Dahlem Workshop Reports
Life Sciences Research Report 34
The Molecular Mechanism of Photoreception

The goal of this Dahlem Workshop is:
to understand the molecular
events linking photon absorption
to membrane conductance change

Life Sciences Research Reports
Editor: Silke Bernhard

Held and published on behalf of the
Stifterverband für die Deutsche Wissenschaft

Sponsored by:
Senat der Stadt Berlin
Stifterverband für die Deutsche Wissenschaft
Stiftungsfonds Schering AG

The Molecular Mechanism of Photoreception

H. Stieve, Editor

Report of the Dahlem Workshop on
The Molecular Mechanism of Photoreception
Berlin 1984, November 25 – 30

Rapporteurs:
M. L. Applebury · W. H. Miller · W. G. Owen
E. N. Pugh, Jr.

Program Advisory Committee:
H. Stieve, Chairperson · M. L. Applebury
M. D. Bownds · M. Chabre · K. Kirschfeld
T. D. Lamb · B. Minke

Springer-Verlag
Berlin Heidelberg New York Tokyo

Copy Editors: K. Geue, J. Lupp
Text Preparation: J. Lambertz, D. Lewis
Photographs: E. P. Thonke

With 4 photographs, 105 figures, and 6 tables

Dahlem Workshop on the Molecular Mechanism
of Photoreception (1984: Berlin, Germany)
The molecular mechanism of photoreception.
(Life sciences research report; 34)
(Dahlem workshop reports)
Includes bibliographies and indexes.
1. Photoreceptors--Congresses.
2. Photobiology--Congresses.
3. Photochemistry--Congresses.
4. Photoelectricity--Congresses.
I. Stieve, H.
II. Applebury, M.L.
III. Title.
IV. Series.
V. Series: Dahlem workshop reports.
QP481.D34 1984 591.1'823 85-27946

ISBN-13: 978-3-642-70446-8 e-ISBN-13: 978-3-642-70444-4
DOI:10.1007/978-3-642-70444-4

Printing: Color-Druck, G. Baucke, Berlin

Bookbinding: Lüderitz & Bauer, Berlin

2131/3020-5 4 3 2 1 0

Table of Contents

The Dahlem Konferenzen

Founders
Recognizing the need for more effective communication between scientists, especially in the natural sciences, the Stifterverband für die Deutsche Wissenschaft*, in cooperation with the Deutsche Forschungsgemeinschaft**, founded Dahlem Konferenzen in 1974. The project is financed by the founders and the Senate of the City of Berlin.

Name
Dahlem Konferenzen was named after the district of Berlin called "Dahlem", which has a long-standing tradition and reputation in the arts and sciences.

Aim
The task of Dahlem Konferenzen is to promote international, interdisciplinary exchange of scientific information and ideas, to stimulate international cooperation in research, and to develop and test new models conducive to more effective communication between scientists.

Dahlem Workshop Model
Dahlem Konferenzen organizes four workshops per year, each with a limited number of participants. Since no type of scientific meeting proved effective enough, Dahlem Konferenzen had to create its own concept. This concept has been tested and varied over the years, and has evolved into its present form which is known as the *Dahlem Workshop Model*. This model provides the framework for the utmost possible interdisciplinary communication and cooperation between scientists in a given time period.

**The Donors Association for the Promotion of Sciences and Humanities*
***German Science Foundation*

The main work of the Dahlem Workshops is done in four interdisciplinary discussion groups. Lectures are not given. Instead, selected participants write background papers providing a review of the field rather than a report on individual work. These are circulated to all participants before the meeting to provide a basis for discussion. During the workshop, the members of the four groups prepare reports reflecting their discussions and providing suggestions for future research needs.

Topics
The topics are chosen from the fields of the Life Sciences and the Physical, Chemical, and Earth Sciences. They are of contemporary international interest, interdisciplinary in nature, and problem-oriented. Once a year, topic suggestions are submitted to a scientific board for approval.

Participants
For each workshop participants are selected exclusively by special Program Advisory Committees. Selection is based on international scientific reputation alone, although a balance between European and American scientists is attempted. Exception is made for younger German scientists.

Publication
The results of the workshops are the Dahlem Workshop Reports, reviewed by selected participants and carefully edited by the editor of each volume. The reports are multidisciplinary surveys by the most internationally distinguished scientists and are based on discussions of new data, experiments, advanced new concepts, techniques, and models. Each report also reviews areas of priority interest and indicates directions for future research on a given topic.

The Dahlem Workshop Reports are published in two series:
1) Life Sciences Research Reports (LS), and
2) Physical, Chemical, and Earth Sciences Research Reports (PC).

Director
Silke Bernhard, M.D.

Address
Dahlem Konferenzen
Wallotstrasse 19
1000 Berlin (West) 33

The Molecular Mechanism of Photoreception, ed. H. Stieve, pp. xi-xiii. Dahlem Konferenzen 1986. Berlin, Heidelberg, New York, Tokyo: Springer-Verlag.

Preface

H. Stieve
Institut für Biologie II
RWTH Aachen
5100 Aachen, F.R. Germany

During the last 10-15 years the investigation of the molecular mechanism of visual transduction has made such considerable progress that a workshop of the Dahlem Konferenzen type seemed desirable. This workshop provided a unique opportunity to review and discuss both recent and old results in order to come up with a reasonable description of the state of the art, a platform to aid the planning of future research in this rapidly developing field.

Phototransduction research has become very multidisciplinary. A researcher competent in one area cannot competently judge the validity of results reported in another discipline which may be very relevant for his own work. Therefore, a panel of scientists competent in the different disciplines involved was needed. Because this workshop had to deal with much more information than a single person can handle, appreciate, and evaluate, we had to discuss in a joint effort the significance of the observations and their implications.

Good discussions (i.e., the exchange of arguments and their evaluation, including the consideration of all arguments for their own value and not so much for the reputation of their authors) are very rare in scientific meetings. Such discussions can be an intellectual pleasure but usually they are flawed by competitiveness.

The quality and intensity of discussion in this Dahlem Workshop was better than in any other scientific meeting I have experienced so far; it included an actual priority of scientific criteria above competitive tactics. The great majority of participants discussed and quoted all publications which seemed relevant to the topics considered.

The format provided a relatively rigid framework for the workshop including restriction of the number of participants to 40 (plus eight young German scientists), a subdivision into four discussion groups, the tasks of providing all participants in advance with manuscripts containing necessary background information for the group discussions, and of producing group reports which describe both the outcome and the highlights of the group discussions.

The program advisory committee chose the four topics for the group discussions on which we felt discussion was needed and which could be most fruitful at our present state of knowledge: a) triggering and amplification, b) internal messengers, c) light-modulated channels, pumps, and carriers, and d) adaptation were selected from a large range of possibilities. They differed largely in level of complexity and in the degree of our present understanding. We restricted our discussion to phototransduction in animals and did not include phototransduction in higher plants, fungi, algae, and bacteria.

There was overlap among the topics of the four discussion groups to various degrees. This led to a joint discussion for one afternoon of the nature and action of the terminal excitatory transmitter in vertebrate photoreceptors by two of the groups (see Miller et al. and Owen et al., both this volume).

The program advisory committee also had the responsibility for choosing the persons invited as participants and their function at the workshop. This was difficult since we wanted to have enough expertise available for discussion of the topics and controversies and in addition some input from related fields. However, there are about three times as many laboratories in the world which have made important contributions to the topics we discussed as there were openings for potential participants. Moreover, the meeting of the program advisory committee was held about a year before the workshop. During that year the progress made in this field was extremely fast. Looking back, we might have made a few different choices.

When planning this workshop in 1983 we did not foresee the exciting timing of the workshop with respect to revisions of the calcium transmitter hypothesis of the vertebrate photoreceptor (2, 5). The recent application of new electrophysiological techniques, especially those using patch-clamp electrodes and suction electrodes has mainly contributed to such progress. Due to a number of independent experimental studies on the action of calcium ions and cyclic GMP on the light-modulated sodium conductance of the plasma membrane of the vertebrate rod outer segment, which were at that time being prepared for publication, we were able to include the most up-to-date research in our discussions. For this reason we particularly regret that neither W.A. Hagins nor S.H. Yoshikami were able to accept our invitation to participate in the workshop.

The debate of the calcium hypothesis obviously was the most spectacular part of the workshop (see also (1, 3, 4)). However, there were a number of other discussions the results of which in my opinion seem also to have been very important, especially the comparison of vertebrate and invertebrate transduction mechanisms.

The workshop was very successful and the outcome of the discussions proved it worth the effort. To no small extent has that success been made possible by Dr. Silke Bernhard who with a combination of authority and charm together with her extremely efficient and dedicated staff organized this workshop, providing the conditions and framework for a scientific debate of outstanding quality in a friendly and pleasant atmosphere. The great majority of participants were also very committed to making this workshop successful.

Besides the reports of the four discussion groups, this publication contains the background papers which were revised by the authors partly as a result of suggestions of some participants.

I hope this book will give a fair overview of the state of our knowledge of research in visual transduction. It was a pleasure to edit, especially because of the friendly and very efficient commitment of K. Geue, J. Lupp, and A. Eckert and the cooperativeness of most of the contributors. Particularly I would like to acknowledge gratefully the extensive efforts and patience of the four rapporteurs, M.L. Applebury, W.H. Miller, W.G. Owen, and E.N. Pugh, Jr., in compiling, writing, and revising the group reports.

REFERENCES

(1) Altman, J. 1985. Sensory transduction, new visions in photoreception. Nature **313**: 264-265.

(2) Hagins, W.A. 1972. The visual process: Excitatory mechanisms in the primary receptor cells. Ann. Rev. Biophys. Bioeng. **1**: 131-158.

(3) Lewin, R. 1985. Unexpected progress in photoreception. Science **227**: 500-503.

(4) Vines, G. 1985. New light on vision. New Sci. **1 August (1467)**: 40-43.

(5) Yoshikami, S., and Hagins, W.A. 1971. Light, calcium and the photocurrent of rods and cones. Biophys. Soc. Ann. Meet. Abstr. **11**: 47a.

The Molecular Mechanism of Photoreception, ed. H. Stieve, pp. 1-10. Dahlem Konferenzen 1986. Berlin, Heidelberg, New York, Tokyo: Springer-Verlag.

Introduction

H. Stieve
Institut für Biologie II
RWTH Aachen
5100 Aachen, F.R. Germany

The aim of the study of visual transduction is to understand the mechanism by which the photoreceptor cell translates the light stimulus into the language of the nervous system. Nature has solved the task of photoreception by using the typical tools of a cell (such as enzymes, membranes, and ion channels) and modifying and adapting them for the visual function. Therefore, besides being a value in itself, our understanding of the mechanism of visual transduction also contributes to general neurobiology and cell biology. For example, the cGMP controlling enzyme cascade in vertebrate visual cells is at present one of the best studied examples of the regulation of cyclic nucleotide as a cellular second messenger (see Applebury et al., this volume).

Moreover, the molecular mechanism of visual transduction is the best understood molecular mechanism of sensory transduction. Compared to other sensory modalities it is relatively well suited to study, because relatively clean photoreceptor cell subunits are available in sufficient amounts and the light stimulus can be manipulated accurately.

A fundamental biological process can often best be studied in very original primitive systems: the study of phages, viruses, and bacteria has greatly advanced our knowledge of genetics, and the study of the fly Drosophila revealed basic principles of genetics which are also valid for vertebrates including humans. The same approach was used for the study of visual transduction but with varying success: the study of the light response of the fungus Phycomyces (6) brought more understanding in terms of developmental morphogenesis and growth processes than of vision. The study of Halobacterium considerably advanced our understanding of the structure of the rhodopsin molecule but has as yet not supplied important clues for the transduction process of visual cells such as the interaction of the photopigment with other types of protein molecules.

A considerable part of our knowledge of the phototransduction process has been derived from experiments on seemingly simple systems such as the ventral photoreceptor of the "old-timer" Limulus or the "primitive" photoreceptor of the barnacle. Closer inspection has shown, however, that these systems are not nearly as simple and primitive as they originally appeared. Phototransduction can be studied surprisingly well in the highly specialized photoreceptor cells of vertebrates, the other main source of our recent gain in knowledge.

In the last ten years a genetical approach using mutants of the fly Drosophila with defects in visual transduction has been applied ((14-17, 25); also see Wong, this volume). This line of research makes use of the vast experience of Drosophila genetics since the work of T.H. Morgan, combined with recent methods of molecular genetics such as gene cloning. After a "warming-up" period of about ten years, the great potential of this approach has recently become evident. It provides, for instance, the potential for specifically varying the structure of a protein molecule in order to investigate in vitro or in vivo its effect on specific functions. This new molecular genetical approach might drastically transform neurobiological research from work in small laboratories to that of huge teams.

Much to our surprise, it became evident in 1965 (21) that nature has solved the task of visual transduction differently for vertebrates than for invertebrates. Some conspicuous differences in the mechanisms of the two groups are well established and include the following:

1 - In invertebrates the visual pigment is converted by light to a thermostable meta-state which is not, or is only very slowly, metabolically regenerated.

2 - The electrical signal of visual excitation in invertebrates is based upon an increase in membrane conductance as opposed to the conductance decrease in vertebrates, resulting in membrane voltage signals (receptor potentials) of opposite polarity.

These apparent distinctions may indicate fundamental differences in the transduction mechanism of the two groups of animals. However, it is most reasonable to assume that vertebrate vision is not a new "invention" of early vertebrates but has developed from an ancestral mechanism of ancient invertebrates which might have been more original than the visual transduction mechanism which we find in recent invertebrates (see also (18)). Therefore, I feel quite strongly that the study of recent invertebrate mechanisms may also give us a clue to a better understanding of the mechanism of vertebrates. At present some aspects of the transduction process are better understood or are easier to study in the photoreceptors of vertebrates, others in those of invertebrates.

Thus, it was of special interest to me to consider throughout the workshop the comparative aspect of the transduction process in photoreceptor cells of vertebrates and invertebrates. Therefore the workshop is reviewed below under this comparative aspect.

In a visual cell of a vertebrate or an invertebrate animal the absorption of a single photon by a molecule of the visual pigment rhodopsin causes, after a delay (latency) of 10-500 ms*, a largely amplified transient change in membrane conductance: the elementary excitatory response. The sum of such elementary responses of a photoreceptor cell forms a macroscopic response. The size and time course of the response are modulated by adaptation, which enables the visual cell to adjust its sensitivity according to ambient illumination.

The process of visual transduction starts with the absorption of a light quantum by a rhodopsin molecule. For the vertebrate photoreceptor it has been shown that light induces conformational changes of the rhodopsin molecule which trigger the following steps in the transduction process. In these cells at least a substantial part of the amplification involved in transduction is supplied by an enzyme cascade which is started by a special type of light activation: a light-activated rhodopsin molecule transiently binds a GTP-binding protein (transducin) which activates the enzyme phosphodiesterase. A single activated rhodopsin molecule transiently binds several G-protein molecules sequentially; this provides a first amplification step in the transduction process. The activated enzyme phosphodiesterase catalyzes the hydrolysis of cyclic GMP, a second stage of amplification. In invertebrates much less is known about these processes. It seems plausible that the early processes are similar, and there is evidence for the involvement of a GTP-binding protein in invertebrates which is activated on illumination.

The rhodopsin molecules display large rotational and translational mobility within the vertebrate photosensory membrane (4), in contrast to the invertebrate photoreceptor membrane in which the rhodopsin molecules are virtually immobile.

The discussion group on triggering and amplification (see Applebury et al., this volume) discussed step-by-step the properties of this enzyme cascade,the molecular interactions, amplification, its control, and the role of cyclic GMP, and dealt with a relatively young but clearly described field which from year to year provides greater understanding of molecular details. The data are clear, the open questions well-defined and seemingly approachable with methods that are already available. The discussion of this

---

* The long latency of the visual response is certainly not desirable from a functional point of view, but it may be an unavoidable property of the transduction mechanism which is optimized for the detection of single photons without much error (false alarm).

group delved into the molecular details of rhodopsin structure and structural changes that constitute transduction. Part of the discussion dealt with amplification and control in the cGMP cascade and with "switch-off" (or turn-off) mechanisms. The latter are probably based on the phosphorylation of rhodopsin that enables the binding of a 48K-protein and the subsequent "breaking" of the cascade process. Finally, the group dealt with more unknown territory - what is the functional significance of this cascade and the role of cGMP in photoreceptors? Does rhodopsin trigger other important biochemical pathways?

Although much less is yet known about transduction processes at the molecular level in invertebrates, the recent cloning of the rhodopsin gene of Drosophila (15, 25) opens exciting new aspects for study: soon it will be possible to change the sequence of the amino acids at will and insert such modified rhodopsin, through the use of Drosophila genetics, as the only rhodopsin in visual cells and thus to investigate the significance of the rhodopsin structure for defined steps in the transduction process.

It has been shown for the light response of invertebrates that substantial amplification cannot take place in the earlier steps of the transduction process, i.e., during latency. This makes it improbable that the same model of amplification formulated for vertebrate photoreceptors (the enzyme cascade which already exhibits amplification in the first stages) can be used without modification for the invertebrate photoreceptor. The delay between photon absorption and the transient change in membrane conductance in invertebrates is not correlated with the size of the single-photon response. Substantial amplification does not take place during the silent delay time (latency) (see Stieve, Schnakenberg and Keiper, and Pugh et al., all this volume).

There is evidence that the excitation spreads in space from the location of photon absorption by the rhodopsin molecule over a relatively large area, in both vertebrate and invertebrate photoreceptors. The only plausible explanation (5) which is still consistent with the present data is the assumption of diffusing molecules, internal messengers or transmitters. It became clear during the workshop that more than one species of molecules may be responsible for the spatial spread of excitation. Part of this spread may be based on diffusion of molecules from the enzyme cascade.

Discussions of the group on internal messengers (see Miller et al., this volume) concentrated on the evidence concerning the nature of the terminal (excitatory) internal transmitter or gating signal molecule whose binding controls the closure or opening of the light-modulated ion channels in the plasma membrane of the vertebrate rod. The experimental evidence favoring cyclic GMP is increasing whereas the calcium hypothesis as originally formulated by Yoshikami and Hagins (24) and Hagins (9) seems to be untenable ((8, 10, 13, 23); also see McNaughton et al., this volume). This latter hypothesis has stimulated important investigations which have demonstrated the complex role of calcium ions in the rod outer segment.

Originally it seemed to be an ideal, clear working hypothesis, and one could think of a number of experiments for critically testing it. However, these experimental tests proved very hard to accomplish and did not provide definite experimental support for the calcium hypothesis. It took more than 15 years to obtain decisive evidence, experimental findings which clearly do not agree with the calcium transmitter hypothesis.

Cyclic GMP, the level and turnover of which is greatly modulated by illumination in photoreceptors of vertebrates, seems to act directly on the light-modulated ion channels in the plasma membrane of these photoreceptors in a manner expected of a terminal excitatory transmitter (8, 10, 23). Its involvement in the transduction process of invertebrates has not yet been clearly demonstrated.

The nature of the excitatory internal transmitter in the invertebrate photoreceptor is still not known. Calcium, which besides its desensitizing effect (see below) also has an excitatory effect in the photoreceptor of Limulus (3, 19, 20), is one of the candidates. Inositol polyphosphate injected intracellularly mimics the effect of light in these receptors (3, 7). It may act indirectly by causing intracellular calcium release. Inositol polyphosphate is unlikely to be the terminal excitatory transmitter in invertebrate photoreceptors because its injection evokes single events (bumps) of similar shape as the photon-evoked responses (Miller et al. and Owen et al., both this volume).

Since not only excitation but also light-induced desensitization (adaptation) shows a limited spatial spread, the existence of an adaptational (desensitizing) transmitter has also been proposed. In invertebrate photoreceptors calcium ions have been identified as an adaptational transmitter (see Brown, this volume), whereas the nature of an adaptational transmitter in vertebrate photoreceptors is still not known. Here, too, calcium might be a candidate (Lamb, this volume).

The plasma membrane enclosing the photoreceptor cell is basically a phospholipid bilayer which is almost impermeable to polar molecules. Protein molecules which form aqueous (or polar-lined) channels through the membrane allow the (passive) transport of ions down the electrochemical gradients. Diffusing carrier molecules may serve the same function. Protein complexes in the membrane form pumps which restore and sustain the electrochemical gradients by active transport coupled with an exergonic chemical reaction. The conductance of a subpopulation of ion channels in the photoreceptor cell membrane is modulated as a consequence of the light stimulus. Following the absorption of a single photon many of these channels open (invertebrates) or close (vertebrates) almost simultaneously, resulting in a concerted conductance change, the single photon-evoked event.

In the group on light-modulated channels, pumps, and carriers (see Owen et al., this volume), single photon-evoked events were also discussed. The molecular nature of the light-modulated channels in the cell (plasma) membrane of the photoreceptor cell and the stoichiometry of its ligand binding are not known. The gating mechanism may be influenced by calcium or may involve a light-induced calcium affinity change. The single-channel conductance of the light-modulated ion channels in the plasma membrane of vertebrate rods is very small, and/or the channel lifetimes are much shorter than in invertebrate photoreceptor cells (see Owen, this volume). In Limulus photoreceptors distinct single-channel events of ca. 20 pS have been observed (2). The data available do not give much support to the assumption that in vertebrate rods the light-modulated membrane current is carried by a light-modulated carrier mechanism (Owen, this volume).

A coupled sodium/calcium exchange transport (antiport) has been shown in vertebrate rods (22) and there are also strong indications for such a process in invertebrate photoreceptors (1, 11).

Single photon-evoked events which seem to be due to a concerted action (the opening in invertebrates or the closing in vertebrates) of many light-modulated ion channels are very different in size and time course characteristics for invertebrate and vertebrate photoreceptors. These conspicuous differences in the characteristics of the single-photon responses may indicate fundamental differences in the mechanism of their generation. In vertebrates single-photon events evoked under identical conditions have a standard size, a standard latency, and a standard time course, all three parameters showing relatively small variations. Responses to dim flashes containing several photons vary only by an amplitude scaling factor, but not in latency and time course. In invertebrates, however, single-photon responses (bumps) vary greatly in delay (latency), time course, and size. These fluctuations of the bump parameters of invertebrate photoreceptors seem to reflect early processes in the causal chain of transduction.

The macroscopic response of vertebrates has a shape and time scale that is similar to the single photon-evoked event, whereas in invertebrates the macroscopic response is much longer than a single-photon response. The latency of the macroscopic light response of invertebrates is shortened much more with increasing intensity of the light stimulus than is that of vertebrate photoreceptors.

The underlying mechanism of invertebrate bumps was discussed in some detail, including the discussion of the use of mathematical models (see also Schnakenberg and Keiper, this volume). There are two kinds of mathematical models. One type is very useful only for describing quantitative experimental data in a handy way. The other is a tool for studying the mechanism underlying the experimental observations; it therefore should be biophysically meaningful. It should be designed in such a way that the stipulations and assumptions made are evident and that the model can be treated rigorously. In order to be useful the model should lead to critical experimental tests which would either disprove it or make it more probable. A model which cannot be critically tested by experiments (and potentially refuted) is useless. At its best a model should lead to unambiguous statements such as the exclusion of certain classes of mechanisms discussed, e.g., the involvement of substantial amplification in the early stages of the bump-generating process (see above).

The power and limitations of noise analysis for the determination of parameters of single-photon responses (bumps) were also discussed (see Schnakenberg and Wong, this volume): noise analysis reveals properties of the noise underlying bumps only at low or medium densities of photon absorption in the photoreceptor. For higher intensities of the light stimulus this still holds, at least as long as the bumps are triggered and formed independently from each other. At very high photon densities where the photoreceptor is virtually flooded with excitatory transmitter, the bump concept is no longer applicable because individual bumps are no longer generated. Under these conditions an individual ion channel is gated by many transmitter molecules which cannot be associated with the activation of one particular rhodopsin molecule. There seems to be no direct experimental clue which allows the reliable determination of the highest intensity of the light stimulus up to which the bump approach is valid.

Adaptation is the control of the sensitivity of the photoreceptor cell. It involves changes in the size and time scale of the light response. Light adaptation increases the intensity range of response by reducing the sensitivity, and the frequency range of response by accelerating the kinetics of the light response.

The group on adaptation probably had the most difficult task. The mechanism of adaptation is only little understood. Much more is known about the mechanism of adaptation in invertebrate photoreceptors than in vertebrate photoreceptors. In both invertebrate and vertebrate photoreceptors the single photon-evoked events become smaller (in size) and faster due to light adaptation. Obviously changes in several steps of the transduction process can influence the sensitivity. In invertebrates the following mechanisms of light adaptation are known:

1 - A very important mechanism of adaptation in invertebrate photoreceptors is a feedback mechanism regulating the light sensitivity of the cell by controlling the intracellular concentration of free calcium ions (see Brown, this volume). This rise in calcium most probably causes the diminution of bump size by control of the amplification.

2 - In addition there is a mechanism of light adaptation in invertebrate photoreceptors which is not controlled by calcium. It can be observed at very low desensitizations (see Minke and Pugh et al., both this volume).

3 - After very strong bleaches which convert larger amounts of visual pigment from rhodopsin to metarhodopsin, a prolonged depolarizing afterpotential (PDA) is evoked which is accompanied by a long lasting desensitization (Minke, this volume).

4 - Migration of screening pigment which is regulated according to the illumination of the photoreceptor cell changes the probability of photon absorption.

In vertebrate photoreceptors the adaptation mechanisms are much less well described. Operationally, two cases are distinguished: desensitization due to background light and desensitization following larger bleaches. Calcium ions may also contribute to the desensitization of the vertebrate photoreceptor (Lamb, this volume). In the photoreceptors of vertebrates, the mechanism of the so-called "bleaching adaptation" may be similar or related to that of the desensitization of invertebrate visual cells accompanying the PDA (see Minke, this volume).

A molecular understanding of most of the phenomena observed in adaptation is still missing. One can speculate that in the case of the sensitivity control by intracellular calcium in the invertebrate photoreceptor internal calcium might regulate the lifetime of an activated enzyme which in turn controls amplification.

In the case of strong bleaches of visual pigment it may be that backward reactions in the chain of the light-induced pigment reactions become prominent because of the enormous quantities of photoproduct. This would lead to the reactivation of inactivated rhodopsin molecules (see Lamb, this volume). The proposed backward reactions give a plausible explanation for the increased "dark light" which is observed after strong bleaches in psychophysical measurements.

Although it is most probable that vertebrate vision has developed from ancient invertebrate vision during evolution (see above), the key to our understanding of how vertebrate vision could be derived from the invertebrate mechanism is still missing. The interesting hypothesis of Kramer and Widman (12) can formalistically explain a number of the observed phenomena: it assumes that in the vertebrate rod the spontaneous bump rate in the dark is so high that the cell is virtually flooded with excitatory transmitter and thus steadily depolarized. A light stimulus then causes hyperpolarization by a mechanism homologous to light adaptation of invertebrates, namely, a reduction of transmitter concentration. Vertebrate visual excitation would thus correspond to visual light adaptation in invertebrates. However, this aspect has to be investigated much further to test whether this is a clue as to the path evolution has taken.

We can compare the investigation of photoreception with the solving of a jigsaw puzzle. This game is far from being finished: some regions are still terra incognita, some we know rather well; in some regions we already know the substructure of building pieces but we do not know where their place is, in some fields we still miss several building pieces. In some areas of the phototransduction research we are close to an understanding on a molecular level, in others we do not even have a quantitatively exact description of the overall phenomenology. In the future, besides pursuing those lines of research which have been fruitful in the past, new techniques will be applied which will lead to a deeper understanding of the scientific landscape of photoreception. Among these, the use of strategically designed antibodies, gene cloning, and other methods of molecular genetics for studying the structure-function relation of a protein appear to me very promising.

Acknowledgements. I would like to thank M.L. Applebury, M.D. Bownds, T.D. Lamb, B. Minke, and U. Thurm for their valuable suggestions which helped me in writing this article; K. Geue and T. Stieve for helping me express my views in passable English; and A. Eckert for typing the manuscript.

REFERENCES

(1) Armon, E., and Minke, B. 1983. Light activated electrogenic Na+-Ca-exchange in fly photoreceptors: modulation by Na^+/K^+-pump activity. Biophys. Struct. Mech. **9:** 349-357.

(2) Bacigalupo, J., and Lisman, J.E. 1983. Single-channel currents activated by light in Limulus ventral photoreceptors. Nature **304:** 268-270.

(3) Brown, J.E.; Rubin, J.L.; Traver, A.P.; Ghalayini, A.J.; Irvine, R.F.; and Anderson, R.E. 1984. Myo-inositol polyphosphate may be a messenger for visual excitation in Limulus photoreceptors. Nature **311:** 160-163.

(4) Cone, R.A. 1972. Rotational diffusion of rhodopsin in the visual receptor membrane. Nat. New Biol. **236:** 39-43.

(5) Cone, R.A. 1973. The internal transmitter model of visual excitation, some quantitative implications. In Biochemistry and Physiology of Visual Pigments, ed. H. Langer, pp. 275-282. Berlin: Springer-Verlag.

(6) Delbrück, M. 1974. Anfänge der Wahrnehmung. Karl-August-Forster Lectures Akademie der Wissenschaften und Literatur (Mainz) **10:** 8-48.

(7) Fein, A.; Payne, R.; Corson, D.; Berridge, M.J.; and Irvine, R.F. 1984. Photoreceptor excitation and adaptation by inositol 1,4,5-triphosphate. Nature **311:** 157-160.

(8) Fesenko, E.E.; Kolesnikov, S.S.; and Lyubarsky, A.L. 1984. Induction by cyclic GMP of cationic conductance in plasma membrane of retinal rod outer segment. Nature **313:** 310-313.

(9) Hagins, W.A. 1972. The visual process: excitatory mechanism in the primary receptor cells. Ann. Rev. Biophys. Bioeng. **1:** 131-158.

(10) Haynes, L., and Yau, K.-W. 1985. Cyclic GMP-sensitive conductance in outer segment membrane of catfish cones. Nature, in press.

(11) Ivens, I., and Stieve, H. 1984. Influence of the membrane potential on the intracellular light induced Ca^{2+}-concentration change of the Limulus ventral photoreceptor monitored by Arsenazo III under voltage clamp conditions. Z. Naturforsch. **39c:** 986-992.

(12) Kramer, L., and Widman, T. 1977. Quantitative model for the electric response of invertebrate and vertebrate photoreceptors. Biophys. Struct. Mech. **2:** 333-336.

(13) Matthews, H.R.; Torre, V.; and Lamb, T.D. 1985. Effects on the photoresponse of calcium buffer and cyclic GMP incorporated into the cytoplasm of retinal rods. Nature **313:** 582-584.

(14) Minke, B. 1977. Drosophila mutant with a transducer defect. Biophys. Struct. Mech. **3:** 59-64.

(15) O'Tousa, J.E.; Baehr, W.; Martin, R.L.; Hirsh J.; Pak, W.L.; and Applebury, M.L. 1985. The Drosophila ninaE gene encodes an opsin. Cell **40:** 839-850.

(16) Pak, W.L.; Conrad, S.K.; Kremer, N.E.; Larrivee, D.C.; Schinz, R.H.; and Wong, F. 1980. Photoreceptor function. In Development and Neurobiology of Drosophila, eds. O. Siddiqu, P. Babu, L.M. Hall, and J.C. Hall, pp. 331-346. New York: Plenum Publishing Corp.

(17) Pak, W.L.; Ostroy, S.E.; Deland, M.C.; and Wu, C.F. 1976. Photoreceptor mutant of Drosophila: is a protein involved in intermediate steps of phototransduction? Science **194:** 956-959.

(18) Stieve, H. 1977. Thoughts on the comparative biology of photosensory function. Verh. Dtsch. Zool. Ges. 1-25. Stuttgart: Gustav-Fischer-Verlag.

(19) Stieve, H., and Bruns, M. 1980. Dependence of bump rate and bump size in Limulus ventral nerve photoreceptor on light adaptation and calcium concentration. Biophys. Struct. Mech. **6:** 271-285.

(20) Szuts, E.Z.; Reid, M.S.; Payne, R.; Corson, D.W.; and Fein, A. 1985. Physiology and biochemistry evidence for a role for inositol 1,4,5-triphosphate in visual transduction. Suppl. Inv. Ophthal. Vis. Sci. **26:** 167.

(21) Tomita, T. 1965. Electrophysiological study of the mechanisms subserving color coding in the fish retina. Cold S.H. Symp. Quant. Biol. **30:** 559-566.

(22) Yau, K.-W., and Nakatani, K. 1985a. Light-induced reduction of cytoplasmic free calcium in retinal rod outer segment. Nature **313:** 579-582.

(23) Yau, K.-W., and Nakatani, K. 1985b. Light-suppressible cyclic GMP-sensitive conductance in the plasma membrane of truncated rod outer segment. Nature, in press.

(24) Yoshikami, S., and Hagins, W.A. 1971. Light, calcium and the photocurrent of rods and cones. Biophys. Soc. Ann. Meet. Abstr. **11:** 47a.

(25) Zuker, C.S.; Cowman, A.F.; and Rubin, G.M. 1985. Isolation and structure of a rhodopsin gene from D. melanogaster. Cell **40:** 851-858.

The Molecular Mechanism of Photoreception, ed. H. Stieve, pp. 11-30. Dahlem Konferenzen 1986. Berlin, Heidelberg, New York, Tokyo: Springer-Verlag.

The Biosynthetic, Functional, and Evolutionary Implications of the Structure of Rhodopsin

J.B.C. Findlay
Dept. of Biochemistry University of Leeds
Leeds LS2 9JT,
England

Abstract. Until fairly recently, some of the fundamental characteristics of rhodopsin were much better described than its structure. Now, however, amino acid and DNA sequence data, the use of chemical probes, and the application of structure prediction algorithms have allowed a much better appreciation of the disposition of the polypeptide chain in the bilayer. The structure which emerges identifies seven transmembrane helices, some of which may contain distortions or irregularities. Computer graphics suggest that most of the hydrophilic side chains are not exposed to the lipid phase. From this structure, it is possible to make tentative speculations as to which regions of the molecule may participate in interactions with the chromophore and which with other proteins in the system. In general, the structural features of opsin have much in common with integral membrane proteins which have a transport function. Vertebrate rhodopsins exhibit greater than 90% identity in their amino acid sequences. The receptor proteins involved in color vision may be a little more distant. In contrast, invertebrate opsins show marked reductions in sequence homology as compared with the vertebrate proteins but the general 3-dimensional features of the molecule appear similar. This burst of new information has accentuated the need for redoubled efforts in two directions - investigations into the full functional capabilities of rhodopsin and the elucidation of the detailed 3-D structure of the protein.

INTRODUCTION

Although the determination of the primary and 3-dimensional structures of a protein are major goals in themselves, the underlying significance of such studies more properly lies in the appreciation which the structural information imparts to the biochemical activity of the molecule and ultimately to the biological role which it fulfils. This review will attempt to place more emphasis on understanding structure in the context of the function of opsin rather than concentrating on the methods which have been used in its elucidation. This is not to belittle the ingenious practical approaches which have been used in the pursuit of this elusive goal, for unravelling the structure of such integral membrane proteins has been and continues to be one of the most challenging problems confronting protein chemists. No attempt will be made here to relate the historical development of our current knowledge, but some indication is given as to the future direction of work in this area. The term opsin is used to denote all those proteins which act as primary light receptors in the visual process, irrespective of the source. Most of our information, however, is derived from the visual pigment from rod cells which, when dark-adapted, is known as rhodopsin.

SYNTHESIS AND DISTRIBUTION

Opsin is essentially a plasma membrane protein whose synthesis appears to follow, at least superficially, the conventional route. Its early association with the membrane and ultimate organization in the cell, however, illustrate fundamental and as yet only imprecisely understood intracellular phenomena. The "Signal Hypothesis" (30) and its numerous refinements go some way toward providing a mechanistic explanation as to how proteins are synthesized on membrane-bound ribosomes and cotranslationally secreted across the bilayer. But is this process readily adaptable to integral membrane proteins whose polypeptide chain traverses the bilayer numerous times?

The first observation to be made is that opsin (possibly like many proteins of this type, a notable exception being bacteriorhodopsin) does not possess at its N-terminus, a conventional signal which is subsequently removed by proteolysis (23). It is relevant to identify two "functions" for these signal peptides, the ability to be recognized by a membrane-bound receptor and the ability to be cleaved by the signal peptidase. The absence of a cleavable signal does not, however, necessarily imply that the recognition site also does not exist. It may well be that a sequence is present somewhere in the first third of the nascent polypeptide chain which allows opsin to become associated with the endoplasmic reticulum.

Thereafter, the situation is complicated by the clustered distribution of hydrophobic amino acid residues in the polypeptide chain. Such distributions do not coexist comfortably with the concept of insertion into the bilayer via a protein-lined pore in a manner similar to that proposed for secreted proteins. Although various "stop-start" mechanisms could be envisaged, the alternative mechanism for association with the membrane should not be too quickly discarded - namely, that long sequences containing mainly hydrophobic residues may allow direct integration into the bilayer in the form of hairpin-like structures. In such a scenario, theoretical considerations suggest that it is not energetically too unfavorable to pull the more hydrophilic portions through the lipid phase of the bilayer provided that sufficient hydrophobic character is also present (10). The final topography of the molecule will probably be dependent on co-translational insertion, not on complete synthesis followed by integration into the bilayer.

Once in the plasma membrane, segregation events influence the final cellular distribution of opsin. The protein is usually restricted to the morphologically characterized outer segment region, either in the form of pinched-off intracellular discs (rods), multiple invaginations of the plasma membrane (cones), or microvilli (invertebrate visual cells). Until the full functional abilities of rhodopsin are described, the advantages of one morphological form over another will remain unclear. Although still a very controversial issue, it may be that the integral membrane protein (and possibly also lipid) composition of the vertebrate rod disc membrane varies significantly from that of the corresponding plasma membrane (e.g., (19)). In such a situation, some form of regulatory mechanism in the cell is capable of confining opsin to certain membranes or membrane regions of the outer segment, and in so doing, exclude most of the other integral proteins. The corollary may be less true - namely, the plasma membrane of the rod cell may not rigorously banish opsin. If the reasons for this possible segregation in rods lie in cellular efficiency and economy, it seems less critical for there to be any separation in the cone and microvillar systems - if only because no alteration will have taken place in the location of those membrane proteins whose functions depend on their presence in the plasma membrane.

The nature of the segregation mechanism in rod cells is entirely obscure. It is facile to say that opsin may contain defined recognition sites, but for what and where and how do they operate? Possibly an instructive example of these basic dilemmas will emerge from a study of the squid visual pigments which, unlike most others, are not free to rotate in the plane of the membrane (31). One can speculate in these instances that the opsins may contain in their structure the entire apparatus which allows them to remain associated with the submembranous cytoskeletal systems which stabilize the microvilli. It is important to recognize in this discussion, however, that "lateral" as well as "vertical" protein-protein association could have a role in these processes.

Thus, even before considering the relationship between structure and primary function, opsin clearly presents in its synthesis and distribution facets of some of the most intriguing enigmas taxing cell biologists. Their resolution may require complete deployment of the full armory of the molecular biologist.

PRIMARY STRUCTURE AND ITS IMPLICATIONS

The quest for the primary structure of opsin has occupied several research groups and cost nearly a hundred man years, but it has been an instructive exercise. The sequences of the bovine (16, 22) and ovine (4, 12) proteins from rod cells have been elucidated independently using a variety of orthodox and novel procedures (Fig. 1). In all instances a common strategy was employed, proteolytic cleavage in situ followed by fractionation of the fragments and subdigestion to smaller peptides. This approach reduced the complexity of the peptide mixture, so facilitating subsequent purification using unusual chromatography systems (4, 12). More conventional peptide procedures were also employed for peptide isolation, but at the cost of considerable losses of material. Final success was critically dependent on the development of automated solid and liquid phase sequencers. The

```
    1 CHO                                   10                                        CHO                                       20                         25
B Ac-Met-Asn-Gly-Thr-Glu-Gly-Pro-Asn-Phe-Tyr-Val-Pro-Phe-Ser-Asn-Lys-Thr-Gly-Val-Val-Arg-Ser-Pro-Phe-Glu-
O     * - * - * - * - * - * - * - * - * - * - * - * - * - * - * - * - * - * - * - * - * - * - * - * - * -
P     * - * - * - * - * - * - * - * - * - * - * - * - * - * - * - * - * - * - * - * - * - * - * - * - * -
E

    26              30                                      40                                      50
B   Ala-Pro-Gln-Tyr-Tyr-Leu-Ala-Glu-Pro-Trp-Gln-Phe-Ser-Met-Leu-Ala-Ala-Tyr-Met-Phe-Leu-Leu-Ile-Met-Leu-
O    * - * - * - * - * - * - * - * - * - * - * - * - * - * - * - * - * - * - * - * - * - * - * -Val- * -
P   Tyr- * - * - * - * - * - * - * - * - * - * - * - * - * - * - * - * - * - *
E

    51                                  60                                      70              75
B   Gly-Phe-Pro-Ile-Asn-Phe-Leu-Thr-Leu-Tyr-Val-Thr-Val-Gln-His-Lys-Lys-Leu-Arg-Thr-Pro-Leu-Asn-Tyr-Ile-
O    * - * - * - * - * - * - * - * - * - * - * - * - * - * - * - * - * - * - * - * - * - * - * - * - * -
P
E

    76              80                                      90                                      100
B   Leu-Leu-Asn-Leu-Ala-Val-Ala-Asp-Leu-Phe-Met-Val-Phe-Gly-Gly-Phe-Thr-Thr-Thr-Leu-Tyr-Thr-Ser-Leu-His-
O    * - * - * - * - * - * - * - * - * - * - * - * - * - * - * - * - * - * - * - * - * - * - * - * - * -
P
E

    101                                 110                                     120V            125
B   Gly-Tyr-Phe-Val-Phe-Gly-Pro-Thr-Gly-Cys-Asn-Leu-Glu-Gly-Phe-Phe-Ala-Thr-Leu-Gly-Gly-Glu-Ile-Ala-Leu-
O    * - * - * - * - * - * - * - * - * - * - * - * - * - * - * - * - * - * - * - * - * - * - * - * - * -
P
E

    126             130                                     140                                     150
B   Trp-Ser-Leu-Val-Val-Leu-Ala-Ile-Glu-Arg-Tyr-Val-Val-Val-Cys-Lys-Pro-Met-Ser-Asn-Phe-Arg-Phe-Gly-Glu-
O    * - * - * - * - * - * - * - * - * - * - * - * - * - * - * - * - * - * - * - * - * - * - * - * - * -
P
E

    151                                 160                                     170             175
B   Asn-His-Ala-Ile-Met-Gly-Val-Ala-Phe-Thr-Trp-Val-Met-Ala-Leu-Ala-Cys-Ala-Ala-Pro-Pro-Leu-Val-Gly-Trp-
O    * - * - * - * - * - * - * - * - * - * - * - * - * - * -Leu- * - * - * - * - * - * - * - * - * - * -
P
E

    176V            180                                     190                                     200
B   Ser-Arg-Tyr-Ile-Pro-Glu-Gly-Met-Gln-Cys-Ser-Cys-Gly-Ile-Asp-Tyr-Tyr-Thr-Pro-His-Glu-Glu-Thr-Asn-Asn-
O    * - * - * - * - * -Gln- * - * - * - * - * - * - * -Leu-Ala- * -Phe- * -Leu-Lys-Pro- * -Ile- * - * -
P
E
```

```
    201                                  210                                     220             225
B   Glu-Ser-Phe-Val-Ile-Tyr-Met-Phe-Val-Val-His-Phe-Ile-Ile-Pro-Leu-Ile-Val-Ile-Phe-Phe-Cys-Tyr-Gly-Gln-
O    * - * - * - * - * - * - * - * - * - * - * - * -Ser- * - * - * - * - * - * - * - * - * - * - * - * -
P
E

    226             230     V                        240                                             250
B   Leu-Val-Phe-Thr-Val-Lys-Glu-Ala-Ala-Ala-Gln-Gln-Gln-Glu-Ser-Ala-Thr-Thr-Gln-Lys-Ala-Glu-Lys-Glu-Val-
O    * - * - * - * - * - * - * - * - * - * - * - * - * - * - * - * - * - * - * - * - * - * - * - * - * -
P                                                         * - * - * - * - * - * - * - * - * - * - * - * -
E                                                         * - * - * - * - * - * - * - * - * - * - * - * -

    251                                  260                                     270             275
B   Thr-Arg-Met-Val-Ile-Ile-Met-Val-Ile-Ala-Phe-Leu-Ile-Cys-Trp-Leu-Pro-Tyr-Ala-Gly-Val-Ala-Phe-Tyr-Ile-
O    * - * - * - * - * - * - * - * -Ile- * - * - * - * - * - * -Leu- * - * - * -Gly- * - * - * - * - * -
P    * - * - * - * - * - * - * - * -Val- * - * - * - * - * - * -Leu- * - * - * -Ser- * - * - * - * - * -
E    * - * - * - * - * - * - * - * -Ile- * - * - * - * - * - * -Val- * - * - * -Ser- * - * - * - * - * -

    276             280                          290                                  300
B   Phe-Thr-His-Gln-Gly-Ser-Asp-Phe-Gly-Pro-Ile-Phe-Met-Thr-Ile-Pro-Ala-Phe-Phe-Ala-Lys | Thr-Ser-Ala-Val |
O    * - * - * - * - * - * -Asp- * - * - * - * - * - * - * -Ile- * - * - * - * - * - * | Ser-Ser-Ser-Val |
P    * - * - * - * - * - * -Asp- * - * - * - * - * - * - * -Ile- * - * - * - * - * - * | Ser-Ala-Ser-Ile |
E    * - * - * - * - * - * -Asn- * - * - * - * - * - * - * -Leu- * - * - * - * - * - * | Ser-Ala-Ala-Ile |

    301                                  310      V                              320             325
B   Tyr-Asn-Pro-Val-Ile-Tyr-Ile-Met-Met-Asn-Lys-Gln-Phe-Arg-Asn-Cys-Met-Val-Thr-Thr-Leu-Cys-Cys-Gly-Lys-
O    * - * - * - * - * - * - * - * - * - * - * - * - * - * - * - * - * -Leu- * - * - * - * - * - * - * -
P    * - * - * - * - * - * - * - * - * - * - * - * - * - * - * - * - * -Leu- * - * - * - * - * - * - * -
E    * - * - * - * - * - * - * - * - * - * - * - * - * - * - * - * - * -Ile- * - * - * - * - * - * - * -

    326             330                          340                                 348
B   Asn-Pro-Leu-Gly-Asp-Asp-Glu-Ala-Ser-Thr-Thr-Val-Ser-Lys-Thr-Glu-Thr-Ser-Gln-Val-Ala-Pro-Ala-OH
O    * - * - * - * - * - * - * - * - * - * - * - * - * - * - * - * - * - * - * - * - * - * - *
P    * - * - * - * - * - * - * - * - * - * - * - * - * - * - * - * - * - * - * - * - * - * - *
E    * - * - * - * - * - * - * - * - * - * - * - * - * - * - * - * - * - * - * - * - * - * - *
```

FIG. 1 - Sequences of opsins. B = bovine, O = ovine, P = porcine, E = equine, V = intron, * denotes same residue.

primary structure of the bovine protein was confirmed by sequencing both the cDNA and the gene (21). This DNA sequence information supported the absence of an N-terminal signal and revealed the intron/exon organization of the gene, an interesting feature to which we will return.

Evolutionary Aspects

Visual inspection of these sequences quickly establishes their near identity, indeed, such variations as do exist - mostly conservative -account for only about 3% to 5% of the total structure of the molecule. The sequences of the C-terminal 110 residues of the porcine (P) and equine (E) proteins reinforce this low mutation rate. On account of the similarity of the four species, this observation is not at first sight too surprising, except that the primary structure of the equivalent region of chicken rhodopsin again demonstrates this remarkable conservation among farmyard cohabitants (Medina and Findlay, unpublished).

Does this immutability extend to the vertebrate cone pigments thought to be responsible for color vision and for vision under conditions of bright light? First impressions suggest that the two classes of protein, although very similar, are significantly less homologous to one another. The cone and rod visual pigments from chickens, for example, which both utilize 11-cis retinal, exhibit slightly different affinities for Concanavalin A-Sepharose and DEAE-Sepharose, migrate a little differently on SDS-polyacrylamide gels, and are not equally stable in detergent (11). Unpublished sequence information of color pigments seems to substantiate these observations.

And how about the pigments of invertebrates where the structure of the eye and the membranes containing the photoreceptor proteins show the greatest diversity (35)? 11-cis retinal again seems to be the usual (but not the only) chromophore and the initial photochemical events following illumination appear roughly similar to vertebrate systems (see Kirschfeld, this volume). Thereafter, however, differences become apparent. The metarhodopsin spectral intermediate is photostable, the all-trans chromophore remaining bound to most invertebrate opsins. There are also gross structural variations. Insect opsins appear a little smaller (Mr 33-39,000) than their vertebrate counterparts while the octopus and squid proteins are certainly larger (Mr 43,000-51,000). This variation is confirmed by sequence studies on the protein from Drosophila (Applebury, personal communication) and from Loligo (Grant-Morris, Pappin, Medina, Saibil, and Findlay, unpublished) which reveal, in contrast to the almost complete conservation noted above, very significant differences from the mammalian rod opsins. Nevertheless, the Drosophila protein is clearly recognizable as belonging to the opsin family and, although the data from Loligo is not yet definitive, the distribution pattern of residues may again be similar. In addition, the squid protein responds to incubation with the S. aureus V8 proteinase in manner reminiscent of the other opsins. It is probable, therefore, that there is pronounced conservation of the major elements of 3-dimensional structure but not of amino acid sequence. In summary, the emerging picture indicates that all the visual light receptors are structurally related and conform to a general 3-dimensional pattern. The linear sequences of the vertebrate rod proteins appear remarkably conserved but exhibit greater variation with respect to the vertebrate cone proteins and differ substantially from the primary structure of the invertebrate polypeptides. Whether there is tight conservation within the cone, insect, and cephalopod families remains to be seen.

The exciting prospect inherent in this mosaic of structures is that comparison may well reveal useful insights into the functional domains of opsin, particularly those associated with interaction with the chromophore. Surprises can be expected! One intriguing example which has already emerged concerns a small mutational hot spot located in the four residues C-terminal to the Schiff base attachment point for retinal (residues 297-300; Fig. 1). In all cases so far examined, the molecular volume occupied by the corresponding amino acids remains relatively constant, and so one would not expect major alterations in conformation (25). Conventional evolutionary wisdom would expect, however, significant effects on some functional parameter since these rapid changes are confined to what one would imagine is a very sensitive region of the molecule. No firm explanation has yet emerged, but we will return to this region in another context.

Role of Lipid

When considering integral membrane proteins, it has been a common supposition that the intramembranous hydrophobic residues exposed to the lipid milieu are subject to few constraints. The implication is that one might expect to see fairly widespread mutational changes in these residues, the only proviso being that the substitutions must involve amino acids of similar character (i.e., hydrophobic). That such a frequent substitution rate is not seen with the opsins suggests that the constraints are greater than supposed and evokes questions as to the part played by the lipid bilayer in the activity of the molecule.

Some indication that a hydrophobic environment is a necessary but not sufficient criterion for the activity of integral proteins has come from reconstitution studies with a variety of transport proteins and from the inhibitory influence which low concentrations of some hydrophobic and hence lipid-soluble substances (e.g., general anesthetics) can have on maximal transport activity. We have monitored the effects of certain inhibitory hydrophobic anesthetics and of cholesterol on rhodopsin function using the rate and extent of phosphorylation following illumination as the criterion of activity (37). In no case can we detect any significant effect, suggesting that the conformational changes which follow illumination and allow phosphorylation are not sufficiently hindered to change significantly this somewhat inadequate criterion of activity. The study is necessarily incomplete, however, and one looks forward to some assay, say, for transport which may present a more satisfactory means of assessing the influence of the hydrophobic phase.

It must be remembered in this context that the fatty acid content of mammalian rod disc membranes is highly unusual, possessing about 50% unsaturated species (particularly the rare 22:6 fatty acid). Moreover, the composition is maintained, at least in the rat, when other bilayers are changing in response to dietary deficiencies in unsaturated fatty acids (7). Just how widespread this unique composition is, remains to be seen, invertebrate and cone systems being of particular interest. One explanation for the high proportion of unsaturated fatty acids is that it makes for a low viscosity environment which facilitates the biological activity of rhodopsin by permitting rapid translational mobility. This in turn allows increased collision frequencies with other proteins in the membrane, most notably, the G-protein (there is no convincing evidence that opsin-opsin interactions are an essential part of the biological function of the molecule). There are, however, a number of discordant observations relevant to this hypothesis, two of which are worthy of comment. The overall composition of the rhodopsin-containing membranes in cone cells (i.e., the plasma membrane) is likely to be different from that of rod discs, if only because of the so far unviolated observation that plasma membranes, in contrast to most intracellular membranes, contain large amounts (up to 50%) of cholesterol (see also (19)). This should significantly alter the overall viscosity of the bilayer. Second, rhodopsin appears to be relatively fixed in the microvillar membrane of at least some invertebrates ((31) and Saibil, personal communication). Both of these elements tend to argue against the indispensability of rapid translational mobility. With such thoughts in mind, therefore, one ought to widen the role of the lipid fraction to include the possibility that the conformational changes undergone by rhodopsin require a fluid environment (see below).

DISPOSITION OF THE POLYPEPTIDE CHAIN IN THE MEMBRANE

Although the primary structure of a protein is vital as a construction template, the information obtained has its limitations. One can appreciate the intramembranous nature of opsin and to some extent its evolutionary history and individual characteristics, but much more data is required before one can begin to understand the 3-dimensional organization of the molecule so crucial for any meaningful perception of mechanism. Many of the results described below take us some way along this path and could only have been obtained once good methods for the dissection of opsin and the isolation of the peptides had been developed.

Early studies quickly established that the protein was transmembranous, the carbohydrate chain near the N-terminus (14, 15) being sequestered inside the rod disc (and presumably outside the cone and invertebrate cells) and the phosphorylated sites in the C-terminal region facing the cytosolic compartment (15, 33, 38). But how else is the polypeptide chain packaged so as to give the compact structure suggested by X-ray and neutron diffraction studies (8, 32)? The answer came from the use of water soluble, essentially impermeant chemical probes (1) and in situ proteolytic dissection. The results from these approaches are summarized in Fig. 2.

The data clearly and unambiguously establish, for the first time with any integral membrane protein of this type, the convoluted path of the polypeptide chain across the lipid bilayer. This kind of study is a much more definitive indicator of polypeptide chain disposition in the membrane than the use of predictive hydrophobic indices, for the latter approach is liable to faulty interpretation when applied to proteins, such as those of the porin family, which appear to traverse the membrane several times yet have no clearly defined, extended hydrophobic segments.

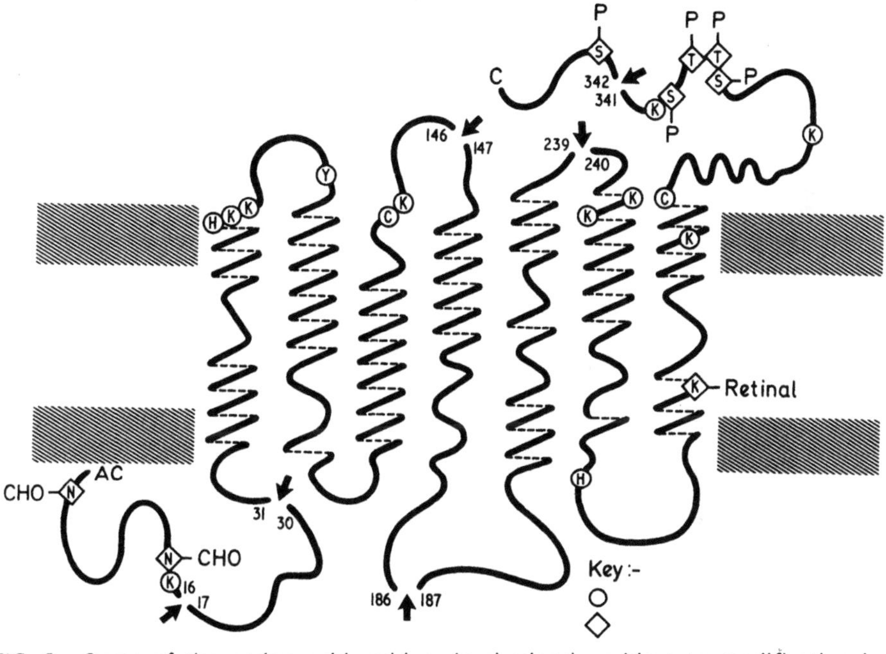

FIG. 2 - Some of the amino acid residues in rhodopsin subject to modification by hydrophilic chemical probes (O), phosphorylation (P), and glycosylation (CHO) and to proteolytic cleavage (↑). (1, 4, 12-14, 22, 25, 37).

CONFORMATION

Having obtained a quasi-2-dimensional impression of opsin which incidentally places not inconsiderable numbers of polar residues within the limits of the membrane, it is important to push on towards a 3-D reconstruction. Unfortunately, conventional X-ray diffraction studies, although showing some promise, will take some time yet to reach fruition. In the meantime, a number of different approaches have led to the emergence of a low-resolution but useful prototype for the monomeric protein. Circular dichroic data (36), the infrared spectrum (5), energy transfer measurements (26), and low angle X-ray diffraction studies on native (8) and 2-D crystalline systems (6) can be amalgamated in the construction of a somewhat irregular, ellipsoid molecular model, 7-8 nm long by 2-3 nm wide, whose transmembrane sections (long axis) are largely α-helical. To this picture we can now begin to add the orientation of some of the amino acid side chains as revealed with the use of the photosensitive hydrophobic probe, [^{125}I] iodo-azidobenzene (Davison and Findlay, unpublished). The results have proved very instructive because they begin to fill in that "big black box," conventionally called the apolar region of the membrane. Interpreted in conjunction with the water soluble probes, they help us to define much more exactly the membrane-aqueous phase boundaries.

As well as the many specific interpretations one can make from this work, some of which are mentioned later, a few more general conclusions can be drawn. At least some part of all the transmembrane segments are exposed to the hydrophobic milieu; most tyrosine, cysteine, and tryptophan residues, along with many of the more conventional hydrophobic side chains, are oriented towards the apolar phase. The intramembranous domain is not "held together" by disulfide bonds; indeed, the only possible interaction of this type involves residues 110 and 187 at the intradiscal face of the protein. Most intriguingly, all the transmembrane helices ought not to be regarded as completely regular or uninterrupted. This last observation emerged from a rather involved prediction study carried out on the whole molecule (Fig. 3) whose main conclusions received unexpectedly firm support from the chemical data (24). The labelling studies suggest that no arrangement of the modified residues in at least two of the transmembrane segments is consistent with both their availability to the probe and their presence in continuous and regular α-helices (assuming, of course, that at least some part of the helix face is involved in protein-protein interaction and hence inaccessible to the hydrophobic probe). The availability of these residues could only arise if distortions or discontinuities are present in these helices. Just as didactic for this model-building exercise is the disposition of those residues which are potential targets for the hydrophobic probe but which clearly are not reactive. Any H-bonding not fulfilled as a result of these irregularities in the helices would be satisfied by interactions with amino acid side chains such as those of serine and threonine. It will be fascinating to see eventually what relationships exist between the functional parameters of opsin and these distorted regions.

In this discussion, it is very important to note that the proportion of polar residues which make up the intramembranous regions of opsin (about 20%) is not significantly different from that found in the buried hydrophobic cores of water soluble, globular proteins. Extensive modeling studies using computer graphics suggest that these hydrophilic side chains are oriented away from the lipid-exposed surface (Pappin and Findlay, unpublished). Integral membrane proteins ought not, however, to be regarded as "inside-out" polypeptides; rather, it is the distribution of the nonpolar residues and the interhelical associations which allow this class of protein to integrate into the bilayer. Since they already are present in a hydrophobic environment, there will be a substantial diminution in the entropic effect which normally

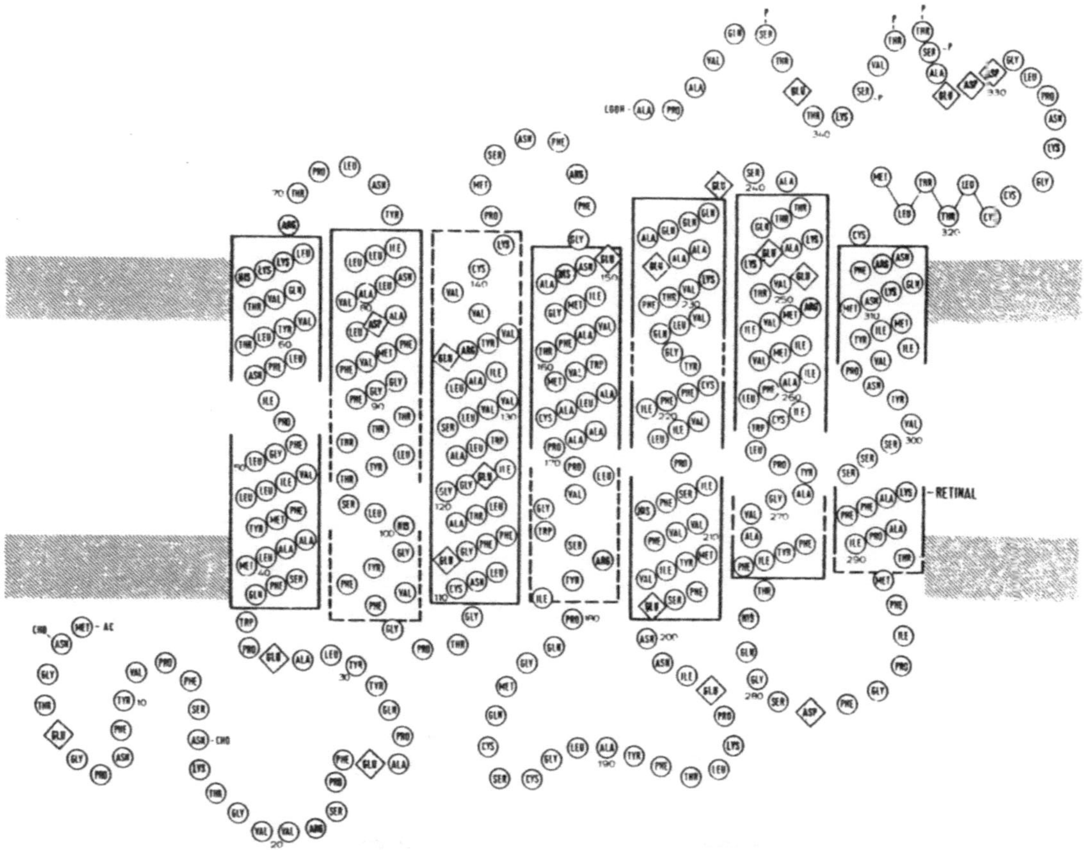

FIG. 3 - Predicted structure for ovine rhodopsin (24). Solid and broken lines represent probable regular helices. Unboxed regions represent irregular secondary structure. Negative charged residues are shown as stippled diamonds, positively charged residues as stippled circles.

has a profound influence on the folding of water soluble proteins. There remains to be described, therefore, the principal influences which combine to fashion the polypeptide into a coherent and cohesive 3-D structure. In this respect it is interesting to note that the proteolysed molecule, apparently stable either in the membrane in both the bleached and unbleached states or in detergent as rhodopsin (25), nevertheless dissociates into its two major fragments when the detergent-solubilized species is exposed to light (27).

FUNCTIONAL DOMAINS

At this point, one can begin the process of assigning, with widely varying degrees of confidence, known and conjectured functional abilities of rhodopsin to defined areas of structure.

Retinal Binding Site

At the essential core of its biological function is the ability to absorb a photon of light energy via its 11-cis retinal chromophore. The critical "bathochromic shift" seen on association of prosthetic group and protein could arise from more than one structural scenario but in the most popular hypothesis it is thought to be derived in part from the formation of a protonated Schiff base between the aldehyde of retinal and the ε-amino of Lys-296, and in part from the influence of a strategically placed negative charge (18). The exact spatial relationship of chromophore to this negative region may well govern the precise absorbance maximum of the pigment complex. How, then, does the putative structure relate to these characteristics?

The first observation is that 11-cis retinal must penetrate deep within the protein in order to form its association with Lysine 296, clearly identified as the attachment point. Some kind of pocket or cleft must, therefore, exist which, because of the dimensions and relative orientation of the chromophore, will extend a good way across a transverse section of the protein. It is interesting that, among the possible candidates for the negative charge, the side chains of potentially buried Glu's 113, 122, and 134 are clearly not exposed to the lipid phase since conditions which give rise to covalent modification by dicyclohexylcarbodiimide of critical glutamic acid residues in several proteins, including the proteolipid protein of myelin and the membane-bound b-subunit in the Fo portion of the ATPase synthase complex (34), do not lead to any reaction in opsin (Thompson and Findlay, unpublished). Two other types of residue, cysteine and tryptophan, have also been implicated in interactions, perhaps with the beta-ionone moiety of the chromophore. In view of their availability to various probes and their location in the 3-dimensional structure, one can reasonably eliminate all but Trp 126 as active candidates. (In this latter respect, it is presumed that the hydrophobic probe does not readily occupy the chromophore binding site since it does not inhibit the process of regeneration. Notice must be taken, however, of the low levels of radiolabel associated with Lys-296).

Since the Schiff base is not available for reduction in unbleached mammalian rhodopsins, it is assumed to reside in a highly protected environment whose characteristics stabilize the protonated group, perhaps by a second negative moiety 3Å away, or by the inclusion of bound water (28). If we then fold the protein up into a squashed barrel approximating the situation seen with bacteriorhodopsin (17), it is possible to imagine that the high proportion of polar hydroxyl-containing amino acids in the more irregular (lower) half of helix 2 could contribute to the environment of the Schiff base.

Capture of the photon of incident light is associated with conformational changes presumably originating from within the bilayer and being transmitted to the extramembranous C-terminal region. The initiating event involves the 11-cis to all-trans isomerization of retinal. One can envisage three scenarios in this process: movement of the ionone ring, movement of the conjugated carbon chain, or both. The Schiff base linkage suddenly becomes available for reduction and the conformational changes that occur are sufficiently marked to create a binding site for the G-protein in the C-terminal region, or more probably, the extended cytoplasmic loop between about residues 230 and 250 (20). It is interesting to note that the structure prediction has this region projecting far up above the surface of the membrane in the form of a twin helix. Previously inaccessible serine and threonine side chains near the C-terminus have now been positively identified as targets for the kinase (see Fig. 2) (37).

The transmission of both a change in configuration and an elongation of the chromophore is not a trivial episode for the protein. In this context one could envisage an important role for the extended loop proposed as lying adjacent to Lys-296, which we have already identified as a mutational hot spot (9). The rapid interchange amongst the polar residues occupying this site tends to suggest that a specific spatial configuration of the moieties making up their respective side chains is not a critical feature for the protein. Could it be, then, that this region functions as a rather flexible loop which can respond to the effective elongation of the chromophore on isomerization and, in so doing, force an alteration in the relative orientation of the blocks of regular helix? This reorientation would in turn lead to a pronounced reorganization of the aqueous C-terminal domain, thereby creating the necessary and new functional binding sites. This mechanism would not involve any change in the total content of α-helix as the molecule is converted from the unbleached to the activated state, just an alteration in their relative spatial relationships.

Transport

So far, this discussion has concerned known activities of rhodopsin which have, to varying extents, some structural validity. It is useful, however, to take a more long-distance overview of the structure which has emerged from the studies outlined above.

Integral membrane proteins fall, roughly speaking, into two main categories. One of these classes is characterized by having the bulk of the polypeptide chain in the aqueous phase, with one or two transmembrane segments anchoring the molecule in the bilayer. The parts of the structure primarily involved in the biological action of the protein reside in the aqueous domain(s). In most instances (one or two receptors may be more complex), the hydrophobic intramembranous domain acts mainly as a tether. Examples of such proteins include enzymes with a variety of hydrolase activities and receptors such as those involved in hormone action, chemotaxis, and endocytosis.

The second group is quite different in that a very substantial proportion of the molecule is embedded in the membrane, the polypeptide chain making repeated sallies across the bilayer. Bacteriorhodopsin, the anion and glucose transport proteins of the erythrocyte membrane, the Ca^{2+} and Na^+/K^+-ATPases, the proteolipid protein of myelin, the lactose and K^+-transport proteins of E. coli, and the neural sodium gate are all members of this second group. The pertinent unifying characteristic of this heterogeneous collection of examples is that they are all thought to be involved in mediating the movement of material across an otherwise impermeable barrier. The mechanisms by which they do this are extremely obscure. Some seem to act as permanent molecular assemblies encircling selective pores, but most have properties in common with enzymes, only involved in physical translocation rather than chemical transformation. In these instances, interactions with substrates are highly specific and transport may involve transient conformational changes which subtly alter the environment of the substrate. No fixed "hole" down through the protein need be envisaged, nor is there any obligatory requirement for substantial changes in conformation or orientation.

Since the structure of opsin clearly assigns it to this second category of integral membrane proteins, one cannot help but be drawn to the possibility that it too fulfils a transport role perhaps similar to that originally proposed by Hagins (13). It may be of some relevance that the 3-dimensional computer modelling mentioned earlier orients virtually all the hydrophilic and charged intramembranous residues away from the lipid environment and in clusters which extend from one membrane surface to the other, through one half of the opsin molecule. A supportive argument could also be made that if rhodopsin were solely involved in activating peripheral G-protein, as a prelude to the aqueous cascade mechanism which results in the turnover of cGMP, it could achieve this via a "lollipop"-type structure akin to the first group of intrinsic membrane proteins. Indeed, such a configuration could also be more convenient for its interactions with the kinase and the 48K protein (see Chabre and Applebury, this volume). In contrast to the water soluble, serum retinol-binding protein (Mr 21,000) which, as well as binding hydrophobic retinol, also interacts with prealbumin and a receptor, rhodopsin would need a membrane-bound anchor for correct orientation of the protein/chromophore with respect to incident light.

These comments are all controversial and speculative inferences from the structure of the molecule. Attempts to demonstrate a convincing transport activity for rhodopsin in either discs or reconstituted systems have not been uniformly successful. It could be that the experimental conditions for such a demonstration have not been optimized, and the time may well be approaching when further investigation is warranted.

Conclusions

From the above discussion, it seems obvious that rhodopsin may be capable of a variety of conformational alterations which are effective over different time scales. On absorption of light, major unheavals take place which may propel the protein simultaneously into two activated modes via a series of spectral intermediates which bear witness to these structural rearrangements. One mode is clearly involved in the activation of a cascade-like process via protein-protein interactions. The second could be concerned with transient conformational changes which are associated with its putative transport function - a function, it should be noted, which will reside with the monomer rather than result from any aggregation phenomena.

Thereafter a slower series of structural events occur, indicative of which is the time-dependent decay in its ability to be phosphorylated. These events will ultimately return the protein to a quiescent but primed state. But even in this "recovery phase," the protein behaves uniquely, for one can regenerate phosphorylated opsin even though the original rhodopsin was not phosphorylated. Moreover, it is then possible to bleach in an apparently "normal" fashion this unusual hybrid species. The inference is that one can, to some extent, uncouple the C-terminal extramembranous region of the protein from events in the intramembranous domain. In such a complex system, one is reminded of the unusual fatty acid content of the lipid bilayer and the role it may play in facilitating conformational change. It is worth noting in this context that as much as 40% of the lipid content of the membrane may be relatively immobilized in the immediate vicinity of opsin and that the influence of the protein may be experienced further out into the bulk phase (39).

EVOLUTIONARY ASSEMBLY

As more protein structures are elucidated, a picture is emerging of families of proteins interrelated by the possession of similar functional capacities. This may be reflected in either sequence or, in "older" proteins, conformational homology. These homologies may extend over much of the molecule or be restricted to defined domains. The utilization of the various retinoids and their close cousins, the carotenoids, presents an obvious case for closer examination.

Bacteriorhodopsin, the $[H^+]$ pump from the membranes of Halobacterium halobium, is light-sensitive by virtue of a bound retinal moiety. Both the low-resolution (17) and predicted (9) "heptahelical" structures for this protein have much in common with that proposed for opsin (24), lending some support to the rather hopeful hypothesis that photosensitivity was conferred on primitive host cells by a symbiotic association with the antecedents of the halobacteria. If such an event did take place, the extended time scale of subsequent evolutionary development would account for the absence of any sequence homology between the two proteins. It must be borne in mind, however, that all transport proteins may turn out to conform to this general structural prototype. In such a case, convergent evolution starting from related transport functions and incorporating the utilization of a light-sensitive pigment may be an equally valid interpretation.

Of more immediate interest, therefore, is the nature of the retinal binding site which in both proteins causes the pronounced bathochromic shifts seen on association with the chromophore. The hypothetical structure postulated for this site is equally suited to either system, and the guess is that the conformation in this region may be similar. The disappointment is that no clear sequence homology is apparent to help us pinpoint the essential residues involved. One can only surmise that some vague similarities exist in the site earlier identified in rhodopsin. Important information may come from photosensitive retinal analogs, but more definitive answers will have to await solution of the 3-dimensional structure. As an indication of the generality of the "point charge" hypothesis, it will be interesting to see what 3-D structure emerges from a study of crustacyanin, a protein from lobster carapace, whose partial amino acid sequence is known (Zagalski and Findlay, unpublished). The interesting feature of this system is that a pronounced bathochromic shift occurs on combination of the oligomeric protein with its carotenoid prosthetic group, astaxanthin (40).

The cellular retinol and retinoic acid binding proteins, although exhibiting clear sequence homology with one another, do not allow any convincing or even useful analogies to be drawn with the opsins. Surprisingly, the serum retinol binding protein, whose molecular weight (21,000) is not too dissimilar from the cellular species (15,000), appears unrelated to either group in terms of its sequence, but its conformation is very much like that of the retinol binding milk-protein β-lactoglobulin. The 11-cis retinal binding protein (Mr 33,000) is different yet again. Clearly, then, this whole group of proteins, although possibly related, have diverged so far that only detailed 3-D information will reveal any domain interrelationships.

With the discovery of the intron/exon structure of the eukaryotic gene, interest has been renewed into the evolutionary assembly of "modern" proteins. Amino acid sequences and 3-dimensional analysis have previously revealed the presence of easily defined functional domains which could be arranged in a number of different permutations to give proteins of different functions and specificities. At first, it was envisaged that these functional domains would correlate with the exons seen in the structure of the gene. As more DNA sequences have become available, however, it is now clear that this correlation is untenable. In a significant number of cases, residues intimately involved in both catalytic and regulatory activities are associated with different exons. One can surmise that opsin also conforms to this newly emergent property, for it seems quite probable that the binding site for retinal will be shared by at least two of the exons which make up the opsin gene. On this evidence, therefore, the migrations of functional domains, including their intron/exon structures, would appear to be a rather later evolutionary phenomenon.

The exon itself may reflect an even earlier assembly process. The latest suggestion is that exons may represent structurally stable peptide entities which may not necessarily possess any functional capacity (2). Although this suggestion sits uncomfortably with the tenets of Darwinian evolution, it may be that we are examining a precellular era to which these concepts are less relevant. These stable structural elements average about 50 amino acids or multiples thereof and can often be readily detected in the 3-D structures of proteins. It is rather exciting that opsin may be an obvious example of this new idea. All but one of the intron/exon boundaries map in regions delineated as extramembranous loops by the structural studies (V symbol, Fig. 1). Most of the exons themselves can be envisaged as coding for stable hydrophobic helical hairpins. As vestigial remnants, these hairpins may still play an important role during the biosynthesis of opsin, particularly in the folding of the polypeptide chain and in its insertion into the lipid bilayer (see above).

EPILOGUE

On consideration of the various structural and functional features of rhodopsin discussed above, the protein appears as a multifaceted system. Disguised by a superficially simple structure, rhodopsin nevertheless emerges as a molecular entity capable of complex conformational changes which may operate over several time scales and give rise to several functional abilities. Although we can now appreciate so much more about the protein, the adage,

> "Many dogs have little fleas
> up their backs to bite 'em
> and little fleas have lesser fleas
> and so ad infinitum,"

remains true when applied to rhodopsins. The future holds more fascinating and fundamental insights, particularly if a high-resolution 3-dimensional structure can be determined.

Acknowledgements. I gratefully acknowledge the legion of scientists who have contributed to our present understanding of rhodopsin. In our own laboratory substantial contributions have been made by M. Brett, D.J.C. Pappin, P. Thompson, P. Barclay, M. Davison, and G. Medina. Our work has been made possible by grants from the M.R.C. and S.E.R.C.

REFERENCES

(1) Barclay, P.L., and Findlay, J.B.C. 1984. Labelling of the cytoplasmic domains of ovine rhodopsin with hydrophilic chemical probes. Biochem. J. **220:** 75-84.

(2) Blake, C.C.F. 1983. Exons - present from the beginning? Nature **306:** 535-537.

(3) Brett, M., and Findlay, J.B.C. 1979. Investigation of the organization of rhodopsin in sheep photoreceptor membranes using cross-linking reagents. Biochem. J. **177:** 31-44.

(4) Brett, M., and Findlay, J.B.C. 1983. Isolation and characterisation of the CNBr peptides from the proteolytically derived N-terminal fragment of ovine opsin. Biochem. J. **211:** 661-670.

(5) Chabre, M. 1981. **In** Membranes and Intracellular Communication, ed. R. Balian, pp. 251-256. Amsterdam: North Holland Publishing.

(6) Corless, J.M.; McCaslin, D.R.; and Scott, B.L. 1982. Two-dimensional rhodopsin crystals from disk membranes of retinal rod outer segments. Proc. Natl. Acad. Sci. USA **79:** 1116-1120.

(7) Daemen, F.J.M. 1973. Vertebrate rod outer segment membranes. Biochim. Biophys. Acta **300:** 255-288.

(8) Dratz, E.A.; Miljanick, G.P.; Nemes, P.P.; Gaw, J.E.; and Schwartz, S. 1979. The structure of rhodopsin and its disposition in the rod outer segment disc membrane. Photochem. Photobiol. **29:** 661-670.

(9) Eliopoulos, E.; Geddes, A.J.; Brett, M.; Pappin, D.J.C.; and Findlay, J.B.C. 1982. A structural model for the chromophore binding domain of ovine rhodopsin. Intl. J. Biol. Macromol. **4:** 263-268.

(10) Engelman, D.M., and Steitz, T.A. 1981. The spontaneous insertion of proteins into and across membranes. The Helical Hairpin Hypothesis. Cell **23:** 411-422.

(11) Fager, L.Y., and Fager, R.S. 1982. Chromatographic separation of rod and cone pigments from chicken retinas. **In** Methods in Enzymology, ed. L. Packer, vol. 81(H), pp. 160-166. New York: Academic Press.

(12) Findlay, J.B.C.; Brett, M.; and Pappin, D.J.C. 1981. Primary structure of the C-terminal functional sites in ovine rhodopsin. Nature **293:** 314-316.

(13) Hagins, W.A. 1972. The visual process: Excitatory mechanisms in the primary receptor cells. Ann. Rev. Biophys. Bioeng. **1:** 131-158.

(14) Hargrave, P.A. 1977. The amino terminal tryptic peptide of bovine rhodopsin. Biochim. Biophys. Acta **492:** 83-94.

(15) Hargrave, P.A., and Fong, S.-L. 1977. The amino and carboxy-terminal sequences of bovine rhodopsin. J. Supramol. Struct. **6:** 559-570.

(16) Hargrave, P.A.; McDowell, J.H.; Curtis, D.R.; Wang, J.K.; Juszczak, E.; Fong, S.-L.; Mohanna Rao, J.K.; and Argos, P. 1983. The structure of bovine rhodopsin. Biophys. Struct. Mech. **9:** 235-244.

(17) Henderson, R. 1977. The purple membrane from Halobacterium halobium. Ann. Rev. Biophys. Bioeng. **6:** 87-109.

(18) Honig, B.; Dinur, U.; Nakanishi, K.; Balogh-Nair, V.; Gawinowicz, M.H.; Arnaboldi, M.; and Motto, M.G. 1979. An external point-charge model for wavelength regulation in visual pigments. J. Am. Chem. Soc. **101:** 7084-7086.

(19) Kamps, M.P.; DeGrip, W.J.; and Deamen, F.J.M. 1982. Use of a density modification technique for isolation of the plasma membrane of rod outer segments. Biochim. Biophys. Acta **687**: 296-302.

(20) Kühn, H., and Hargrave, P.A. 1981. Light-induced binding of GTPase to bovine photoreceptor membranes. Effect of limited proteolysis of the membranes. Biochemistry **20**: 2410-2417.

(21) Nathans, J., and Hogness, D.S. 1983. Isolation, sequence analysis, and intron-exon arrangement of the gene encoding bovine rhodopsin. Cell **34**: 807-814.

(22) Ovchinnikov, Y.A.; Abdulaev, N.G.; Feigina, M.Y.; Artomonov, I.D.; Zolotarev, A.S.; Moroshnikov, A.I.; Martynow, V.I.; Kostina, M.B.; Kudelin, A.B.; and Bogachuk, A.S. 1982. The complete amino acid sequence of visual rhodopsin. Biorg. Khim. **8**: 1424-1427.

(23) Papermaster, D.S.; Burstein, Y.; and Schechter, I. 1980. Opsin mRNA isolation from bovine retina and partial sequence of the in vitro translation product. Ann. NY Acad. Sci. **343**: 347-355.

(24) Pappin, D.J.C.; Eliopoulos, E.; Brett, M.; and Findlay, J.B.C. 1984. A structural model for ovine rhodopsin. Intl. J. Biol. Macromol. **6**: 73-76.

(25) Pappin, D.J.C., and Findlay, J.B.C. 1984. Sequence variability in the retinal attachment domain of mammalian rhodopsins. Biochem. J. **217**: 605-613.

(26) Pober, J.S.; Iwarij, V.; Reich, E.; and Stryer, L. 1978. Transglutaminase-catalysed insertion of a fluorescent probe into the protease sensitive region of rhodopsin. Biochemistry **17**: 2163-2169.

(27) Pober, J.S., and Stryer, L. 1975. Light dissociates enzymatically cleaved rhodopsin into two different fragments. J. Molec. Biol. **95**: 477-481.

(28) Rafferty, C.N., and Shichi, H. 1981. The involvement of water at the retinal binding site in rhodopsin and early light-induced intramolecular proton transfer. Photochem. Photobiol. **33**: 229-234.

(29) Rohlich, P. 1976. Photoreceptor membrane carbohydrate on the intradiscal surface of retinal rod discs. Nature **263**: 789-791.

(30) Sabatini, D.D.; Kreibich, G.; Morimoto, T.; and Adesnik, M. 1982. Mechanisms for the incorporation of proteins in membranes and organelles. J. Cell Biol. **92**: 1-22.

(31) Saibil, H. 1982. An ordered membrane-cytoskeleton network in squid photoreceptor microvilli. J. Molec. Biol. **158**: 435-456.

(32) Saibil, H.; Chabre, M.; and Worcester, D. 1976. Neutron diffraction studies of retinal rod outer segment membranes. Nature **262**: 266-270.

(33) Sale, G.T.; Towner, P.; and Akhtar, M. 1978. Topography of the rhodopsin molecule. Identification of the domain phosphorylated. Biochem. J. **175**: 421-430.

(34) Solioz, M. 1984. Dicyclohexylcarbodiimide as a probe for proton translocating enzymes. Trends Biochem. Sci. **July**: 309-312.

(35) Stavenga, D.G., and Schwemer, J. 1983. Visual pigments of invertebrates. In Photoreception and Vision in Invertebrates, ed. M.A. Ali. New York: Plenum Press.

(36) Stubbs, G.W.; Smith, H.G.; and Litman, B.J. 1976. Alkyl glucosides as effective solubilising agents for bovine rhodopsin. A comparison with several commonly used detergents. Biochim. Biophys. Acta **426:** 46-56.

(37) Thompson, P., and Findlay, J.B.C. 1984. Phosphorylation of ovine rhodopsin: identification of the phosphorylation sites. Biochem. J. **220:** 773-780.

(38) Virmaux, N.; Weller, M.; Mandel, P.; and Trayhurn, R. 1975. Localisation of the major site of light stimulated phosphorylation in a region of rhodopsin distinct from the chromophore binding site. FEBS Lett. **53:** 320-323.

(39) Watts, A.; Volotovski, I.D.; and Marsh, D. 1979. Rhodopsin-lipid associations in bovine rod outer segment membranes. Identification of immobilized lipid by spin labels. Biochemistry **18:** 5006-5013.

(40) Zagalski, P. 1976. Carotenoid-protein complexes. Pure Appl. Chem. **47:** 103-120.

The Molecular Mechanism of Photoreception, ed. H. Stieve, pp. 31-49. Dahlem Konferenzen 1986. Berlin, Heidelberg, New York, Tokyo: Springer-Verlag.

Activation of Visual Pigment: Chromophore Structure and Function

K. Kirschfeld
Max-Planck-Institut für biologische Kybernetik
7400 Tübingen 1, F.R. Germany

Abstract. Besides the "classical" chromophores retinal (visual pigment: rhodopsin) and 3-dehydroretinal (visual pigment: porphyropsin), recently a new chromophore has been found in several insect groups: 3-hydroxyretinal (visual pigment: xanthopsin). Evolutionary aspects are considered - the first interaction of light with the photoreceptor must not necessarily take place at the Schiff base-linked chromophore. In many photoreceptors, e.g., of many fly species, light can be absorbed by a sensitizing pigment which then transfers energy (Förster mechanism) to the Schiff base-linked chromophore. This chromophore is then isomerized and leads to excitation of the receptor. The sensitizing pigment in higher flies is identified as 3- hydroxyretinol, and in one more primitive fly species (Simuliid) most likely as retinol. Functional consequences of sensitization are illustrated.

INTRODUCTION

Activation of visual pigment usually begins with the absorption of a quantum of light. The absorption leads to an isomerization of the chromophoric group, followed by a sequence of sterical changes of the protein moiety. Since Wald's (48) discovery of retinal as the chromophoric group of the visual pigment in most vertebrates, the only other naturally occurring chromophore that has been described is 3-dehydroretinal which is found in some freshwater fishes and amphibia. Details of these findings are discussed in several reviews (28, 30, 36). I will describe here some recent findings which show that in many insect species neither retinal nor 3-dehydroretinal is the visual pigment chromophore, but that there is another, third chromophore. I will then show that the interaction of light with the visual pigment need not happen directly at the normal, Schiff base-linked chromophore, but that photoreceptor function can be improved by sensitization. Some aspects of the problem of photooxidation in photoreceptors will also be illustrated, and finally, similarities and differences between kinetic behavior in vertebrate and invertebrate visual pigments will be discussed.

THE CHROMOPHORIC GROUP IN FLY VISUAL PIGMENT: 3-HYDROXYRETINAL

The visual pigment of flies has been believed to be a rhodopsin, i.e., to have retinal as the chromophoric group. This was first claimed by Wolken et al. (49) who found that the $SbCl_3$ product of an acetone extract of Musca heads has an absorbance peak (665 nm) fitting to the $SbCl_3$ product of retinal. Furthermore, the reaction of the visual pigment of Drosophila with hydroxylamine yields a compound with an absorption peak near 360 nm (32) in accordance with the spectrum of retinaloxime. However, with thin-layer chromatography of retinal extracts of Calliphora, instead of retinal (or retinaloximes after NH_2OH treatment), Vogt (40) found a considerably more polar aldehyde or its corresponding oximes. That these compounds originate from the chromophoric group of the visual pigment was shown by the occurrence of different stereoisomers in extracts, depending on the color of light with which the eyes were previously illuminated: red illumination leads to one isomer of the oximes, whereas blue illumination reduces this isomer to some 20%, the rest being changed into a new isomer (Fig. 1). This behavior corresponds to the ratios of visual pigment and metapigment that are created in living flies, depending upon illumination with red and blue light, respectively. Furthermore, flies raised on a carotenoid-deficient diet are known to have drastically reduced concentrations of visual pigment (2, 8, 38). They also do not have this polar aldehyde, which by means of biochemical methods was specified as 3-hydroxyretinal (Fig. 2, (46)). In support of this result is the finding that a diet with the 3-hydroxyxanthophyll lutein leads to flies with higher visual pigment content than a diet of β-carotene (40).

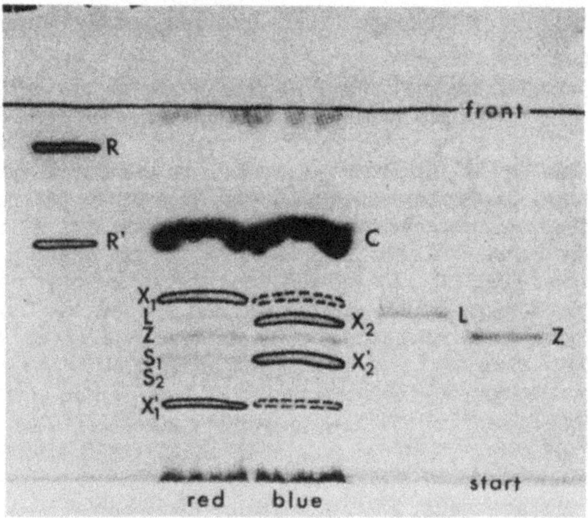

FIG. 1 - Thin-layer chromatogram of extracts from NH_2OH-treated retinae of Calliphora, previously adapted with red or blue light. Retinae extracts: C -cholesterol; L - lutein; Z - zeaxanthin; S_1, S_2 - 3-hydroxyretinol; X_1, X_1', X_2, X_2' - 3-hydroxyretinal-oximes. References: R,R' - all-trans-retinal oximes; L - lutein; Z - zeaxanthin (from (40)).

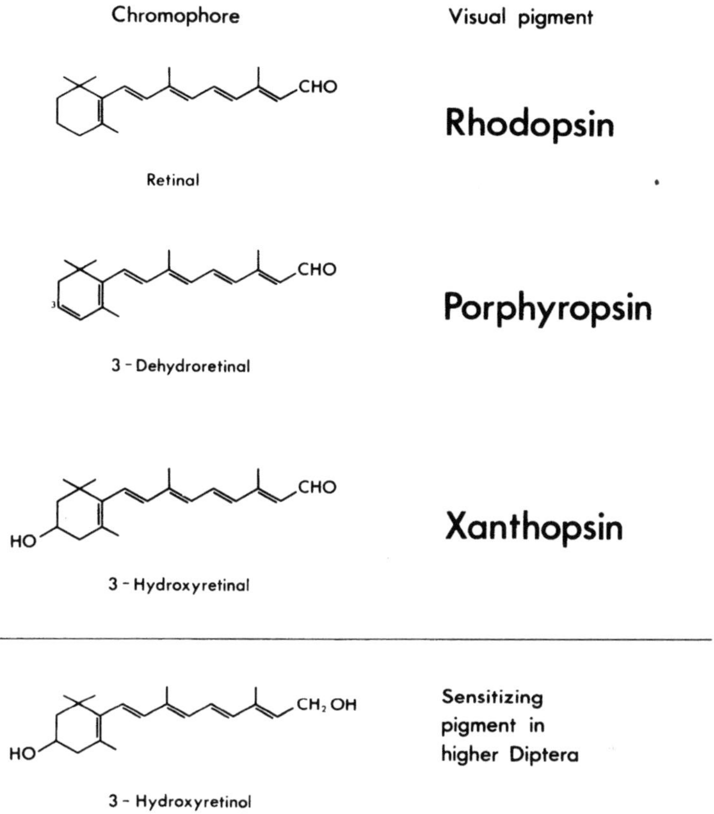

FIG. 2 - Nomenclature of visual pigments and their chromophores.

NOMENCLATURE OF VISUAL PIGMENTS

A simple way to characterize visual pigments as suggested by Dartnall (4) is by their absorption maxima, with an index indicating the chromophore: VP 523_2 means a visual pigment with λ_{max} = 523 nm, based on vitamin A_2 (3-dehydroretinol). In this way the fly visual pigment of the most common receptor (type R1-6) would be VP 490_3, whereby "3" indicates 3-hydroxyretinal; the corresponding alcohol consequently should be called vitamin A_3.

Another nomenclature goes back to Wald, who called visual pigments based on retinal "rhodopsins" and those based on 3-dehydroretinal "porphyropsins." Because the chromophore of fly visual pigments can be derived from oxygenated carotenoids, i.e., xanthophylls, Vogt (40) suggested that 3-hydroxyretinal-derived visual pigments should be called "xanthopsins" (Fig. 2).

SENSITIZATION OF VISUAL PIGMENTS

In the compound eye of the fly the most common receptor type (R1-6) has a receptor potential action spectrum with two maxima: one close to 500 nm, the other, usually still higher one, in the near-ultraviolet, close to 350 nm (3, 19). Dual peak sensitivity of this type cannot be explained on the basis of extinction spectra of known visual pigments. These pigments have only a small peak (β-peak) at shorter wavelengths, on the order of 25% of the maximum. For a long time this dual peak spectral sensitivity was a matter of debate. Suggested explanations have included two different pigments in one and the same photoreceptor (19), b) waveguide effects that can selectively enhance short wavelength extinction of light (37), or c) electrical coupling of receptors with different spectral sensitivities (35). More recently it has been suggested that the high UV-sensitivity might be due to an unusually enhanced β-peak of the visual pigment (34).

With electrophysiological techniques it can be excluded that there is substantial coupling between photoreceptors with different spectral sensitivities. Older experiments in which the spectral sensitivity of white-eyed mutants was measured by means of the ERG exclude the waveguide concept as a possible explanation (20). Experiments described later show that the two-pigment as well as the enhanced β-peak hypotheses cannot explain the high UV-sensitivity in photoreceptors of the fly.

A direct method of investigating visual pigments is by means of microspectrophotometry. In Musca photoreceptors R1-6, a difference spectrum can be measured which has a minimum at about 470 nm, an isosbestic point at 510 nm, and a maximum at 570 nm (Fig. 3). This difference spectrum is similar to that measured in Calliphora (13, 39) and corresponds to a visual pigment with maximal absorption at 490 nm and a metapigment with an absorption maximum at 560 nm.

It is possible to show that UV light is capable of shifting the visual pigment with high efficiency into the same metapigment as blue light and that there is, nevertheless, no significant decrease of extinction in the UV but only in the blue spectral range. This means that there is only one pigment system present in these rhabdomeres that can be shifted back and forth between visual pigment and metapigment. This observation, together with the fact that in the absolute extinction spectrum there is stable high extinction in the UV (Fig. 3), led us to the concept that there could be an energy transfer from a photostable, UV-absorbing pigment onto the blue light-absorbing visual pigment. This sensitizing pigment hypothesis can be formulated as follows:

$$X + h\gamma \rightarrow X^*$$
$$X^* + P \rightarrow P^* + X$$
$$P^* \rightarrow M,$$

where X is the sensitizing pigment, X* its activated state, P and P* are the visual pigment and its activated state, respectively, and M is the visual

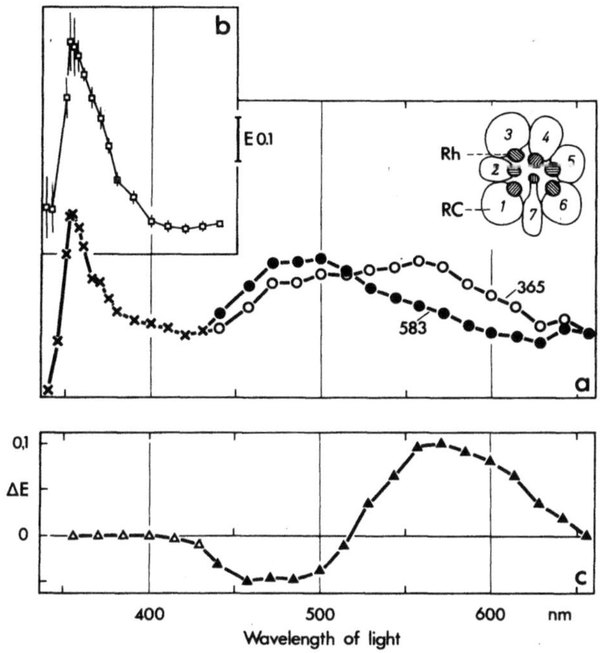

FIG. 3 - Extinction spectrum of rhabdomeres type R1-6 of Musca as measured in a microspectrophotometer. During the measurement the preparation at regular intervals was illuminated either with strong orange light (λ = 583 nm) in order to shift most of the pigment into the xanthopsin state, or with ultraviolet light (λ = 365 nm) in order to shift some xanthopsin into metaxanthopsin. Inset: Cross section through ommatidium indicating receptor cells RC and rhabdomeres Rh. b) Mean extinction spectrum in the ultraviolet of 6 ommatidia. c) Difference spectrum of a). Redrawn from Kirschfeld et al. (24).

pigment in the meta-state. In the meantime, we have collected a variety of experimental evidence that unanimously supports the sensitizing pigment hypothesis.

Evidence for a sensitizing function comes, for instance, from experiments in which flies are reared on a carotenoid-deficient diet (8-10, 38). These flies lose absolute sensitivity, which indicates that vitamin A or a derivative might be essential for their visual pigment. Unexpectedly, however, the loss in sensitivity in the UV is much stronger than in the visible. This is not to be expected if the sensitivity in the UV and visible is due to a single pigment. The sensitizing pigment concept, however, allows an easy explanation of this finding: the sensitizing pigment in conditions of vitamin A deprivation is no longer present or at least not capable of transferring energy.

The photoreceptors R1-6 exhibit polarization-sensitivity (PS), however, only in the visible and not in the UV (9, 10, 14). Presumably the dipoles, responsible for UV-sensitivity, are aligned in a different way compared to the normal chromophores of the visual pigment. If we measure the polarization-sensitivity in the UV in receptors of carotenoid-deprived flies, we expect again polarization-sensitivity, since now the β-band absorption should be responsible for the remaining UV-sensitivity. That this is actually the case is illustrated in Fig. 4. Quantitative data are discussed by Vogt and Kirschfeld (45). We can conclude from these results that the visual pigment in fly rhabdomeres is quite normal with respect to the height of the β-band absorption, and hence that an unusual β-band is not the explanation for the high UV-sensitivity.

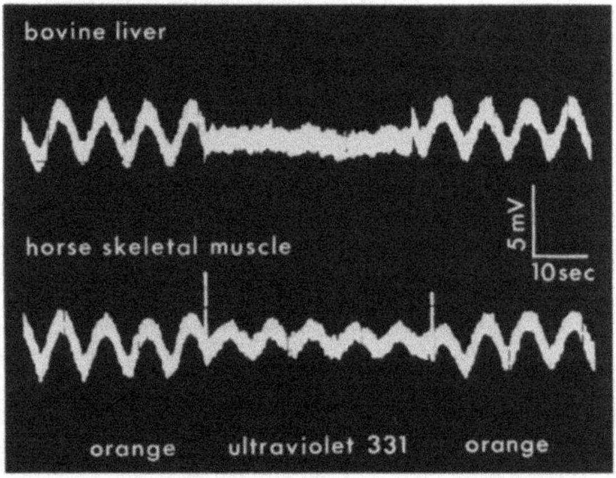

FIG. 4 - Response of receptor cells of type R1-6 to light delivered through a continuously rotating polarization filter. First orange light was applied, and then the color was switched to UV light adjusted in intensity to evoke approximately the same mean response. Then again orange light was applied. While there is no modulation during the UV in the fly raised on bovine liver (upper trace), a response to the angle of the e-vector of the light is obvious in the carotenoid-deprived fly, raised on horse skeletal muscle (from (45)).

A prediction from the sensitizing pigment concept is that energy transfer from the sensitizing pigment is to be expected not only onto the visual pigment but also onto the metapigment. This follows, at least, if we assume that the energy transfer occurs according to Förster's theory (5): the fluorescence spectrum of the sensitizing pigment should overlap not only with the visual pigment but also with the metapigment absorption spectrum, because both are close together on the wavelength scale. And this overlap is one important factor for the energy transfer. This prediction has been confirmed by determining the photosensitivity spectrum of the metapigment which also has a high UV maximum (29).

The quantum efficiency of energy transfer to the visual pigment was estimated by electrophysiological methods (43). The remarkably high value of

≥0.8 infers a distance of less than 25 Å between the sensitizing pigment and the normal chromophore if the energy transfer is mediated according to Förster-type dipole-dipole interaction. Since this distance is considerably smaller than the diameter of the visual pigment molecule, it is likely that the sensitizing pigment is directly attached to the visual pigment protein ((44), see Fig. 7).

The Sensitizing Pigment in Fly Photoreceptors: 3-hydroxyretinol

In order to recognize the sensitizing pigment spectrophotometrically, a finding by Gemperlein et al. (7) was important. These authors developed a special method (Fourier spectroscopy) that allows the rapid measurement of spectral sensitivity with high spectral resolution. They found that the UV peak in Calliphora exhibits a vibrational fine structure with peaks at 332, 350, and 369 nm. We improved our microspectrophotometer for better spectral resolution and were also able to measure the vibrational fine structure in the UV in the extinction spectrum of individual rhabdomeres of receptors R1-6. If flies are reared on a carotenoid-deprived diet, however, extinction in the UV can no longer be detected (Fig. 5; (23)). In thin-layer chromatography Vogt (40) has shown that there is, besides the 3-hydroxyretinal which represents the chromophore of the visual pigment, a still more polar compound exhibiting strong fluorescence (Fig. 1: S_1, S_2 = isomers of the sensitizing pigment). This compound is also considerably reduced in flies reared on a carotenoid-deprived diet and therefore fulfills one of the prerequisites of the sensitizing pigment. This compound was shown to be 3-hydroxyretinol (Fig. 2; (46)).

At first sight it seemed unlikely, however, that this alcohol could represent the sensitizing pigment because the peak absorbance of 3-hydroxyretinol in

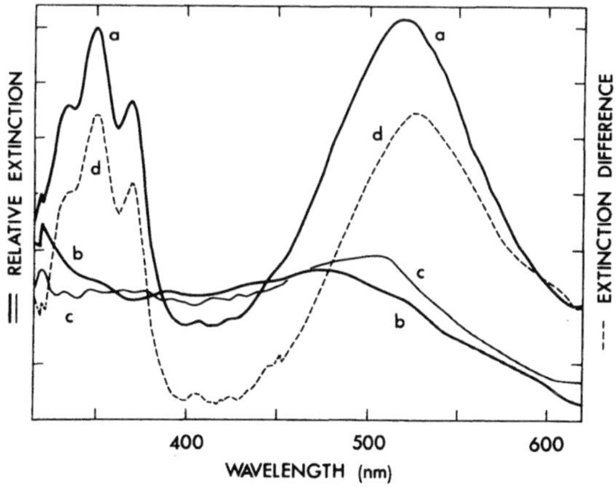

FIG. 5 - Extinction spectra of receptors type R1-6 of Musca. a: Fly grown on normal diet. b, c: Two spectra from two different flies, grown on a carotenoid-deprived medium. d: Difference between a and c (from (23)).

ethanol is at 325 nm and has no vibrational fine structure, while the UV-sensitivity in flies is at 350 nm with a fine structure (Fig. 6). This discrepancy may be explained, however, by analogy with retinol: if in retinol the ionon ring is fixed relative to the side chain at the C6-C7 bond and approximately coplanar, then a shift of the extinction maximum of some 25 nm and a fine structure become obvious (31). Such a fixation and coplanarity can be achieved either by a retrostructure (C6-C7 double bond) or by binding of retinol to a protein. For reasons discussed in detail by Vogt and Kirschfeld (46), we tentatively suggest a structure of the visual pigment complex as drawn in Fig. 7a (46).

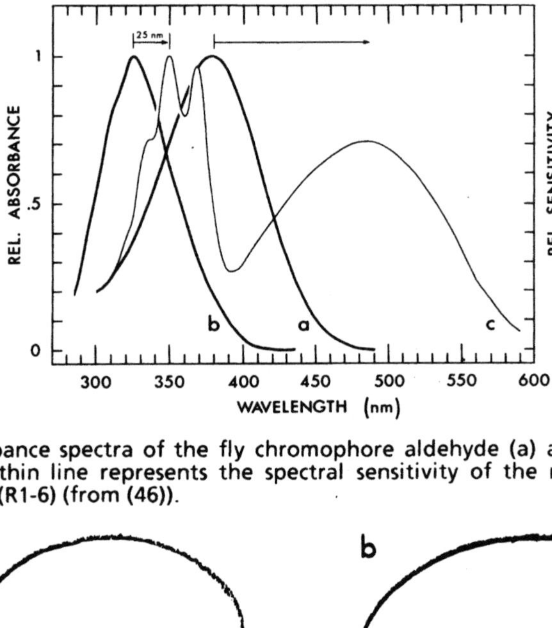

FIG. 6 - Absorbance spectra of the fly chromophore aldehyde (a) and alcohol (b) in ethanol. The thin line represents the spectral sensitivity of the most common fly photoreceptor (R1-6) (from (46)).

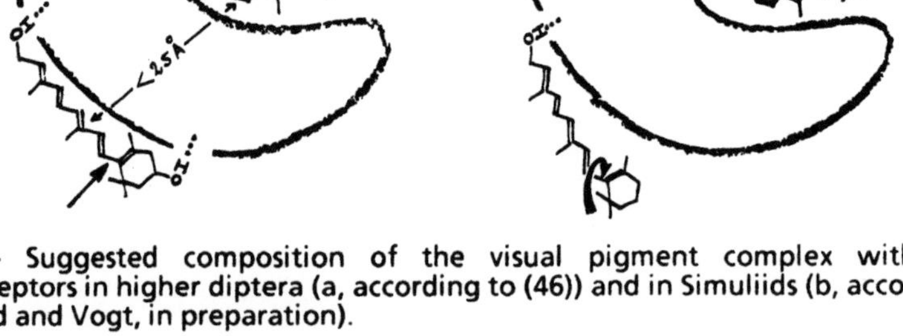

FIG. 7 - Suggested composition of the visual pigment complex within the photoreceptors in higher diptera (a, according to (46)) and in Simuliids (b, according to Kirschfeld and Vogt, in preparation).

Other Photoreceptors in the Fly with a Sensitizing Pigment
Up to now I have discussed properties of receptors R1-6 from the fly. These receptors, though anatomically distinct, all have the same spectral sensitivity and, as we have seen, a sensitizing pigment. Besides the six receptors type R1-6, there are two more receptors in each ommatidium, called R7 and R8. These receptors are not a homogeneous population: in the dorsal marginal eye region they are anatomically distinct from normal R7 and R8 (15, 47), and even in the regular ommatidia, in which they look structurally rather similar, they are functionally different (18).

Finally, the ocellar receptors have to be taken into account. As can be seen from Table 1, most of these receptors (eleven out of fifteen) have a UV-sensitizing pigment as characterized from the vibrational fine structure in the UV. In the case of R7y, it was shown with microspectrophotometry in addition that the visual pigment has λ_{max} = 430 nm.

Evolutionary Aspects
The existence of an as yet unknown visual pigment (xanthopsin) in flies leads to the question of where else in the animal kingdom xanthopsin can be found. It has been shown that, in insects, the neuropter Ascalaphus has retinal as a chromophore (33). The same chromophore was found in groups such as Caelifera, Heteroptera, Coleoptera, and Hymenoptera, whereas such groups as Lepidoptera and Diptera have 3-hydroxyretinal (41). This result supports the view that the new chromophore has a monophyletic origin and has been "invented" only once, some 300 million years ago, during the Carboniferous. Actually, because of its monophyletic origin the chromophore can probably be used as an indicator for phylogenetic relationships. For instance, the presence of either retinal or 3-hydroxyretinal supports the view that Neuropteroidea and Mecopteroidea together are a monophyletic group. This is contrary to the view favored today ((26); Fig. 8).

The next obvious question concerns the problem of which insect groups have a sensitizing pigment. The outcome of the comparative study (Vogt and Kirschfeld, in preparation) is a) the insect groups with <u>rhodopsin</u> never have UV-sensitivity with the vibrational fine structure. b) In the groups with <u>xanthopsin</u> there are two different cases: either there is UV-sensitivity with vibrational fine structure, indicative of a sensitizing function, or a UV-vibrational fine structure is lacking. Actually, higher diptera such as Bibioniformia, Empidiformia, Acalyptratae, and Calyptratae have a vibrational fine structure, whereas the more primitive diptera such as Tipuloidea and the Lepidoptera do not (Vogt and Kirschfeld, in preparation; examples in Fig. 9). The latter lack a sensitizing pigment. This result indicates that in order to exhibit a vibrational fine structure, the sensitizing pigment seems to need the second OH-group at the ionon ring: only then is there a fixation of the C6-C7 double bond, as indicated in Fig. 7a (46). A further conclusion from this study is that it took some 80 million years after the occurrence of xanthopsin in insects until the sensitizing pigment had been "invented."

TABLE 1 - Presence of sensitizing pigment in different receptor types of the fly (Musca) compound eye.

Receptor Type	Anatomical Location	UV-sensitizing Pigment	Visual - Pigment max (nm)	References	Remarks
R1-6	regular and marginal ommatidia	+	490	(23, 24)	specialized for high absolute sensitivity
R7y	in 2/3 of the ommatidia statisically distributed	+	430	(16, 18)	
R7p	in 1/3 of the regular ommatidia	-	335	(18)	
R7r	"love spot" of males	+	490	(6, 17)	
R8y	as R7y	+	515	(18)	
R8p	as R7p	-	460	(18)	
R8r	"love spot" of males	+	490	(17)	
R7marg	dorsal marginal eye region	-	335	(15)	specialized for high polarization sensitivity
R8marg	as R7marg	-	335		
ocelli receptors	ocelli	+	430	(Kirschfeld and Vogt, in preparation)	

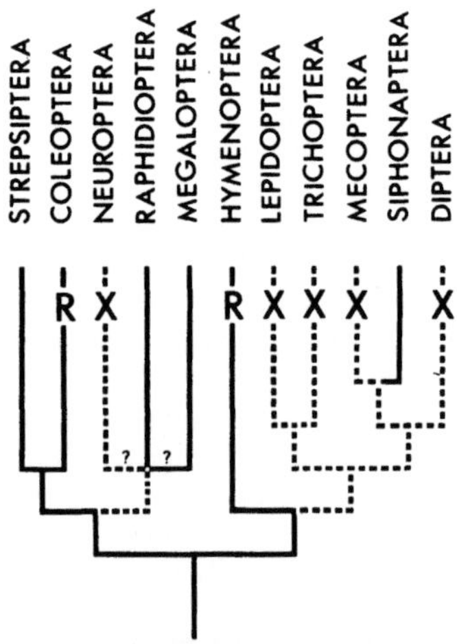

FIG. 8 - Family tree of holometabolic insects according to Kristensen (26). The letters R and X indicate in which orders rhodopsin or xanthopsin were found, respectively (according to Vogt (42)).

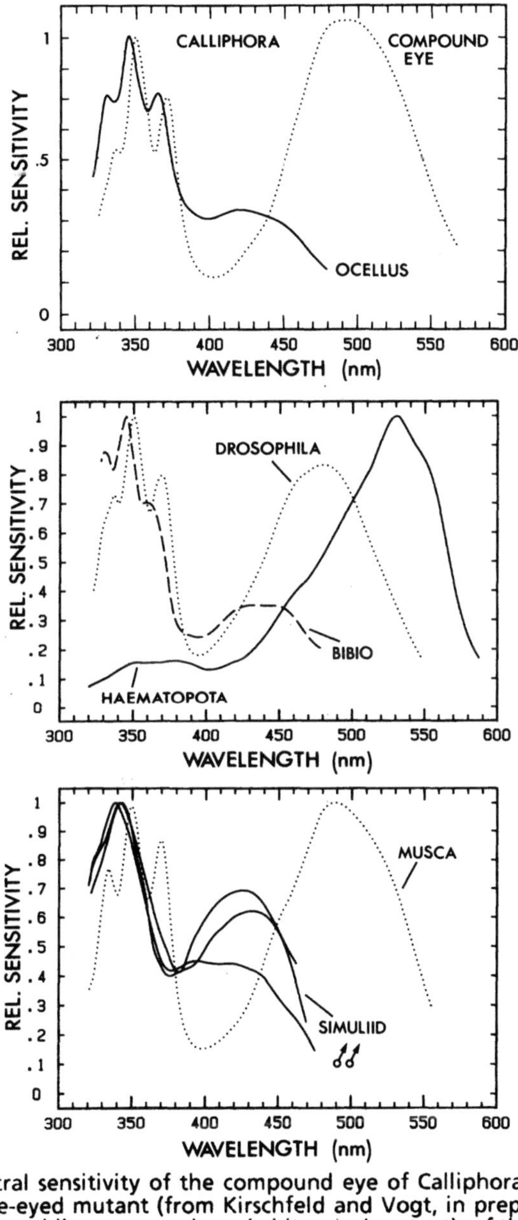

FIG. 9 - a) Spectral sensitivity of the compound eye of Calliphora and of the ocellus. Data from white-eyed mutant (from Kirschfeld and Vogt, in preparation). b) Spectral sensitivity of Drosophila compound eye (white-eyed mutant), of the dorsal eye of Bibio male, and of the tabanid Haematopota compound eye (from Vogt and Kirschfeld, in preparation). c) Spectral sensitivity of Musca compound eye (white-eyed mutant) and of the dorsal eye of three different Simuliids (from Kirschfeld and Vogt, in preparation.

Is 3-hydroxyretinol the Only Sensitizing Pigment?

In most cases, one of the characteristic properties of spectral sensitivity in photoreceptors with a sensitizing pigment is high sensitivity in the UV combined with a second peak, or at least a shoulder, at longer wavelengths. During the measurements we found one group of flies, Simuliids, which also have this general property. However, the UV-sensitivity in Simuliids is unusual in that a) it does not show the vibrational fine structure but only one maximum, and b) this maximum is shifted significantly to shorter wavelengths, i.e., at 340 instead of at 350 nm (Fig. 9c). Several observations are indicative of a UV-sensitizing pigment combined with a visual pigment of λ_{max} = 430 nm; there is a considerable variability between the heights of the two peaks in the UV and at 430 nm, respectively (as if the diet of the individuals had been different), and furthermore, it is not possible to modify the relative heights of these two peaks by means of selective chromatic adaptation.

We analyzed the visual pigment in the rhabdomeres by means of microspectrophotometry and found, indeed, a visual pigment with λ_{max} = 430 nm which fits with the position of the peak of sensitivity in the visible, the metapigment having a λ_{max} = 530 nm. The situation insofar is quite similar to that in receptors R1-6. However, why is there no vibrational fine structure as in all other sensitizing pigments analyzed up to now?

A possible explanation comes from the biochemical analysis of heads of Simuliid flies (Vogt and Kirschfeld, in preparation). Here we found - the only exception in diptera up to now - retinal instead of 3-hydroxyretinal. If our explanation of the UV vibrational fine structure is correct, namely, that it depends upon the fixation of the C6-C7 bond due to two hydrogen bonds (Fig. 7a), then we do not expect such a vibrational fine structure in the Simuliids, because the second hydrogen bond, responsible for the fixation of the C6-C7 bond in the other flies, is lacking. If retinol is the sensitizing pigment, then the spectral sensitivity in the UV should be similar to the absorption spectrum of retinol in solution, which it actually is. These data suggest that not only 3-hydroxyretinol can act as a sensitizing pigment, but that there is still at least one other molecule, which is most likely retinol (Kirschfeld and Vogt, submitted).

FUNCTIONAL CONSEQUENCES OF SENSITIZATION

Highly evolved photoreceptors should absorb light with a high probability. This can be important for an animal for several reasons: in order to allow vision at low ambient intensities where only a few quanta are available, or when the animal has to detect small optical signals (small modulation of intensity) in a short time. The detection of such signals can be a problem even in bright light because the light-quantum noise must be smaller than the signal. The only way to achieve a low quantum noise is to absorb many quanta per unit time. One of the functional consequences of sensitizing pigments is that they increase the rate of absorbed quanta. This is so because sensitizing pigments allow the overcoming of the limitation as far as absorption probability is concerned, given by membrane density, visual pigment concentration, and length of the absorbing photoreceptor organelle. The advantage of a sensitizing pigment is obvious: these molecules, since they do not directly mediate transduction, need not be large and hence can be incorporated into the photoreceptor membranes along with the visual pigment without demanding much space.

Another functional consequence of sensitizing pigments is that in principle they may extend the spectral range over which the receptor is sensitive. That is, besides the gain in number of collected quanta, patterns can also be seen (e.g., structures present in the ultraviolet spectral range only) that otherwise would be undetectable.

It is surprising that sensitization of visual pigment that is so common in insects has not yet been described in vertebrates. At the beginning of our work we thought that the reason might be that vertebrates do not have 3-hydroxyretinol, which could have been a prerequisite for sensitization. Since the Simuliids do not have this chromophore but nevertheless have a sensitizing pigment, we have to take into account that animals with rhodopsin or porphyropsin in principle could also make use of sensitization to improve their photoreceptors.

THE PROBLEM OF PHOTOOXIDATION IN PHOTORECEPTORS

Light is necessary for vision, but light is also potentially dangerous: it is capable of destroying cells or their components by photooxidation. Many different mechanisms have been developed in evolution to overcome this problem, whereby a most efficient method is just to cover cells with an integument that absorbs dangerous wavelengths of light. Since photoreceptors must be accessible to light for functional reasons, special mechanisms have to be developed. The situation is comparable to that in photosynthetic plant cells: here also light must enter the cells to reach the sites of chlorophyll, and free oxygen is produced by photosynthesis. The major function of carotenoid pigments seems to be to protect such cells from photosensitized oxidation (25). Actually, there are also photoreceptors which incorporate C_{40}-carotenoids into the membranes. The most sophisticated type is the type of receptor called R7y: in these cells there is a UV-sensitizing pigment that sensitizes a 430 nm xanthopsin (18). This blue-absorbing visual pigment itself is protected from direct access of light by means of a C_{40}-carotenoid (22) -identified as zeaxanthin (44) - that acts as a blue-absorbing light filter. By this means a UV-sensitive photoreceptor is created in which part of the energy of the UV quanta is dissipated as heat during the energy transfer to the normal chromophore. In addition, the C_{40}-carotenoid most likely acts as a "quencher" for dangerous activated states of molecules. The complex setup leads to a receptor that is extremely resistant to short wavelength irradiation (21, 50). The photoreceptors of the dorsal eye of Simuliids (Fig. 9c) apparently have the same organization, the only difference being that rhodopsin rather than xanthopsin is the visual pigment, and retinol (vitamin A_1) rather than 3-

hydroxyretinol (vitamin A_3) is probably the sensitizing pigment. There is, as in R7y, also a C_{40}-carotenoid incorporated into these photoreceptors (Kirschfeld and Vogt, in preparation). High concentrations of C_{40}-carotenoids cannot be incorporated into all types of photoreceptors because of the strong absorption of C_{40}-carotenoids in the blue spectral range. Hence, only photoreceptors which need not have high blue-sensitivity can make use of C_{40}-carotenoids. Other photoreceptors may make use of different mechanisms for overcoming the problem of photooxidation, such as, e.g., repair mechanisms (21).

FROM OPTICAL INTERMEDIATES TO EXCITATION

The immediate consequence of absorption of a light quantum by a visual pigment molecule is the isomerization of the chromophore from the 11-cis to the all-trans conformer. This change can be followed optically with picosecond technology, and it seems as if this isomerization is a necessary step in order to excite the cell: if 11-cis-locked analogs of retinal are incorporated into opsin, an unbleachable pigment is formed (1). Following initial isomerization a sequence of spectral intermediates can be identified at different temperatures. These changes can also be followed at room or body temperature if instruments with sufficient time resolution are used. Two schemes that illustrate the temporal sequence of events for vertebrate rhodopsin and for Calliphora xanthopsin are given in Fig. 10.

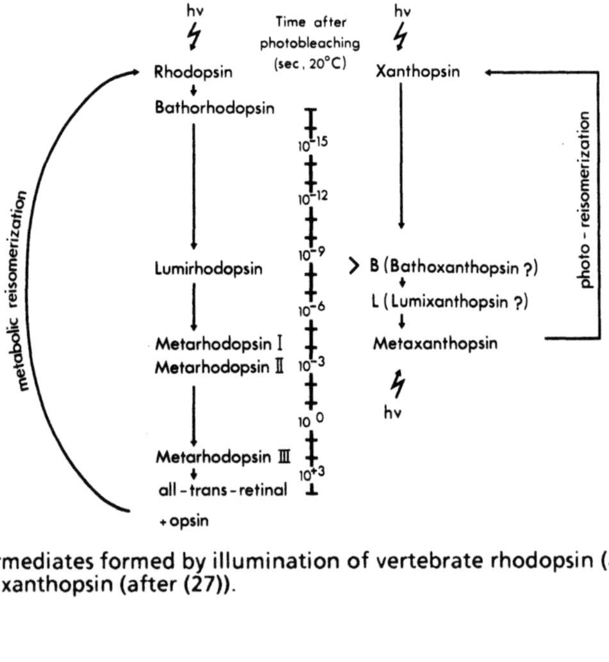

FIG. 10 - Intermediates formed by illumination of vertebrate rhodopsin (after (36)) and of Calliphora xanthopsin (after (27)).

It has been a matter of debate as to which one of the spectrally defined products or transitions is necessary for the excitation of the receptor. It is clear that in vertebrates, substances such as metarhodopsin III that occur only after the onset of the receptor potential cannot be necessary to induce it. Because of the temporal coincidence of the Meta I-to-Meta II transition with the receptor potential this step has been considered as a possible source of excitation.

From our present knowledge of the structure of the rhodopsin molecule, it follows that those parts of the opsin moiety that are important for the induction of the enzyme cascade which is considered as a causal step in the transduction process (see Chabre and Applebury, this volume) are quite distant from the chromophoric group. Hence it cannot necessarily be expected that these processes are reflected in absorption changes.

Actually, there is an interesting difference between vertebrate and invertebrate visual pigment: whereas the isomerized chromophore in vertebrates eventually comes off of the opsin, this is not so in invertebrates: reisomerization in vertebrates usually is a metabolic process, whereas reisomerization in invertebrates is usually performed by light (review, (11)). Thus, invertebrate visual pigments such as xanthopsin of flies can be considered as bistable visual pigments, whereas in vertebrates they are monostable.

The bistability of xanthopsin suggests a method of analyzing the question of whether it is sufficient for excitation to isomerize xanthopsin (X), or whether metaxanthopsin (MX) must exist for some time in order to trigger a response. The experiment makes use of the fact that the latency of the receptor potential is significantly longer (≈ 10 ms) than it takes to form MX (≈ 100 µs). Hence it is possible, first by means of a colored (blue) flash that is absorbed mainly by X, to convert X to MX, and then, within the latency of the response, to reconvert MX to X by an orange flash that is mainly absorbed by MX. The question is whether a receptor potential is induced even if MX has disappeared before the occurrence of the late receptor potential (LRP). It was found that the occurrence of the LRP can be stopped by the second flash if this flash is given 3 to 5 ms after the first one (12). For practical reasons, the reduction of the response is not "all-or-none" as might be expected for an ideal case. In fact, the response is only reduced by the second flash (Fig. 11), but nevertheless, a blue flash followed by a red flash gives a smaller response than a blue flash alone! The response cannot be cancelled completely because a) not all MX molecules created by the first flash can be reconverted into R by the second flash, and b) the second flash itself, due to the overlap of the absorption spectra of X and MX, also converts some X to MX. Quantitative estimates show that the reduced response amplitude corresponds to a reduction of the stimulus intensity to some 25%. Parallel spectrophotometric measurements have shown that some 25% of MX remains after the second flash. Hence the amplitude reduction corresponds to that expected.

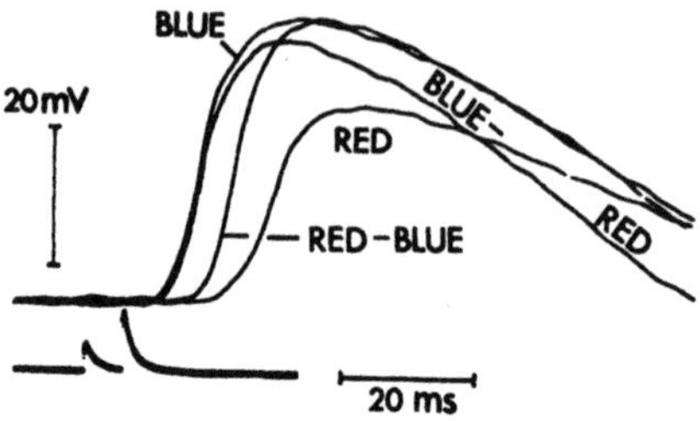

FIG. 11 - Intracellularly recorded responses of a photoreceptor type R1-6 of Calliphora to light flashes (the lower trace shows the time course of the light intensity for the case of a double flash stimulus). Either a blue flash, a red flash, or two flashes with a delay of 5 ms in the sequence blue-red or red-blue, respectively, were given. The response to the sequence blue-red was significantly smaller than that to a blue flash alone. Note that the longer latency of the sequence red-blue is due to the fact that the blue flash occurs second. Temperature, 10C (redrawn from (12)).

This result shows that, following the conversion of X into MX, there remains a period of 3 to 5 ms during which reconversion of MX to X more or less completely abolishes the effect of the initial X-to-MX conversion. During this period no irreversible transmitter release or chemical chain reaction can have been initiated. The MX formed must exist for a few milliseconds in order to trigger the response.

Activation of the visual pigment by light is only a first step in photoreception. Analysis of the following steps probably will be more easy in vertebrate photoreceptors which are more easily accessible with biochemical methods.

Acknowledgements. I thank R. Hardie and K. Vogt for discussing and reading the manuscript.

REFERENCES

(1) Akita, H.; Tanis, S.P.; Adams, M.; Balogh-Nair, V.; and Nakanishi, K. 1980. Nonbleachable rhodopsin retaining the full natural chromophore. J. Am. Chem. Soc. **102:** 6370-6372.

(2) Boschek, C.B., and Hamdorf, K. 1976. Rhodopsin particles in the photoreceptor membrane of an insect. Z. Naturforsch. **31c:** 763.

(3) Burkhardt, D. 1962. Spectral sensitivity and other response characteristics of single visual cells in the arthropod eye. Symp. Soc. Exp. Biol. **16:** 86-109.

(4) Dartnall, H.J.A. 1952. Visual pigment 467, a photosensitive pigment present in tench retinae. J. Physiol. **116:** 257-289.

(5) Förster, T. 1951. Fluoreszenz organischer Verbindungen. Göttingen: Vandenhoeck und Ruprecht.

(6) Franceschini, N.; Kirschfeld, K.; and Minke, B. 1981. Fluorescence of photoreceptor cells observed in vivo. Science **213:** 1264-1267.

(7) Gemperlein, R.; Paul, R.; Lindauer, E.; and Steiner, A. 1980. UV fine structure of the spectral sensitivity of flies visual cells. Naturwiss. **67:** 565-566.

(8) Goldsmith, T.H.; Barker, R.J.; and Cohen, C.F. 1964. Sensitivity of visual receptors of carotenoid-depleted flies: a vitamin A deficiency in an invertebrate. Science **146:** 65-67.

(9) Guo, A.K. 1980. Elektrophysiologische Untersuchungen zur Spektral- und Polarisations-Empfindlichkeit der Sehzellen von Calliphora erythrocephala II. Sci. Sin. **XXIII:** 1461-1468.

(10) Guo, A.K. 1981. Elektrophysiologische Untersuchungen zur Spektral- und Polarisationsempfindlichkeit an den Sehzellen von Calliphora erythrocephala III. Sci. Sin. **XXIV:** 272-286.

(11) Hamdorf, K. 1979. The physiology of invertebrate visual pigments. In Handbook of Sensory Physiology, ed. H. Autrum, vol. VII/6A, pp. 145-224. Berlin, Heidelberg, New York: Springer-Verlag.

(12) Hamdorf, K., and Kirschfeld, K. 1980. Reversible events in the transduction process of photoreceptors. Nature **283:** 859-860.

(13) Hamdorf, K.; Paulsen, R.; and Schwemer, J. 1973. Photoregeneration and sensitivity control of photoreceptors of invertebrates. In Biochemistry and Physiology of Visual Pigments, ed. H. Langer, pp. 155-166. Berlin, Heidelberg, New York: Springer-Verlag.

(14) Hardie, R.C. 1978. Peripheral visual function in the fly. Ph.D. Thesis, Australian National University, Canberra.

(15) Hardie, R.C. 1984. Properties of photoreceptors R7 and R8 in dorsal marginal ommatidia in the compound eyes of Musca and Calliphora. J. Comp. Physiol. A **154:** 157-165.

(16) Hardie, R.C.; Franceschini, N.; and McIntyre, P.D. 1979. Electrophysiological analysis of fly retina. II. Spectral and polarisation sensitivity in R7 and R8. J. Comp. Physiol. **133:** 23-39.

(17) Hardie, R.C.; Franceschini, N.; Ribi, W.; and Kirschfeld, K. 1981. Distribution and properties of sex-specific photoreceptors in the fly Musca domestica. J. Comp. Physiol. **145:** 139-152.

(18) Hardie, R.C., and Kirschfeld, K. 1983. Ultraviolet sensitivity of fly photoreceptors R7 and R8: evidence for a sensitising function. Biophys. Struct. Mech. **9:** 171-180.

(19) Horridge, G.A., and Mimura, K. 1975. Fly photoreceptors. I. Physical separation of two visual pigments in Calliphora retinula cells 1-6. Proc. Roy. Soc. Lond. B **190:** 211-224.

(20) Kirschfeld, K. 1979. The function of photostable pigments in fly photoreceptors. Biophys. Struct. Mech. **5:** 117-128.

(21) Kirschfeld, K. 1982. Carotenoid pigments: their possible role in protecting against photooxidation in eyes and photoreceptor cells. Proc. Roy. Soc. Lond. B **216:** 71-85.

(22) Kirschfeld, K.; Feiler, R.; and Franceschini, N. 1978. A photostable pigment within the rhabdomere of fly photoreceptors no. 7. J. Comp. Physiol. **125:** 275-284.

(23) Kirschfeld, K.; Feiler, R.; Hardie, R.; Vogt, K.; and Franceschini, N. 1983. The sensitizing pigment in fly photoreceptors. Properties and candidates. Biophys. Struct. Mech. **10:** 81-92.

(24) Kirschfeld, K.; Franceschini, N.; and Minke, B. 1977. Evidence for a sensitising pigment in fly photoreceptors. Nature **269:** 386-390.

(25) Krinsky, N.I. 1968. The protective function of carotenoid pigments. Photophysiology **3:** 123-195.

(26) Kristensen, N.P. 1975. The phylogeny of hexapod "orders". A critical review of recent accounts. Z. Zool. Syst. Evol. Forsch. **13:** 1-44.

(27) Kruizinga, B.; Kamman, R.L.; and Stavenga, D.G. 1983. Laser induced visual pigment conversions in fly photoreceptor. Measured in vivo. Biophys. Struct. Mech. **9:** 299-307.

(28) Lythgoe, J.N. 1972. The adaptation of visual pigments to the photic environment. **In** Handbook of Sensory Physiology, ed. H.J.A. Dartnall, vol. VII/I, pp. 566-603. Berlin, Heidelberg, New York: Springer-Verlag.

(29) Minke, B., and Kirschfeld, K. 1979. The contribution of a sensitizing pigment to the photosensitivity spectra of fly rhodopsin and metarhodopsin. J. Gen. Physiol. **73:** 517-540.

(30) Morton, R.A. 1972. The chemistry of the visual pigments. **In** Handbook of Sensory Physiology, ed. H.J.A. Dartnall, vol. VII/I, pp. 33-68. Berlin, Heidelberg, New York: Springer-Verlag.

(31) Ong, D.E., and Chytil, F. 1978. Cellular retinol-binding protein from rat liver. J. Biol. Chem. **253:** 828-832.

(32) Ostroy, S.E. 1978. Characteristics of Drosophila rhodopsin in the wild-type and norpA vision transduction mutants. J. Gen. Physiol. **72:** 714-732.

(33) Paulsen, R., and Schwemer, J. 1972. Studies on the insect visual pigment sensitive to ultraviolet light: retinal as the chromophoric group. Biochim. Biophys. Acta **283:** 520-529.

(34) Paulsen, R., and Schwemer, J. 1979. Vitamin A deficiency reduces the concentration of visual pigment protein within blowfly photoreceptor membranes. Biochim. Biophys. Acta **557:** 385-390.

(35) Shaw, S.R. 1968. Organization of the locust retina. Symp. Zool. Soc. Lond. **23:** 35-163.

(36) Shichi, H. 1983. Biochemistry of Vision. New York: Academic Press.

(37) Snyder, A.W., and Pask, C. 1973. Spectral sensitivity of dipteran retinula cells. J. Comp. Physiol. **84:** 59-76.

(38) Stark, W.S., and Zitzmann, W.G. 1976. Isolation of adaptation mechanisms and photopigment spectra by vitamin A deprivation in Drosophila. J. Comp. Physiol. **105:** 15-27.

(39) Stavenga, D.G.; Zantema, A.; and Kuiper, J.W. 1973. Rhodopsin processes and the function of the pupil mechanism in flies. **In**

Biochemistry and Physiology of Visual Pigments, ed. H. Langer, pp. 175-180. Berlin, Heidelberg, New York: Springer-Verlag.

(40) Vogt, K. 1983. Is the fly visual pigment a rhodopsin? Z. Naturforsch. **38c:** 329-333.

(41) Vogt, K. 1984. The chromophore of the visual pigment in some insect orders. Z. Naturforsch. **39c:** 196-197.

(42) Vogt, K. 1984. Zur Verteilung von Rhodopsin und Xanthopsin bei Insekten. (Verh. Dtsch. Zool. Ges.) Stuttgart: Gustav Fischer.

(43) Vogt, K., and Kirschfeld, K. 1982. Die Quantenausbeute der Energieübertragung von Photorezeptoren von Fliegen, p. 337. (Verh. Dtsch. Zool. Ges.) Stuttgart: Gustav Fischer.

(44) Vogt, K., and Kirschfeld, K. 1983. C40 carotinoide in Fliegenaugen. (Verh. Dtsch. Zool. Ges.) Stuttgart: Gustav Fischer.

(45) Vogt, K., and Kirschfeld, K. 1983. Sensitizing pigment in the fly. Biophys. Struct. Mech. **9:** 319-328.

(46) Vogt, K., and Kirschfeld, K. 1984. Chemical identity of the chromophores of fly visual pigment. Naturwiss. **71:** 211-213.

(47) Wada, S. 1974. Spezielle randzonale Ommatidien von Calliphora erythrocephala Meig. (Diptera: Calliphoridae): Architektur der zentralen Rhabdomeren-Kolumne und Topographie im Komplexauge. Intl. J. Insect Morphol. Embryol. **3:** 397-424.

(48) Wald, G. 1936. Carotenoids and the visual cycle. J. Gen. Physiol. **19:** 351-371.

(49) Wolken, J.J.; Bowness, J.M.; and Scheer, I.J. 1960. The visual complex of the insect: Retinene in the housefly. Biochim. Biophys. Acta **43:** 531-537.

(50) Zhu, H., and Kirschfeld, K. 1984. Protection against photodestruction in fly photoreceptors by carotenoid pigments. J. Comp. Physiol. A **154:** 153-156.

The Molecular Mechanism of Photoreception, ed. H. Stieve, pp. 51-66. Dahlem Konferenzen 1986. Berlin, Heidelberg, New York, Tokyo: Springer-Verlag.

Interaction of Photoactivated Rhodopsin with Photoreceptor Proteins: The cGMP Cascade

M. Chabre* and M.L. Applebury**
*Laboratoire de Biologie Moleculaire et Cellulaire
Département de Recherche Fondamentale
Grenoble 38041, France
**Dept. of Biological Sciences, Purdue University
West Lafayette, IN 47907, USA

Abstract. Photoactivated rhodopsin (R*) triggers a cascade of fast reactions mediated by transducin (T) which lead to the amplified activation of cGMP phosphodiesterase (PDE). The first step is the catalysis by R* of a GRP/GDP exchange in the α-subunit of T: the holoenzyme $T_{\alpha}GDP$-$T_{\beta\gamma}$ binds specifically to R*, and the interaction loosens the nucleotide site, allowing the fast exchange of the bound nucleotide for free GDP or GTP. Upon GTP binding, the R*-T complex dissociates and $T_{\alpha}GTP$ separates from $T_{\beta\gamma}$ and becomes soluble. This three-body dissociation is probably important in pulling the equilibrium toward GTP/GDP exchange. One R* can interact in sequence with hundreds of T at the rate of one per millisecond. Each dissociated $T_{\alpha}GTP$ then interacts with the inhibitor (I) subunit of a PDE and releases the inhibition, thus allowing a very fast cGMP hydrolysis.

The process is reversed upon the spontaneous hydrolysis of the GTP on T_{α}. Then $T_{\alpha}GTP$ ceases to interact with I, which can re-inhibit the PDE, and $T_{\alpha}GDP$ binds back to $T_{\beta\gamma}$. This inactivation process is slow ($\tau_{1/2} \simeq 30$ s.). Another regulation process is initiated by the phosphorylation of R* by an ATP-dependent kinase specific for R*. After multiple phosphorylation of R* on a site different from that of T binding, a soluble "48K" protein binds to R*-P, which results in the quenching of R*-T interaction.

The activation and regulation of the transducin cascade is probably the only function of R*, but the cascade might control another process besides cGMP hydrolysis, namely, the hydrolysis of phosphotidylinositol and release of inositol triphosphate.

Transducin is analogous to the various G proteins of the hormone-controlled adenylate cyclase systems, and the detailed analysis of the transducin cascade and its regulation may provide instructive models for the other systems.

INTRODUCTION

It has become clear that a major role of photoexcited rhodopsin is to initiate rapid cGMP hydrolysis in the vertebrate rod outer segments (ROS). The process entails a cascade of fast reactions that provide high gain (amplification) and that are under tight regulation. Current investigations address the questions of whether or not this is the unique task of photoexcited rhodopsin (R*) and whether or not cGMP hydrolysis is the first amplified signal in the phototransduction signalling process. The identification of the components of this cascade has given us the unique opportunity to explore the molecular basis of a receptor-mediated signalling process in detail.

There are at least four proteins involved with R* in this reaction cascade: the guanine nucleotide binding protein named alternatively "GTPase," "G-protein," or transducin (T), the ATP-dependent kinase (K) specific for R*, the "48K" protein, and the cGMP phosphodiesterase (PDE). All of them are peripherally bound to the disc membrane or soluble in the cytoplasm. The first three (T, K, and "48K") interact directly at some stage with R*.

Three contributions have been essential in laying down the basic scheme of the amplification cascade: a) Bownd's group (56) demonstrated that rapid, light-dependent cGMP hydrolysis occurred in frog rod outer segments (ROS). Liebman's group (35, 58) further characterized the process using their pH titrimetric method and showed that low levels of illumination initiated a rate of up to 10^6 cGMP hydrolyzed R* s in fractionated bovine ROS. The observations emphasized a possible role for rapid light-induced cGMP hydrolysis in visual transduction. b) Wheeler and Bitensky (54) illustrated that a light-activated GTPase (T) activity regulated cGMP phosphodiesterase activity, and Kühn (29) observed a specific binding of T, but not of PDE, to R* and a GTP-dependent dissociation of T from R*. These studies pointed out that T was the first intermediate to interact with R*. c) Godchaux and Zimmerman (23) identified the α and β subunits of GTPase (T) and illustrated a light-catalyzed GTP/GDP exchange function for this protein. The key work of Fung et al. (19, 21) measured the amplified GTP/GDP turnover catalyzed by R* and demonstrated that T, when it binds GTP or a non-hydrolyzable analog of GTP, can activate PDE in the absence of R*. Light-scattering studies (31) have further demonstrated that T binding to R*, as well as release from R*, was fast enough to be involved in the visual transduction process. Thus, the functional protein activities, the sequence of events, and the appropriate time scale for events in the cGMP cascade had been established.

Progress on the biochemical determination of T (3, 4, 18, 19, 23, 29) and analysis of the various interactions between all the proteins involved (30, 44) now lead to the elaborate but still partially tentative scheme of Fig. 3 (below). We shall recapitulate the biochemical data (Table 1 and Fig. 1). Then we will discuss the successive steps which control the triggering, the rate, the gain, and the turnoff of the cascade.

TABLE 1 - The biochemical data for the components of the cascade in vertebrate rods.

Proteins (symbol)	Relation to membrane	Molecular weight (kiloDaltons)	Stoichiometry v Rhodopsin	Equivalent molar concentration in cytoplasm
Rhodopsin (R)	intrinsic	39 + 2 (glycocong.)	1	-
Transducin ($T = T_\alpha + T_\beta + T_\gamma$)	peripheral or soluble	= 80 (39 + 37 + 6)	= 10^{-1}	= 500 μM
"48K"	soluble	= 50	= 3.10^{-2}	= 150 μM
cGMP phos-phodiesterase (PDE) + inhibitor (I)	peripheral	= 180 (83 + 84 + 13)	= 10^{-2}	= 50 μM
ATP-dependent R* Kinase (K)	soluble	68	≤ 10^{-3}	= 5 μM
"Rim Protein" (P)	intrinsic	240	= 3.10^{-3}	-

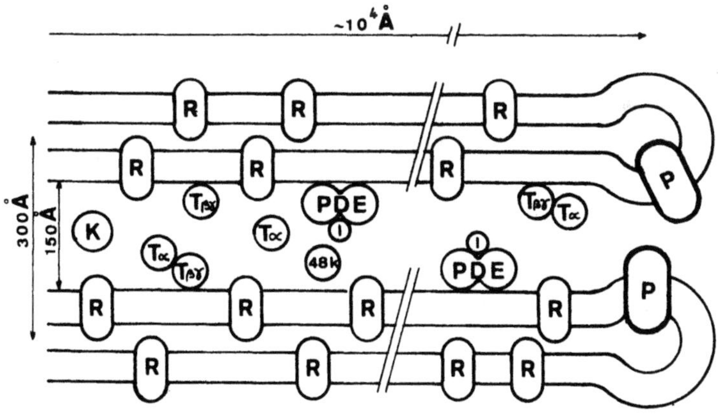

FIG. 1 - The locations of the proteins in the interdisc cleft.

THE PARTNERS

Rhodopsin (R)

Structural modeling from the known amino acid sequence predicts that this integral membrane protein is folded into seven α-helices spanning the disc membrane (14). The model is in agreement with all the previous biochemical and biophysical data. The chromophore site is buried in a central hydrophobic core. All the known sites of interaction with other proteins are on the hydrophilic cytoplasmic face which constitutes no more than 1/4 of the protein. The same proportion of the protein protudes inside the disc. This face has a larger negative charge but has no known function besides its possible contribution to the binding of intradiscal calcium (see Fig. 2).

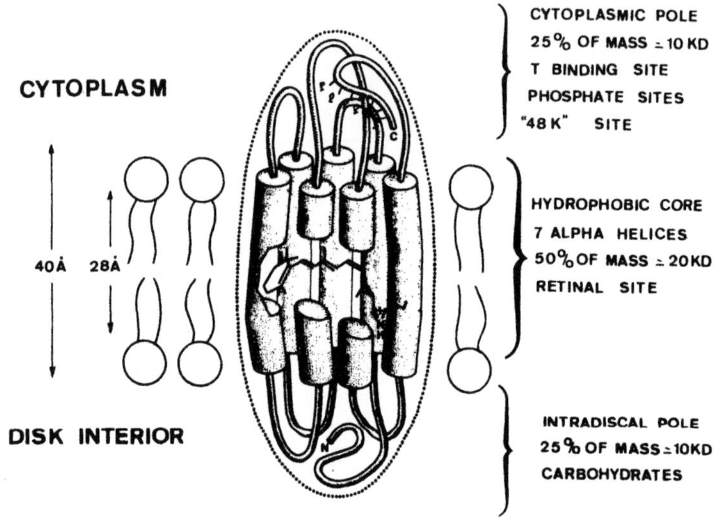

FIG. 2 - The structure of rhodopsin from retinal rods (from a figure given by P. Hargrave).

Transducin (T)

Transducin is made up of two functionally differentiated subunits, T_α[39K], which bears the guanine nucleotide site, and $T_{\beta\gamma}$, made up of two polypeptide chains [37K and 6K]. It is present at about one copy [$T_\alpha \cdot T_{\beta\gamma}$] for every ten rhodopsins. In the dark, at the ionic strength of the cytoplasm, T is peripherally bound to the disc membrane. The binding to the membrane is sensitive to the ionic strength and is probably not specific to rhodopsin, except in that rhodopsin contributes to the electric charge of the membrane. At low ionic strength, T is released as a complex T_αGDP-$T_{\beta\gamma}$. The association between the subunits is not very strong; T_αGDP and $T_{\beta\gamma}$ can be dissociated by various mild treatments (high pH, high Mg^{2+}) and independently isolated (4, 18), but the isolated subunits do not bind anymore to the membrane. The exchange of the GDP for GTP or non-hydrolyzable analogs markedly changes the properties of the T_α subunit; it dissociates from $T_{\beta\gamma}$ at normal pH and ionic strength and becomes markedly soluble.

ATP-dependent Kinase (K)

This protein specifically catalyzes the phosphorylation of photoexcited rhodopsin. It is a soluble protein, apparently present in low concentrations of < 1 per 1000 rhodopsins (though it might be partially lost in the ROS isolation procedure). Upon illumination, K binds to R*, and this property can be used for its isolation. It is composed of a single polypeptide of 68 kD (30).

cGMP Phosphodiesterase (PDE)

PDE is peripherally bound to the disc membrane and present at ~1 copy per 100 rhodopsins. Like T, it can be extracted at low ionic strength, although there is no light-dark dependence. It is composed of two subunits (88 kD and 84 kD) whose peptide maps are remarkably similar and a heat stable, trypsin-sensitive, inhibitory subunit of 13 kD. The soluble enzyme is a stable, 185 kD holoenzyme of these three subunits (3).

48K Protein

48K protein was shown by Kühn to elute from dark-adapted disc membranes and bind to light-adapted membranes. Its binding must be less tight than T, since it can be extracted by 1 mol/l NH_4Cl even under light-adapted conditions. Moreover, in the presence of ATP or GTP the light-induced binding becomes irreversible (28). Because relatively significant quantities of this protein are present ([48K] ≃ [T] in unbroken ROS) and because of the striking light-dependent binding, it is an obvious candidate for a participant in visual transduction. Two recent observations can be added to the story: 48K binds preferentially to phosphorylated, illuminated rhodopsin (Kühn, private communication). Recent work in a complementary field indicates that the major antigen causing the experimental autoimmune uveo retinitis is a 50K polypeptide called "Retinal S Antigen," arising from photoreceptor cells. The polypeptide has been purified to homogeneity. Polyclonal and monoclonal antibodies localize the peptide to ROS (but also to inner segments) (12, 16). Biochemical, immunological, and functional evidence has been recently obtained which identifies the retinal S antigen with the "48K" protein (43).

R → R* Transformation

In the dark, rhodopsin is remarkably inert and does not seem to interact with any other protein. R*, defined functionally as the state of high affinity for T_αGDP-$T_{\beta\gamma}$,has been identified with the state defined spectrally as Meta II rhodopsin (6, 15). Kühn et al. (34) have shown that the C terminal end, which bears the multiple sites of phosphorylation of R*, is made more accessible to proteolytic attack upon photoexcitation. This C terminal end is not critical for the R*-T binding, since binding requires the integrity of other cytoplasmic loops (33). The structural changes in the protein correlated with the R → R* transition are not well understood, and the detailed nature of the newly exposed R* site is yet to be defined.

R*-T^{GDP} Binding

This is the key interaction which mediates the amplifying cascade. The binding constant has not been directly measured but is estimated to be $\sim 10^7$ based on GTP titration of T release and the GTP-T binding constant (4). Rapid T binding can be titrated with formation of stoichiometric amounts of R* (31). This suggests that there is one-to-one binding of T and R*. The binding is not very sensitive to the ionic strength. The specific interaction between R* and T is probably mediated by the T_{α}-subunit, but the presence of stoichiometric amounts of $T_{\beta\gamma}$, and therefore the formation of the T_{α}^{GDP}-$T_{\beta\gamma}$ complex, is necessary for the binding to R*. Under the non-physiological conditions of GTP omission in the media, the binding is long-lived. Besides providing a convenient method of purification of T, this artificial stabilization allows one to study the effect of the binding of both R* and T in reconstituted systems. The nature of the quaternary structural interfaces of this complex is an outstanding question.

Effect of the Binding on T

Binding to R* perturbs the structure of T, resulting in modification of its nucleotide site. Structural perturbations in T are demonstrated by the change of reactivity of SH groups and in exposure to proteolytic cleavage (20, 25). In addition, chemical labeling studies using bacterial toxins have indicated conformational changes. In the presence of GppNHp, T becomes sensitive to ADP-ribosylation by cholera toxin (1); the presence of pre-illuminated membranes is necessary, and only non-hydrolyzable analogs (except $GTP_{\gamma}S$) permit reaction. Fung proposes R*-T_{α}^{GppNHp}-$T_{\beta\gamma}$ as the transient intermediate susceptible to ribosylation (40). It is puzzling, however, that neither $GTP_{\gamma}S$ nor a continuous supply of GTP, which should also form such intermediates, promotes reaction. The transient lifetime of this complex may be one factor that contributes to whether T may be modified by cholera toxin.

In the presence of GTP, T cycles repetitively between its GDP and GTP states and does not remain bound for any significant time. Robinson et al. (45) noted that the amount of R*-T^{GTP} measured is regulated by the GTP/GDP ratio and showed that GDP directly competes with GTP for the light-induced binding site. Bennett and Dupont (7) have shown that in the absence of GTP the rate of GDP/GDP exchange on T, which is very low in the dark, is considerably enhanced upon illumination. The binding of T to R* seems to alter the nucleotide site, thus facilitating exchange with any other guanine nucleotide present. Other support for T transitions comes from work of Vuong, Chabre, and Stryer. They have used fast light-scattering signals on oriented ROS (52) to show that the kinetics of GTP/GDP exchange are slowed by addition of increasing GDP concentrations. The GTP/GDP exchange competes significantly with GDP/GDP exchange, and only the GTP exchanged form contributes to the scattering signal. Like R-R* transitions, little is known about the specific structural changes that take place in T on binding.

Effect of the Binding on R*

Although the binding of T is on the cytoplasmic surface of R*, binding also affects the chromophore environment in the hydrophobic core of opsin. The stable binding of T to R* in the absence of GTP shifts the Meta I - Meta II equilibrium formed upon bleaching rhodopsin towards the Meta II state (6, 15). Furthermore, at physiological temperature and pH in frog rods, strong T binding perturbs the spontaneous decay of Meta II to Meta III and other late photoproducts, apparently blocking it in the Meta II state (44). Another effect of the binding of T on R* is to block the kinase action (see below). Thus transducin binding also appears to alter opsin conformational changes which are reflected by the spectrum of the bound chromophore. Do these subtle reciprocal controls on dynamic transitions have implications for visual triggering?

The GTP/GDP Exchange and Transducin Dissociation

As noted above, upon R*-T interaction GTP rapidly replaces GDP in the now "loosened" nucleotide site of T_α. This exchange dramatically modifies T_α; $T_\alpha GTP$ loses its affinity for R* as well as for $T_{\beta\gamma}$ and becomes soluble. Release is most easily observed with non-hydrolyzable GTP analogs which block T_α in the "active" conformation by preventing normal "decay" due to hydrolysis of GTP by GTPase activity. Upon addition of GTP to illuminated membranes at physiological ionic strength, followed by sedimentation of the membranes, $T_\alpha GDP$ is preferentially collected in the supernatant. Presumably, it must have been eluted as $T_\alpha GTP$ which hydrolyzed to $T_\alpha GDP$ during the centrifugation. Without GTP, $T_\alpha GDP$-$T_{\beta\gamma}$ is not released under these conditions. Non-hydrolyzable analogs of GTP effect release as well as GTP. At low ionic strength with GTP, both T_α and $T_{\beta\gamma}$ are released as soluble entities, though $T_{\beta\gamma}$ has a tendency to aggregate. The dissociation of T_α from $T_{\beta\gamma}$ (ultimately releasing T_α and leaving $T_{\beta\gamma}$ behind peripherally associated with the membrane) may be an important factor leading to GTP/GDP exchange. Although the nucleotide site would have comparable affinities for GDP and GTP, presumably only GTP would lead to release. The sequence of dynamic events and the state of T_α following dissociation need to be addressed.

Activation of the PDE by $T_\alpha GTP$ and Role of $T_{\beta\gamma}$

The soluble $T_\alpha GTP$ is the messenger which activates PDE. Hurley and Stryer (26) have proposed an attractive model in which T_α binds to the 13K subunit, releasing inhibition by the subunit. Yamazaki et al. (57) have proposed that the inhibitory subunit becomes solubilized by $T_\alpha GTP$. Several issues in this model need further confirmation. It seems surprising that $T_\alpha GTP$ becomes soluble and has to shuttle between two proteins, R* and the PDE, which are both bound on the disc membrane. This does make sense when one considers the peculiar morphology of the ROS: the solubilized $T_\alpha GTP$ is entrapped between two discs and can only diffuse across the 150 Å cytoplasmic space between the discs. This ensures a high rate of collisions with the PDE molecules on both disc surfaces limiting the space. It may be significant to point out here that for the analogous β-adrenergic hormone system, the receptor and adenylate cyclase are integral membrane proteins of the cell's plasma membrane which surrounds a large cytoplasmic volume (27). The GTP-dependent protein analog to transducin always remains membrane-bound.

Turnoff Mechanisms for Activated PDE

Once the cGMP cascade has been triggered, it must be turned off to restore resting cGMP cell levels and to accommodate repetitive cell triggering. The hydrolysis of the GTP bound to T_α is the key step in the inactivation of the PDE. This GTPase activity does not require the presence of $T_{\beta\gamma}$. Sometimes there is misunderstanding of this fact because $T_{\beta\gamma}$ is needed to assay a persistent GTPase activity. T_αGDP needs $T_{\beta\gamma}$ to be able to rebind R* and load a new GTP. A GTPase assay measures the repetitive rate of full cycles of GTP/GDP exchange and GTP hydrolysis. The GTPase activity of isolated T_α is not easily measured, because once bound GTP is hydrolyzed, there is no further exchange. Fung (18) has shown that a stoichiometry of one $T_{\beta\gamma}$ for 10 T_α is sufficient to achieve the full GTPase activity measured in an assay with a reconstituted system, giving a maximum value of two to three GTP/T min. This indicates that the rate-limiting step in the complete cycle is not determined by the presence of $T_{\beta\gamma}$ · $T_{\beta\gamma}$ seems to be needed only for the fast binding and exchange steps catalyzed by R*, and not for the slower inactivation steps related to GTP hydrolysis. An alternate role for $T_{\beta\gamma}$ in the inactivation of the PDE may be proposed based on analogies to the inhibitory GTP-dependent protein (Gi) in hormone systems (27). If one assumes that T_αGDP retains a significant affinity for the inhibitor of the PDE but has a higher one for $T_{\beta\gamma}$, then the inhibitor will remain attached to T_α even after the GTP hydrolysis, until it is released by the competition of $T_{\beta\gamma}$ for the same site. In this case, the spontaneous hydrolytic activity of T_α is the rate-limiting step and is necessary but not sufficient for PDE deactivation. In addition, other cascade interfaces must be deactivated, e.g., the triggering R* must be deactivated to prevent further catalytic activation of T (see below). A full understanding of these "turnoff" mechanisms is one of the weakest links in our knowledge of the cascade.

R*-kinase Interaction and R* Phosphorylation

The ATP-dependent kinase can phosphorylate up to nine serine and threonine residues, seven of which are concentrated near the C terminal end of R* (55). A strong inhibition of this phosphorylation is observed under the conditions where all the R* formed are bound to transducin, i.e., when little R is photoexcited, GTP is omitted, but enough ATP is present to keep the kinase active (30, 44). This inhibition suggests that there might be steric hindrance between T and K, both of which have to bind to the same cytoplasmic face of R*. The C terminal phosphorylation sites are not themselves an important part of the transducin binding site (see above), but transducin may mask kinase access to the phosphorylation sites. Conversely, kinase might hinder T binding under conditions where R* is saturated with K. This might be observed in the absence of ATP but with enough GTP present to induce the exchange, but this experiment is not feasible since it would require unrealistic quantities of purified kinase. For the same reason, the interference of the binding of the kinase with the chromophore site has not been investigated. R* becomes sensitive to phosphorylation by the kinase following its transition to metarhodopsin I (42). It is sensitive in the

Meta II state, and the decay of its sensitivity is not strictly correlated with that of Meta II (30). The phosphate groups, once added, do not seem to interfere with the later decay or the spectral regeneration of rhodopsin (39, 55). They do, however, interfere with the functional regeneration (55).

The Role of Phosphorylation

Liebman and Pugh (36) have observed that in vitro, ATP mediates a R* deactivation mechanism for the light-induced PDE activation. In addition, Sitaramayya and Liebman (47) have indicated that phosphorylation of rhodopsin is sufficient in amount and is fast enough to be compatible with PDE quenching. It has not been proven, however, that phosphorylation alone is the deactivation mechanism nor that phosphorylated R* is still able to catalyze the GTP/GDP exchange. There are several other ATP-dependent phosphorylations which take place in ROS: some of them might be important in PDE deactivation or regulation. This topic is more fully explored by Bownds and Brewer (this volume).

Is the "48K" a "Blocking" Protein?

Integration of a role for 48K in visual transduction remains to be elucidated. The interaction of 48K to R* has been puzzling for a long time. 48K binding has proven to be much slower than the transducin binding and requires the presence of ATP or a high concentration of GTP to develop. This suggests that the binding is on a late-intermediate state of R*, different from that which interacts with T (44). Kühn has now demonstrated that the 48K protein binds more specifically to phosphorylated R* and that there is competition between 48K and T for binding on phosphorylated R* (32).

KINETICS, AMPLIFICATION, AND REGULATION

The reaction scheme of the cascade looks quite complete (Fig. 3). The scheme is designed to illustrate an amplificiation loop controlled by two regulatory mechanisms. The amplification loop is the central catalytic turnover of transducin: it is flanked on either side by control schemes for setting the level of the catalyst R* and the level of activated PDE. Both phosphorylation and binding of 48K are proposed here as mechanisms of removing or regulating participating R*. The level of active PDE is set by accessibility of the PDE inhibitor [13K] to PDE and the GTPase activity of T_α; in addition, $T_{\beta\gamma}$ may compete for T_α^{GDP}, thus releasing I. Several considerations are to be noted: the fact that the system may be activated by non-hydrolyzable analogs of GTP often introduces confusion about the requirement of GTP hydrolysis. It should not be considered as a simple turnoff signal. If the hydrolysis is impossible, the system is indeed blocked in the "active" state. But the hydrolysis is required to restore the conformation of T which will be able to interact spontaneously with R* and T. The conformational change of T^{GDP} promoted by binding to R*, which leads to GTP/GDP exchange, is only facilitated and not driven by R*. As a catalyst, R* does not provide any energy to the reaction; it must be released unchanged upon the completion of the exchange reaction, otherwise it would be unable to interact repeatedly with a large number of T^{GDP}

molecules providing amplification. One may assume that TGDP is driving the reaction with energy acquired from GTP hydrolysis in the preceding turn of the cycle or, alternatively, that the new GTP accepted upon the exchange provides energy for the dissociation. Whatever the source of energy driving the cycle, the energy of phosphate bond hydrolysis is required to complete the cycle of information transfer from R* to the PDE.

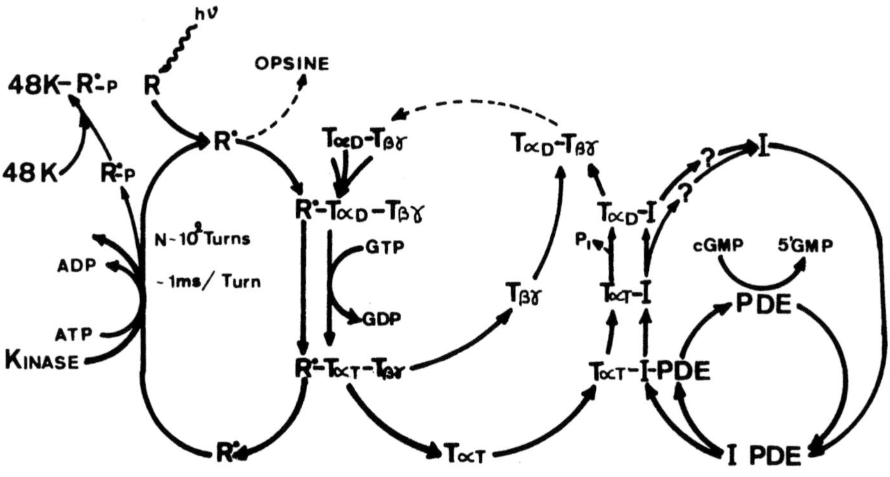

FIG. 3 -The amplifying cascade.

It remains to be shown whether both the activation and the control mechanisms are fast enough for visual signalling. Near infrared light-scattering methods on rods fragments (31) and more recently on structurally intact and oriented ROS (52) have shown that the first stage of amplification, i.e., the rate of release of $T_{\alpha}GTP$, is very fast: at very low levels of illumination ($\simeq 10^{-4}$ R*/rod), one R* sequentially processes hundreds of transducins at the rate of about one per millisecond. This rate is probably controlled by the lateral diffusion constants of both R* and T, respectively, in and on the disc membrane. It may be worthwhile to remark that the lateral diffusion coefficient of rhodopsin, $D = 0.5\ 10^{-9}$ cm^2/s, is the highest ever measured for an intrinsic protein (2). The next stage of the cascade involves the diffusion of $T_{\alpha}GTP$ toward the PDE. The morphology of the interdiscal compartment is probably critical for this action. The confining 150Å width of interdiscal space might be compared to a synaptic cleft. Local concentration of the transmitter molecule $T_{\alpha}GTP$ can reach a very high value; the $T_{\alpha}GTP$ pool amounts to an interdiscal cytoplasmic concentration of about 500 µmol/l; total release might be accomplished at only a few R* per disc to allow $T_{\alpha}GTP$ to diffuse toward the PDE molecules attached to either disc membrane.

The proposed turnoff mechanism of R* involves two soluble proteins, the kinase and the 48K. They are easily lost during ROS preparations and are seldom present in "correct" amounts in biochemical assays. Thus, it has been hard to assess their mechanistic role. Furthermore, measurements of phosphorylation rates are usually performed at high levels of photoexcitation. Given the high level of kinase activity and the high concentration of 48K protein in vivo, one may estimate that the rate of phosphorylation and ensuing blocking of R* by the 48K could be fast enough at the physiological level of illumination to provide a rapid switch-off mechanism. Experimental evidence exists to support the adequate rate of phosphorylation (47).

Experiments addressing the mechanistic role of 48K are in progress (Kühn, Zuckerman, and others) and a discussion of the caveats and possibilities is warranted. Once R* is removed from the cycle, the decay of PDE activity should be correlated with the rate of GTP hydrolysis by T_α, if the model for control is correct. The slow rate measured in vitro remains a problem for this model. Perhaps there exists another in situ soluble factor which is responsible for the inactivation of T_αGTP. Clarification of the role of I is still needed. While the model is attractive, nonstoichiometric excesses of I are needed to inhibit PDE from which I has been removed. As Hurley and Stryer pose (26), it is not clear whether I is ever dissociated from PDE; perhaps T_αGTP simply changes the structure of PDE-I holoenzyme to achieve activation. A subtle question about the cascade is unaddressed: photoreceptors function over a range of photon fluence of 1-10^5 units. Is there regulation of the gain of this cGMP cascade over this range as well? Are there additional modes of functional regulation in this cascade, e.g., other phosphorylations (see Bownds and Brewer, this volume) or methylations (49)? How would such modes of regulation affect the dynamics of this cascade? Informative experiments in this area are still forthcoming.

LESSONS FROM THE FAMILY OF ANALOGOUS GUANINE NUCLEOTIDE-BINDING PROTEINS

Transducin is but one of a family of GTP-binding proteins which generally serve to communicate information from an activated membrane receptor to an intracellular enzymatic function (22). There are now five members of this family. Other members that are well characterized are the G_s (s = stimulatory) and G_i (i = inhibitory) proteins which function in receptor-mediated adenylcyclase activation or inhibition, respectively. Sternweis and Robishaw (48) have recently identified G_o (o = other) from brain tissue and have indicated that it mediates the neural muscarinic receptor action. In addition, there is the ras oncogene relative encoded by RNA tumor viruses; this relative is a GTP-binding protein but is a smaller polypeptide of about 21 kD. With the exception of the ras protein, all G regulatory binding proteins consist of the three α-, β-, γ-subunits. The β- and γ-subunits seem to be nearly identical for the different systems, whereas the α-subunits are homologous but functionally distinct. T has been suggested to be an evolutionary precursor of G_s and G_i (37). G_s can be ADP-ribosylated by cholera toxin: G_i and G_o can be ribosylated by pertussis toxin (IAP); T can be ribosylated by both toxins at independent sites (51). The striking homologies of these subunits give the suggestion that mechanisms of action will be analogous. Indeed, some studies of T were based on known properties of the G-proteins, and vice versa. Support for the dissociation of T_α- and $T_{\beta\gamma}$-subunits upon interaction with activated receptor comes from the properties of G_s and G_i interaction with the β-adrenergic receptor and α-adrenergic receptor, respectively. These receptors mediate dissociation of the subunits of G (27). A tantalizing hypothesis from comparative analyses of the G family functions has been put forth by Gilman in which each of the G-subunits seems to be dually functional. Evidence suggests that while each generally mediates activation or inhibition of a cyclic nucleotide metabolic enzyme, each also seems to mediate control of an ion channel. Is such a dual function realized in the photoreceptor by transducin?

OUTSTANDING QUESTIONS

Is R* Triggering Other Visual Transduction Processes?

Most of the photoinduced reactions documented up to now in ROS are concerned with amplified control of cGMP phosphodiesterase activity by R*. This seems to be the main, if not the exclusive, function of photoexcited rhodopsin. It is hard to imagine that R* could have another function. This modest-sized protein has interfaces for inteacting with three different proteins, T, K, and 48K, and its hydrophobic core is devoted to providing the appropriate environment for the retinal chromophore. The latter provides all the proper interfaces for primary photon energy transduction. Only the intradiscal face of rhodopsin remains to be given a defined function. Rhodopsin is definitely not a calcium channel in the traditional sense of an ion channel, and its is hard to conceive that it has another interface that could drive an independent process for amplified calcium ion release. The analysis of the time course of single-photon responses indicates that many sequential steps of reaction should exist between the fast spectroscopic change in photoexcited rhodopsin and the delayed and slow-rising cellular response (5). There must be an amplification step before the eventual calcium release is triggered. The R*-T-PDE cascade is an excellent candidate to provide this amplification.

Are There Other Roles for G-Proteins?

The provocation given above and recent evidence from study of the role of G-protein in platelets, pancreatic cells, and elsewhere (13, 24, 38, 41) suggest that T might control inositol triphosphate (ITP) release in photoreceptors. In other systems ITP controls Ca^{2+} efflux from intra- or extracellular Ca^{2+} pools (18). ITP injections do hyperpolarize or depolarize photoreceptors, depending upon the species studied (9, 17, 53), but probably not via Ca^{2+} mobilization, at least in Limulus: injection of EGTA together with IP_3 did not change the IP_3 response (J. Brown). Potential roles for T in ITP metabolism should certainly be discussed.

How Universal Is the cGMP Cascade?

Protein elements of the cascade have been identified or intimated in several other species, e.g., squid (46, 50) and Limulus (10, 11). Adaptations of the cascade mechanisms found in other animals may provide exceptional insight into the nature of this R*-triggered event. For example, in invertebrates where rhodopsin has little translational mobility, what changes in the R* T catalytic mechanisms might be expected, and what implication does this hold?

REFERENCES

(1) Abood, M.A.; Hurley, J.; Pappone, M.C.; Bourne, H.R.; and Stryer, L. 1982. Functional homology between signal-coupling proteins. J. Biol. Chem. **257:** 10540-10543.

(2) Axelrod, D. 1983. Lateral motion of membrane proteins and biological functions. J. Membr. Biol. **75:** 1-10.

(3) Baehr, W.; Devlin, M.J.; and Applebury, M.L. 1979. Isolation and characterization of cGMP phosphodiesterase from bovine rod outer segments. J. Biol. Chem. **254:** 11669-11677.

(4) Baehr, W.; Morita, E.A.; Swanson, R.J.; and Applebury, M.L. 1982. Characterization of bovine rod outer segment G-protein. J. Biol. Chem. **257:** 6452-6460.

(5) Baylor, D.A.; Hodgkin, A.L.; and Lamb, T.D. 1974. The electrical response of turtle cones to flashes and steps of light. J. Physiol. **242:** 685-727.

(6) Bennett, N.; Michel-Villaz, M.; and Kühn, H. 1982. Light-induced interaction between rhodopsin and the GTP-binding protein. Metarhodopsin II is the major photoproduct involved. Eur. J. Biochem. **127:** 97-103.

(7) Bennett, N., and Dupont, Y. 1985. The G protein of retinal rod outer segments (Transducin): Mechanism of interaction with rhodopsin and nucleotides. J. Biol. Chem. **260:** 4156-4168.

(8) Berridge, M.J., and Irvine, R.F. 1984. Inositol triphosphate, a novel second messenger in cellular signal transduction. Nature **312:** 315-320.

(9) Brown, J.E.; Rubin, L.J.; Ghalayini, A.J.; Tarver, A.P.; Irvine, R.F.; Berridge, M.J.; and Anderson, R.E. 1984. Myo-inositol polyphosphate may be a messenger for visual excitation in Limulus photoreceptors. Nature **311:** 160-163.

(10) Corson, D.W., and Fein, A. 1983. Chemical excitation of Limulus photoreceptors I. Phosphatase inhibitors induce discrete-wave production in the dark. J. Gen. Physiol. **82:** 639-657.

(11) Corson, D.W.; Fein, A.; and Walthall, W.W. 1983. Chemical excitation of Limulus photoreceptors II. Vanadate, GTP-y-S, and fluoride prolong excitation evoked by dim flashes of light. J. Gen. Physiol. **82:** 659-677.

(12) Das, N.D.; Ulshafer, R.J.; Zam, Z.S.; Leverenz, V.R.; and Shichi, H. 1984. Radioimmunocytochemical localization of retinal S-antigen with monoclonal-antibodies. J. Histochem. Cytochem. **32:** 834-838.

(13) Downes, C.P. 1983. Inositol phospholipids and neurotransmitter receptor signalling mechanisms. TINS **5:** 313-316.

(14) Dratz, E.A., and Hargrave, P.A. 1983. The structure of rhodopsin and the rod outer segment disk membrane. TIBS **8:** 128-131.

(15) Emeis, D.; Kühn, H.; Reichert, J.; and Hofmann, K.P. 1982. Complex formation between metarhodopsin II and GTP-binding protein in bovine photoreceptor membranes leads to a shift of the photoproduct equilibrium. FEBS Lett. **143:** 29-34.

(16) Faure, J.P.; Mirshahi, M.; Dorey, C.; Thillaye, B.; DeKozak, Y.; and Boucheix, C. 1984. Production and specificity of monoclonal antibodies to retinal S antigen. Curr. Eye Res. **3:** 867-872.

(17) Fein, A.; Payne, R.; Corson, D.W.; Berridge, M.J.; and Irvine, R.F. 1984. Photoreceptor excitation and adaptation by inositol 1,4,5-triphosphate. Nature **311:** 157-160.

(18) Fung, B.K.K. 1983. Characterization of transducin from bovine retinal rod outer segments. I. Separation and reconstitution of the subunits. J. Biol. Chem. **258:** 10495-10502.

(19) Fung, B.K.K.; Hurley, J.B.; and Stryer, L. 1981. Flow of information in the light-triggered cyclic nucleotide cascade of vision. Proc. Natl. Acad. Sci. USA **78:** 152-156.

(20) Fung, B.K.K., and Nash, C.R. 1983. Characterization of transducin from bovine retinal rod outer segments. II. Evidence for distinct binding sites and conformational changes revealed by limited proteolysis with trypsin. J. Biol. Chem. **258:** 10503-10510.

(21) Fung, B.K.K., and Stryer, L. 1980. Photolyzed rhodopsin catalyzes the exchange of GTP for bound GDP in retinal rod outer segments. Proc. Natl. Acad. Sci. USA **77:** 2500-2504.

(22) Gilman, A.G. 1984. G proteins and dual control of adenylate cyclase. Cell **36:** 577-579.

(23) Godchaux, W., and Zimmerman, W.F. 1979. Membrane dependent guanine nucleotide binding and GTPase activities of soluble protein from bovine rod cell outer segments. J. Biol. Chem. **254:** 7874-7884.

(24) Gomperts, B.D. 1983. Involvement of guanine nucleotide binding protein in the gating of Ca^{++} by receptors. Nature **306:** 65-66.

(25) Ho, Y.K., and Fung, B.K.K. 1984. Characterization of transducin from bovine retinal rod outer segments. The role of sulfhydryl groups. J. Biol. Chem. **259:** 6694-6699.

(26) Hurley, J.B., and Stryer, L. 1982. Purification and characterization of the gamma regulatory subunit of the cyclic GMP phosphodiesterase from retinal rod outer segments. J. Biol. Chem. **257:** 11094-11099.

(27) Katada, T.; Northup, J.K.; Bokoch, G.M.; Michio, U.; and Gilman, A.G. 1984. The inhibitory guanine nucleotide-binding regulatory component of adenylate cyclase. Subunit dissociation and guanine nucleotide-dependent hormonal inhibition. J. Biol. Chem. **259:** 3578-3585.

(28) Kühn, H. 1980. Light-induced, reversible binding of proteins to bovine photoreceptor membranes. Neurochem. Int. **1:** 269-285.

(29) Kühn, H. 1981. Light- and GTP-regulated interaction of GTPase and other proteins with bovine photoreceptor membranes. Nature **283:** 587-589.

(30) Kühn, H. 1984. Interactions between photoexcited rhodopsin and light-activated enzymes in rods. In Progress in Retinal Research, eds. N. Osborne and J. Chader, vol. 3, pp. 123-156. Oxford: Pergamon Press.

(31) Kühn, H.; Bennett, N.; Michel-Villaz, M.; and Chabre, M. 1981. Interactions between photoexcited rhodopsin and GTP-binding protein: Kinetic and stoichiometric analysis from light-scattering changes. Proc. Natl. Acad. Sci. USA **18:** 6873-6877.

(32) Kühn, H.; Hall, S.W.; and Wilden, U. 1984. Light-induced binding of 48-kDa protein to photoreceptor membranes is highly enhanced by phosphorylation of rhodopsin. FEBS Lett. **176:** 473-478.

(33) Kühn, H., and Hargrave, P.A. 1981. Light-induced binding of guanosine triphosphatase to bovine photoreceptor membranes: Effect of limited proteolysis of the membranes. Biochemistry **20:** 2410-2417.

(34) Kühn, H.; Mommertz, O.; and Hargrave, P.A. 1982. Light-dependent conformational change at rhodopsin's cytoplasmic surface detected by increased susceptibility to proteolysis. Biochim. Biophys. Acta **679:** 95-100.

(35) Liebman, P.A., and Pugh, E.N., Jr. 1979. The control of phosphodiesterase in rod disk membranes: kinetics, possible mechanisms and significance for vision. Vision Res. **19:** 375-380.

(36) Liebman, P.A., and Pugh, E.N. 1980. ATP mediates rapid reversal of cyclic GMP phosphodiesterase activation in visual receptor membranes. Nature **287:** 734-736.

(37) Manning, D.R.; Fraser, B.A.; Kahn, R.A.; and Gilman, A.G. 1984. ADP-ribosylation of transducin by islet-activating protein. Identification of asparagine as the site of ADP-ribosylation. J. Biol. Chem. **258:** 749-756.

(38) Michell, B. 1984. Hormone action at membranes. TIBS **9:** 3-4.

(39) Miller, J.A., and Paulsen, R. 1975. Phosphorylation and dephosphorylation of frog rod outer segment membranes as part of the visual process. J. Biol. Chem. **250:** 4427-4432.

(40) Navon, S.E., and Fung, B.K.K. 1984. Characterization of transducin from bovine retinal rod outer segments. Mechanisms and effects of cholera toxin-catalyzed ADP-ribosylation. J. Biol. Chem. **259:** 6686-6693.

(41) Nishizuka, Y. 1974. Protein kinases in signal transduction. TIBS **9:** 163-166.

(42) Paulsen, R., and Bentrop, J. 1984. Activation of rhodopsin phosphorylation is triggered by the lumirhodopsin-metarhodopsin I transition. Nature **302:** 417-419.

(43) Pfister, C.; Dorey, C.; Vadot, E.; Mirschahi, M.; Deterre, P.; Chabre, M.; and Faure, J.P. 1984. Identification of the 48-K protein that interacts with illuminated rhodopsin in vertebrate retinal rods with the retinal S-antigen inducing experimental autoimmune uveoretinitis. Cpt. Rend. Acad. Sci. Paris **299:** 261-266.

(44) Pfister, C.; Kühn, H.; and Chabre, M. 1983. Interaction between photoexcited rhodopsin and peripheral enzymes in frog retinal rods. Influence on the post metarhodopsin II decay and phosphorylation rate of rhodopsin. Eur. J. Biochem. **136:** 489-499.

(45) Robinson, P.R.; Radeke, M.J.; and Bownds, M.D. 1982. Regulation of the flash responses of frog photoreceptor cyclic GMP phosphodiesterase and GTP binding protein. Inv. Ophthalmol. Vis. Sci. **22:** 186.

(46) Saibil, H.R., and Michel-Villaz, M. 1984. Squid rhodopsin and GTP-binding protein crossreact with vertebrate photoreceptor enzymes. Proc. Natl. Acad. Sci. USA **81:** 5111-5115.

(47) Sitaramayya, A., and Liebman, P.A. 1983. Phosphorylation of rhodopsin and quenching of cyclic GMP phosphodiesterase activation by ATP at weak bleaches. J. Biol. Chem. **258:** 12106-12109.

(48) Sternweis, P.C., and Robishaw, J.D. 1984. Isolation of two proteins with high affinity for guanine nucleotides from membranes of bovine brain. J. Biol. Chem. **259:** 13806-13813.

(49) Swanson, R.J., and Applebury, M.L. 1983. Methylation of proteins in photoreceptor rod outer segments. J. Biol. Chem. **258:** 10599-10605.

(50) Vandenberg, C.A., and Montal, M. 1984. Light-regulated biochemical events in invertebrate photoreceptors. Biochemistry **23:** 2339-2352.

(51) VanDop, C.; Yamanaka, G.; Steinberg, F.; Sekura, R.D.; Manclark, C.R.; Stryer, L.; and Bourne, H. 1984. ADP-ribosylation of transducin by pertussis toxin blocks the light-stimulated hydrolysis of GTP and cGMP in retinal photoreceptors. J. Biol. Chem. **259:** 23-26.

(52) Vuong, T.M.; Chabre, M.; and Stryer, L. 1984. Millisecond activation of transducin in the cyclic nucleotide cascade of vision. Nature **311:** 659-661.

(53) Waloga, G.; Anderson, R.E.; and Irvine, R.F. 1985. Modulation of vertebrate photoreceptor potentials by injection of inositol triphosphate. Biophys. J. **47:** 37a.

(54) Wheeler, G.L., and Bitensky, M.V. 1977. A light-activated GTPase in vertebrate photoreceptors: Regulation of light-activated cyclic-GMP phosphodiesterase. Proc. Natl. Acad. Sci. USA **74:** 4238-4242.

(55) Wilden, U.A., and Kühn, H. 1982. Light-dependent phosphorylation of rhodopsin: Number of phosphorylation sites. Biochemistry **21:** 3014-3022.

(56) Woodruff, M.L.; Bownds, D.; Green, S.H.; Morrisey, J.L.; and Shedlovsky, A. 1977. Guanosine 3',5'-cyclic monophosphate and the in vitro physiology of frog photoreceptor membranes. J. Physiol. **69:** 667-679.

(57) Yamazaki, A.; Stein, P.J.; Chernoff, N.; and Bitensky, M. 1983. Activation mechanism of rod outer segment cyclic GMP phosphodiesterase. Release of inhibitor by the GTP/GDP-binding protein. J. Biol. Chem. **258:** 8188-8194.

(58) Yee, R., and Liebman, P.A. 1978. Light-activated phosphodiesterase of the rod outer segments. Kinetics and parameters of activation and deactivation. J. Biol. Chem. **253:** 8902-8909.

The Molecular Mechanism of Photoreception, ed. H. Stieve, pp. 67-77. Dahlem Konferenzen 1986. Berlin, Heidelberg, New York, Tokyo: Springer-Verlag.

Evidence for a Role of Messenger Substances in Phototransduction

G.L. Fain
Dept. of Ophthalmology
Jules Stein Eye Institute
UCLA School of Medicine
Los Angeles, CA 90024, USA

Abstract. This paper is a review of the evidence for internal ("second") messengers in the photoreceptors of vertebrate rods and cones and Limulus ventral photoreceptors.

INTRODUCTION

Fifteen years ago, when Baylor and Fuortes (5) and Hagins (27) suggested that the response of vertebrate photoreceptors was produced by the release of some internal transmitter substance, the mediation of a cytosolic messenger in excitation seemed like a novel idea. At that time we were used to thinking of electrical events as exclusively membrane phenomena. The squid axon action potential had been shown only a few years earlier to remain intact even after the removal of all of the axoplasm (2). The only demonstrated effects of synaptic transmitters were thought to be direct effects of binding of the transmitters to extracellular membrane receptors (32).

Now "internal transmitters," or second messengers, as they are usually called, are a pervasive feature of the neurobiological landscape. There are many examples of effects of small-molecular-weight intracellular substances such as cAMP on the membrane potential and action potential of neurons (16, 53, 54). Though some synaptic receptors now have been demonstrated to function just as we originally thought they did, by directly regulating the flow of current through the plasma membrane (41, 43, 50), many others appear to produce their responses only by the intervention of second messengers (18, 25, 51).

The notion of a second messenger was first introduced to explain the responses of cells to hormones (46). Hormones bind to external receptors which activate other, usually membrane-bound, proteins. These lead eventually to a change in the concentration of a small-molecular-weight substance which diffuses at least to some extent within the cytosol to produce the response of the cell. The hormone can be thought of as a first messenger, and the diffusible, small-molecular-weight substance in the cytosol as a second messenger or "internal transmitter." In some cases, more than one messenger substance appears to be generated during the response to a single hormone or growth factor (42). As the mechanisms of these responses and the interaction of the various messengers are elucidated, it may be possible eventually to categorize them and place them in their proper order, as second messengers, or third, or fourth. For the present, I shall refer to them all as internal messengers, abandoning the term "internal transmitter," now obsolete.

INTERNAL MESSENGERS IN VISUAL EXCITATION

The first realization that messengers might play some role in the generation of the photoresponse came from considerations of the sensitivity of photoreceptors. In a turtle cone, a single, activated pigment molecule (Rh*) can change the magnitude of the light-dependent conductance by about one part in 10^3 (5). This is much greater than one would expect if each pigment molecule were itself a membrane channel (5, 14). For Limulus photoreceptors, a single quantum can cause a change in membrane conductance of tens of nS (14, 19). However, the single-channel conductance of the light-activated channel is orders of magnitude smaller (about 25 pS -see (1, 56)). Clearly thousands of channels must open as the result of a single Rh*.

In retrospect, these considerations no longer seem compelling evidence for a role for messenger substances, since we could easily imagine other means of amplification. There could, for example, be diffusion of enzymes or other proteins within the plasma membrane of the cone invaginations, as has been postulated for the proteins in the electron transport chain of mitochondria (26). The diffusion of small-molecular-weight proteins might also occur within the cytosol of cones or Limulus receptors between different parts of the photoreceptive membrane. Such proteins would not count as messenger substances according to the usual definition, though in practice they would function in a nearly identical fashion. One might also imagine a mechanism whereby some ion or other small-molecular-weight substance entered the cell from the extracellular space, as Ca was postulated to do for cones in Yoshikami and Hagins' original version of the calcium hypothesis (57). Although it now seems unlikely that Ca actually enters cones in this fashion (8), models of this kind cannot yet be completely excluded. Such substances would again not strictly count as second messengers.

A better case for messengers in excitation can be made for the rods. Here most of the photosensitive pigment is contained within the membrane of the disks, which are physically separate from the plasma membrane (see Fig. 1). The fibrillar connections between the plasma membrane and the disks (Fig. 1, arrows), which have recently been demonstrated in transmission electron micrography (EM) of fast-frozen rods by Schröder and myself, and previously in deep-etched fast-frozen rods by others (47, 55), seem only to serve to hold the disks in place. There seems to be no anatomical (12, 35) or electrical (24, 44) connection between the disks and the extracellular space.

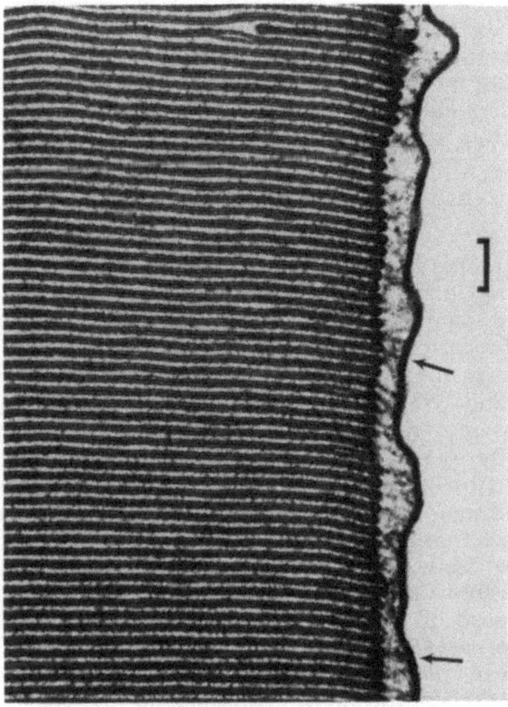

FIG. 1 - Transmission electronmicrograph of outer segment from quick- frozen toad rod from retina freeze-substituted in acetone and 0.5% OsO_4 and embedded in Epon-Araldite. Calibration bar: 100 nm. Arrows: fibrillar connections between plasma membrane and disks.

The quantum sensitivity measurements of Baylor and his colleagues (7) clearly show that rods use all, or nearly all, of the photopigment in their outer segments, including that within the disks. If the absorption of a photon in a disk is to cause a change in the light-dependent conductance, some substance must diffuse from the disks to the plasma membrane. Although one could again imagine that what is diffusing is a protein rather than a small-molecular-weight messenger, there is now so much evidence for effects of Ca^{2+} and cyclic GMP on rods (see McNaughton et al., Miller, and Pugh and Cobbs, this volume) that a role for internal messengers seems rather likely. In my opinion, small-molecular-weight messengers probably also contribute to excitation in cones and invertebrate photoreceptors.

INTERNAL MESSENGERS IN ADAPTATION

In order to describe evidence for a role of messenger substances in the regulation of photoreceptor sensitivity, it is useful to distinguish two kinds of sensitivity changes which occur during or after exposure to light. I shall call light adaptation the decrease in the sensitivity of the receptor in the presence of a maintained background or ambient illumination. The term dark adaptation will be used for the receeovery of sensitivity which occurs after exposure to bright illumination. I shall not distinguish between the various forms of dark adaptation such as neural and photochemical, or fast and slow, since the state of our knowledge about messengers is at present not sufficiently detailed to warrant this.

Light Adaptation

The first evidence that an internal messenger might be involved in light adaptation came again from measurements of the sensitivity of vision. In the Limulus ventral photoreceptor, for example, decreases in sensitivity can occur with the absorption of fewer than 100 photons (36). Since a single photoreceptor has on the order of 10^9 pigment molecules (37), the absorption of light by these 100 molecules must somehow be communicated to the rest of the photoreceptor, in order that the sensitivity of the transduction process initiated by photon absorption in the other 999,999,900 rhodopsins can be reduced. Although this could occur by diffusion of proteins along the microvillar membrane, this seems unlikely, given the slow diffusion of rhodopsin itself in arthropod microvilli (23). A large body of work by Brown, Fein, Lisman, and their co-workers has shown that light produces an increase in the cytosolic Ca^{2+} concentration in Limulus receptors probably in large part by release of Ca^{2+} from internal stores, and that injection of Ca^{2+} produces decreases in sensitivity which mimic those produced by background light. A more thorough discussion of Ca^{2+} as a messenger in invertebrate photoreceptor adaptation is contained in the paper by Brown (this volume).

Messenger substances may also play a role in the regulation of sensitivity during the light adaptation of vertebrate rods. As we (3) and Donner and Hemila (17) first pointed out, the sensitivity of rods to background lights is so high that single disks cannot be the locus of light adaptation. A light bleaching only four rhodopsins per rod per second reduces the sensitivity of a rod (in toad) by a factor of two (3). Since the sensitivity of the rod reaches steady state in only a few seconds, such a background light would bleach rhodopsins in only a small fraction of the 1000-2000 disks in a rod. The disks which absorb photons must somehow be able to communicate with the other disks in the rod, so that desensitization spreads throughout the whole of the photoreceptor. Lamb, McNaughton, and Yau (34) have actually measured the spread of desensitization in toad rods and shown that it decays approximately exponentially on either side of a localized background light, with a space constant (in toad) of about 6 μm (see also (29)).

At present, we know virtually nothing about how this spread of desensitization occurs. It is possible that disks containing a bleached rhodopsin release a messenger which diffuses to neighboring disks and alters the gain (and the kinetics of decay) of the transduction process. There is evidence suggesting that the messenger is neither Ca^{2+} (3, 4) nor cyclic GMP (10). It is in any case unlikely that it is the same substance as the messenger for excitation (13). It may not be a messenger at all, in the usual sense of this term, but rather a protein which during the process of transduction is released from the disks and diffuses between them, perhaps modulating the rate of re-uptake of the excitatory messenger.

Dark Adaptation

As single pigment molecules are not solely responsible for excitation or light adaption, neither are they the locus of the changes in sensitivity which occur after bleaching. It is well-known that rods are desensitized by bleaching to a much greater extent than expected simply on the basis of the reduction in the number of pigment molecules in the receptor available to be bleached (49). There must be some intermediary process which is activated by pigment bleaching and which somehow greatly decreases the gain of transduction. This process could occur entirely within the disks by depletion of internal transmitter or by modification of the light-activated enzymes, or it could occur by the intervention of a messenger substance diffusing within the cytosol along or between the disks. The only evidence we have on this question comes from attempts to measure the spread of desensitization in rods produced by local exposure to a bright bleaching illumination (6, 15, 29, 31). Baylor and Lamb (6) used transverse slits 7 μm wide to measure sensitivity up and down the rod outer segment after exposure to lights bleaching less than 10% of the pigment in a localized region of the outer segment. Their results failed to reveal any spread in desensitization other than that which would be expected from light scatter. Not only was the desensitization localized, but the increase in rod membrane current noise following bleaching, first described by Lamb (33), appeared also to be localized to the region of the rod exposed to the bleaching light. Because of the difficulties of measuring with precision the extent of scatter of the bleaching light and the exact area of outer segment over which current was recorded by their pipettes, it is not possible to exclude some spread of desensitization in Baylor and Lamb's experiments (for example, see (29)). However, their results provide no evidence in favor of a role for a messenger in the spread of dark adaptation in rods.

Similar results have been obtained by Cornwall, Fein, and MacNichol (15), using more extensive bleaches. In these experiments, local illumination bleaching as much as 95% of the pigment in a 1-2 μm wide slit along the outer segment produced desensitization which was mostly confined to the region of the bleaching light. Although a small spread of desensitization could be measured, it could be wholly accounted for by light scatter.

In Limulus, Fein and DeVoe (21) showed that there is no relationship between recovery of sensitivity after bright illumination and the time course of recovery of the early receptor potential. This suggests that sensitivity in invertebrates is also largely independent of the concentration of rhodopsin during dark adaptation (see also (38)). One interpretation of this result (21, 38) is that sensitivity during dark adaptation is controlled by some intermediate substance, perhaps Ca^{2+}. Though there is little spread of desensitization in Limulus following local illumination (20), this does not necessarily argue against a role for a messenger such as Ca^{2+}, since the diffusion of Ca^{2+} in ventral photoreceptors is very restricted (22). One possibility is that Ca^{2+} is the messenger for sensitivity regulation in Limulus during both light and dark adaptation, in such a way that the desensitization in the presence of backgrounds is produced by a continuous release of Ca^{2+} from internal stores, and the recovery of sensitivity during dark adaptation by a slow re-uptake or removal of Ca^{2+} from the cytosol. The reader is again referred to Brown (this volume) for a fuller discussion of these matters.

SOME FURTHER THOUGHTS ON MESSENGER SUBSTANCES

The search for messengers has had an extremely important influence on photoreceptor biochemistry and physiology during the last ten to fifteen years. There is reason to think that messengers play some role in vertebrate and invertebrate phototransduction, but exactly what this role is and what these messengers are cannot as yet be said with certainty. One thing which can be said, however, in itself rather remarkable, is that whatever the messenger substances may be, they must diffuse rather slowly. The longitudinal spread of excitation in rods is so small that it has not as yet been measured with certainty (28, 31, 34). Desensitization in both rods and Limulus photoreceptors is highly localized (20, 34). The effective diffusion constant for the adaptation messenger of rods has been estimated to be on the order of 10^{-7} cm^2 s^{-1} (34), two orders of magnitude smaller than for Na^+ or Ca^{2+} in aqueous solution. The effective diffusion constant for the excitatory messenger may be even smaller.

How can we explain this very slow rate of diffusion? One possibility (34), at least for rods, is that the disks present a barrier to longitudinal diffusion. One might think that diffusion would be restricted, since substances would be able to diffuse longitudinally through the cytosol only at the disk borders. It is important to realize, however, that disks are often not circular but are penetrated by incisures, which invaginate at the disk rim and can reach nearly to the center of the disk. The number and length of the incisures appears to increase as the diameter of the disk increases, and this has the effect of greatly increasing the area for diffusion. Recent calculations (S. Hart and R.E. Cone, personal communication) indicate that the incisures have a dramatic effect on the rate of longitudinal diffusion, so much so that the diffusion constant of a small ion or molecule would from these considerations alone be expected to be only about five times smaller longitudinally down an outer segment than for the same ion or molecule in aqueous solution. Thus, the obstruction produced by the disks cannot by itself account for the slow longitudinal spread observed in the physiological measurements.

A more likely possibility for the reduction of the rate of diffusion is that the excitatory or adaptation messengers are slowed either by rapid degradation of the messenger or by binding of the messenger to phospholipids or proteins on the external surface of the disks of rods, or on the internal surface of membranes and within the cytosol of Limulus receptors or cones. This is easy to imagine if Ca^{2+} were a messenger, since Ca^{2+} has been shown to diffuse slowly in Limulus (22) and other cells (30, 48). In most tissues, diffusion of Ca^{2+} is probably slowed by uptake into mitochondria and smooth endoplasmic reticulum (ER) (39, 48). In rod outer segments, which have no ER or mitochondria, McLaughlin and Brown (40) have proposed that Ca^{2+} diffusion is retarded by adsorption of Ca^{2+} to phospholipids in the disk membrane.

For putative messengers other than Ca^{2+}, some explanation must be given of the slow spread of phototransduction. This may be difficult to do for substances such as Mg^{2+} or ATP, which would be expected to diffuse with little impediment in nerve cells. It may be particularly difficult to do for H^+, whose diffusion constant in water is especially high, five to ten times greater than that for Na^+ or Ca^{2+}. The diffusion of protons is unlikely to be retarded by binding to phospholipids, since the pK_a's of phosphatidylethanolamine (52) and phosphatidylserine (11) are too far from neutral pH to produce significant buffering of H^+ under physiological conditions. Protons could bind to rhodopsin or other proteins in the disk membrane, but cytosolic constituents such as ATP and GTP (9, 45) and other small-molecular-weight compounds probably provide the bulk of the buffering power in rods. The binding of protons to these molecules would reduce their diffusion constant to some extent, but not by a large enough amount to produce the small rate of longitudinal diffusion demonstrated physiologically. In my opinion, it is highly unlikely that H^+ is an internal messenger in rods.

In conclusion, I have reviewed the evidence for messenger substances in photoreceptors. The evidence I have presented indicates that messengers are of considerable importance in the physiology of transduction. The papers by McNaughton et al., Miller, and Pugh and Cobbs (this volume) evaluate the possible roles of individual putative messenger substances in more detail.

Acknowledgements. I am grateful to H. Kühn and J. Lisman for their thoughtful criticisms of an earlier draft of this manuscript. I also thank W.H. Schröder for permission to use Fig. 1. Work in my laboratory is supported by NIH grants EY 01844, EY 05234, and EY 00331.

REFERENCES

(1) Bacigalupo, J., and Lisman, J.E. 1983. Single-channel currents activated by light in Limulus ventral photoreceptors. Nature **304:** 268-270.

(2) Baker, P.F.; Hodgkin, A.L.; and Shaw, T.I. 1962. Replacement of the axoplasm of giant nerve fibers with artificial solutions. J. Physiol. **164:** 330-337.

(3) Bastian, B.L., and Fain, G.L. 1979. Light adaptation in toad rods: requirement for an internal messenger which is not calcium. J. Physiol. **297:** 493-520.

(4) Bastian, B.L., and Fain, G.L. 1982. The effects of low calcium and background light on the sensitivity of toad rods. J. Physiol. **330:** 307-329.

(5) Baylor, D.A., and Fuortes, M.G.F. 1970. Electrical responses of single cones in the retina of the turtle. J. Physiol. **207:** 77-92.

(6) Baylor, D.A., and Lamb, T.D. 1982. Local effects of bleaching in retinal rods of the toad. J. Physiol. **328:** 49-71.

(7) Baylor, D.A.; Lamb, T.D.; and Yau, K.-W. 1979. Responses of retinal rods to single photons. J. Physiol. **288:** 613-634.

(8) Bertrand, D.; Fuortes, M.G.F.; and Pochobradsky, J. 1978. Actions of EGTA and high calcium on the cones in the turtle retina. J. Physiol. **275:** 419-437.

(9) Biernbaum, M.S., and Bownds, M.D. 1985. Frog rod outer segments with attached inner segment ellipsoids as an in vitro model for photoreceptors of the retina. J. Gen. Physiol. **85:** 83-105.

(10) Capovilla, M.; Cervetto, L.; and Torre, V. 1983. The effect of phosphodiesterase inhibitors on the electrical activity of toad rods. J. Physiol. **343:** 277-294.

(11) Ceve, G.; Watts, A.; and Marsh, D. 1981. Titration of the phase transition of phosphatidylserine bilayer membranes. Effects of pH, surface electrostatics, ion binding, and head-group hydration. Biochemistry **20:** 4955-4965.

(12) Cohen, A.I. 1968. New evidence supporting the linkage to extracellular space of outer segment saccules of frog cones but not rods. J. Cell Biol. **37:** 424-444.

(13) Coles, J.A., and Yamane, S. 1975. Effects of adapting lights on the time course of the receptor potential of the anuran retinal rod. J. Physiol. **247:** 189-207.

(14) Cone, R.A. 1973. The internal transmitter model for visual excitation: some quantitative implications. In Biochemistry and Physiology of Visual Pigments, ed. H. Langer, pp. 275-282. Berlin: Springer-Verlag.

(15) Cornwall, M.C.; Fein, A.; and MacNichol, E.F., Jr. 1983. Spatial localization of bleaching adaptation in isolated vertebrate rod photoreceptors. Proc. Natl. Acad. Sci. USA **80:** 2785-2788.

(16) DePeyer, J.; Cachelin, A.B.; Levitan, I.B.; ad Reuter, H. 1982. Ca^{2+}-activated K^{+} conductance in internally perfused snail neurons is enhanced by protein phosphorylation. Proc. Natl. Acad. Sci. USA **79:** 4207-4211.

(17) Donner, K.O., and Hemila, S. 1978. Excitation and adaptation in the vertebrate rod photoreceptor. Med. Biol. **56:** 52-63.

(18) Drummond, G.I. 1983. Cyclic nucleotides in the nervous system. Adv. Cyc. Nucl. Res. **15:** 373-494.

(19) Fain, G.L., and Lisman, J.E. 1981. Membrane conductances of photoreceptors. Prog. Biophys. Molec. Biol. **37:** 91-147.

(20) Fein, A., and Charlton, J.S. 1975. Local adaptation in the ventral photoreceptors of Limulus. J. Gen. Physiol. **66:** 823-836.

(21) Fein, A., and DeVoe, R.D. 1973. Adaptation in the ventral eye of Limulus is functionally independent of the photochemical cycle, membrane potential, and membrane resistance. J. Gen. Physiol. **61:** 273-289.

(22) Fein, A., and Lisman, J. 1975. Localized desensitization of Limulus photoreceptors produced by light or intracellular calcium ion injection. Science **187:** 1094-1096.

(23) Goldsmith, T.H., and Wehner, R. 1977. Restrictions on rotational and translational diffusion of pigment in the membranes of a rhabdomeric photoreceptor. J. Gen. Physiol. **70:** 453-490.

(24) Goldstein, E.B. 1967. Early receptor potential of the isolated frog (Rana pipiens) retina. Vision Res. **7:** 837-845.

(25) Greengard, P. 1978. Cyclic Nucleotides, Phosphorylated Proteins, and Neuronal Function. New York: Raven.

(26) Hackenbrock, C.R. 1981. Lateral diffusion and electron transfer in the mitochondrial inner membrane. Trends Biochem. Sci. **6:** 151-154.

(27) Hagins, W.A. 1972. The visual process: excitatory mechanisms in the primary receptor cells. Ann. Rev. Biophys. Bioeng. **1:** 131-158.

(28) Hagins, W.A.; Penn, R.D.; and Yoshikami, S. 1970. Dark current and photocurrent in retinal rods. Biophys. J. **10:** 380-412.

(29) Hemila, S., and Reuter, T. 1981. Longitudinal spread of adaptation in the rods of the frog's retina. J. Physiol. **310:** 501-528.

(30) Hodgkin, A.L., and Keynes, R.D. 1957. Movements of labelled calcium in squid giant axons. J. Physiol. **138:** 253-281.

(31) Jagger, W.S. 1979. Local stimulation and local adaptation of single isolated frog rod outer segments. Vision Res. **19:** 381-384.

(32) Katz, B. 1966. Nerve, Muscle, and Synapse. New York: McGraw-Hill.

(33) Lamb, T.D. 1980. Spontaneous quantal events induced in toad rods by pigment bleaching. Nature **287:** 349-351.

(34) Lamb, T.D.; McNaughton, P.A.; and Yau, K.-W. 1981. Spatial spread of activation and background desensitization in toad rod outer segments. J. Physiol. **319:** 463-496.

(35) Laties, A.M., and Liebman, P.A. 1970. Cones of living amphibian eye: selective staining. Science **168:** 1475-1477.

(36) Lisman, J.E. 1971. An electrophysiological investigation of the ventral eye of the horseshoe crab Limulus polyphemus. Ph.D. Thesis, Massachusetts Institute of Technology, Cambridge, MA, USA.

(37) Lisman, J.E., and Bering, H. 1977. Electrophysiological measurement of the number of rhodopsin molecules in single Limulus photoreceptors. J. Gen. Physiol. **70:** 621-633.

(38) Lisman, J.E., and Sheline, Y. 1976. Analysis of the rhodopsin cycle in Limulus ventral photoreceptors using the early receptor potential. J. Gen. Physiol. **68:** 487-501.

(39) Martonosi, A.N. 1983. The regulation of cytoplasmic Ca^{2+} concentration in muscle and nonmuscle cells. In Muscle and Non-muscle Motility, ed. A. Stracher, vol. 1, pp. 233-357. New York: Academic.

(40) McLaughlin, S., and Brown, J. 1981. Diffusion of calcium ions in retinal rods. A theoretical calculation. J. Gen. Physiol. **77:** 475-487.

(41) Nelson, N.; Anholt, R.; Lindstrom, J.; and Montal, M. 1980. Reconstitution of purified acetylcholine receptors with functional ion channels in planar lipid bilayers. Proc. Natl. Acad. Sci. USA **77:** 3057-3061.

(42) Nishizuka, Y. 1984. The role of protein kinase C in cell surface signal transduction and tumor promotion. Nature **308:** 693-698.

(43) Nowak, L.; Bregestovski, P.; Ascher, P.; Herbet, A.; and Prochiantz, A. 1984. Magnesium gates glutamate-activated channels in mouse central neurones. Nature **307:** 462-465.

(44) Penn, R.D., and Hagins, W.A. 1972. Kinetics of the photocurrent of retinal rods. Biophys. J. **12:** 1073-1099.

(45) Robinson, W.E., and Hagins, W.A. 1979. GTP hydrolysis in intact rod outer segments and the transmitter cycle in visual excitation. Nature **280:** 398-400.

(46) Robison, G.A.; Butcher, R.W.; and Sutherland, E.W. 1971. Cyclic AMP. New York: Academic.

(47) Roof, D.J., and Heuser, J.E. 1983. Surfaces of rod photoreceptor disk membranes: integral membrane components. J. Cell Biol. **95:** 487-500.

(48) Rose, B., and Lowenstein, W.R. 1975. Calcium ion distribution in cytoplasm visualized by acquorin: diffusion in cytosol restricted by energized sequestering. Science **190:** 1204-1206.

(49) Rushton, W.A.H. 1956. The rhodopsin density in human rods. J. Physiol. **134:** 30-46.

(50) Schindler, H., and Quast, U. 1980. Functional acetylcholine receptor from Torpedo marmorata in planar membranes. Proc. Natl. Acad. Sci. USA **77:** 3052-3056.

(51) Siegelbaum, S.A.; Camardo, J.S.; and Kandel, E.R. 1982. Serotonin and cAMP close single K^+ channels in Aplysia sensory neurones. Nature **299:** 413-417.

(52) Szabo, G.; Eisenman, G.; McLaughlin, S.G.A.; and Krasne, S. 1972. Ionic probes of membrane structures. Ann. NY Acad. Sci. **195:** 273- 290.

(53) Treistman, S.N. 1980. Effect of adenosine 3'-5'-monophosphate on neuronal pacemaker activity: a voltage clamp analysis. Science **211:** 59-61.

(54) Tsien, R.W. 1973. Adrenaline-like effects of intracellular iontophoresis of cAMP in cardiac purkinje fibers. Nat. New Biol. **245:** 120-122.

(55) Usukura, J., and Yamada, E. 1981. Molecular organization of the rod outer segment. A deep-etching study with rapid freezing using unfixed frog retina. Biomed. Res. **2:** 177-193.

(56) Wong, F. 1978. Nature of light-induced conductance changes in ventral photoreceptors of Limulus. Nature **276:** 76-78.

(57) Yoshikami, S., and Hagins, W.A. 1971. Light, calcium, and the photocurrent of rods and cones. Biophys. Soc. Abstr. **15:** 47a.

The Molecular Mechanism of Photoreception, ed. H. Stieve, pp. 79-92. Dahlem Konferenzen 1986. Berlin, Heidelberg, New York, Tokyo: Springer-Verlag.

Evaluation of Internal Transmitter Candidates: Ca

P.A. McNaughton, B.J. Nunn, and A.L. Hodgkin
Physiological Laboratory, University of Cambridge
Cambridge CB2 3EG, England

Abstract. This paper reviews the evidence for and against the "calcium hypothesis," which proposes that light releases Ca from the discs of vertebrate photoreceptors and that the light-sensitive channels are closed by combination with this Ca. In favor of this hypothesis are the observations that when Ca is applied either externally or internally the light-sensitive current is reduced, and that a net efflux of Ca from the cell occurs after a flash of light, as one would expect if Ca is released by light and is subsequently pumped from the cell. More recent observations show that an abrupt rise in internal Ca does not produce an abrupt closure of channels. The slow step intervening between a rise in internal Ca and channel closure is inconsistent with a simple form of the Ca hypothesis but is consistent with a model where channels are maintained open by cyclic GMP, and where Ca inhibits the cyclase responsible for producing cyclic GMP. In this model light closes channels by activating the phosphodiesterase responsible for breaking down cGMP, while internal Ca closes channels by inhibiting the production of cGMP.

INTRODUCTION

No one now doubts the importance of calcium as an intracellular messenger in biological systems. Students in the advanced class in Cambridge hear so much about calcium that their reaction is often to groan and roll their eyes to the ceiling on being told of yet another calcium-dependent cellular function. Yet in contrast with other systems (muscle, synapse, Ca-dependent K channels), firm evidence for a role of Ca in the response of photoreceptors to light is still lacking. Partly this is because of the small size of photoreceptors, partly it reflects the complexity of the system, and partly it is due to the light-sensitive nature of the cells themselves, which makes the usual array of Ca-sensitive dyes and photoproteins useless for most purposes. Nonetheless, in the last few years many interesting results have been published, and at the present rate of progress we may soon expect to understand the function of Ca. This review briefly presents some of the evidence for and against the argument that Ca is the primary internal transmitter in the light response. We have concentrated mainly on recent work, and for a more detailed discussion of earlier studies we refer the reader to reviews by Hagins (15) and by Fain and Lisman (8), and to papers in the volume edited by Miller (21).

THE Ca HYPOTHESIS

The "Ca hypothesis," as it is often called, was first proposed by Yoshikami and Hagins (29) and by Hagins (15). Like muscle, the outer segment of a vertebrate photoreceptor contains an extensive system of vesicles (the discs); as in muscle, the signal for excitation is transmitted from these vesicles to another structure, myofilaments in muscle and the surface membrane in the photoreceptor. There are also significant differences. In the rod the discs are formed by invagination of the surface membrane, while in muscle they are true intracellular organelles, a specialization of the endoplasmic reticulum. In muscle there is a visible lacuna inside the sarcoplasmic reticulum, while in the rod little space would appear to be available inside the disc to store a large quantity of calcium. Nevertheless, it was attractive to suppose that photoreceptors, by analogy with muscle, released Ca from discs which had absorbed a photon, and that this Ca diffused to the surface membrane where it closed off light-sensitive channels. On the simplest level one might suppose that an isomerized rhodopsin molecule itself formed a Ca-permeable channel in the disc membrane. However, now that more is known of the structure of rhodopsin (for a review see (7)) it seems difficult to imagine how ionic channels could form in the transmembrane part of rhodopsin, which consists largely of hydrophobic α-helices. But even if rhodopsin itself does not form a Ca-permeable pore, the signal might well be transmitted from an isomerized rhodopsin to some other molecule which either forms a pore in the disc membrane or releases Ca bound to the surface of the disc.

Support for the Ca hypothesis came from the experiments of Yoshikami and Hagins (30), who showed that externally applied Ca suppressed the light-sensitive current. Bertrand, Fuortes, and Pochobradsky (2) showed that when external Ca is lowered the light-sensitive current increases but the fractional suppression of light-sensitive conductance by a dim flash remains the same, which would be consistent with the idea that Ca closes channels by an external action without affecting the mechanism of transduction. The experiments of Bertrand et al. (2) were performed on cones, where the lumen of the discs is thought to be freely accessible to the external medium; their observation that the sensitivity to light was little affected by changes in external Ca was not promising for the Ca hypothesis but might be explained by the difficulty of changing the solution bathing the photoreceptor outer segments in an intact eyecup preparation. Substantial reductions in sensitivity were observed by Yoshikami and Hagins (30), Yau, McNaughton, and Hodgkin (27), and Bastian and Fain (1) in situations where rod outer segments are exposed directly to a flowing solution.

More direct support for the Ca hypothesis was provided by Brown, Coles, and Pinto (3), who found that injecting Ca into rod outer segments caused a hyperpolarization, while the Ca chelator EGTA had the opposite effect. Making the membrane more permeable to Ca by applying the ionophore X537A caused a suppression of light-sensitive current at normal extracellular calcium (Ca_o) but had much less effect at low Ca_o (16). Yet even here one could argue that changing intracellular calcium (Ca_i) alters the light-sensitive conductance for reasons which are irrelevant to the normal light response of the cell, in which the conductance is suppressed by some quite separate means.

LIGHT-DEPENDENT Ca EFFLUX FROM PHOTORECEPTORS

A central role for Ca in the light response was made more probable by the discovery that light causes a Ca efflux from the outer segment, consistent with the idea that the primary effect of light is to cause a Ca release from discs, and that part of this Ca is then extruded from the rod outer segment. Gold and Korenbrot (11-13) detected the efflux by draping a toad retina, receptor-side down, over a large Ca-sensitive electrode. They measured a Ca efflux of 2 x 10^4 Ca^{2+} released per photon absorbed. The efflux time course was slowed by reducing Na or elevating Ca in the external medium, suggesting that an Na:Ca exchange pump was responsible for at least part of the efflux. Although the temporal resolution of their method was reduced by the distance of the sensing electrode from the outer segments, they were able to conclude that the time course of the recorded efflux was consistent with a rise in Ca_i being responsible for the suppression of the light-sensitive current.

In an independent series of experiments Yoshikami, George, and Hagins (28) detected the Ca efflux by inserting a Ca electrode between the outer segments of a rat retina. They measured the smaller release of 400 Ca^{2+} per isomerization, a difference which may be due to the lower Ca_o and higher Na_o used in their experiments or to a species difference. With the improved temporal resolution achieved with the electrode positioned alongside the outer segment, they concluded that the Ca efflux did not lag the suppression of light-sensitive current by more than 100 ms.

These observations have recently been confirmed in an elegant experiment by Schröder and Fain (24), who measured the intracellular calcium concentration by laser micro-mass analysis of shock-frozen and freeze-substituted sections of frog retina. The total calcium concentration inside the rod outer segment (not distinguishing between disc and cytoplasm) decreased from about 4.5 mmol/l in the dark to about 2 mmol/l in bright light. The authors concluded that the decline was not due to a shutting off of Ca influx, as they were able to distinguish between Ca influx and efflux by replacing the external Ca with the heavier isotope ^{44}Ca.

Contrasting results have since been reported by Walz and Somlyo (25), using electron-probe analysis of sections shock-frozen in a manner apparently similar to those of Schröder and Fain and subsequently freeze-dried. These authors report a low total Ca concentration in outer segments, 0.3 mmol/l expressed relative to cell water, a high concentration of 1.4 mmol/l in the inner segment, and no significant effect of light on either. The only significant difference from the experiments of Schröder and Fain (24) appears to be a different method for extracting water from the frozen tissue. We await with interest the resolution of the discrepancies between the results of these two groups.

RELEASE OF Ca FROM DISCS BY LIGHT

If a Ca release from discs initiates the light response of the cell, then it is clearly important to measure this release in isolated discs. Initial results on this front were disappointing (for example, see (26)), with isolated discs showing little Ca-accumulating activity in the dark and releasing only one or two Ca ions per photon absorbed even when loaded with Ca. By suspending discs in a medium containing ATP and GTP, George and Hagins (10) were able to obtain much larger Ca releases of up to 10,000 per photoisomerization. Adding cGMP to disc suspensions promoted an uptake of Ca. The Ca release observed by George and Hagins (10) was too slow to be a plausible trigger for a normal light response, but it is possible that some factor missing from their suspension medium speeds up the in vivo release.

Caretta and Cavaggioni (5) and Kaupp and Koch (this volume) have studied Ca movements across disc membranes which are apparently quite different from those reported by George and Hagins (10). They observed Ca movements which are passive and unaffected by the presence of bleached rhodopsin molecules. The channels responsible for the Ca permeability pass Na, K, and Rb ions but exclude Cl (4). The channels are gated open by cyclic GMP; the action is apparently direct as it takes place when the membranes have been well washed to remove soluble proteins, and in the absence of nucleotide triphosphates. It is still unclear quite what the role is of these channels in the function of the discs, but these reports gain topicality in view of the recent demonstration by Fesenko, Kolesnikov, and Lyubarsky (9) of similar channels in the surface membrane (see below).

EFFECTS OF ELEVATED Ca ON THE ROD LIGHT-SENSITIVE CURRENT

One way of testing the Ca hypothesis is to introduce a pulse of Ca into the cell; the effects should be exactly the same as those of a flash of light. We have investigated this question by exposing the rod to brief pulses of elevated external Ca. The light-sensitive current is recorded by drawing the inner segment of an isolated rod into a suction pipette connected to a sensitive current-to-voltage converter; the outer segment projects into a flowing test solution (see (19, 27)). The solution bathing the outer segment can be changed in about 50 ms by transferring the rod across the boundary between two flowing solutions (6).

Figure 1 shows the effects of briefly exposing a rod to an elevated Ca of 10 mmol/l. In the top traces the net light-sensitive inward current is plotted downward from the zero level shown by the dotted line. The traces have been obtained by subtracting the current observed in bright light from the current observed in darkness. The procedure is explained more fully in the legend to Fig. 5 (and see (18)). In this rod a current of 25 pA flowed into the

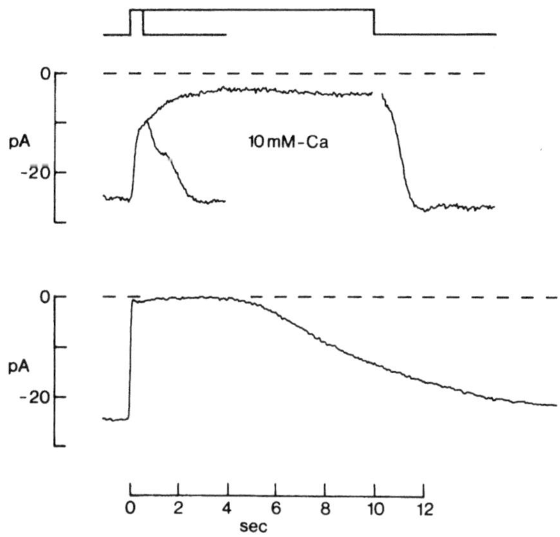

FIG. 1 - Comparison between the effects of elevated external Ca and a flash of light on the light-sensitive current of a rod from Bufo marinus. The top trace shows the duration of two applications of 10 mmol/l Ca, and the middle trace shows the effects of these two pulses of high Ca on the light-sensitive current flowing in through the membrane of the rod outer segment (see text). The lower trace compares the effect of a bright flash of light (6320 isomerizations) on the current recorded from the same cell. In this and subsequent figures the light-sensitive current is shown in pA (1 pA = 10^{-12} A); inward current is plotted downwards according to the usual convention. The zero of light-sensitive current is the level observed in a steady bright light. See (18) for further details.

rod outer segment in darkness and in normal Ringer's solution. Elevating the Ca for 0.5 s and for 10 s had a strong effect in suppressing the light-sensitive current.

The effects of high Ca can be resolved into two components: a rapid and a delayed suppression of current. The rapid suppression is effectively instantaneous within the time resolution of our experiments and may reflect a rapid external competition between Na and Ca ions for channels. The delayed suppression is substantially slower than the solution change rate and probably reflects an internal action of Ca in suppressing channels.

On return to normal Ringer's with 1 mmol/l Ca, the recovery of current is also resolvable into a rapid and a delayed recovery, consistent with removal of a rapid blocking action followed by a slower transport of Ca ions from the cell.

From this and other experiments we conclude that Ca ions inside the cell suppress light-sensitive channels. Unfortunately, this does not prove that

Ca ions are involved in the normal light response. In fact, this experiment rather argues against that, since the recovery from a Ca load is more rapid than the recovery from a light response, as shown in the bottom panel of Fig. 1. It is clear that the recovery of current after a flash of light is not determined by the rate of extrusion of Ca ions, which occurs on the much shorter time scale shown above.

However, under some circumstances the response to a flash of light and a pulse of Ca can be remarkably similar. Figure 2 shows the effect of flashes and a pulse of high Ca on a rod which is maintained in 1 μmol/l Ca. The effect of the low Ca is to increase the light-sensitive current: this loads the cell with Na, and it seems reasonable to suppose that Ca extrusion via Na:Ca exchange has been slowed by this load.

Perhaps because the extrusion of Ca has been slowed down so that it is now the slowest step in the recovery from a flash of light, the recoveries from a 0.5 s pulse of Ca and from a flash are now remarkably similar. On the face of it, this seems more promising for the view that transduction involves the closure of channels by a pulse of Ca released inside the cell. However, closer inspection shows that the suppression of current is much slower when induced by external Ca than when induced by light, even though the Ca entry during the pulse is sufficient to cause a suppression of channels which greatly outlasts the response to a flash of light.

The suppression caused by Ca could be slower because Ca has to cross the membrane before it acts. However, this does not seem to be the whole explanation, for two reasons: first, the suppression does not begin until after the high Ca has been removed, and second, the delay is strikingly reduced by even quite weak background lights (Hodgkin, McNaughton, and Nunn, unpublished results).

These experiments show that there is a slow step between a rise of internal Ca and the closure of channels. This slow step, which is not present in the

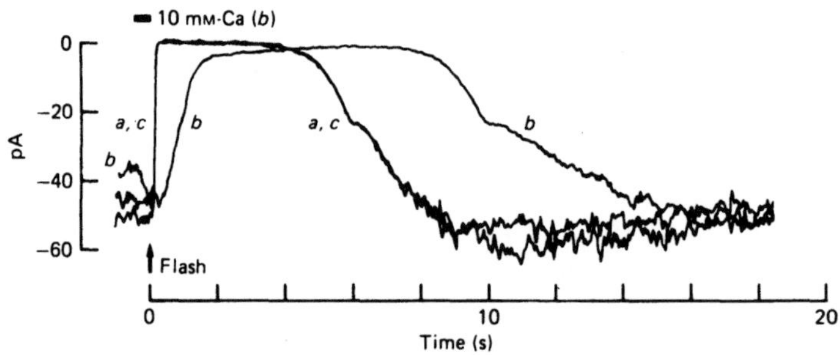

FIG. 2 - Effect on current of a rod from Ambystoma tigrinum of a 0.5 s pulse of 10 mmol/l Ca (b) and of flashes (a, c) each delivering 2.6 x 10^4 isomerizations. The rod was in 110 mmol/l Na, 1 μmol/l Ca except during the pulse. Reprinted with permission from (17).

light response, is speeded by background lights. The speeding action comes on rapidly after exposure to light, since weak flashes delivered coincident with the Ca pulse are also effective.

Another experiment which makes this point in a more striking way is shown in Fig. 3. Here the Ca concentration was abruptly elevated from 1 µmol/l to 74 mmol/l. A large current flows which must be carried almost entirely by Ca ions. Integrating the area under this current transient, we find that the Ca concentration inside the rod must have increased by about 1.4 mmol/l. In spite of this large Ca load, the closure of channels is still slow: a flash of light delivered at any point during this transient shuts off channels at least ten times faster and, therefore, on a simple version of the Ca hypothesis should release the unlikely amount of 14 mmol/l Ca.

The lower panel of Fig. 3 shows that weak background light has the effect of speeding the closure of channels and reducing the Ca influx necessary to produce a complete shutoff of current.

We conclude from these experiments that the Ca hypothesis in its simple form is unlikely to be correct. A rise in internal Ca acts far too slowly to be underlying the light response.

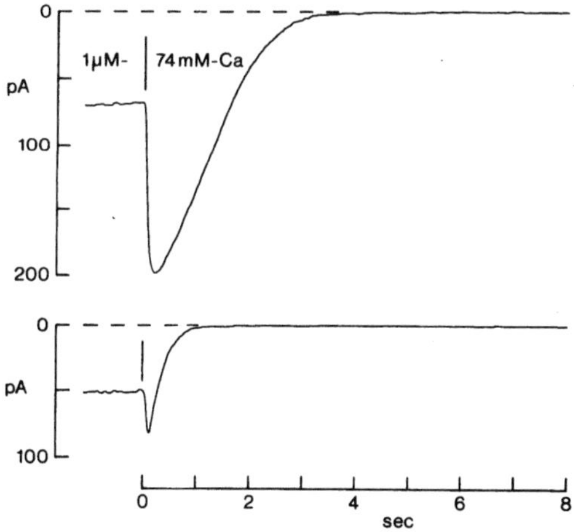

FIG. 3 - Large inward currents carried by Ca. In each record the outer segment was initially in 35 mmol/l Na, 1 µmol/l Ca solution and Ca was raised to 74 mmol/l (and Na removed) as shown. Upper record in dark, lower record in dim background light (2.4 isomerizations s^{-1}). Modified from (18).

EVIDENCE FOR A Na:Ca EXCHANGE MECHANISM

Several different lines of evidence suggest that an Na:Ca exchange mechanism is important in maintaining a low Ca_i in photoreceptors (1). The Ca efflux following a light flash is slowed by lowering the external Na (12, 13,

28). In isolated rod outer segments the efflux of ^{45}Ca is stimulated by Na, Ca, and Sr (23).

Indirect evidence that Na:Ca exchange is important in maintaining the light- sensitive current comes from experiments in which the Na is replaced by other cations. In Fig. 4 (taken from (18)), the Na in Ringer's solution was replaced by other monovalent cations for 6 s. From the initial rapid change in current we conclude that the permeability of the light-sensitive channels to the various ions is Li > Na > K > Rb > Cs > choline. Other experiments have shown that the current recorded in choline is carried by Ca, since it is abolished when Na and Ca are removed together, and that choline is therefore effectively impermeable (18).

Removing Na causes the current to relax to a low level in spite of the relatively high permeability of the membrane to Li, K, and Rb. The probable reason is the inability of all the monovalent cations apart from Na to participate in an Na:Ca exchange mechanism which keeps light-sensitive channels open by maintaining Ca_i at a low level. Clearly the Na:Ca exchange and the light-sensitive channel have very different ionic selectivities and are therefore likely to be separate entities.

Recently a light-insensitive current has been recorded which probably reflects an electrogenic extrusion of Ca in exchange for Na. The current is seen after a large Ca load (22) or after a period in low Na (6). Nakatani and Yau (22) have found the integral of the current to be proportional to the Ca load and to be consistent with a 3:1 exchange of Na for Ca.

Figure 5 shows a recording of both the light-sensitive current (top) and the light-insensitive exchange current (bottom) in an experiment in which the rod outer segment was exposed to a zero-Na solution for various periods of time. On removing Na, the light-sensitive current is rapidly reduced to about 20% of its value in Ringer's solution; as noted above this residual current is carried principally by Ca. On return to Ringer's the light-sensitive current recovers in two phases, the rapid recovery reflecting the passage of Na ions through open channels and the delayed recovery being due to the reopening of channels which have closed in the absence of Na. As the exposure period lengthens, the fraction of channels remaining open declines, as seen in the decline both of the residual current and of the rapid recovery phase on return of Na.

The light-insensitive current (bottom) is the difference between a trace in which a bright flash was given coincident with the return to full Na and a trace taken in bright light throughout. The difference between the traces is truly light-insensitive, as it was not abolished by giving a still brighter flash nor by turning on a bright light a few seconds before the return to full Na.

We think that the light-insensitive current arises because there is a Ca influx in the dark, seen as the residual current in the absence of Na, and this Ca load must be extruded by an Na:Ca exchange on return to normal Na.

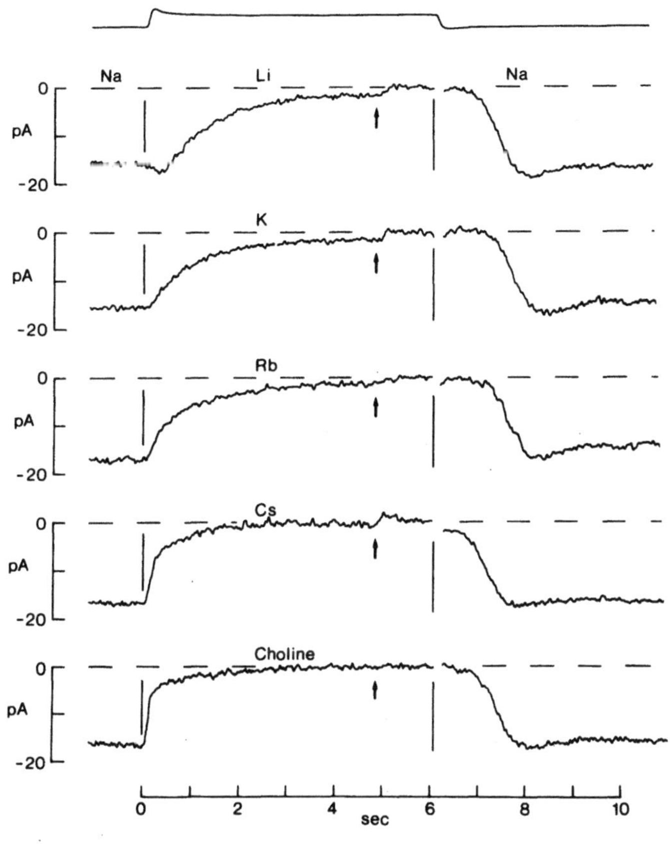

FIG. 4 - Na was replaced by other monovalent ions for 6 s as shown. A saturating light was turned on at the arrow shortly before the end of each step; recovery of light-sensitive current after exposure to zero-Na was recorded in separate exposures without the step of light. Reprinted with permission from (18).

For this reason we refer to the current as a pumping current. This view is supported by the observation that the total charge which flows during the pumping current is proportional to the preceding Ca influx, measured as the integral of the residual current. The fact that the pumping current is inward means that at least 3 Na^+ ions must exchange with each Ca^{2+}, but the exact stoichiometry of the exchange pump cannot be determined from experiments such as this since other ions may participate to some degree in the residual current. Figure 5 shows that the pumping current appears to saturate at about 5 pA in this rod and declines with a time constant of about 0.5 s.

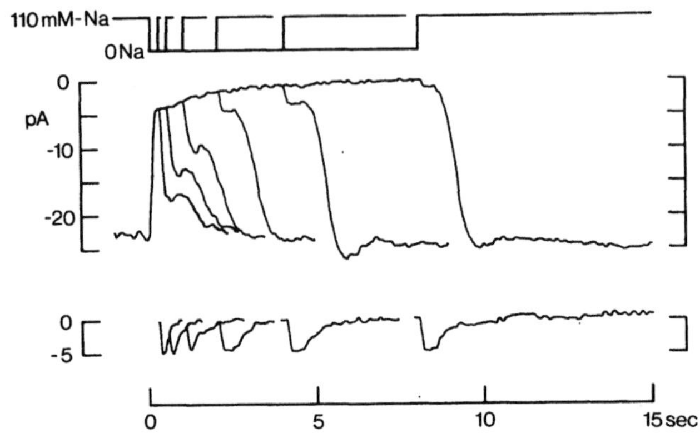

FIG. 5 - Light-sensitive (top) and light-insensitive (bottom) currents recorded during exposure of a toad rod outer segment to a zero-Na solution for various times. Na replaced by choline. Solution monitor shown at top. Currents during the exposure to zero-Na are the difference between the current recorded in darkness and in continuous bright light. On return to full Na the light-sensitive current was computed as the difference between a trace in the dark and a trace with a bright, 20 ms flash delivered coincident with the return to Na; light-insensitive current is the difference between the trace with a bright light flash and the trace taken in continuous bright light (unpublished results of Hodgkin, McNaughton, and Nunn).

If the calcium pumped out of the cell after a flash of light is transported by an electrogenic Na:Ca exchange, then the electrical effects of the operation of the pump should be apparent in the light response. In fact, there is a small transient inward current after a bright flash - an example is shown in the lower trace in Fig. 1. In this experiment the inward charge transferred in this small "notch" following the flash was about 1 pC, or 1000 Ca^{2+} ions per photoisomerization if the pump operates as a $3Na^{+}:1Ca^{2+}$ exchange.

DIRECT EFFECTS OF CYCLIC GMP ON THE PHOTORECEPTOR MEMBRANE

No review of the Ca hypothesis would be complete without at least a mention of the recent paper by Fesenko et al. (9), published since the above was written, in which a direct action of cyclic GMP on the surface membrane of the rod outer segment has been reported. Cyclic GMP increases the conductance of excised patches of the rod outer segment; the conductance has the ionic selectivity and underlying unit event expected from studies of the light-sensitive current in intact rods. The conductance is little affected by calcium ions, is half-activated at a cyclic GMP concentration of 30 µmol/l, and seems to require two molecules of cGMP to activate it. There is a striking resemblance to the cGMP-activated conductance in the disc membranes (4, 5) which are, after all, formed by pinching off invaginations of outer segment membrane. This report, taken together with the more indirect arguments outlined above, probably means that the Ca hypothesis as originally proposed can be laid to rest.

CONCLUSION

It is still unclear exactly which role Ca plays in the light response. We know that Ca inside the cell has a strong effect in suppressing the light-sensitive current, but the slow rate of its action compared with the light response suggests that some other substance controlled by light is responsible for shutting channels directly. The recent report by Fesenko et al. (9) makes it likely that this substance is cyclic GMP. Calcium is transported from the cell after a light flash, probably by an electrogenic Na:Ca exchange, but we still do not know whether this Ca efflux occurs as a result of a Ca release inside the cell, whether it results from the activation of the Ca pump by the flash of light, or whether it is due to the shutting off of the Ca influx through light-sensitive channels. Of these three possibilities, the first implies a rise in free Ca inside the cell, while the last two actually predict a fall in free Ca after a flash of light. It seems unlikely that the complex mechanism for controlling the breakdown of cyclic GMP which we know exists within the cell does not have an important role in the light response; equally, it seems unlikely that the cell would devote so much energy to pumping Ca for no good reason. It looks increasingly likely that the control of channels proceeds via some sort of dual control system in which cyclic GMP is the primary transmitter and Ca has a regulatory function.

It may be worth spelling out a speculative model in more detail (see Fig. 6). We assume that Ca_i does not interact directly with channels, but instead modulates the production of cyclic GMP, which is the primary transmitter responsible for maintaining channels in the open state (9). In this model the rate of closure of channels would not depend on the rate of rise of Ca_i, provided this was significantly faster than the rate β at which cGMP is broken down. Background light which stimulates PDE, and IBMX which inhibits it, would produce the observed speeding and slowing of channel closure rate by modulating β . The guanylate cyclase in rat rods appears to be inhibited by Ca in the range 10^{-9} mol/l to 10^{-7} mmol/l (20), a range of sensitivity which makes control of Ca_i quite plausible. In the experiment of Fig. 5, for instance, in which the residual current of about 5 pA is largely

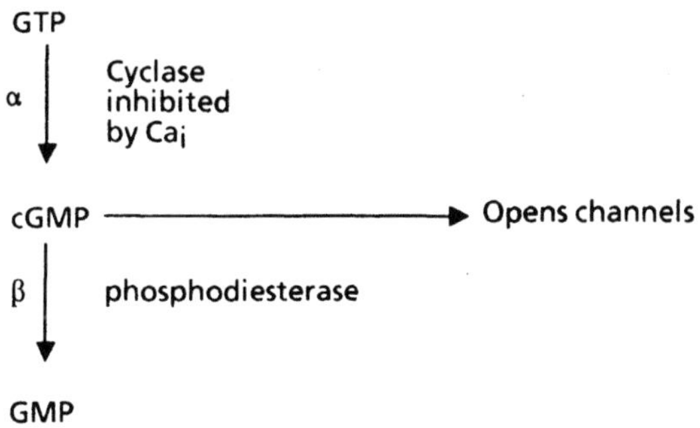

FIG. 6 - Possible model for the control of light-sensitive channels by Ca and cGMP.

carried by Ca, enough Ca enters in the first 20 ms of exposure to zero-Na to raise Ca_i by 10^{-6} mol/l.

The light-sensitive channels are quite permeable to Ca (18), and their closure by light will tend to reduce the Ca influx into the cell. If Ca is not directly released inside the cell by the action of light, then it is quite possible that Ca_i will decrease in light, in contrast to the proposal of the original Ca hypothesis. A shutting off of Ca influx would be consistent with the rise in external Ca after a flash of light (11-13, 28). In the model of Fig. 6 a fall in Ca_i in light has the additional advantage that the production of cGMP will be speeded as well as its breakdown; this may explain the puzzling observation (14) that the rate of turnover of cGMP is more sensitive to light than is the actual concentration of cGMP.

Acknowledgements. We thank L. Cervetto, T.D. Lamb, and V. Torre for their comments on this manuscript.

REFERENCES

(1) Bastian, B.L., and Fain, G.L. 1982. The effects of low calcium and background light on the sensitivity of toad rods. J. Physiol. **330:** 307-329.

(2) Bertrand, D.; Fuortes, M.G.F.; and Pochobradsky, J. 1978. Actions of EGTA and high calcium on the cones in the turtle retina. J. Physiol. **275:** 419-437.

(3) Brown, J.E.; Coles, J.A.; and Pinto, L.H. 1977. Effects of injection of calcium and EGTA into the outer segment of retinal rods of Bufo marinus. J. Physiol. **269:** 707-722.

(4) Caretta, A., and Cavaggioni, A. 1983. Fast ionic flux activated by cyclic GMP in the membrane of cattle rod outer segments. Eur. J. Biochem. **132:** 1-8.

(5) Caretta, A.; Cavaggioni, A.; and Sorbi, R.T. 1979. Cyclic GMP and the permeability of the disks of the frog photoreceptors. J. Physiol. **295:** 171-178.

(6) Cook, R.H.; Hodgkin, A.L.; McNaughton, P.A.; and Nunn, B.J. 1984. Rapid change of solutions bathing a rod outer segment. J. Physiol. **357:** 2P.

(7) Dratz, E.A., and Hargrave, P.A. 1983. The structure of rhodopsin and the rod outer segment disc membrane. Trends Biochem. Sci. **8:** 128-131.

(8) Fain, G.L., and Lisman, J.E. 1981. Membrane conductances of photoreceptors. Prog. Biophys. Molec. Biol. **37:** 91-147.

(9) Fesenko, E.E.; Kolesnikov, S.S.; and Lyubarsky, A.L. 1985. Induction by cyclic GMP of cationic conductance in plasma membrane of retinal rod outer segment. Nature **313:** 310-313.

(10) George, J.S., and Hagins, W.A. 1983. Control of Ca^{2+} in rod outer segment disks by light and cyclic GMP. Nature **303:** 344-348.

(11) Gold, G.H., and Korenbrot, J.I. 1980. Light induced Ca efflux from intact rod cells in living retinas. Fed. Proc. **39:** 1814.

(12) Gold, G.H., and Korenbrot, J.I. 1980. Light-induced Ca release by intact retinal rods. Proc. Natl. Acad. Sci. USA **77:** 5557-5561.

(13) Gold, G.H., and Korenbrot, J.I. 1981. The regulation of calcium in the intact retinal rod: a study of light induced calcium release by the outer segment. In Molecular Mechanisms of Photoreceptor Transduction, ed. W.H. Miller. New York and London: Academic.

(14) Goldberg, N.D.; Ames, A.; Gander, J.E.; and Walseth, T.F. 1983. Magnitude of increase in retinal cGMP metabolic flux determined by ^{18}O incorporation into nucleotide-phosphoryls corresponds with intensity of photic stimulation. J. Biol. Chem. **258:** 9213-9219.

(15) Hagins, W.A. 1972. The visual process: excitatory mechanisms in the primary receptor cells. Ann Rev. Biophys. Bioeng. **1:** 131-158.

(16) Hagins, W.A., and Yoshikami, S. 1974. A role for Ca^{2+} in excitation of retinal rods and cones. Exp. Eye Res. **18:** 299-305.

(17) Hodgkin, A.L.; McNaughton, P.A.; and Nunn, B.J. 1984. Comparison between the effects of flashes of light and brief pulses of calcium on the current of toad and salamander rods. J. Physiol. **357:** 10P.

(18) Hodgkin, A.L.; McNaughton, P.A.; and Nunn, B.J. 1985. The ionic selectivity and calcium dependence of the light-sensitive pathway in toad rods. J. Physiol. **358:** 447-468.

(19) Hodgkin, A.L.; McNaughton, P.A.; Nunn, B.J.; and Yau, K.-W. 1984. Effects of ions on retinal rods from Bufo marinus. J. Physiol. **350:** 649- 680.

(20) Lolley, R.N., and Racz, E. 1982. Calcium modulation of cyclic GMP synthesis in rat visual cells. Vision Res. **22:** 1481-1486.

(21) Miller, W.H. 1981. Current Topics in Membranes and Transport. Molecular Mechanisms of Photoreceptor Transduction, ed. W.H. Miller, vol. 15. New York and London: Academic Press.

(22) Nakatani, K., and Yau, K.W. 1984. Measurement of Na-Ca exchange in toad after Ca loading. J. Physiol. **353:** 77P.

(23) Schnetkamp, P.P.M. 1980. Ion selectivity of the cation transport system of isolated intact cattle rod outer segments: evidence for a direct communication between the rod plasma and the rod disk membranes. Biochem. Biophys. Acta **595:** 66-90.

(24) Schröder, W.H., and Fain, G.L. 1984. Light-dependent calcium release from photoreceptors measured by laser micro-mass analysis. Nature **309:** 268-270.

(25) Somlyo, A.P., and Walz, B. 1985. Elemental distribution in Rana pipiens retinal rods: quantitative electron probe analysis. J. Physiol. **358:** 183-195.

(26) Szuts, E. 1981. Calcium trace exchange in the rods of excised retinas. In Molecular Mechanisms of Photoreceptor Transduction, ed. W.H. Miller. New York and London: Academic.

(27) Yau, K.-W.; McNaughton, P.A.; and Hodgkin, A.L. 1981. Effect of ions on the light-sensitive current in retinal rods. Nature **292:** 502-505.

(28) Yoshikami, S.; George, J.S.; and Hagins, W.A. 1980. Light-induced calcium fluxes from outer segment layer of vertebrate retina. Nature **286:** 395-398.

(29) Yoshikami, S., and Hagins, W.A. 1971. Light, calcium and the photocurrent of rods and cones. Abstract. 15th Annual Meeting of the Biophysical Society. Biophys. J. **11:** 47a.

(30) Yoshikami, S., and Hagins, W.A. 1973. Control of the dark current in vertebrate rods and cones. In Biochemistry and Physiology of Visual Pigments, ed. H. Langer, pp. 245-255. Berlin: Springer-Verlag.

The Molecular Mechanism of Photoreception, ed. H. Stieve, pp. 93-107. Dahlem Konferenzen 1986. Berlin, Heidelberg, New York, Tokyo: Springer-Verlag.

Significance of Changes in Intracellular Ca^{2+} for the Mechanism of Signal Transduction in Vertebrate Rod Cells

U.B. Kaupp and K.-W. Koch
Abteilung Biophysik Universität Osnabrück
4500 Osnabrück, F.R.Germany

Abstract. Light and cyclic GMP stimulate the flux of Ca^{2+} ions across the plasma and disk membrane in rod cells of the vertebrate retina. Ca^{2+} enters the cytosol through the light-sensitive channels in the plasma membrane and a cyclic GMP-regulated conductance in the disk membrane. Ca^{2+} is extruded from the cell by an Na^+/Ca^{2+} exchange mechanism. The existence of an active, ATP-dependent uptake of Ca^{2+} into disks is likely; a Ca^{2+}-transport ATPase in the plasma membrane, however, has not yet been identified. Some of these transport systems may be directly or indirectly regulated by light. The individual contributions of each transport system to the maintenance of the cytosolic Ca^{2+} concentration in the dark and its change by light have not yet been delineated. In particular, attempts to detect the rapid injection of Ca^{2+} (<100 ms) from inside disks into the cytosol - a crucial component of the "Ca^{2+} hypothesis" - have been unsuccessful. A final decision on the relevance of changes in $Ca^{2+}{}_i$ for the generation of the electrical signal must be postponed until the much needed measurements of intracellular Ca^{2+} in these cells are available.

INTRODUCTION

Vertebrate photoreceptors respond to stimulation by light with a transient hyperpolarization of their plasma membrane. It has been suggested that light changes the cytosolic concentration of an internal transmitter substance, thereby closing the light-regulated ion channels in the plasma membrane (3). Ca^{2+} ions and cyclic GMP have been tentatively proposed as intracellular messengers in the process of phototransduction (34, 59).

Necessary conditions consistent with Ca^{2+} being the excitatory messenger are:

1) Light causes a transient change of sufficient speed and magnitude in the cytosolic concentration of Ca^{2+}. The rise in the concentration of Ca^{2+} in the vicinity of the channel and the change in membrane potential must be of similar velocity. The absorption of one photon causes many ion channels (30-300) to close. Therefore, one photon must release at least several hundreds of Ca^{2+} ions.

2) Changes in the cytosolic concentration of Ca^{2+} control the dark current across the plasma membrane and the responses to light.

There is satisfactory evidence on the second condition being fulfilled (reviewed in (25)), even though Ca^{2+} may not control the light-regulated conductance directly but rather regulates the synthesis or removal of another excitatory transmitter.

This paper reviews the evidence for the first condition, namely, a rapid, light-stimulated rise in Ca^{2+}_i. Unfortunately, attempts to measure changes in the intracellular concentration of Ca^{2+} in a living rod have been unsuccessful. Therefore, Ca^{2+} fluxes either were followed extracellularly in the photoreceptor layer of a retina (19, 58) or were studied in suspensions of "intact" or fragmented rod outer segments (e.g., (28, 51, 54)). Both approaches have their virtues and disadvantages. Working with rod outer segments or isolated disks may suffer from the loss of cellular components that are relevant for the mechanism of the release, whereas changes in Ca^{2+} concentration measured extracellularly are difficult to interpret in terms of changes in the cytosolic Ca^{2+} concentration. Two agents stimulate Ca^{2+} fluxes in rod cells: light and cyclic GMP. These two conditions are discussed separately below.

THE RAPID REDISTRIBUTION OF Ca^{2+} BY LIGHT IN SUSPENSIONS OF BROKEN ROD CELLS

Most attempts have failed to detect the rapid release of a large number of Ca^{2+} ions in rod outer segments with an intact plasma membrane, in isolated disks, in sonicated disks, or in rhodopsin-phospholipid recombinants (reviewed in (27)). Kaupp et al. (28) detected a rapid, light-stimulated redistribution of Ca^{2+} between binding sites inside disks and the external medium in the presence of the Ca^{2+} ionophore A23187. The appearance of the released Ca^{2+} ions in the outer medium occurred rapidly in fragmented, sonicated disks ($t_{1/2}$ = 15 ms) but was retarded in "intact" rod outer segments (~300 ms), probably due to diffusional barriers existing in an intact stack of disks (29, 30). The metarhodopsin II transition and the light-stimulated uptake of protons into the disk membrane have similar time courses to the rapid release of Ca^{2+} (Fig. 1). This kinetic coincidence suggests that either conformational changes within the rhodopsin molecule or changes in the electrostatic properties of the membrane surface during the proton uptake are responsible for the redistribution between bound and free Ca^{2+} ions. Maximally 1 Ca^{2+} ion/photoisomerization (Rh*) was released. This release stoichiometry is far less than the several hundred transmitter molecules required to close 3% of the open channels in the plasma membrane. Therefore, we do not consider this rapid Ca^{2+} redistribution as a crucial step in the amplification sequence of phototransduction. However, it may be an indicator of some important structural reorganization at the lumenal surface of the disk membrane during the metarhodopsin II formation.

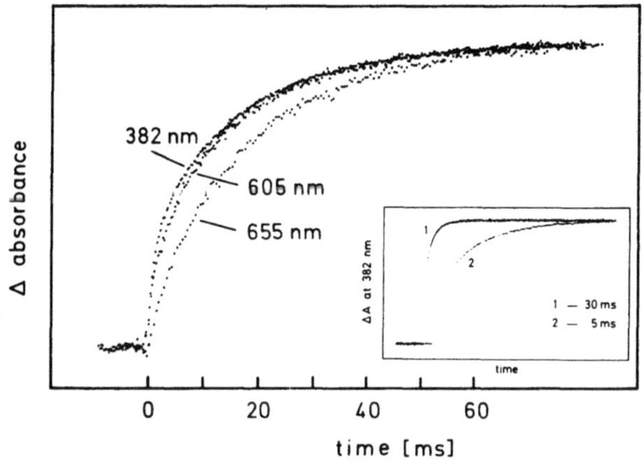

FIG. 1 - Time courses of light-induced absorption changes at 382 nm (MI/MII), at 605 nm in the presence of bromocresol purple (proton uptake), and at 655 nm in the presence of arsenazo III (calcium release).

The release by light of Ca^{2+} ions was only observed in the presence of A23187. We could not detect a light-stimulated translocation of Ca^{2+} across the disk membrane, although the disks were filled with 1-3 mol Ca^{2+}/mol Rh, and hence a Ca^{2+} gradient existed across the disk membrane. These experiments argue against the release of Ca^{2+} by a simple mechanism such as the direct opening of Ca^{2+} channels by rhodopsin itself. The above results were obtained with rod outer segments whose plasma membrane had been made permeable to small solutes by freeze-thawing or washing with tris-buffer. Some cellular component(s) which is (are) of critical importance for the intactness of the Ca^{2+} release mechanism may have been lost or damaged, hence the failure to detect a rapid, light-stimulated translocation of Ca^{2+} should be interpreted with caution. Recently, George and Hagins (18) reported a light-stimulated release of Ca^{2+} from frog rod outer segments whose plasma membrane had been made permeable by a simple sedimentation step. The release required the presence of high concentrations of nucleotide triphosphates. Although the release stoichiometry (28,000 Ca^{2+} ions released per photoisomerization) was sufficient to account for the amplification in phototransduction, the apparent velocity of the Ca^{2+} efflux was too slow (half rise-time approximately 30-40 s at saturating light levels) to be regarded as a precursor of the receptor potential.

Ca^{2+} EFFLUX BY LIGHT IN THE INTACT RETINA

A large number of Ca^{2+} ions per photoactivated rhodopsin molecule (Rh*) were extruded from a rod in a living retina (19, 58). The release occurred rapidly (within 10 s) at high stoichiometry (400 Ca^{2+}/Rh*, see (58); and 10,000 Ca^{2+}/Rh*, see (19)). These experiments provided the first convincing evidence that light causes a redistribution of intracellular Ca^{2+} in an intact cell. Schröder and Fain (50), employing a different technique, reported that a rod outer segment loses an appreciable fraction of its total Ca^{2+} content (50-66%) in bright light. This Ca^{2+} loss was slower (2-5 min) than the Ca^{2+} loss reported by Gold and Korenbrot. Most likely, Ca^{2+} is extruded via a Ca^{2+}/Na^{+} exchange mechanism because the velocity of the efflux is dependent on the extracellular concentrations of both Ca^{2+} and Na^{+} (19, 20).

A crucial question is: does the appearance of Ca^{2+} in the extracellular medium quantitatively indicate a transient change in the cytosolic concentration of Ca^{2+}? The simplest view assumes that light causes the cytosolic concentration of Ca^{2+} to drop owing to the active extrusion of Ca^{2+} from the cell. This interpretation implies that the Ca^{2+} extruded is predominantly of cytosolic origin and that the Ca^{2+} buffering capacity in the cytosol is large enough to account for all of the Ca^{2+} that appears outside the cell. A Ca^{2+} efflux from the cell without a rise in cytosolic concentration of Ca^{2+} could result from a) activation of a Ca^{2+} transport-ATPase by light, b) a more negative membrane potential during the photoresponse, c) a decrease in the Na^{+} concentration inside the cell, or d) the inhibition of the Ca^{2+} influx by light.

No light-regulated Ca^{2+}-ATPase activity has been detected in the outer segment of rods. A change in the membrane potential from -30 mV to -60 mV could stimulate the efflux rate approximately fivefold, assuming an electrogenic exchange of four Na^{+} ions against one Ca^{2+}. The voltage-dependence of the Na^{+}/Ca^{2+} exchange (41) can be described by

$$\frac{[Ca^{2+}]_o}{[Ca^{2+}]_i} = \left(\frac{[Na^{+}]_o}{[Na^{+}]_i}\right)^n exp\left(-\frac{re\Delta\phi}{kT}\right) \qquad (1)$$

Although the contribution of the membrane potential may be significant under normal conditions, it is negligible in the experiments by Gold and Korenbrot (20) because changes in membrane voltage were abolished by working in a low Na^{+} medium. Intracellular Na^{+} inhibits the extrusion of Ca^{2+} in other cells (41). Less Na^{+} enters the rod during a photoreceptor potential and the concentration of Na^{+} may therefore decrease, thereby removing the inhibitory effect of Na^{+}.

Alternatively, it has been proposed that Ca^{2+} is extruded as a result of a rise in the concentration of Ca^{2+} in the cytosol. Accordingly, active removal of Ca^{2+} across the plasma membrane represents one route whereby the

* Symbols: $[Ca^{2+}]_o$, $[Ca^{2+}]_i$, $[Na^{+}]_o$, $[Na^{+}]_i$ are the respective activities of calcium and sodium ions outside and inside the cell. e, elementary charge of an electron; $\Delta\phi$, membrane potential; n, number of Na^{+} ions per turnover; r, number of elementary charges transported per turnover; k, Boltzman constant; T, absolute temperature.

elevated Ca^{2+} level is returned to its original value in the dark. The 1,000-10,000 Ca^{2+} ions/Rh* that appear outside the cell may be sufficient to account for the closing of several hundred light-regulated channels per absorbed photon. The considerations which follow, however, illustrate that the extrusion of Ca^{2+} from the cell could be a mere epiphenomenon of a much larger change in Ca^{2+} concentration inside the cell.

1,000-10,000 messenger molecules/Rh* are sufficient to close several hundred channels in the plasma membrane. This number of transmitter molecules is sufficient only if the light-regulated channels are the only or the predominant binding component in the cytosol, i.e., the buffering capacity for the excitatory transmitter is determined by the concentration of the light-regulated channels. The number of light-regulated channels in the outer segment of a rod is on the order of 10^{-6} mol/l (~3,000-10,000 channels per outer segment (14, 60)); volume of a bovine outer segment v = 30 fl; volume of a frog outer segment v = 1.5 pl) and, therefore, these values are a lower limit for the cytosolic buffering capacity. This estimate may be wrong by several orders of magnitude if the cytosolic space has a significant binding capacity for the excitatory transmitter in excess to that of the light-regulated channels. This statement applies in particular to Ca^{2+} ions and protons: the buffering capacity of cells for protons can be several 10 mmol/l. To our knowledge, Ca^{2+} buffering capacities of cells have not been quantitatively determined, but a value β = 1 mmol/l may be realistic. A list of several Ca^{2+} binding components of cells are given by Campbell (7). Phospholipids of the disk membrane may already contribute β = 100-500 µmol/l. Ca^{2+} buffering capacities can be determined from EGTA injections into cells. A rise in Ca^{2+} or the triggering of a Ca^{2+}-dependent response commences to be suppressed by EGTA when $\beta_{EGTA} \simeq \beta_t$. Experiments on three different types of cells suggest that the injection of EGTA to a final concentration of 1-10 mmol/l is needed to affect either the cellular responses or the changes in Ca^{2+} (1, 36, 42).

The incremental change in the free concentration of Ca^{2+}, $dpCa_i$, is proportional to the change in total Ca^{2+}, dCa_t:

$$dCa_t = -\beta_t \, dpCa_i. \tag{2}$$

dCa_t is also the number of Ca^{2+} ions that have been injected into the cytosol by light. The constant β_t is the total buffering capacity of the cytosol for Ca^{2+}. It is composed of instantaneous, passive binding of Ca^{2+} to binding sites in the cytosol and of the active uptake into subcellular stores. A 1,000-fold larger release stoichiometry would be required to give the same change in the free concentration of Ca^{2+} if β = 1 mmol/l instead of β = 1 µmol/l (the concentration of channels is $\lesssim$ 1 µmol/l). In conclusion, the number of Ca^{2+} ions that are needed to close 3% of the channels that were still open in the dark could easily amount to 100,000-1,000,000 Ca^{2+} ions/Rh*.

The above arguments are only valid if the released Ca^{2+} rapidly equilibrates between the free pool and binding sites in the cytosol. If the site of the Ca^{2+} ejection into the cytosol is in close juxtaposition to the channels, or if specialized osmotic compartments exist ("localized calcium"), equilibrium binding does not apply and, consequently, the intrinsic buffering groups in the cytosol are not "seen" by the Ca^{2+} ions. There is indeed good evidence that the excitation of a rod cell by light is approximately localized to the area of illumination (33). The above is a rather general consideration of the intrinsic buffering capacity owing to the lack of sufficiently detailed information on the concentration and subcellular distribution of Ca^{2+} buffering groups. It illustrates, however, that Ca^{2+} is a candidate for the excitatory transmitter only if the changes in Ca^{2+} by light are highly localized and not spread over the larger portion of the cell.

The buffering capacity has two important consequences. It determines the effective velocity of Ca^{2+} diffusion in the cytosolic matrix and the size of changes in free Ca^{2+}. It can explain the slow rise-time of the photoresponses to low levels of light (39) at the expense, however, of much larger release stoichiometries needed to raise Ca^{2+} significantly. In our opinion, the diffusion of a soluble component of small molecular weight is unlikely to be the rate-limiting step in the transduction process. Diffusion in an aqueous medium is little dependent on the temperature ($E_A \sim RT \sim 4$ kJ/mol). In contrast, the rising phase of the photoresponse exhibits an E_A = 40-50 kJ/mol (e.g., (32)). The lateral and rotational diffusion of proteins in the matrix of the disk membrane has a higher $Q_{10} \sim 1.7$-3.5 (~40-80 kJ/mol) (11, 45, 55), similar to the Q_{10} of the photoresponse.

Immobilized Ca^{2+} buffering groups at the surface of the disk membrane could contribute to the decrease in diffusional mobility of cytosolic Ca^{2+}. Soluble and small Ca^{2+}-binding molecules, however, do not slow down the diffusion of Ca^{2+}. Therefore, an increase in the buffering capacity of the cytosol, i.e., by the injection of EGTA or Quin 2 into a rod cell does not necessarily impede the mobility of Ca^{2+} ions. Instead, it causes a redistribution of Ca^{2+} between binding sites at the surface of membranes and the additional buffer that has been introduced into the cytosol. Consequently, more Ca^{2+} ions are bound to a molecule which is less restricted in its diffusional mobility than the binding sites at the surface of the membrane. As a result, the velocity of the Ca^{2+} flux from the site where it was released to the plasma membrane through the cytosolic matrix is expected to increase despite a larger buffering capacity.

STIMULATION OF Ca^{2+} FLUXES BY CYCLIC GMP

Manipulating the intracellular concentrations of either Ca^{2+} or cyclic GMP affects the light-sensitive conductance in the plasma membrane (see McNaughton et al. and Miller, both this volume). The biochemical pathways of Ca^{2+} and cyclic GMP in rods are intimately coupled (10, 37). Since Ca^{2+} can regulate the activity of the enzyme that synthesizes cyclic GMP, it is difficult to interpret the results of experiments that involved manipulating the intracellular levels of cyclic GMP and Ca^{2+} independently of each other. Some effects that have been taken as supporting the "Ca^{2+} transmitter hypothesis" can be equally well explained as the result of the effects of Ca^{2+} on the cyclic GMP level. It is, therefore, important to understand the interaction between cyclic GMP and Ca^{2+} in greater detail.

Cyclic GMP regulates one or several Ca^{2+} transport systems in the disk membrane and probably also in the plasma membrane. Three different research groups reported two different results - unfortunately of conflicting nature. Caretta, Cavaggioni, and Sorbi (8, 9, 52) and Koch and Kaupp (26, 31) reported a cyclic GMP-sensitive release of Ca^{2+} from disks, whereas George and Hagins (18) found an ATP-dependent stimulation of Ca^{2+} uptake by cyclic GMP into some intracellular compartment in either the inner or outer segment. We have not yet followed the experimental procedures as published by George and Hagins (18) and, therefore, do not know why cyclic GMP stimulates the sequestration of Ca^{2+} into an intracellular store under some conditions and induces the release of Ca^{2+} under other conditions. We emphasize, however, that the results obtained by the Parma and Osnabrück groups are in good quantitative agreement (8, 9, 26, 31, 52).

Stoichiometry. The release of Ca^{2+} by cyclic GMP occurs with a high stoichiometry of about 11,000 Ca^{2+} ions/disk (26).

Velocity of the release. The velocity of the Ca^{2+} efflux depends on the concentration of cyclic GMP and the transmembrane Ca^{2+} gradient (Fig. 2) (8, 31). At saturating concentrations of cyclic GMP ($\gtrsim$100 µmol/l) it is sufficiently rapid to be of significance for phototransduction even though the time course is slower in fragmented outer segments ($\lesssim$20 s (26, 31)) than in sonicated disk vesicles ($\lesssim$100 ms (8)).

Dissociation constant. The Ca^{2+} permeability is activated by cyclic GMP concentrations between 25 and 100 µmol/l, i.e., the concentration range of cyclic GMP in a living rod (Fig. 2) (10). The apparent Michaelis constant is K_D = 50-70 µmol/l, as determined from the rate of the Ca^{2+} efflux at different concentrations of cyclic GMP (8, 31). Ca^{2+} binding sites at the surface of the disk membrane are not affected by cyclic GMP (31).

Mechanism of activation. The ion conductance is activated by simple binding of cyclic GMP to a specific receptor at the channel protein. Therefore, cyclic GMP may be the natural agonist for this chemically gated channel or carrier (31).

The hydrolysis of cyclic GMP is not a necessary condition for the permeability increase because poorly hydrolyzable analogues of cyclic GMP are even more efficient (Fig. 3) (31). The release is not inhibited by high concentrations of protonophores (e.g., FCCP) and therefore does not require a protonmotive force across the disk membrane. It is almost certain that the activation mechanism also does not involve a phosphorylation step (31). We proposed that disk membranes contain a receptor for cyclic GMP which is part of a regulatory site of an ion channel (31).

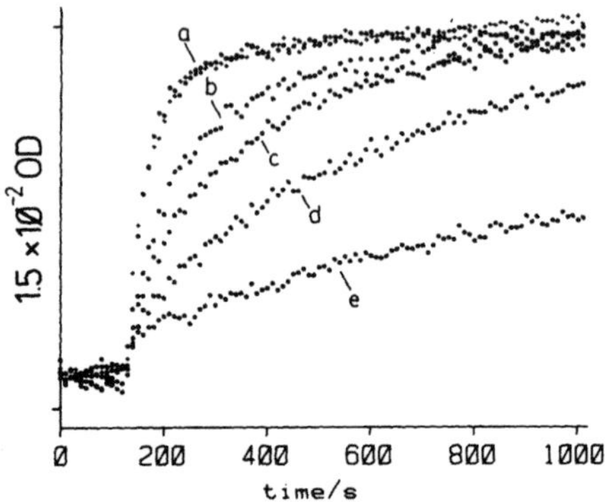

FIG. 2 - Ca2+ release by different concentrations of cyclic GMP. The time course of the Ca2+-indicating absorption changes was recorded in a spectrophotometer (Aminco DW-2a) in the double wavelength mode. The difference in absorbance at 650 nm and 730 nm was recorded with a bandwidth of 10 nm. Recordings were stored in a transient recorder (Tracor TN-1500) and transferred to a PDP 11/34 computer. Intact ROS were suspended in 20 volumes of distilled water (final volume 2.1 ml). After 20-30 s the lysis process was stopped by the addition of 100 mmol/l KCl and 10 mmol/l Hepes (pH 7.4). Final concentrations in the cuvette were: 100 mmol/l KCl, 10 mmol/l Hepes at pH 7.4, 30 mmol/l sucrose, 50 µmol/l arsenazo III, and 5 µmol/l rhodopsin. Ca2+ release was started by the injection of a small volume of cyclic GMP (2-10 µl) into the cuvette with a Hamilton syringe. Concentrations of cyclic GMP: (a), (o) 250 µmol/l and (+) 150 µmol/l; (b) 100 µmol/l; (c) 75 µmol/l; (d) 50 µmol/l; and (e) 25 µmol/l. Calibration pulse: 1 µmol/l Ca2+ = 1.5 x 10-2 A.

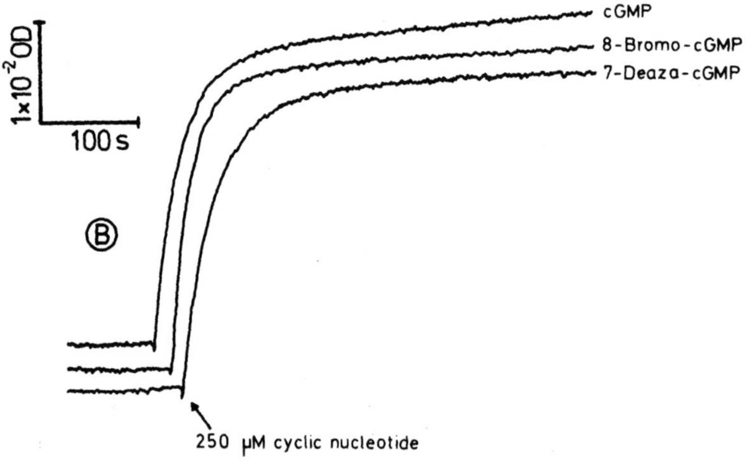

FIG. 3 - Cyclic nucleotide specificity of the Ca2+ release. Comparison of the Ca2+ responses stimulated by 250 µmol/l of three different analogues of cyclic GMP: calibration pulses, 1 µmol/l Ca2+ = 1.7 x 10-2 A (cGMP and 8-Br- cGMP) and 1.5 x 10-2 A (7-deaza-cGMP); rhodopsin concentration, 7 µmol/l.

Nature of the cyclic GMP-sensitive conductance. Both the light-sensitive conductance in the plasma membrane (35, 38, 44) and this cation conductance in the disk membrane are activated by cyclic GMP and are permeable to most monovalent cations as well as to Ca^{2+}, Ba^{2+}, Sr^{2+}, and Mn^{2+} (8, 24, 31, 56). The permeability for monovalent cations of the light-sensitive conductance in the plasma membrane decreases in the following sequence (17, 56):

$Li^+ \sim Na^+ > K^+ > Rb^+ > Cs^+$.

Permeability ratios for the cyclic GMP-activated conductance in the disk membrane are not yet available. These similarities suggest that the two ion conductances may be similar if not identical. The increase in membrane permeability induced by cyclic GMP may reflect the opening of a cation channel or the activation of a carrier-type transport system. Noise analysis of current fluctuations across the plasma membrane, however, exclude a shuttle-type carrier as the translocation mechanism (17).

REGULATION OF Ca^{2+} - A SUMMARY

Four Ca^{2+}-pathways have been described in the literature (Fig. 4):

1) cyclic GMP-regulated Ca^{2+} conductance in the disk membrane (8, 31);
2) light-regulated Ca^{2+} conductance (24, 56);

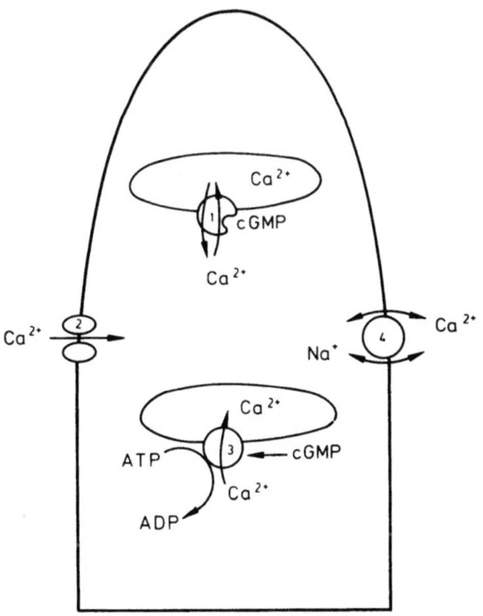

FIG. 4 - Ca^{2+} transport systems or Ca^{2+} permeabilities in rod cells of the vertebrate retina: 1) cGMP-activated Ca^{2+} permeability; 2) light-regulated conductance; 3) ATP-dependent Ca^{2+} uptake into disks; 4) Na^+/Ca^{2+} exchange carrier.

3) ATP-dependent uptake of Ca^{2+} into disks (46) - the uptake may be stimulated by cyclic GMP (18);
4) Na^{+}/Ca^{2+} exchange system in the plasma membrane (2, 24, 49, 57).

This list is tentative and reflects the uncertainty in our knowledge of the molecular identity of some of these transport systems. For example, the cyclic GMP-regulated Ca^{2+} conductance in the disk membrane may be identical with the light-regulated conductance in the plasma membrane. The cyclic GMP-stimulated transport system described by George and Hagins (18) may occur in some subcellular organelles other than disks. With this proviso in mind, the following general picture can be drawn:

In the dark, Ca^{2+} enters the cytosol through the light-regulated conductance (24, 56) and possibly also via the cyclic GMP-activated channels in the disk membrane. The influx of Ca^{2+} is compensated for by removal of Ca^{2+} by the Na^{+}/Ca^{2+} exchanger and possibly by an ATP-dependent "Ca^{2+} pump" in the disk membrane. A Ca^{2+} transport-ATPase has not yet been identified in the plasma membrane of the outer segment. Stimulation by light closes the light-sensitive conductance and decreases the concentration of cyclic GMP in the cytosol, thereby curtailing the Ca^{2+} influx into the cytosol. Consequently, the cytosolic concentration of Ca^{2+} is expected to fall during illumination. So far the intracellular free Ca^{2+} has not been measured in rod cells, and therefore it is unknown on which time scale these different pathways individually contribute to the maintenance of Ca^{2+} in the dark and of its modulation by light.

CONCLUSION

We reviewed the evidence for changes in the concentration of Ca^{2+} inside the outer segment of a rod cell and the possible mechanisms underlying these changes. Several aspects of Ca^{2+} metabolism may be regulated by cyclic GMP, including a slow uptake into subcellular stores and a rapid release from disks within seconds. Light regulates the concentration of cyclic GMP and also releases Ca^{2+} from internal stores, possible disks. Both observations suggest that light may control Ca^{2+} via a cyclic GMP pathway. Ca^{2+} movements in other cellular systems are controlled either by the voltage across the membrane, by cyclic nucleotides (47), by the inositol trisphosphate pathway (IP_3) (4), or by ion antiports. The photoisomerization of a single rhodopsin molecule can probably change the potential across the disk membrane by no more than a few nanovolts and, therefore, we have to envisage the possible existence of a "first messenger" which could trigger the release of Ca^{2+} from inside the disks. Rhodopsin itself is probably not a channel-forming protein (22). Recently it has been demonstrated that inositol polyphosphates stimulate the efflux of Ca^{2+} from intracellular sites in a variety of cells (4, 53), including photoreceptors of the ventral eye of Limulus (5, 12). A light-regulated IP_3 pathway may also be involved in the redistribution of Ca^{2+} by light in vertebrate rod cells.

In the past, much importance has been attributed to a rapid light-stimulated rise in Ca^{2+}. In our opinion, even though a rise in Ca^{2+} certainly is a crucial component of the "Ca^{2+} transmitter hypothesis," it is not sufficient to warrant acceptance of this hypothesis. Why? We expect in the near future that discussions on this issue will replicate the debates that have been provoked by the seminal paper of Michell (40) on the role of a Ca^{2+} influx for hormone secretion or hormone excitation coupling. This controversy focused on the question as to whether a rise in Ca^{2+} is a necessary step for secretion or excitation or simply a consequence of the processes elicited by a hormone. There is a growing body of evidence that a rise in Ca^{2+} is sufficient to initiate some cellular responses but is not a necessary condition, because hormone secretion or excitation may take place even in the absence of a change in Ca^{2+} (e.g., (15)). These observations are suggestive of at least two different excitatory pathways, one employing Ca^{2+} as an essential ingredient. It comes then as no surprise that research on the excitatory messenger substance in the ventral eye of Limulus has already arrived at this stage (6, 16, 43). We conclude this brief survey by quoting a sentence by Chargaff which expresses some of the flavor of photoreceptor research: "One of the most insidious and nefarious properties of scientific models is their tendency to take over, and sometimes supplant, reality" (quoted from a review by Hawthorne (23)).

Acknowledgements. This work has been supported by the Deutsche Forschungsgemeinschaft (SFB 171-C6). We thank W. Junge for helpful comments and reading of the manuscript.

REFERENCES

(1) Ashley, C.C. 1967. The role of cell calcium in the contraction of single cannulated muscle fibers. Am. Zool. **7:** 647.

(2) Bastian, B.L., and Fain, G.L. 1982. The effects of sodium replacement on the responses of toad rods. J. Physiol. **330:** 331-347.

(3) Baylor, D.A., and Fuortes, M.G.F. 1970. Electrical responses of single cones in the retina of the turtle. J. Physiol. **207:** 77-92.

(4) Berridge, M.J., and Irvine, R.F. 1984. Inositol trisphosphate, a novel second messenger in cellular signal transduction. Nature **312:** 315-321.

(5) Brown, J.E., and Rubin, L,J. 1984. A direct demonstration that inositol trisphosphate induces an increase in intracellular calcium in Limulus photoreceptors. Biohem. Biophys. Res. Commun. **125:** 1137-1142.

(6) Brown, J.E.; Rubin, L.J.; Ghalayini, A.J.; Tarver, A.P.; Irvine, R.F.; Berridge, M.J.; and Anderson, R.E. 1984. Evidence that myo-inositol polyphosphate may be a messenger for visual excitation in Limulus photoreceptors. Nature **311:** 160-163.

(7) Campbell, A.K. 1983. Intracellular Calcium. Chichester: John Wiley.

(8) Caretta, A., and Çavaggioni, A. 1983. Fast ionic flux activated by cyclic GMP in the membrane of cattle rod outer segment. Eur. J. Biochem. **132:** 1-8.

(9) Cavaggioni, A., and Sorbi, R.T. 1981. Cyclic GMP releases calcium from disk membranes of vertebrate photoreceptors. Proc. Natl. Acad. Sci. USA **78:** 3964-3968.

(10) Cohen, A.I. 1981. The use of incubated retinas in investigating the effects of calcium and other ions on cyclic-nucleotide levels in photoreceptors. Curr. Top. Membr. Trans. **15:** 215-229.

(11) Cone, R.A. 1972. Rotational diffusion of rhodopsin in the visual receptor membrane. Nat. New Biol. **236:** 39-43.

(12) Corson, D.W.; Fein, A.; and Payne, R. 1985. Detection of an inositol 1,4,5-triphosphate-induced rise in intracellular free Ca^{2+} with aequorin in Limulus ventral photoreceptors. Biol. Bull., in press.

(13) Cote, R.H.; Biernbaum, M.S.; Nicol, G.D.; and Bownds, M.D. 1984. Light-induced decreases in cGMP concentration precede changes in membrane permeability in frog rod photoreceptors. J. Biol. Chem. **259:** 9635-9641.

(14) Detwiler, P.B.; Conner, J.A.; and Bodoia, R.D. 1982. Gigaseal patch-clamp recordings from outer segments of intact retinal rods. Nature **300:** 59-62.

(15) Di Virgilio, F.; Lew, D.P.; and Pozzan, T. 1984. Protein kinase C activation of physiological processes in human neutrophils at vanishingly small cytosolic concentrations. Nature **310:** 691-693.

(16) Fein, A.; Payne, R.; Corson, D.W.; Berridge, M.J.; and Irvine, R.F. 1984. Photoreceptor excitation and adaptation by inositol 1,4,5 trisphosphate. Nature **311:** 157-160.

(17) Fesenko, E.E.; Kolesnikov, S.S.; and Lyubarsky, A.L. 1985. Induction by cyclic GMP of cationic conductance in plasma membrane of retinal rod outer segment. Nature **313:** 310-313.

(18) George, J.S., and Hagins, W.A. 1983. Control of Ca^{2+} in rod outer segment disks by light and cyclic GMP. Nature **303:** 344-348.

(19) Gold, G.H., and Korenbrot, J.I. 1980. Light-induced calcium release by intact retinal rods. Proc. Natl. Acad. Sci. USA **77:** 5557-5561.

(20) Gold, G.H., and Korenbrot, J.I. 1981. The regulation of calcium in the intact retinal rod: A study of light-induced calcium release by outer segment. Curr. Top. Membr. Trans. **15:** 307-330.

(21) Goldberg, N.D.; Ames, A. III; Gander, J.E.; and Walseth, T.F. 1983. Magnitude of increase in retinal cGMP metabolic flux determined by ^{18}O incorporation into nucleotide-phosphoryls corresponds with intensity of photic stimulation. J. Biol. Chem. **258:** 9213-9219.

(22) Hanke, W., and Kaupp, U.B. 1984. Incorporation of ion channels from bovine rod outer segments into planar lipid bilayers. Biophys. J. **46:** 587-595.

(23) Hawthorne, J.N. 1983. Polyphosphoinositide metabolism in excitable membranes. Biosci. Rep. **3:** 887-904.

(24) Hodgkin, A.L.; McNaughton, P.A.; Nunn, B.J.; and Yau, K.-W. 1984. Effect of ions on retinal rods from Bufo marinus. J. Physiol. **350:** 649-680.

(25) Kaupp, U.B. 1984. The role of calcium in visual transduction. In Information and Energy Transduction in Biological Membranes, eds. E.J. Helmreich, pp. 325-339. New York: A.R. Riss.

(26) Kaupp, U.B., and Koch, K.-W. 1984. Cyclic GMP releases calcium from leaky rod outer segments. Vision Res. **24:** 1477-1479.

(27) Kaupp, U.B., and Schnetkamp, P.P.M. 1982. Calcium metabolism in vertebrate photoreceptors. Cell Calcium **3:** 83-112.

(28) Kaupp, U.B.; Schnetkamp, P.P.M.; and Junge, W. 1979. Light-induced calcium release in intact rod outer segments upon photoexcitation of rhodopsin. Biochim. Biophys. Acta **552:** 390-403.

(29) Kaupp, U.B.; Schnetkamp, P.P.M.; and Junge, W. 1981a. Rapid calcium release and proton uptake at the disk membrane of isolated cattle rod outer segments. 1. Stoichiometry of light-stimulated calcium release and proton uptake. Biochemistry **20:** 5500-5510.

(30) Kaupp, U.B.; Schnetkamp, P.P.M.; and Junge, W. 1981b. Rapid calcium release and proton uptake at the disk membrane of isolated cattle rod outer segments. 2. Kinetics of light-stimulated calcium release and proton uptake. Biochemistry **20:** 5511-5516.

(31) Koch, K.-W., and Kaupp, U.B. 1985. Cyclic GMP directly regulates a cation conductance in membranes of bovine rods by a cooperative mechanism. J. Biol. Chem. **260:** 6788-6800.

(32) Lamb, T.D. 1984. Effects of temperature changes on toad rod photocurrents. J. Physiol. **346:** 557-578.

(33) Lamb, T.D.; McNaughton, P.A.; and Yau, K.-W. 1981. Spatial spread of activation and background desensitization in toad rod outer segments. J. Physiol. **319:** 463-496.

(34) Liebman, P.A., and Pugh, E.N., Jr. 1981. Control of rod disk membrane phosphodiesterase and a model for visual transduction. Curr. Top. Membr. Trans. **15:** 157-170.

(35) Lipton, S.A. 1983. cGMP and EGTA increase the light-sensitive current of retinal rods. Brain Res. **265:** 41-48.

(36) Lisman, J.E., and Brown, J.E. 1975. Effects of intracellular injection of calcium buffers on light adaptation in Limulus ventral photoreceptors. J. Gen. Physiol. **66:** 489-506.

(37) Lolley, R.N., and Racz, E. 1982. Calcium modulation of cyclic GMP synthesis in rat visual cells. Vision Res. **22:** 1481-1486.

(38) MacLeish, P.R.; Schwartz, E.A.; and Tachibana, M. 1984. Control of the generator current in solitary rods of the Ambystoma tigrinum retina. J. Physiol. **348:** 645-664.

(39) McLaughlin, S., and Brown, J.E. 1981. Diffusion of calcium ions in retinal rods. J. Gen. Physiol. **77:** 475-487.

(40) Michell, R.H. 1975. Inositol phospholipids and cell surface receptor function. Biochim. Biophys. Acta **415:** 81-147.

(41) Mullins, L.J. 1981. Ion Transport in Heart. New York: Raven Press.

(42) Palade, P., and Vergara, J. 1982. Arsenazo III and Antipyrilazo III calcium transients in single skeletal muscle fibers. J. Gen. Physiol. **79:** 679-707.

(43) Payne, R.; Fein, A.; and Corson, D.W. 1985. A rise in intracellular Ca^{2+} is necessary and perhaps sufficient for photoreceptor excitation and adaptation by inositol 1,4,5-triphosphate. Biol. Bull., in press.

(44) Pinto, L.H., and Brown, J.E. 1984. Pressure injection of 3',5'-cyclic GMP into solitary rod photoreceptors of the tiger salamander. Brain Res. **304:** 197-200.

(45) Poo, M.-M., and Cone, R.A. 1974. Lateral diffusion of rhodopsin in the photoreceptor membrane. Nature **247:** 438-441.

(46) Puckett, K.L.; Aronson, E.T.; and Goldin, S.M. 1985. ATP-dependent calcium uptake activity associated with a disk membrane fraction isolated from bovine retinal rod outer segments. Biochemistry **24:** 390-400.

(47) Reuter, H. 1983. Calcium channel modulation by neurotransmitters, enzymes and drugs. Nature **301:** 569-574.

(48) Robinson, P.R.; Kawamura, S.; Abramson, B.; and Bownds, M.D. 1980. Control of the cyclic GMP phosphodiesterase of frog photoreceptor membranes. J. Gen. Physiol. **76:** 631-645.

(49) Schnetkamp, P.P.M. 1979. Calcium translocation and storage of isolated intact cattle rod outer segments. Biochim. Biophys. Acta **554:** 441-459.

(50) Schröder, W.H., and Fain, G.L. 1984. Light-dependent calcium release from photoreceptors measured by laser micro-mass analysis. Nature **309:** 268-270.

(51) Smith, H.G., Jr.; Fager, R.S.; and Litman, B.J. 1977. Light-activated calcium release from sonicated bovine retinal rod outer segment disks. Biochemistry **16:** 1399-1405.

(52) Sorbi, R.A. 1981. Modulation of sodium conductance in photoreceptor membranes by calcium ions and cGMP. Curr. Top. Membr. Trans. **15:** 331-338.

(53) Streb, H.; Irvine, R.F.; Berridge, M.J.; and Schulz, I. 1983. Release of Ca^{2+} from a nonmitochondrial intracellular store in pancreatic acinar cells by inositol-1,4,5-trisphosphate. Nature **306:** 67-69.

(54) Szuts, E. 1981. Calcium tracer exchange in the rods of excised retina. Curr. Top. Membr. Trans. **15:** 291-305.

(55) Wey, C.-L.; Cone, R.A.; and Edidin, M.A. 1981. Lateral diffusion of rhodopsin in photoreceptor cells measured by fluorescence photobleaching and recovery. Biophys. J. **33:** 225-232.

(56) Yau, K.-W., and Nakatani, K. 1984. Cation selectivity of light-sensitive conductance in retinal rods. Nature **309:** 352-354.

(57) Yau, K.-W., and Nakatani, K. 1984. Electrogenic Na-Ca exchange in retinal rod outer segment. Nature **311:** 661-663.

(58) Yoshikami, S.; George, J.S.; Hagins, W.A. 1980. Light-induced calcium fluxes from rod outer segment layer of vertebrate retina. Nature **286:** 395-398.

(59) Yoshikami, S., and Hagins, W.A. 1971. Ionic basis of dark current and photocurrent of retinal rods. Biophys. J. **10:** 60a.

(60) Yoshikami, S., and Hagins, W.A. 1973. Control of the dark current in vertebrate rods and cones. In Biochemistry and Physiology of Visual Pigments, ed. H. Langer, pp. 245-255. Berlin: Springer-Verlag.

The Molecular Mechanism of Photoreception, ed. H. Stieve, pp. 109-125. Dahlem Konferenzen 1986. Berlin, Heidelberg, New York, Tokyo: Springer-Verlag.

Evaluation of Rod Internal Transmitter Candidates: Cyclic GMP

W.H. Miller
Dept. of Ophthalmology and Visual Science
Yale Medical School
New Haven, CT 06510, USA

Abstract. Several systems in the rod outer segment (ROS) are affected by light (7, 8, 20, 30, 41, 59) and may prove to be either in the direct chain of command of or modulators of transduction. Two, rhodopsin kinase (8, 30) and the cyclic nucleotide enzymatic cascade, are activated by rhodopsin. But the cyclic nucleotide cascade is the only system with the speed, amplification, and physiological control capable of mediating transduction that is known to be directly activated by photolyzed rhodopsin (R*). One R* can catalyze the exchange of GTP for GDP on hundreds of molecules of the guanine-nucleotide binding regulatory "G-type" (47) protein, transducin (T) (19), which in turn activates phosphodiesterase (PDE) (37). Dim bleaches can lead to the hydrolysis of more than 10^6 molecules of cyclic GMP/R* s (32, 53). For bright flashes the latency of decrease of cyclic GMP concentration is shorter than the latency of the light-dependent conductance decrease (13). Cyclic GMP pulses injected into ROS cause reversible depolarizations (38) that, in correspondence with the biochemical evidence of light-activated PDE, are antagonized by light. Pulses of cyclic GMP transiently block the receptor potential and increase latency (43). Thus the hydrolysis of cyclic GMP appears necessary but not sufficient for transduction since the receptor potential would not be blocked by cyclic GMP plus light if cyclic GMP hydrolysis per se mediates transduction. Another event, perhaps a decrease in cyclic GMP concentration, is therefore necessary in addition to cyclic GMP hydrolysis. The input to the cyclic nucleotide cascade is coupled to the transduction mechanism by R* activation, and the output is coupled to the transduction mechanism by control of receptor potential latency (38, 43) and kinetics (9, 38). The mechanism of coupling to the input is understood (reviewed in (49)); the mechanism of coupling to the output is not understood. Nevertheless, cyclic GMP appears to be directly in the chain of events initiated by Rh* that leads to control of ROS phototransduction. (As this manuscript goes to press, two publications have appeared that give strong evidence based on patch-clamping of the plasma membrane that cyclic GMP is the internal messenger that controls the light-sensitive conductance of ROS (17, 42); see Miller et al. (this volume) for discussion of the data.)

INTRODUCTION

The search for increased understanding of how photolyzed rhodopsin initiates the receptor potential focused on the cyclic nucleotide enzymatic cascade because of evidence that the cascade is regulated by light (7). That light activates a ROS phosphodiesterase (PDE) (37) to cause the hydrolysis of cyclic GMP is not in dispute. There is, however, controversy as to whether high cyclic GMP levels in the ROS associated with darkness and the cyclic GMP hydrolysis caused by light are epiphenomena or whether they are causally related to the regulation of phototransduction.

Inference from the fact that light activates PDE leads to the suggestion for a model of transduction in which cyclic GMP mediates the inward current that keeps the ROS depolarized at about -35 mV in darkness. In this model photolyzed rhodopsin (R*), through transducin (T), activates phosphodiesterase (PDE). Cyclic GMP hydrolysis causes a decrease in dark current to produce the hyperpolarizing receptor potential. Such a minimum model postulates nothing about how cyclic GMP and its hydrolysis mediate their effects or whether other processes contribute to modulate adaptation and excitation. This model does, however, postulate that cyclic GMP is a negative transmitter and that cyclic GMP is a direct link in the chain of molecular events connecting photon capture by rhodopsin with control of ROS plasma membrane permeability.

Some important criteria to be met if cyclic GMP is a transmitter are:

1. Cyclic GMP must be present in the ROS in darkness.
2. Cyclic GMP hydrolysis must be triggered by R*.
3. One R* must lead to the hydrolysis of at least 10^3 cyclic GMP for a detectable physiological response in the amphibian ROS (36).
4. The latency of the hydrolysis must be shorter than that of the receptor potential.
5. Cyclic GMP must control the latency of the light response. Control is used here to mean directly control by being in a linear chain of command between rhodopsin and the light-sensitive conductance. If calcium is regulated by the sodium-calcium exchange and influences the light-sensitive conductance by exerting its effect through binding or concentration of cyclic GMP, that would be considered here to be an indirect influence rather than direct control. The possibilities that cyclic GMP and calcium could both control directly are not mutually exclusive if it turns out that they are both in chains directly linking rhodopsin with the light-sensitive conductance.
6. Cyclic GMP must control the light-sensitive conductance.
7. Cyclic GMP must control the kinetics and amplitude of the light response.

DARK ROS CONCENTRATION IS 60 μmol/l

ROS cyclic GMP concentration is roughly 60 μmol/l in the dark (54). The dark steady state is maintained by background PDE and cyclase activities (22). About half of the cyclic GMP may be bound (53). Illumination of ROS leads to the activation of PDE, a peripheral protein with 88, 84, and 11 kD subunits that is present at about 1% of rhodopsin's concentration (3). PDE K_m is roughly 50 μmol/l cyclic GMP (25, 37).

PHOTOLYZED RHODOPSIN (R*) ACTIVATES MANY PDE'S

The activation of PDE is mediated by the guanine nucleotide-binding (19, 21, 28) regulatory protein, transducin (T) ((18, 19), and Chabre and Applebury, this volume). R* catalzyes the exchange of GTP for GDP bound to T in darkness by lowering the activation barrier for that exchange (49). The amplification derives from the fact that R* diffusing in the disk membrane (58) serially activates hundreds of T's (18) (see Fig. 1). The T_α GTP subunit activates PDE by relieving the inhibitory constraint exerted by PDE's 11 kD subunit (15, 25, 56). T_α has GTPase activity. This exchange of GTP for GDP proceeds in darkness to inactivate PDE (18).

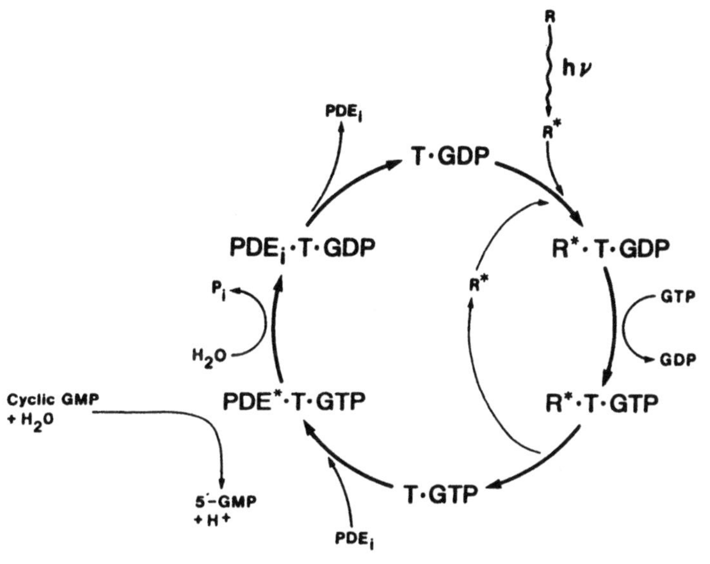

FIG. 1 - Light amplification cycle by which R* leads to PDE activation (from Fung and Stryer (19)).

Since PDE activation involves two stages, not only T but also R* must be inactivated. R* is probably inactivated by rhodopsin kinase (Bownds and Brewer and Chabre and Applebury, both this volume). Gamma phosphate transfer from ATP to R* following a light flash can quench in vitro light-activated cyclic GMP hydrolysis within seconds after the introduction of ATP to a rod disk membrane preparation as shown in Fig. 2 (31, 48).

THERE IS ENOUGH AMPLIFICATION

The single-photon physiological response with a time-to-peak in the one second range in the Bufo rod involves about a 4% reduction in dark current, which implies a decrease of a minimum of 800 "negative" transmitter

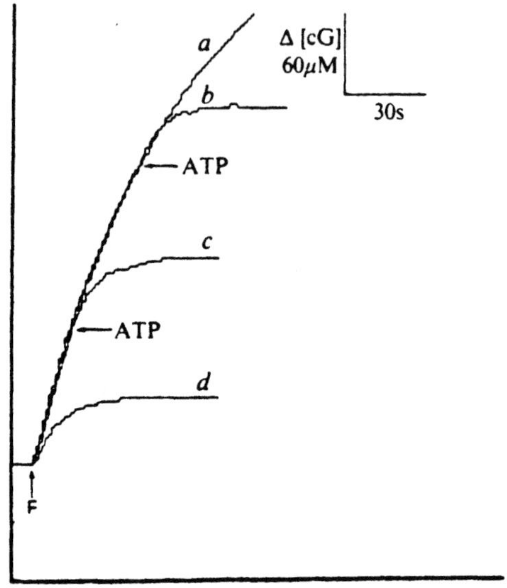

FIG. 2 - Cyclic GMP hydrolysis assayed by measuring proton release following a 1 ms duration light flash, F, that bleaches 24 million rhodopsins present, showing short latency of PDE activation by light and rapid turnoff caused by ATP. Curves b-d: 8 μmol/l ATP injected at 25 s, 15 s, and 5 s into rod disk membrane preparation in presence of 60 μmol/l GTP (from Liebman and Pugh (31)).

particles (36). Thus one R* should be capable of causing the hydrolysis of at least 10^3 cyclic GMP's per second. Amplification is greatest in the physiological range of weak bleaches because the ratio of R*'s to PDE's is lowest (58). The minimum criterion for amplification is easily met in an amphibian isolated ROS or outer-inner segment preparation (13, 53), and also in an amphibian disk preparation (31, 58) where more than 10^6 cyclic GMP's are hydrolyzed per second per R* using dim bleaches.

CYCLIC GMP HYDROLYSIS PRECEDES RECEPTOR POTENTIAL

Does PDE Activation Reduce Cyclic GMP Levels?

The latency of light-activated cyclic GMP hydrolysis must be less than the latency of the receptor potential if a causal relationship is to be established. The importance of establishing the temporal relation of these events was recognized early and resulted in studies with conflicting results both with regard to whether the latency of cGMP hydrolysis is shorter than that of the receptor potential and whether the activation of PDE results in a reduction of ROS cyclic GMP levels. From monitoring PDE activity by ^{18}O labeling of guanine nucleotide α-phosphoryls it appears that the dark flux of cyclic GMP synthesis and hydrolysis corresponds to a $t_{1/2}$ of 1.5 seconds for turnover of the entire pool of ROS cyclic GMP (22). With maximal illumination the hydrolytic rate rises to 5 pools per second. It was concluded (22) that there is normally no decrease in ROS cyclic GMP levels except at the highest (non-physiologic) light intensities when synthetic activity lags. While it is argued that cyclic GMP flux is the important variable (22), this conclusion appears to be premature since the study does not rule out the possibility that a small pool of free cyclic GMP controls transduction. The same can be said of other studies that do not reveal changes in concentration of cyclic GMP with light. No measurable cyclic GMP decrease within 0.5 seconds after a flash bleaching 2000 rhodopsins per ROS is found when the whole retina is illuminated and cyclic GMP measurements are made on microdissected outer segments (24). A marked decrease in cyclic GMP is observed only after 3 s of continuous illumination bleaching 10^7 rhodopsins per ROS (24). In another study (23) weak illumination caused a more rapid decrease in cyclic GMP levels than strong illumination. Investigations on whole retinas suggest that ROS cyclic GMP levels are decreased by light, but the detectable decrease comes too late to mediate the receptor potential (23, 27). Outer segment cyclic GMP changes may be obscured in these studies by changes in other retinal layers (1). Nevertheless, a reduction in ROS cyclic GMP levels can occur rapidly in the isolated retina using bright light and EGTA Ringer's (27). EGTA lowers rod Ca^{2+}, depolarizes and increases the amplitude of the voltage response to bright light (4). Low Ca^{2+} also increases ROS cyclic GMP concentration, possibly by activating cyclase (12).

Cyclic GMP levels decrease 50% with a half-time of about 125 ms with strong illumination in low Ca^{2+} Ringer's using isolated ROS (53) or disk membranes ((58); see Fig. 2). The short latency of PDE activation is confirmed by light-scattering experiments (6, 29). Photolyzed rhodopsin catalyzes the activation of a molecule of T in about 1 ms (50). The latency of the decrease in cyclic GMP in response to light using isolated outer-inner segments in 1 mmol/l Ca^{2+} is shorter than the latency of the decrease in dark current (13). As shown in Fig. 3, the light-induced cyclic GMP decrease at 8×10^3 R*/ROS s is completed within 50 ms and before the start of the light-induced decrease in dark current (13).

Two groups have evidence that could be interpreted as indicating that cyclase is either directly or indirectly stimulated as a result of light-induced PDE activation (1, 22). This synthetic power of the light-stimulated cyclase (1, 22) is insufficient to prevent a light-induced decrease in the concentration of cyclic GMP in these isolated "intact" photoreceptors. But since the experiment of Fig. 3 uses bright light, the question of whether the cyclic GMP hydrolysis caused by dim bleaches reduces cyclic GMP levels (22) remains unresolved. For the physiological range of dim bleaches at which the rods normally function in night vision, it is still important to find out whether cyclic GMP levels actually decline. Although dim bleaches may cause local depletion of cyclic GMP, this has yet to be proven. There seems to be a well developed homeostasis with regard to cyclic GMP levels as measured by the ROS response to cyclic GMP pulses (Fig. 4, below).

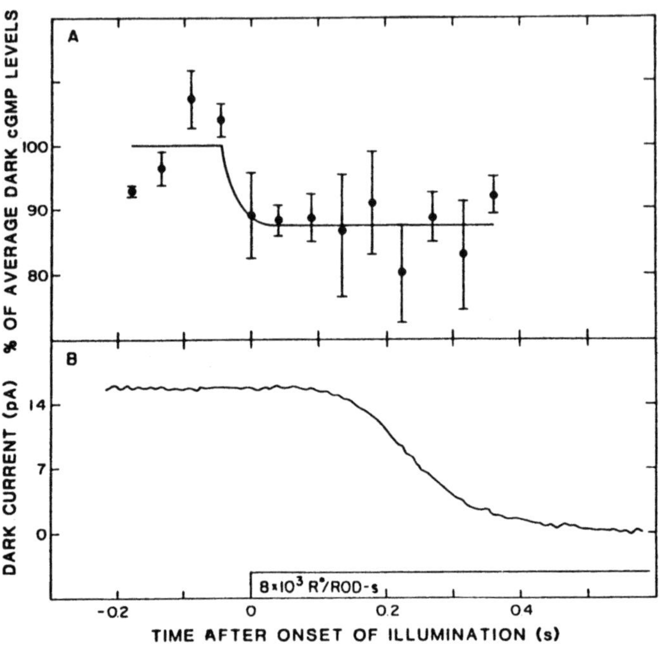

FIG. 3 - Comparison of the time course of cyclic GMP (A) and dark current (B) decreases in isolated outer-inner segments in 1 mmol/l Ca^{2+} Ringer's (from Cote et al. (13)).

CYCLIC GMP AFFECTS RECEPTOR POTENTIAL LATENCY

Immediate Effect of Increasing Cyclic GMP Is Reversible

A pulse of cyclic GMP injected into a ROS depolarizes for a time directly proportional to the amount injected and inversely proportional to light intensity (38, 43). Uniform ionophoretic 5 pC pulses on Fig. 4 indicated by the downward spikes on the traces deliver 3 x 10^7 molecules of cyclic GMP at most. This reversible depolarization is not caused by the injection current since pressure injections have the same result. Nor does the membrane respond transiently while cyclic GMP remains intact. The cyclic GMP response is antagonized by light (Fig. 4) which corresponds with the biochemical evidence that cyclic GMP is rapidly hydrolyzed by light (13) (Fig. 3). As weak illumination that hyperpolarizes only slightly is no impediment to antagonism of the cyclic GMP response, the antagonism is not caused by the Cs^+-sensitive conductance (16). Ca^{2+} does not cause the antagonism because the kinetics of the cyclic GMP response are not speeded by low Na^+ Ringer's that might be expected to raise internal ROS Ca^{2+} levels (39).

Ten pC pulses of cyclic GMP that produce cyclic GMP responses before a bright flash (Fig. 4A) cause no cyclic GMP responses for more than half a minute after the flash. Five pC pulses briefly restore the membrane potential following the longer flash at one-tenth the intensity in Fig. 4B, but as the light remains on, activating more and more PDE, the cyclic GMP response is increasingly antagonized. Figure 4 illustrates that the cyclic GMP response can be titrated with light intensity.

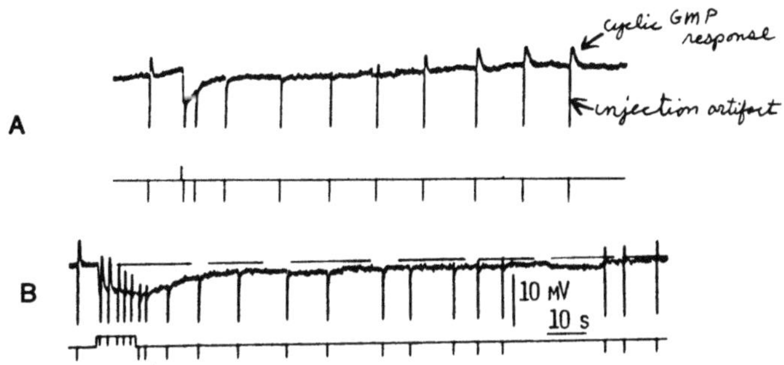

FIG. 4 - A: Train of 10 pC pulses of cyclic GMP injected into ROS. Up on signal trace indicates 0.1 s light flash that contains about 100 effective photons (from Miller and Nicol (40)). B: Train of 5 pC injections of cyclic GMP into single ROS indicated by downward spikes on traces. Ten-second light flash of about 100 photons/ROS s (one-tenth intensity of flash in A) indicated by up on signal trace (from Miller (38)).

That membrane potential can be briefly restored by cyclic GMP after light should be a reflection of the resultant of cyclase and PDE activities and the buffering action of cyclic GMP stores. Thus the response to uniform pulse injections of cyclic GMP should provide an index of the effect of PDE activity relative to cyclase and other cyclic GMP buffering actions. The time course of PDE activity following illumination as measured by the cyclic GMP response corresponds more closely with the photocurrent response than the photovoltage response.

Are Irreversible Depolarization and Increased Conductance Secondary Effects of Cyclic GMP?

Because the cyclic GMP voltage response is rapidly reversible, it appears that the concentration of cyclic GMP can only be raised transiently in a normal ROS. Irreversible depolarization following cyclic GMP injection is likely to be a result of secondary effects. Measurements of conductance in that condition cannot be guaranteed to represent a primary effect of cyclic GMP. The 5 pC bursts of cyclic GMP delivered at a rate of 1 pC/ms in the experiment of Fig. 4 have a reversible effect, but when cyclic GMP is continuously ionophoresed at a rate twenty times slower, the ROS has been shown to depolarize irreversibly (38, 51), Fig. 5, top, or slowly increase its dark current (33, 35) under voltage-clamp, Fig. 5, bottom. In Fig. 5, bottom, the continuous injection of cyclic GMP is at a rate of 50 pC/s. The transference number is estimated at 10^{-3} or less (33), which means that less than 300,000 molecules/s of cyclic GMP are injected. A single isomerization of rhodopsin can lead to hydrolysis of more than 10^6 cyclic GMP/s (32), and the dark turnover number is 10^8 cyclic GMP/s by one estimate (22). The system may adjust to a slow continuous injection 1/300 of the dark rate with no change in cyclic GMP levels or with an elevation and new steady state in which physiological changes caused by the injection may not be a direct effect of cyclic GMP. Cyclic GMP pulses do not cause an irreversible depolarization ((38, 43); see Fig. 4, above).

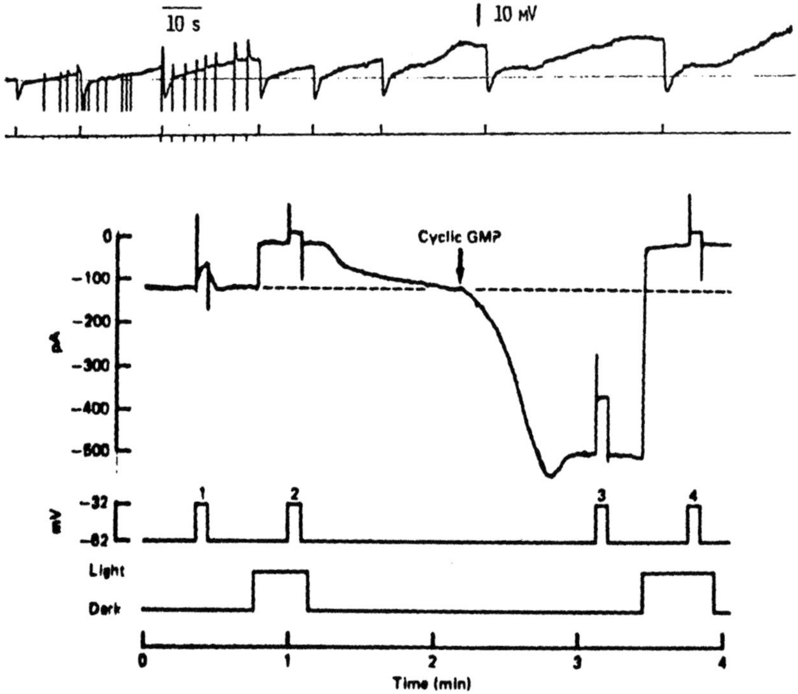

FIG. 5 - Top: All light flashes about 20 photons/ROS. Apparently irreversible depolarization follows cyclic GMP injections (from Miller, 38)). Bottom: Continuous injection of cyclic GMP into isolated rod at -50 pA starting at arrow causes increase in dark current from -100 pA to -500 pA measured with membrane voltage-clamped at -60 mV. Top signal trace indicates clamped voltage, bottom signal trace, light. Irreversible increase in current and irreversible depolarization in Fig. 5, top, may be secondary effects (from MacLeish et al. (35)).

The immediate effects on membrane potential of a cyclic GMP pulse in the millisecond range may be ascribed to the primary consequences of cyclic GMP alone because those consequences are reversible. Effects that have not been shown to be reversible may be at least in part secondary. For example, cyclic GMP may stimulate Ca^{2+} release from disks (10), which may imply that ROS-free Ca^{2+} concentration decreases with illumination. Ca^{2+} (55) and other substances (44) have been shown to alter cyclic GMP levels independent of membrane potential as shown in Fig. 6.

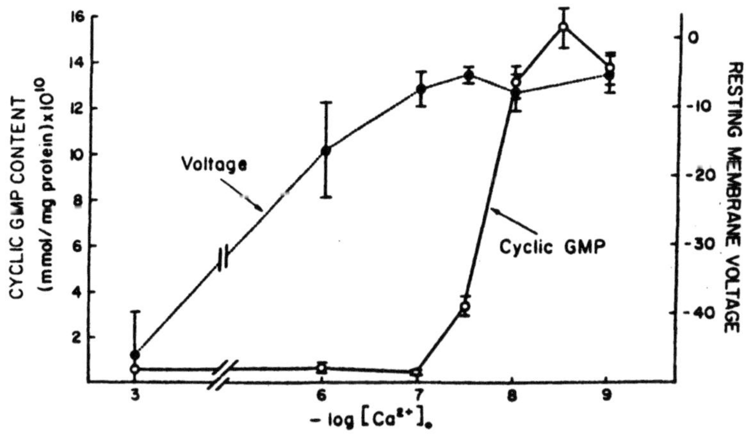

FIG. 6 - ROS cyclic GMP content and membrane potential in the dark as a function of $[Ca^{2+}]_o$ (from Woodruff and Fain (55)).

Receptor Potential Latency Increased by Cyclic GMP

When a pulse of cyclic GMP which causes a reversible depolarization is followed by a light flash, the receptor potential is transiently blocked (43), increasing the latency of the receptor potential for a time that is directly proportional to the amount injected and inversely proportional to the light intensity. This effect of cyclic GMP on receptor potential latency appears to be unique. Other agents that depolarize, such as cyclic AMP, 5' GMP, and EGTA do not have much effect on latency, (Fig. 10 (40), Fig. 12 (38)), presumably because they do not directly control the light-sensitive conductance. Increased latency must reflect local ROS effects since coupling of responses from neighboring rods is inimical to this observation. In the example of Fig. 7, cyclic GMP not only increases latency more than fivefold over the control but also speeds the kinetics of the hyperpolarization and prolongs the recovery phase.

Three studies in which cyclic GMP was injected into ROS show no evidence of increased latency in the published records (26, 35, 51), and in one of these (51) it is stated that no increase in latency was observed. The light flashes were delivered in all of these experiments (26, 35, 51) after cyclic GMP had resulted in irreversible depolarization. No increase in latency would be expected as shown in the responses of Fig. 5, top. On the other hand, a pulse of cyclic GMP that causes reversible physiological effects delays the receptor potential in proportion to the amount injected and inversely proportionally to the light intensity (38, 43), even in the face of increased light-induced cyclic GMP hydrolysis while the ROS is "clamped" with excess cyclic GMP. This contrasts with the effect of clamping calcium inasmuch as when calcium is clamped by the injection of calcium buffers, there is little change in latency (38). Therefore, cyclic GMP appears directly to control latency.

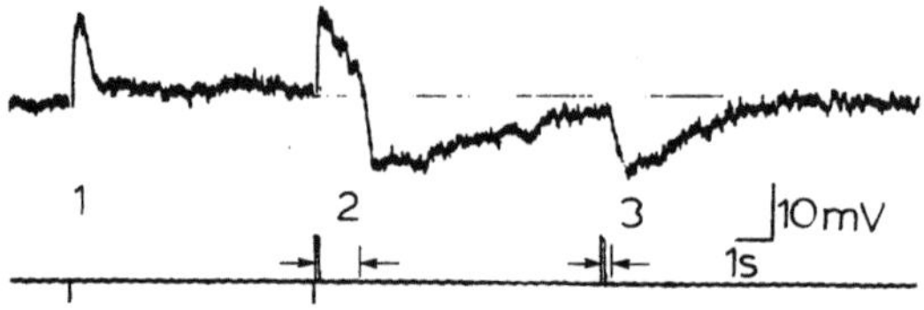

FIG. 7 - 12 pC pulses of cyclic GMP indicated by down on signal trace. Injection artifact removed from record trace. Up on signal trace indicates flash containing about 20 absorbed photons per ROS. Response 1 to cyclic GMP alone, 3 to light alone, and 2 is voltage response to cyclic GMP pulse followed by a light flash. Latency of receptor potential (arrows) following cyclic GMP is 1.3 s compared to 0.24 s for the control, response 3. Noisy staircase effect preceding delayed response often observed. Cyclic GMP not only increases latency of response to light but also speeds kinetics of hyperpolarization and prolongs recovery (from Miller (38)).

DOES CYCLIC GMP CONTROL LIGHT-SENSITIVE CONDUCTANCE?

Whether cyclic GMP affects the light-sensitive conductance has not been determined using voltage-clamp with the ROS reversibly under the influence of cyclic GMP, though less critical tests (e.g., (26, 40)) indicate that cyclic GMP causes a conductance increase. Nevertheless, the increased inward dark current from the continuous injection of cyclic GMP in the clamped cell reverses near 0 mV at the same membrane potential that the outward light-induced current reverses (33) (Fig. 8, below). Time and voltage-sensitive conductances were pharmacologically suppressed in the experiment of Fig. 8 (33). When these conductances were not suppressed, the inward current following cyclic GMP injection reversed when the voltage was clamped between +20 and +30 mV, whereas the light response failed to reverse at depolarizations of up to +46 mV from the resting potential (45). These results (45) are difficult to interpret since the light response has been shown to reverse near zero (2, 5, 14, 52), which is expected since the light-dependent conductance is nonselective (57). Though the result of Fig. 8 (33) is convincing, it may not be a primary consequence of cyclic GMP for the reasons discussed above. The evidence is quite suggestive but not proof that cyclic GMP affects the light-dependent conductance. Other indirect evidence that cyclic GMP affects the light-dependent conductance derives from the fact that cyclic GMP controls the latency (Fig. 7, above) and kinetics (Fig. 9, below) of the light response.

CYCLIC GMP CONTROLS LIGHT-RESPONSE KINETICS

When the membrane is irreversibly depolarized and dark current is increased following injection of cyclic GMP, light responses can be elicited but the membrane potential and current do not recover to the pre-injection level (Fig. 5). As discussed above, it is by no means clear that these effects are primarily due to cyclic GMP. When large amounts of cyclic GMP are injected with the ROS irreversibly depolarized, the cyclic GMP does not cause further depolarization but slows the kinetics of the light response (Fig. 10B). Non-hydrolyzable analogues of cyclic GMP such as 8-bromo cyclic GMP and

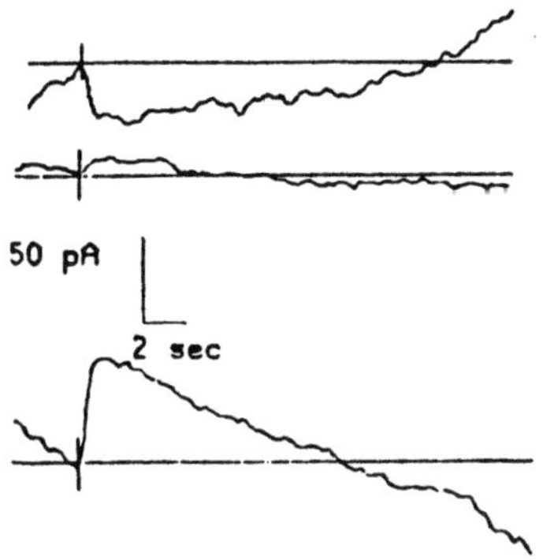

FIG. 8 - After injection of cyclic GMP has increased inward dark current, voltage is stepped from -30 mV to 0 and 10 mV. Vertical line on each trace indicates 10 ms light flash. Both the inward current following cyclic GMP injection and the outward current which is caused by light reverse between 0 and + 10 mV (from Lipton (33)).

IBMX, which act as competitive inhibitors of PDE and would be expected to interfere with cyclic GMP hydrolysis, also slow the kinetics of the receptor potential (34, 51) and light-induced reduction in dark current (35). Very small doses of the PDE inhibitor IBMX cause slowing of the kinetics of the voltage light response which can be reversed by light (9). Thus effects of light on kinetics may be mediated by cyclic GMP hydrolysis (9). Five µmol/l IBMX increases the amplitude and slows the recovery of the voltage response to a dim flash (9). The effect on kinetics is reversed by background light, though the intensity of background light needed to restore kinetics is about twice that needed to restore sensitivity (9) (Fig. 9).

The conclusion that cyclic GMP hydrolysis controls receptor potential kinetics is also supported by comparison of the kinetics of the voltage responses to cyclic GMP hydrolysis and a just-preceding light response (38) (Fig. 10). Illumination antagonizes the cyclic GMP voltage response as noted in connection with Fig. 4, above, and the cyclic GMP response may be an index of effective PDE activity. The cyclic GMP response just after the initial hyperpolarizing phase of the light response changes only slowly with time (Fig. 4), as if after activation by light, PDE activity is relatively stable near its maximum and where it begins to decline slowly. If the light response is controlled by cyclic GMP hydrolysis, the kinetics of the light response and the voltage signal of a just-following hyperpolarization of a cyclic GMP response should be similar. The kinetics of the voltage signal of cyclic GMP hydrolysis are compared to those of a just-preceding hyperpolarizing phase of the light response both in the normal ROS (Fig. 10A) and when light-response kinetics are slowed when the ROS is overwhelmed with injected cyclic GMP (Fig. 10B). That a) background light can

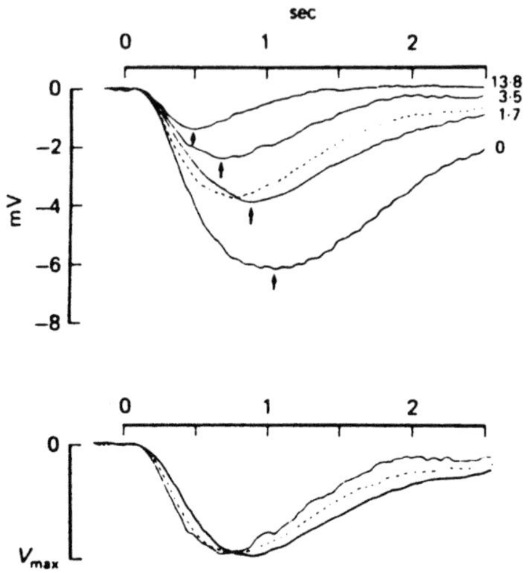

FIG. 9 - Voltage responses to four absorbed photons per rod. Dashed line, control. Solid lines in presence of 5 µmol/l IBMX and background light equivalent to number of absorbed photons/ROS s indicated by numbers to right of curves. Lower panel: Control response (dashed line) and responses 3.5 (thick trace) and 1.7 (thin trace) scaled to same peak response to show that background light can restore changes in kinetics of the voltage response caused by IBMX (from Capovilla et al. (9)).

reverse effects of IBMX on voltage-response kinetics and b) kinetics of the voltage signal of cyclic GMP-induced hyperpolarization just after light are similar to those of a just-preceding light-induced hyperpolarization support the conclusion that cyclic GMP hydrolysis is one of the factors controlling light-response kinetics.

MECHANISM OF ACTION OF CYCLIC GMP

Excess cyclic GMP injected into a ROS transiently blocks the receptor potential. Since both injected cyclic GMP and light should increase cyclic GMP hydrolysis, hydrolysis by itself cannot control transduction. Thus the increase in latency caused by cyclic GMP (43) is consistent with cyclic GMP hydrolysis being necessary but not sufficient for transduction. The results of two other experiments support that conclusion. First, in ROS irreversibly depolarized following continuous injection of cyclic GMP the further injection of cyclic GMP both increases latency and slows the kinetics of the light response (38), as if cyclic GMP concentration must be reduced for transduction. Second, the injection of active PDE in the dark when cyclic GMP levels are high does not hyperpolarize immediately (but cf. (11)), though it prolongs the response to a subsequent light flash for minutes (46). That cyclic GMP hydrolysis is not sufficient for transduction (43) appears to be a) against the hypothesis that protons from cyclic GMP hydrolysis can mediate transduction (41), b) against the hypothesis that transduction is controlled

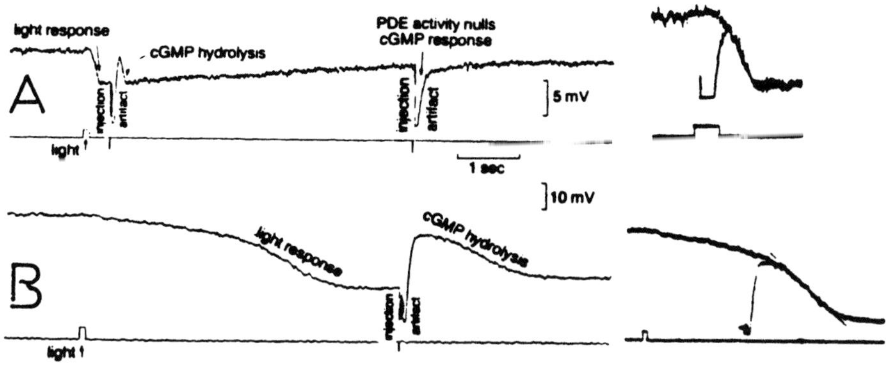

FIG. 10 - Kinetics of cyclic GMP hydrolysis response are similar to kinetics of just-preceding initial hyperpolarizing phase of light response in both normal ROS (A) and ROS in which kinetics are slowed by overloading with cyclic GMP (B). Intensity of all light flashes about 20 photons/ROS. Voltage responses to light and cyclic GMP superimposed on right-hand records (from Miller (38)).

by cyclic GMP flux (22), and c) consistent with transduction being mediated by free cyclic GMP concentration.

CONCLUSION

Cyclic GMP is present in high concentration in ROS. Its hydrolysis is triggered by R* with sufficiently short latency and great amplification to mediate transduction. Cyclic GMP and its hydrolysis appear to control light-response kinetics and latency, all of which support the conclusion that the cyclic nucleotide enzymatic cascade through its link with R* designates cyclic GMP as a messenger molecule in a direct chain linking rhodopsin with the light-sensitive conductance. However, the phototransduction system is complex. The mechanisms by which cyclic GMP and its hydrolysis exert their effects are largely unknown. Other molecular systems have been implicated in transduction, including calcium (59), rhodopsin kinase (8, 30), protons (41), and the phosphatidylinositol system (20). One or more of these may communicate with R*, the light-dependent channel, or elements of the chain in between. Controversy is bound to remain as to the role of its individual elements until the entire transduction system is reconstructed. Because of its direct connection with R*, its biochemical and its physiological effects, cyclic GMP is certainly one of the key elements in the transmitter system.

Acknowledgement. Supported by U.S. Public Health Service grant EY 03196.

REFERENCES

(1) Azeredo, A.M.; de Lust, W.D.; and Passonneau, J.V. 1981. Light-induced changes in energy metabolites, guanine nucleotides, and guanylate cyclase within frog retinal layers. J. Biol. Chem. **256:** 2731-2735.

(2) Bader, C.R.; MacLeish, P.R.; and Schwartz, E.A. 1978. Responses to light of solitary rod photoreceptors isolated from tiger salamander retina. Proc. Natl. Acad. Sci. USA **75:** 3507-3511.

(3) Baehr, W.; Devlin, M.J.; and Applebury, M.L. 1979. Isolation and characterization of cGMP phosphodiesterase from bovine rod outer segments. J. Biol. Chem. **254:** 11669-11677.

(4) Bastian, B.L., and Fain, G.L. 1982. The effects of low calcium and background light on the sensitivity of toad rods. J. Physiol. **330:** 307-329.

(5) Baylor, D.A., and Nunn, B. 1983. Voltage dependence of the light-sensitive conductance of salamander retinal rods. Biophys. J. **41:** 125a.

(6) Bennett, N. 1982. Light-induced interactions between rhodopsin and the GTP-binding protein; relation with phosphodiesterase action. Eur. J. Biochem. **123:** 133-139.

(7) Bitensky, M.W.; Gorman, R.E.; and Miller, W.H. 1971. Adenyl cyclase as a link between photon capture and changes in membrane permeability of frog photoreceptors. Proc. Natl. Acad. Sci. USA **68:** 561-562.

(8) Bownds, D.; Dawes, J.; Miller, J.; and Stahlman, M. 1972. Phosphorylation of frog photoreceptor membranes induced by light. Nature **236:** 125-127.

(9) Capovilla, M.; Cervetto, L.; and Torre, V. 1983. The effect of phosphodiesterase inhibitors on the electrical activity of toad rods. J. Physiol. **343:** 277-294.

(10) Cavaggioni, A., and Sorbi, R.T. 1981. Cyclic GMP releases calcium from disc membranes of vertebrate photoreceptors. Proc. Natl. Acad. Sci. USA **78:** 3964-3968.

(11) Clack, J.W.; Oakley, B. II; and Stein, P.J. 1983. Injection of GTP-binding protein or cyclic GMP phosphodiesterase hyperpolarizes retinal rods. Nature **305:** 50-52.

(12) Cohen, A.I.; Hall, I.A.; and Ferrendelli, J.A. 1978. Calcium and cyclic nucleotide regulation in incubated mouse retinas. J. Gen. Physiol. **71:** 595-612.

(13) Cote, R.H.; Biernbaum, M.S.; Nicol, G.D.; and Bownds, M.D. 1984. Light-induced decreases in cGMP concentration precede changes in membrane permeability in frog rod photoreceptors. J. Biol. Chem. **259:** 9635-9641.

(14) Detwiler, P.B.; Conner, J.D.; and Bodois, R.D. 1982. Gigaseal patch clamp recordings from outer segments of intact retinal rods. Nature **300:** 56-61.

(15) Dumler, I., and Etingof, R. 1976. Protein inhibitor of cAMP phosphodiesterase in retina. Biochim. Biophys. Acta. **429:** 474-484.

(16) Fain, G.L.; Quandt, F.N.; Bastian, B.L.; and Gerschenfeld, H.M. 1978. Contribution of a caesium-sensitive conductance increase to the rod photoresponse. Nature **272:** 467-469.

(17) Fesenko, E.E.; Kolesnikov, S.S.; and Lyubarsky, A.L. 1985. Induction by cyclic GMP of cationic conductance in plasma membrane of retinal rod outer segment. Nature **313:** 310-313.

(18) Fung, B.K.-K.; Hurley, J.B.; and Stryer, L. 1981. Flow of information in the light-triggered cyclic nucleotide cascade of vision. Proc. Natl. Acad. Sci. USA **78:** 152-156.

(19) Fung, B.K.-K., and Stryer, L. 1980. Photolyzed rhodopsin catalyzes the exchange of GTP for GDP in retinal rod outer segment membranes. Proc. Natl. Acad. Sci. USA **77:** 2500-2504.

(20) Ghalayini, A.J., and Anderson, R.E. 1984. Brief light exposure stimulates phosphatidylinositol-4,5-bisphosphate breakdown in frog retina. Inv. Ophthalmol. Vis. Sci. **25:** 61.

(21) Godchaux, W., and Zimmerman, W.F. 1979. Membrane-dependent guanine nucleotide binding and GTPase activities of soluble protein from bovine rod cell outer segments. J. Biol. Chem. **254:** 7874-7884.

(22) Goldberg, N.D.; Ames, A. III; Gander, J.E.; and Walseth, T.F. 1983. Magnitude of increase in retinal cGMP metabolic flux determined by ^{18}O incorporation into nucleotide-phosphoryls corresponds with intensity of photic stimulation. J. Biol. Chem. **258:** 9213-9219.

(23) Goridis, C.; Urban, P.F.; and Mandel, P. 1977. The effect of flash illumination on the endogenous cyclic GMP content of isolated frog retinae. Exp. Eye Res. **24:** 171-177.

(24) Govardovskii, V.I., and Berman, A.L. 1981. Light-induced changes in cyclic AMP content in frog retinal rod outer segments measured with rapid freezing and microdissection. Biophys. Struct. Mech. **7:** 125-130.

(25) Hurley, J.B., and Stryer, L. 1982. Purification and characterization of the γ regulatory unit of the cyclic GMP phosphodiesterase from rod outer segments. J. Biol. Chem. **257:** 11094-11099.

(26) Kawamura, S., and Murakami, M. 1983. Intracellular injection of cyclic-GMP increases sodium conductance in gecko photoreceptors. Jpn. J. Physiol. **33:** 789-800.

(27) Kilbride, P. 1980. Calcium effects on frog retinal cyclic guanosine 3', 5'-monophosphate levels and light-initiated rate of decay. J. Gen. Physiol. **75:** 457-465.

(28) Kühn, H. 1980. Light- and GTP-regulated interaction of GTPase and other proteins with bovine photoreceptor membranes. Nature **283:** 587-589.

(29) Kühn, H.; Bennett, N.; Michel-Villaz, M.; and Chabre, M. 1981. Interactions between photoexcited rhodopsin and GTP-binding protein: Kinetic and stoichiometric analyses from light-scattering changes. Proc. Natl. Acad. Sci. USA **78:** 6873-6877.

(30) Kühn, H., and Dreyer, W.J. 1982. Light dependent phosphorylation of rhodopsin ATP. FEBS Lett. **20:** 1-6.

(31) Liebman, P.A., and Pugh, E.N., Jr. 1980. ATP mediates rapid reversal of cyclic GMP phosphodiesterase activation in visual receptor membranes. Nature **287:** 734-736.

(32) Liebman, P.A., and Pugh, E.N., Jr. 1982. Gain, speed and sensitivity of GTP binding vs. PDE activation in visual excitation. Vision Res. **22:** 1475-1480.

(33) Lipton, S.A. 1982. cGMP and EGTA increase the light-sensitive current of retinal rods. Brain Res. **265:** 41-48.

(34) Lipton, S.A.; Rasmussen, H.; and Dowling, J. 1977. Electrical and adaptive properties of rod photoreceptors in Bufo marinus. J. Gen. Physiol. **70:** 771-791.

(35) MacLeish, P.R.; Schwartz, E.A.; and Tachibana, M. 1984. Control of the generator current in solitary rods of the Ambystoma Tigrinum retina. J. Physiol. **348:** 645-664.

(36) Matthews, G., and Baylor, D.A. 1981. The photocurrent and dark current of retinal rods. Curr. Top. Membr. Trans. **15:** 3-18.

(37) Miki, N.; Keirns, J.J.; Marcus, F.R.; Freeman, J.; and Bitensky, M.W. 1973. Regulation of cyclic nucleotide concentrations in photoreceptors: An ATP-dependent stimulation of cyclic nucleotide phosphodiesterase by light. Proc. Natl. Acad. Sci. USA **70:** 3820-3824.

(38) Miller, W.H. 1982. Physiological evidence that light-mediated decrease in cyclic GMP is an intermediary process in retinal rod transduction. J. Gen. Physiol. **80:** 103-123.

(39) Miller, W.H., and Laughlin, S.B. 1983. Light-mediated cyclic GMP hydrolysis controls important aspects of kinetics of retinal rod voltage response. Biophys. Struct. Mech. **9:** 269-276.

(40) Miller, W.H., and Nicol, G.D. 1981. Cyclic GMP-induced depolarization and increased response latency of rods: antagonism by light. Curr. Top. Membr. Trans. **15:** 417-437.

(41) Mueller, P., and Pugh, E.N., Jr. 1983. Protons block the dark current of isolated retinal rods. Proc. Natl. Acad. Sci. USA **80:** 1892-1896.

(42) Nakatani, K., and Yau, K.-W. 1985. cGMP opens the light-sensitive conductance in retinal rods. Biophys. J. **47:** 356a.

(43) Nicol, G.D., and Miller, W.H. 1978. Cyclic GMP injected into retinal rod outer segments increases latency and amplitude of response to illumination. Proc. Natl. Acad. Sci. USA **75:** 5217-5220.

(44) Ostroy, S.E.; Wilson, M.J.; and Meyertholen, E.P. 1984. cGMP concentration changes in the physiological range do not exhibit the properties of a transduction, adaptation or calcium controlling component. Inv. Ophthalmol. Vis. Sci. **25:** 236.

(45) Pinto, L.H., and Brown, J.E. 1984. Pressure injection of 3', 5'-cyclic GMP into solitary rod photoreceptors of the tiger salamander. Brain Res. **304:** 197-200.

(46) Shimoda, Y.; Hurley, J.B.; and Miller, W.H. 1984. Rod light response augmented by active phosphodiesterase. Proc. Natl. Acad. Sci. USA **81:** 616-619.

(47) Shinozawa, T.; Sen, I.; Wheeler, G.; and Bitensky, M.W. 1979. Predictive value of the analogy between hormone-sensitive adenylate cyclase and light-sensitive photoreceptor cyclic GMP phosphodiesterase: A specific role for a light-sensitive GTPase as a component in the activation sequence. J. Supramol. Struct. **10:** 185-190.

(48) Sitaramayya, A., and Liebman, P.A. 1983. Phosphorylation of rhodopsin and quenching of cyclic GMP phosphodiesterase activation by ATP at weak bleaches. J. Biol. Chem. **258:** 12106-12109.

(49) Stryer, L. 1983. Transducin and the cyclic GMP phosphodiesterase: amplifier proteins in vision. Cold S. H. Symp. Quant. Biol. **48:** 841-852.

(50) Vuong, T.M.; Chabre, M.; and Stryer, L. 1984. Millisecond activation of transducin in the cyclic nucleotide cascade of vision. Nature **311:** 659-661.

(51) Waloga, G. 1983. Effects of calcium and guanosine-3':5'-cyclic-monophosphoric acid on receptor potentials of toad rods. J. Physiol. **341:** 341-357.

(52) Werblin, F.S. 1978. Light, voltage, and time-dependent components of the rod response. Sens. Proc. **2:** 306-315.

(53) Woodruff, M.L., and Bownds, M.D. 1979. Amplitude, kinetics and reversibility of a light induced decrease in 3',5'-cyclic monophosphate in frog photoreceptor membranes. J. Gen. Physiol. **73:** 629-653.

(54) Woodruff, M.L.; Bownds, M.D.; Green, S.H.; Morrisey, J.L.; and Shedlovsky, A. 1977. Guanosine 3',5'-cyclic monophosphate and the in vitro physiology of frog photoreceptor membranes. J. Gen. Physiol. **69:** 667-679.

(55) Woodruff, M.L., and Fain, G. 1982. Ca^{2+}-dependent changes in cyclic GMP levels are not correlated with opening and closing of the light-dependent permeability of toad photoreceptors. J. Gen. Physiol. **80:** 537-555.

(56) Yamazaki, A.; Stein, P.J.; Chernoff, N.; and Bitensky, M.W. 1983. Activation mechanism of rod outer segment cyclic GMP phosphodiesterase. J. Biol. Chem. **258:** 8188-8194.

(57) Yau, K.-W., and Nakatani, K. 1984. Cation selectivity of light-sensitive conductance in retinal rods. Nature **309:** 352-354.

(58) Yee, R., and Liebman, P.A. 1978. Light activated phosphodiesterase of the rod outer segments: kinetics and parameters of activation and deactivation. J. Biol. Chem. **253:** 1802-1809.

(59) Yoshikami, S., and Hagins, W.A. 1971. Light, calcium, and the photocurrent of rods and cones. Biophys. Soc. Abstr.: Biophys. J. **11:** 47a.

The Molecular Mechanism of Photoreception, ed. H. Stieve, pp. 127-158. Dahlem Konferenzen 1986. Berlin, Heidelberg, New York, Tokyo: Springer-Verlag.

Properties of Cytoplasmic Transmitters of Excitation in Vertebrate Rods and Evaluation of Candidate Intermediary Transmitters

E.N. Pugh, Jr.*, and W.H. Cobbs**
*Dept. of Psychology, University of Pennsylvania
Philadelphia, PA 19104
**Dept. of Neurology, University of Pennsylvania Medical School
Philadelphia, PA 19104, USA

Abstract. A cytoplasmically diffusing substance or transmitter must carry the message of excitation from the vertebrate rod disk membrane to the rod plasma membrane, there effecting a decrease in the light-sensitive membrane current. A number of general properties of the transmitter molecule that communicates excitation to the rod plasma membrane either can be deduced from or are strongly constrained by facts of rod physiology. Here we analyze seven general properties of an excitational transmitter: a) transmitter sign (positive or negative concentration change induced by light); b) multi-order sequence of events in production/destruction; c) numerical gain in production; d) restricted longitudinal diffusion along the outer segment; e) buffering effects on gain and diffusion coefficient; f) limited transmitter lifetime; and g) linearity of transmitter production/reduction with light intensity. Although only cGMP and calcium have been hypothesized to be the molecule communicating excitation to the plasma membrane, other substances have been hypothesized to serve as intermediary transmitters in excitation. We examine the following five intermediary transmitter candidates in the light of the seven general properties: (i) G-protein; (ii) protons; (iii) 5'GMP; (iv) cGMP-dependent protein kinase; and (v) inositol-1,4,5-trisphosphate.

INTRODUCTION

Requirement for an Internal Transmitter

The insight that visual transduction in vertebrate rods requires an internal transmitter between disk membranes and plasma membrane follows from two premises: first, the rhodopsin molecules that trigger the normal light response are integrally embedded in disk membranes; second, rod disk membranes are (with the exception of a few basal disks) not continuous with the outer segment plasma membrane. The first premise is supported by a variety of biochemical and biophysical evidence, such as the requirement for detergent to solubilize rhodopsin, the relatively slow lateral diffusion of rhodopsin, X-ray diffraction studies of rods, etc. (see (66) for a fairly recent review of much of this evidence). The second premise is supported by the ultrastructural evidence of electron micrographs of many laboratories (e.g., (18)), by the lack of contribution of most rat rod disks to the rod membrane capacitance (67, 73), and by the inability of the fluorescent lipophilic dye di-dansyl-cystine to access the disk membranes without prior plasma membrane rupture (83). Given these two structural premises, the deduction that in a rod a messenger must carry information about photon absorption from the disk membranes to plasma membrane is immediate. That it is the outer segment plasma membrane whose conductance is altered by light and is therefore the receiver of the transmitter signal is supported by the ability of local outer segment illumination to cause local photocurrents (6, 41), as well as by the linearity of maximal photocurrents with the length of outer segment pulled into a recording suction electrode (5).

Transmitter Code: Concentration Change, Flux Change, or Something Else?
The defining property of an internal transmitter in rods is cytoplasmic diffusibility. Perhaps the simplest way in which a transmitter could represent the event of photon absorption is as a change in cytoplasmic concentration (or more properly, in cytoplasmic activity). Another way the message of excitation could be represented is as a change in the enzymic activity of an aqueous-diffusing enzyme.

Yet another possibility for representing the message of excitation is as a flux such as the hydrolytic flux of cGMP. The essential notion of a "flux signal" (as opposed to a "concentration-change signal") is that a species of molecules is altered rapidly without there being much change in the bulk concentration of that species. For there to be a flux of molecules without a concentration change on a given time scale, the removal/restoration mechanism must operate on a faster scale. For a flux unaccompanied by a cytoplasmic concentration change to be a signal, it seems necessary that the mechanism that responds to the flux be located within a few molecular lengths of the flux-generating molecule. A pure flux could conceivably supply energy to do chemical work such as to drive a trans-diskal transport mechanism (51), or to alter rapidly a second species of molecules. It appears to us, however, that any transduction scheme based upon a light-triggered flux at the disk membrane would still require an aqueous messenger whose cytoplasmic concentration or enzymic activity is altered by light. On the other hand, one can imagine a flux at the plasma membrane to be coupled to the conductance change.

This contribution is concerned with examination of the transmitter concept as applied to transduction in rods and with evaluation of the possibility that a substance (or substances) other than calcium or cGMP might serve as transmitter(s) in the outer segment. In the first part of this article we examine some of the properties of any putative transmitter, as constrained by electrophysiological data and as dictated by physical chemistry and ultrastructural features of the outer segment. In the second part of the paper we consider some specific transmitter candidates. We concentrate our attention on candidate transmitters hypothesized to represent the excitational event as a change in cytoplasmic concentration.

Definition of Terms

In this paper excitation will refer to the sequence of events in the rod leading from photon absorption to closure of the light-sensitive conductance of the outer segment. A messenger that carries information from one site in the rod to another as an essential step in the excitational sequence will therefore be called an excitational transmitter. The excitational transmitter that effects the closure of the outer segment light-sensitive conductance may be called "terminal excitational transmitter" when the need arises to distinguish it from other internal transmitters that may play an intermediary role in excitation. In addition to excitation, the rods of most vertebrate species also undergo light adaptation, a process that reduces the effectiveness of an isomerization and usually results in an increase in response bandwidth. An internal transmitter that carries information effecting sensitivity decrease or frequency bandwidth increase will be called an adaptational transmitter. Electrophysiological evidence supports the existence in the rod outer segment of at least one transmitter of each type.

Several general properties of the terminal excitational transmitter to the rod plasma membrane can be deduced from rod structure, principles of physical chemistry, and known facts about rod photocurrents. We will consider these properties: a) direction (positive vs. negative) of light-induced change in transmitter concentration; b) multi-order sequence of events in transmitter production or reduction; c) numerical gain in transmitter production or reduction; d) restricted longitudinal diffusion of transmitter; e) temperature-dependence of diffusion of transmitter; f) effects of buffering on gain and diffusion of transmitter; g) linearity of transmitter production or reduction with light intensity; and h) lifetime of transmitter.

GENERAL ANALYSIS OF OUTER SEGMENT TRANSMITTER PROPERTIES

Transmitter Sign

For a transmitter T that represents the excitational message as a change in its cytoplasmic activity a_T, it is useful to define a property of the transmitter we shall call sign. We define, then, the excitational sign of a transmitter to be positive (+) if increase in a_T is hypothesized to cause a decrease in the light-sensitive conductance, $g_{h\nu}$. Similarly, we define the sign of a putative transmitter to be negative if a decrease in a_T is hypothesized to cause a decrease in $g_{h\nu}$. The calcium hypothesis specifies Ca^{2+} to be a positive transmitter; the cGMP hypothesis specifies cGMP to be a negative transmitter. The sign of a transmitter has some interesting consequences.

Let a transmitter interact with a binding site on $g_{h\nu}$ via a simple binding reaction, and let the conductance have only two states. For a positive transmitter T^+ the reaction is

$$T^+ + (g_{h\nu})\ \mathrm{open} \underset{k_{-1}}{\overset{k_1}{\rightleftharpoons}} (T^+ \cdot g_{h\nu})\mathrm{closed}\ , \tag{1}$$

whereas for a negative transmitter T^-,

$$T^- + (g_{h\nu})\ \mathrm{closed} \underset{k_{-1}}{\overset{k_1}{\rightleftharpoons}} (T^- \cdot g_{h\nu})\mathrm{open}\ . \tag{2}$$

For both reactions, the equilibrium rate constant for a change to a new steady-state level of transmitter cytoplasmic activity a_T is $k_{eq} = (k_1 a_T + k_{-1})$.

Examination of these reaction schemes shows that they make different predictions about the effect of increasingly intense lights on the rate of conductance closure. Clearly, the rate of channel closure according to reaction 1 is driven by a_{T+}. Thus, assuming that increasingly intense lights produce monotonically increasing a_{T+}, under Eq. 1 increasing light flash intensity should cause ever more rapid shutdown of $g_{h\nu}$. On the other hand, even if increasingly intense lights cause a_{T-} to drop with ever greater rapidity, under reaction 2 the $g_{h\nu}$ shutdown is rate-limited by the rate constant k_{-1}. Essentially the same point was made by Lamb (48).

Figure 1 shows some results of Penn and Hagins (67) which place a lower limit on k_{-1} in reaction 2. They found that the rising phase of rat rod photocurrent was rate-limited by cable properties and it follows from their data that if Eq. 2 holds for rat rods, $k_{-1} > 1400\ s^{-1}$. Another interesting consequence follows if the light-sensitive conductance is governed by reaction 2. For any instantaneous step change in T^- the rate constant of equilibration of the conductance to the new level a_{T-} is $(k_{-1} + K_1 a_{T-})$, which cannot be any smaller than k_{-1}. Thus, Penn and Hagins' result requires that the fraction conductance open equilibrate with any change in activity of a negative transmitter in less than 1 ms, and allows one to observe the kinetics of the biochemistry underlying the negative transmitter change without any material delay intervening.

Multi-order Delay in Transmitter Production

Recordings of rod photocurrents with extracellular electrodes in rat retina (41, 67), voltage-clamping microelectrodes in isolated salamander rods (2, 9), and suction electrodes in toad (5) or monkey (64) have found that in healthy rods the linear response has a rising phase incorporating $n = 4$-6 stages of exponential delay (with roughly equal time constants, about $t = t_{max}/(n-1)$), and that 1-5% of the photocurrent is suppressed per isomerization in the linear range of the peak photocurrent amplitude vs. light intensity curve. Based on these findings, it seems reasonable to assume that the linear response is compounded of relatively similar, multi-order

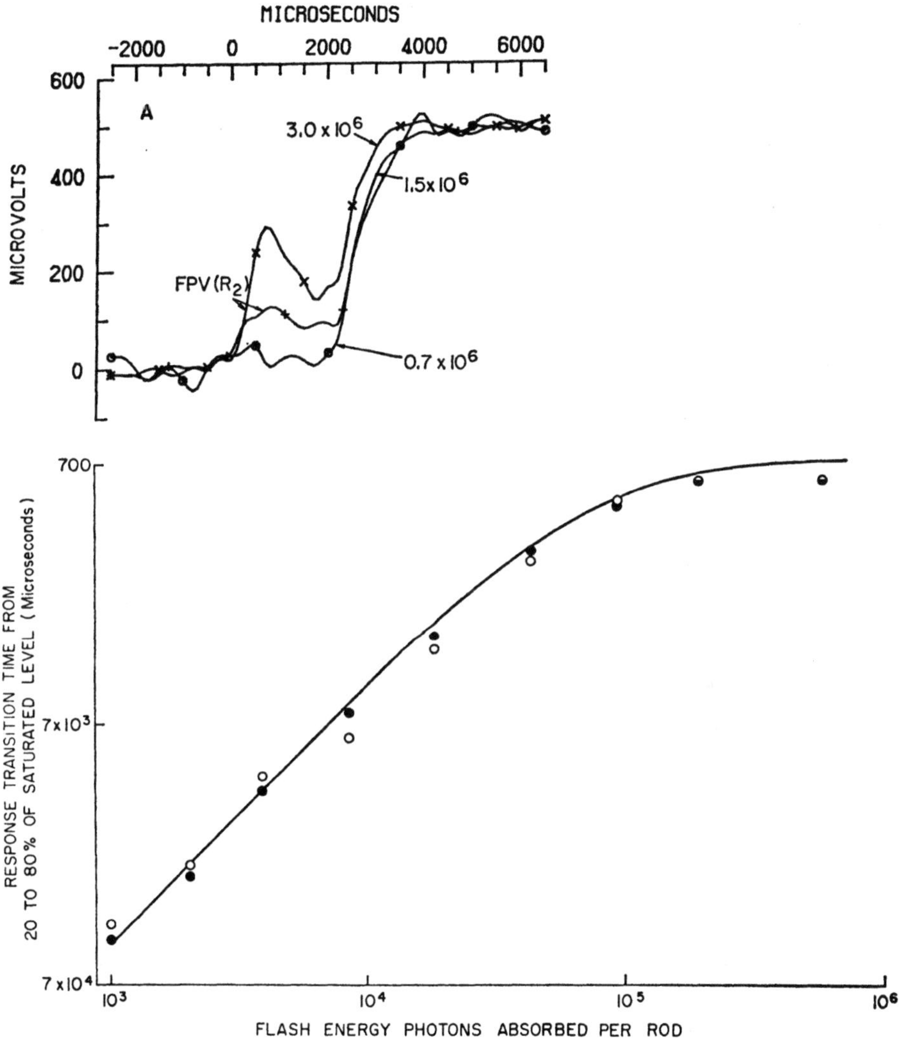

FIG. 1 - Upper panel: Extracellular transreceptor photovoltages from rat retina in response to 10 µs flashes yielding the indicated number of isomerizations/rod. The two responses to the more intense flashes show "fast photo-voltages" (or "early receptor potentials"). Penn and Hagins (67) argue that the rising phase of these responses is rate-limited by rod cable properties (from (67)).

single-isomerization responses. This conclusion is greatly strengthened by examination of responses to single isomerizations.

Figure 2, left panel, shows a sequence of 23 photocurrent "responses" of a frog rod in a suction electrode, stimulated repeatedly every 10 s with a 20 ms, 500 nm flash of nominally identical intensity. Qualitative evidence for photon shot-noise as the principal cause of the response amplitude variation comes from the number of "no-shows," as well as rough grouping of the response-peak histogram (Fig. 1C) into distinct peaks lying at regular amplitude intervals. Quantitative arguments come from computation of the variance/mean ratio of the peak response, analysis of the response peak amplitude histogram, and from computations based on careful light calibration, known receptor size and position in pipette, and rhodopsin concentration and extinction. (These arguments are presented in detail in (6) and will not be reviewed here.)

Having concluded that the major response amplitude variations in data such as those shown in Fig. 2 arise due to Poisson variation in the number of isomerizations, a need for a multi-order sequence of events generating the photocurrent can be argued as follows. All the individual responses in Fig. 2A (indeed, all 125 response records analyzed to form the histogram of Fig. 2C) were fit by least-squares analysis with the normalized average response (inset, Fig. 2C); these fitted curves are shown as the smooth curves in Fig. 2A). One can clearly see that the individual responses are highly stereotypic and well described by the smooth curve fit to them. Furthermore, the idealized single-isomerization waveform is well described as a cascade of four to six reactions with exponential decay (6, 75), in the same manner as the linear response of the rod (to, say, 10-20 isomerizations) has been fit in previous analyses by a number of investigators (cited above). Results such as those shown in Fig. 2 (6, 75) thus exclude the possibility that the photocurrent generated by a single isomerization is a rectangular event, and likewise exclude the idea that the multi-order rising phase of responses to a flash causing $n > 1$ isomerizations is comprised of n temporally distributed rectangular events, one for each photon. The temporal waveform of the single-photon response, however, is almost surely comprised of many hundreds or thousands of much smaller unitary current events (see section on Transmitter Production or Reduction Gain).

Formal analysis of the single-isomerization response into several exponential decay steps does not per se implicate a multi-order process in the production of positive transmitters (1). For example, the formal analysis does not exclude the possibility that one or more of the relatively slow time constant events occurs at the plasma membrane - e.g., as might occur if the transmitter binding had to give rise to a major structural change in the Na^+-conductance, or if the conductance or removal mechanism were an ionophore that had to diffuse through the membrane. The formal analysis can exclude certain hypotheses, however, such as this: at the moment of isomerization (i.e., in a time-window brief relative to the time-to-response

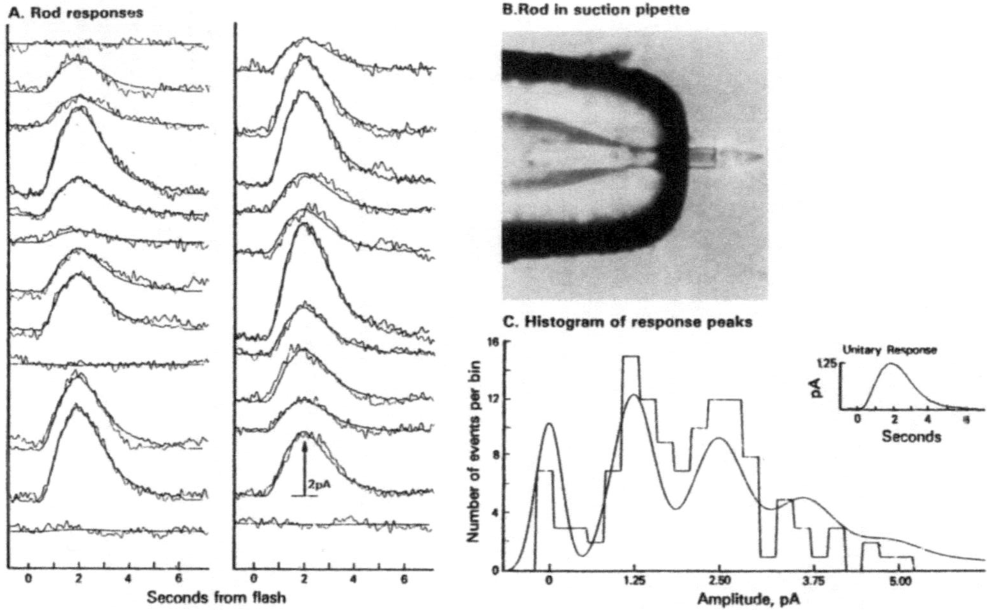

FIG. 2 - A: Photocurrent responses of an isolated toad rod held in a suction pipette (Pugh, 1981, unpublished), measured with the current-to-voltage conversion technique of Baylor, Lamb, and Yau (6). The cell was stimulated at 10 s intervals with nominally identical 20 ms, unpolarized 500 nm flashes of 0.12 quanta/μm^2. The successive traces show fluctuations in amplitude, sometimes showing no clear response at all. The "smooth" curve drawn through each trace is the fit of the normalized average of 125 such responses (inset, Fig. 1C) to the individual traces, scaled by a least-squares fitting procedure. The technique used for the fitting is described by Baylor, Lamb, and Yau (6). The maximal photocurrent was 18 pA; temperature ca. 22°C; sealing resistance of the suction pipette 10 megohm. B: Photograph of a toad rod held with the outer segment in the suction pipette, with the inner segment truncated at the ellipsoid. The cell that gave the responses in A was of the same form and held in the same configuration. C: Histogram of 125 peak amplitudes of the least squares fits to individual traces of responses of the cell of Fig. 1A. The inset shows the average of the 125 responses, scaled to the estimated amplitude of the unitary event. The smooth curve drawn through the histogram is a probability distribution derived from convolution of the Poisson distribution (photon absorption distribution) with Gaussian distributions of dark noise amplitude fluctuations and unitary amplitude fluctuations (6). Estimated parameters: 1.9 events/flash; unitary event amplitude, 1.25 pA; Gaussian standard deviations, 0.2 pA.

peak), a quantity of transmitter particles are simultaneously released, diffusing unhindered in an aqueous phase to the plasma membrane, effecting conductance decrease by a rapid equilibrium binding. Such an hypothesis can be excluded because aqueous diffusion alone cannot account either for the form of the dependence of the rate of rise of photoresponse on light intensity (4) or for the temperature-dependence of the linear response ((8, 47, 67); see section on Temperature-dependence of Diffusion).

The view that most of the multistage delay of the rising phase of the light response is dominated by transmitter production kinetics in the case of a positive transmitter is strengthened by recent evidence from noise power analysis by Bodoia and Detwiler (13), suggesting that the light-sensitive channels have a Lorentzian spectrum with corner frequency (half-power point), f_c, of ca. 400 Hz. For a two-state channel, if the transition rate for closing is a and that for opening is b, then by standard theory (e.g., (24)) $(a + b) = 2\pi f_c$; thus, for Bodoia and Detwiler's data, a + b = 2400/s. Clearly, at least one of the rate constants must be quite large, and the closing of the light-sensitive channels cannot constitute more than one delay stage on the time scale of several hundred ms. (Since only the sum of the opening and closing rates appears in the formula for the corner frequency, one of the time constants could be large, and so channel kinetics could contribute to the multistage delay. See Schnakenberg and Keiper, this volume.)

The case of a negative transmitter is different. Recall that the equilibrium rate for a transmitter satisfying Eq. 2 is constrained by the data of Penn and Hagins to be always greater than 1000 s^{-1}. Thus, for a negative transmitter the multi-order delay in the rising phase of the photoresponse necessarily arises in events leading to the drop in activity of the transmitter, and any activity change is reflected in a few milliseconds by the photocurrent. The outer segment membrane conductance (current) virtually instantly reflects (through some possibly nonlinear binding or other relation) the cytoplasmic concentration of a negative transmitter.

Transmitter Production or Reduction Gain

Analysis of single-isomerization photocurrents has been used to support the conclusion that many hundreds of transmitters are produced per isomerization. In Fig. 2C it may be noted that there is a degree of dispersion about the histogram peaks. As modelled by Baylor et al. (6), the fit to the histogram postulates a minor Gaussian variation (std. dev. $\sigma = 0.2$ pA) in response amplitude, convoluted with the major variation due to the Poisson variation in number of isomerizations. If this minor variation is generated by Poisson variation in the number z of transmitters produced, and if the photocurrent is assumed proportional (with proportionality constant α) to z, then an estimate of the number of transmitters z(tmax) at response peak may be computed (6). This follows, since for an ensemble of single-photon responses, $\{j_1(t)\}$.

mean peak current $= \mu_1 = \langle j_1(tmax) \rangle = \alpha z(tmax)$, and

$$\text{variance} \quad = \sigma^2 = \langle (j_1(tmax) - \alpha z_{peak})^2$$
$$= \alpha^2 z(tmax),$$
$$\text{so that} \quad (\mu_1)^2/\sigma^2 = z(tmax). \tag{3}$$

For the data of Fig. 2, $\sigma = 0.2$ pA and $\mu_1 = 1.25$ pA, giving z(tmax) about 40. The nine cells of Table 1 of Baylor et al. (6) had a mean $\mu_1 = 0.96$ pA, $\sigma = 0.19$ pA, yielding essentially the same estimate for z(tmax). As pointed out by Baylor et al. (6), the total number of positive transmitters produced by a single isomerization will be greater by the ratio of the integrating time of the response divided by the mean lifetime of the transmitter - for a frog or toad rod with single-photon response as in Fig. 2C, inset, this factor should be about three to four, giving an estimate for total transmitter production over the time course of the response of 120-160.

The estimate z(tmax) based upon Eq. 3 applied to homogeneously illuminated rods depends implicitly on the assumption of response isotropy - that single-isomerization responses from different regions of the outer segment produce the same mean number of transmitters. Schnapf (75) has shown that some of the dispersion in the single-photon response in homogeneously illuminated amphibian rods is due to systematic longitudinal variation in response amplitude for rod base to rod tip. When a narrow transverse stimulus is confined to one region, the variance σ^2 above) in single-photon response amplitude is less than that observed for homogenous illumination. From Schnapf's Fig. 4, the number of transmitters at response peak calculated with Eq. 3 for cell A is z(tmax) = 11 at base and 13 at tip; for cell B, z(tmax) = 81 at base and 144 at tip. For the data of Fig. 1, if 50% of the variance is ascribed to longitudinal variation in peak response, the underlying σ for a local region would be 0.14 rather than 2, and we would then estimate $z(tmax) = (125)^2/(0.14)^2 = 80$.

A likely flaw in computations based upon Eq. 3 is the assumption that, at the level of transmitter produced by single-photon responses, peak photocurrent is linearly related to the total number of transmitters produced. Such an assumption could be rationalized if one could assume a) that the rod is effectively a single, well stirred compartment, and b) that the dissociation constant for the transmitter binding is much higher than the effective cytoplasmic concentration of channel-binding sites (see (4), p. 692 for relevant discussion). Assumption a) is certainly false, because the region of the plasma membrane affected by a single isomerization at response peak is a few microns or less ((49); see below, next section). Assumption b), linearity between transmitter concentration and channels closed at the single-isomerization level, is also likely false. If binding and unbinding are rapid, the number of transmitters in the local region whose conductance is totally blocked by one isomerization must equal or exceed the number of ionic conductances blocked. And thus, adding more transmitters to the same locale will not necessarily cause linear variations in response amplitude. It seems unlikely that a linear relation between transmitters and photocurrent at the single-photon level can be rescued within the context of restricted transmitter diffusion, and thus calculations based upon Eq. 3 cannot be rationalized. One should not conclude that variation in transmitter production is not a likely cause of single-photon response variance, however, simply because the model of the rod upon which Eq. 3 seems to rest is false.

In any model of the rod in which a positive transmitter acts by binding as expressed in Eq. 1, the number of transmitters produced must equal or exceed the number of unitary conductances closed at any moment. Assuming a unitary current of ca. 2 fA estimated by Detwiler, Conner, and Bodoia (25), a single-photon peak photocurrent of 1 pA would result from closure of 500 channels and require at least as many transmitter molecules.

Another estimate of the number of transmitters can be made based upon reaction rate theory analysis, and in particular the theory of bimolecular encounters in solution (e.g., (1), p. 905; (27)). Under standard conditions in aqueous solution the diffusion limit for the rate of a bimolecular reaction is about 10^{10} liters/mol. s; this limit is set by the rate at which small molecules in solution can encounter one another, i.e., get within distance of reaction. Evidence reviewed below on the restricted diffusion of transmitter in the outer segment indicates that the transmitter activity change effected by a single isomerization in a toad rod is confined to a length of outer segment of ca. 2 µm and results in virtually complete suppression of the light-sensitive conductance in this region at the time t_{max} of peak photocurrent. Thus, if at t_{max}, z(tmax) transmitters are evenly distributed over a region of Δµm in a toad rod - i.e., in an aqueous volume of $(0.5 \times \pi \times 3^2 \times \Delta)$ μm^3 (see Fig. 2) - then the maximal rate d_+ of transmitter encounters with the channel binding site would be 10^{10} x the concentration of transmitters,

$$d_+ = (3.7 / \Delta)\, z(tmax)\, s^{-1}. \qquad (4)$$

For example, if 100 transmitters were homogeneously spread over an outer segment region of Δ = 2 µm, each channel could expect encounters with transmitters at a maximal rate of 185 per second. If effective transmitter binding has an appreciable activation energy, the rate of channel closures is considerably less than the number of encounters. For a binding activation energy of 5 kcal/mol, the rate of successful encounters would be reduced by a factor $\exp(-5/RT) = 2 \times 10^{-4}$.

In summary, calculations based either upon Detwiler and Bodoia's estimate of the unitary current or upon collision rates set by solution encounter theory, and the assumption of a modest activation energy, require many thousands of positive transmitters to be produced per isomerization at photocurrent peak.

Analysis of reduction gain for a negative transmitter can be done along the same lines as those that give rise to Eq. 3, if one considers, instead of the absolute level of transmitter, the deviation of transmitter concentration from its resting level in the dark. Similarly, an analysis based upon Eq. 4 may be made. For a negative transmitter, Eq. 4 applies in the dark, with z(tmax) equal to the resting number of free transmitter per μm length. For example, suppose a resting concentration of a negative transmitter in the dark in a frog outer segment (diam. 6 μm were 1 μmol/l, so that there are 8500 free transmitter molecules/μm length. Then, based upon Eq. 4, one would expect no more than about 31,500 encounters/s of transmitters with each conductance site. If the binding activation energy were 4 kcal/mol, about 36 of these encounters/s would be successful and open the conductance. If the mean open time (k_{-1} in Eq. 2) were 1 ms, the outer segment conductance in the dark resting state then would be 0.036 its maximal possible value. Presumably the single-photon response would have to result in the destruction of more than 99% of the negative transmitter over a region of ca. 2 μm.

Transmitter Diffusion

In the previous paragraph, in describing encounters between transmitters and channels we made use of physical facts about the diffusion of small molecules in solution. The diffusion of transmitter in the outer segment is an essential step in excitation, and the possible role of transmitter diffusion must be examined in any model of the time course of the photocurrent response (see (4), p. 725). Generally, one can conclude that a single isomerization must produce transmitter that diffuses at least 1-2 μm (30-60 disks) from the disk where absorption occurred (and could diffuse much farther), since a few % of the light-sensitive outer segment conductance is suppressed per isomerization.

A molecule diffusing unhindered with a diffusion coefficient D has an uncertainty in its location at time t relative to its location at t = 0 given by a Gaussian distribution with a standard deviation $\sigma = \sqrt{2Dt}$. From this one might conclude that a transmitter molecule with a diffusion coefficient of $D = 10^{-5}$ cm^2/s = 10^3 μm^2/s could readily diffuse anywhere within an amphibian rod outer segment in 1 s, since at t = 1 s, $\sqrt{2Dt}$ = 44 μm. However, this naive estimate is incorrect by almost two orders of magnitude! The error arises from a failure to take into consideration the boundary conditions - specifically, the hindrance or baffling by the disks. A molecule diffusing in the outer segment spends most of its time wandering between disks rather than moving longitudinally.

Figure 3 shows the model of the outer segment used by Lamb, McNaughton, and Yau (49) to account quantitatively for the baffling effect of the disks on longitudinal diffusion of transmitter in toad rods. They first point out that in this model rod the concentration in an interdiskal space of a small molecule with cytoplasmic diffusion coefficient $D = 10^{-5}$ cm^2/s can be expected to equilibrate in a few ms with the concentration apposite the plasma membrane - this can be arrived at roughly by computing $t = a^2/2D$ = 0.005 s, where a = 3 μm = 3×10^{-4} cm is the rod radius. This value for equilibration time could be off by as much as a factor of 100 for some transmitters ((59); next section). Next, they note that if A is the outer segment cross section, A' is the effective cross-section segment for longitudinal diffusion, V' the aqueous volume and V the total volume, then the effective diffusion coefficient for longitudinal diffusion of a molecule with cytoplasmic diffusion coefficient D is given by $D_{eff} = (A'/A)(V/V')D = (2gs/ac)D = 0.02D$, in terms of the parameters of Fig. 3. In their elegant paper Lamb et al. (49) compare photocurrents in response to narrow transverse slits and diffuse illumination and argue, based upon comparison of their data with solutions to the diffusion equation for longitudinal transmitter diffusion[2], that the spread of excitational transmitter produced by one isomerization is no more than a few microns, and that the effective longitudinal diffusion coefficient is less than 3×10^{-7} cm^2/s. Furthermore, as Lamb et al. point out, the fact that photocurrent vs. light-intensity curves for homogeneous illumination are well described by the equation

$$j = j_{max} [1 - \exp(-\bar{n}\Delta)], \tag{5}$$

where $\bar{n}$ is mean number of isomerizations per unit length, is consistent with the notion that small lengths of the outer segment on the order of 1-2 μm act effectively as isolated compartments with respect to light-released transmitter.

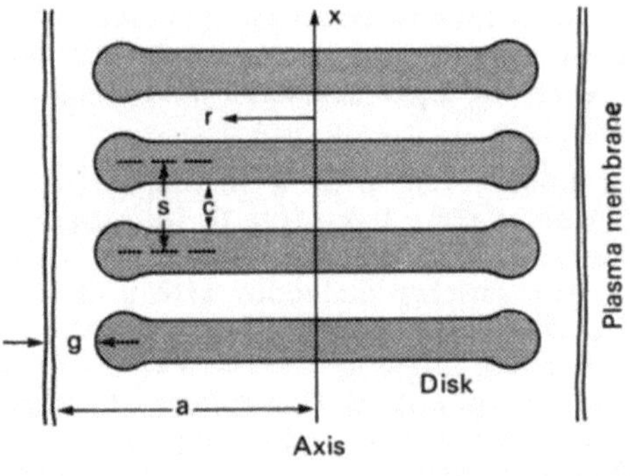

FIG. 3 - Schematic toad outer segment of Lamb, McNaughton, and Yau (49) used for computing longitudinal diffusion in outer segment: a is rod radius, about 3 μm; g is the distance from the edges of the disks to the plasma membrane, ca. 150 Å; s is the center-to-center spacing of the disks, 300 Å; c is the cytoplasmic space between the disks, ca. 150 Å.

Transmitter Buffering and Diffusion

A problem closely related to that of diffusion in the characterization of the outer segment terminal excitational transmitter is the problem of buffering or binding of transmitter to sites other than the channel binding sites. As the number of transmitter binding sites other than the light-sensitive channels increases, then the number of transmitters produced to achieve a given cytoplasmic activity must be increased correspondingly. Furthermore, binding of transmitter to extraneous sites reduces the effective diffusion coefficient: if binding equilibrium is achieved rapidly on the time scale of the light response, then $D_{eff} = D(c_f/c_{tot})$, where c_f is the concentration of free transmitter and c_{tot} the total, averaged over the same space[3] (22). Because of the narrow interdiskal space (ca. 150 Å - see Fig. 20 in (22)), the "diffuse double layer" induced by the negative surface charge of each disk (58) must be taken into consideration in discussion of diffusion and buffering of positively charged transmitter candidates in the outer segment. McLaughlin and Brown (59) examine one case in detail: the effect of the net negatively charged disk phosphotidylserines (outer segment concentration taken to be 50-100 mmol/l) on calcium. They demonstrate that the radial diffusion of calcium would be retarded 7-30 times by the presence of the negative phospholipids and that the buffering would reduce the effectiveness of a given amount of calcium release by a factor in excess of 10. The effect of binding on radial diffusion appears large enough to call into question the assumption of Lamb et al. (49) that radial equilibration occurs very rapidly on the time scale of the linear light response, if Ca^{2+} is the internal transmitter.

Temperature-dependence of Diffusion

On the basis of the Stokes-Einstein relation for the diffusion coefficient, $D = kT/6\pi R\eta$, where R is the hydrodynamic radius and η the solution viscosity, and the relation $\eta = \exp(-E/RT)$, cytoplasmic diffusion of most candidate transmitters would be expected to have a temperature-dependence determined by the temperature-dependence of water viscosity. For water viscosity E = 4.2 kcal/mol, giving $Q_{10} = 1.28$ between 15° and 25°C.

The Q_{10} for the reciprocal time-to-peak of the linear response of vertebrate photoreceptors has been reported in a number of studies (4, 7, 8, 47, 67) to range from 1.8 to 2.7, corresponding to apparent activation energies of between 10 and 17 kcal/mol. This relatively high apparent activation energy appears to exclude aqueous diffusion as playing a significant role in shaping the time course of the linear photocurrent. Baylor, Matthews, and Yau (7) and Lamb (47) have shown that the linear photocurrent at various temperatures scales on the time axis, i.e., linear responses at various temperatures can be made exactly coincident if they are normalized at peak and the time axis is scaled linearly. Lamb (47) also showed that the same scaling applies to the tail phase of the saturated photocurrent. A simple way to account for these findings is to assume that the rate constants of the approximately four reactions that determine the waveform of the linear photocurrent have the same temperature dependence. Lamb (47) points out that this temperature-dependence is approximately the same high temperature dependence as the intra-membrane diffusion of rhodopsin (54), and he discusses a series of four membrane-bound reactions that could conceivably be identified with the hypothetical reactions of the four-stage models.

In summary, the temperature-dependence (Q_{10}) of the photocurrent waveform is about twice that expected if simple aqueous diffusion were the dominant component. The actual temperature-dependence may suggest that transmitter production and removal kinetics are dominated by movement of molecules in the lipid phase of the rod outer segment.

Transmitter Production/Destruction Linearity

Given the apparent compartmentalization of the outer segment due to restricted longitudinal diffusion, photocurrent linearity follows simply because each "compartment" contributes almost identically at low light intensities. Two photons arriving, however, in the same 2 µm "compartment" might not contribute linearly to transmitter production - in the extreme case, one isomerization per compartment might completely inactivate that compartment by releasing or destroying all its transmitter. Evidence that production/destruction of transmitter (at least during the rising phase of the photoresponse) continues linearly well above photocurrent saturation comes from experiments that have shown that the rate of rise of photocurrent is linear up to three log units above the intensity at which linearity of response peak begins to fail. Rat rods' photocurrent peak obeys a hyperbolic saturation law with a half-saturation constant of 30 isomerizations; the rate of rise (10%-90%), however, obeys hyperbolic saturation with a half-saturation constant of 30,000 isomerizations, or ca. 30 isomerizations/disk (67). For turtle cones, the photovoltage semisaturation constant is ca. 1000 isomerizations, but the photovoltage rising phase remains linear over 10^6 isomerizations (4). Penn and Hagins (67) point out that the linearity of the rate of rise excludes the hypothesis that "compartments" - whose size one estimates from the photocurrent vs. light intensity curve (or, for that matter, from analysis showing limited longitudinal diffusion) - act as units, such that one isomerization per compartment completely saturates (inactivates) the compartment. The linearity of the rising phase over a very large range of light intensities does not prove that the underlying transmitter production kinetics operate unchanged over this large range of light intensities, although such global linearity is a key feature of models of photoresponses (4, 67). The obvious difficulty is that as the stimulating flash gets more and more intense the time to response saturation becomes shorter and shorter - and during saturation there is no information about the underlying kinetics. Nonetheless, it is clear that there are no compartments that saturate with only a few (less than 10) isomerizations.

Transmitter Lifetime

The lifetime of an active transmitter molecule under a given set of conditions (e.g., dark adaptation) cannot exceed the time course of the photocurrent response obtained under the same conditions. Although the statement in the last sentence is an obvious truism, the time course of the photocurrent response can serve to winnow transmitter candidates. Lamb (47) showed that the recovery tail phase of the photocurrent response of toad rods to saturating flashes followed a single exponential, with a time constant of 4.8 s in bicarbonate Ringer's at 20°C. Thus, any proposed positive transmitter must come with a mechanism that can remove the transmitter with a time constant no greater than 4.8 s under those conditions. Likewise, any proposed negative transmitter must also have an associated concentration mechanism that can restore the resting level with a time constant of $\leq$ 4.8 s. (If the 4.8 s tail phase is indeed the final removal step of a positive transmitter under those conditions, then the question is pertinent as to why this time constant does not appear in the linear response. The answer may lie in the restricted longitudinal diffusion: at low light levels longitudinal diffusion may provide "sinks" for positive or sources for "negative" transmitter - areas of disks and plasma membrane relatively unstimulated that can participate in removal or restoral at a greater rate than the relatively more stimulated region.)

TRANSMITTER CANDIDATES OTHER THAN cGMP OR CALCIUM

In the context of the general properties of internal transmitters stated above, we shall next discuss five specific rod outer segment excitational transmitter candidates: (i) GTP-binding protein; (ii) protons; (iii) 5' GMP; (iv) cGMP-dependent protein kinase; (v) inositol-1,4,5-trisphosphate. In the discussion above we did not make a distinction between the terminal excitational transmitter - the aqueous diffusible molecule which carries the excitational message to the plasma membrane - and other intermediary excitational transmitters. An intermediary transmitter would carry the message of photon absorption from one site to another in the outer segment in the process leading up to the production of the terminal transmitters. Certainly, the possibility of intermediary transmitters is implicit in the structure of the single-photon response: each sequential step in the inferred multi-order production of terminal transmitter could involve messengers, and the gain of the terminal transmitter production could be achieved by multiplication of gains of cascaded steps of production of intermediaries.

We shall restrict our discussion of specific transmitters to molecules proven or thought to diffuse in the rod cytoplasm. This restriction means that we shall not consider intermediaries that diffuse only in the disk or plasma membrane. Intra-membrane diffusion of intermediaries in transduction, however, could be very important in explaining the complex kinetics of the single-photon response (53) and the temperature-dependence of the response (47).

GTP-binding Protein

GTP-binding protein (or "G-protein") is an approximately 80 kD extrinsic protein of rod disks, with surface density ca. 1 per 10-15 rhodopsins. (A number of laboratories have contributed to its isolation and characterization. Due to our restricted space and aim we shall only refer to material directly pertinent to the transmitter issue.) The most important fact about G-protein is that photolyzed rhodopsin catalyzes GTP binding to it in exchange for priorly bound GDP; GDP remains bound to G-protein in the absence of photoactivated rhodopsin after terminal phosphate hydrolysis (33, 36). A very important feature of the catalytic activity of photolyzed rhodopsin is that at low light levels it shows relatively high gain: one rhodopsin can catalyze hundreds of GDP/GTP exchanges (33, 54). The second important fact is that GTP-activated G-protein (or more specifically, its α-subunit) is the activator of rod disk phosphodiesterase (PDE) (32).

Our first question about G-protein as a hypothetical transmitter candidate is this: does the catalysis by R* of GTP binding to G-protein produce an aqueous diffusible transmitter under intracellular conditions and on the time scale of the single-photon response? In fact, an unequivocal answer to this question cannot be given at this time. Godchaux and Zimmerman (36) showed that protein-bound GTP and GTP-ase activity is released from bleached bovine disk membranes upon addition of 20 μmol/l GTP in a solution containing 100 mol/l NaCl, 1 mol/l $MgCl_2$, 20 mol/l Tris-HCl, pH = 7.5. Kühn (45) showed that the 37 kD α-subunit of G-protein (which initially contains bound GTP and the GTP-ase activity of the protein) is released from bleached bovine membranes suspended in medium containing 130 mol/l KCl, 10 mol/l NaCl, 5 mol/l $MgCl_2$, 8 mol/l Tris-HCl, 2 mol/l ATP, 1 mol/l GTP, pH = 7.5. Although the work just cited does seem adequate proof that the protein is released under cellular ionic conditions in the presence of GTP, no proof is given that such release occurs at low light levels and on the time scale of the single-photon response.

Kühn, Bennett, Michel-Villaz, and Chabre (46) discovered two light-scattering signals of disk membranes which were linearly related to the amount of G-protein added previously to protein-stripped membranes. If their "dissociation" signal can be causally linked to release of GTP-activated G-protein from the membranes, then the half-time of G-protein release from bovine membranes after 3×10^{-5} flash bleach at 20°C in solution containing 100 mol/l KCl, 1 mol/l $MgCl_2$, 10 mol/l Tris-HCl, 17 μmol/l GTP, pH = 7.4, is less than 2 s. In the presence of ATP, the half-time would likely be briefer since the activity and lifetime of R* is reduced (52). Pfister, Kühn and Chabre (68) showed that the "dissociation" signal for frog membranes at 20°C in the presence of 100 μmol/l GTP has a half-time at 2×10^{-4} bleach of ca. 2 s. These observations lend support to the notion, but do not prove, that light-triggered release of G-protein occurs on a time scale appropriate to excitation.

Gain. Since G-protein remains bound to disk membranes in the dark at normal ionic strength (3, 44, 45), the process by which rhodopsin catalyzes G-protein GTP/GDP exchange probably involves mutual lateral diffusion - rhodopsin within the lipid phase and G-protein attached to a molecule (possibly rhodopsin) in the lipid phase. Both by simulation of the random walk of rhodopsin and its targets in the membrane and by calculation based upon "first passage theory" (79), one can demonstrate that one rhodopsin at full catalytic activity can fully activate the G-protein on 1 μm^2 of surface area per second (71). Since there are 25,000-30,000 rhodopsins/μm^2 and 1 G-protein per 10 rhodopsins, the total gain in 1 s would be about 2500-3000 G-proteins activated per R*. However, at low bleaches a rhodopsin is expected to lose its catalytic activity very rapidly after isomerization, due to ATP-dependent quench (52, 76, 77). Thus the gain of the production of activated G-protein is expected to be somewhat lower, ca. 500, at the peak of the single-photon response in an amphibian rod.

Production kinetics. The time course of activated G-protein production should follow closely the time course of R* decay and, according to simulation and calculation, be roughly first-order. Work already mentioned on G-protein-dependent light-scattering is at present the best evidence about G-protein production kinetics.

Aqueous diffusion. Since the α-subunit of G-protein, the subunit thought to be released into the cytoplasm, has a molecular weight of ca. 40,000, as a sphere it would have a radius of ca. 30 Å. Thus, from the Stokes-Einstein formula, its aqueous diffusion coefficient would be about 7×10^{-7} cm^2/s. Diffusing from a point source at the center of a 6 μm diameter disk, the "time constant" for the concentration at the plasma membrane to equilibrate with the source would be ca. 0.1 s, significant on the time scale of the linear response. The effective longitudinal diffusion coefficient calculated with the model rod of Fig. 2 would be only 1.45×10^{-8} cm^2/s = 1.45 $\mu m^2/s$. This would result in a longitudinal space constant at t = 1 s of 0.667 μm ((49), Eq. 14), or about ± 22 disks.

Production linearity. The rate of production of activated G-protein by photolyzed rhodopsin would be expected to continue to rise linearly with light intensity until nearly one rhodopsin were photolyzed per G-protein. Unfortunately, no direct evidence is available on this matter. Kühn et al. (46) show that their "dissociation signal" saturates a ca. 10^{-3} bleach, but this saturation is not to be confused with saturation of the rate of rise.

Decay time. Perhaps the most disturbing fact concerning the hypothesis that G-protein is an intermediary transmitter concerns the rate of GTP-ase activity, which is thought to be the process by which the α-subunit which activates PDE becomes inactive. A number of laboratories have reported GTP-ase rates of ca. 1 min^{-1} (10, 36, 44, 50). This slow turnoff and hence long lifetime of the activated G-protein would seem to exclude G-protein from playing a role as an intermediate excitational transmitter, unless there is some means other than GTP hydrolysis of inactivating it, and thereby PDE. In Fig. 4 we show data from the work of Liebman et al. (51) which shows that some process can inactivate PDE (and presumably G-protein) on a considerably faster time scale.

Upper panel, cGMP hydrolysis: Noisy trace shows the average of five successive responses to flashes that each bleached 5×10^{-6} of the rhodopsin in the cuvette, or about 5 isomerizations per side of each toad disk. At this light level and below the reaction is linear in light, and under these nucleotide conditions the system gives repeatable responses (52, 53). Protons produced per isomerization can be readily computed from the known rhodopsin content and calibration of the buffer responses to known

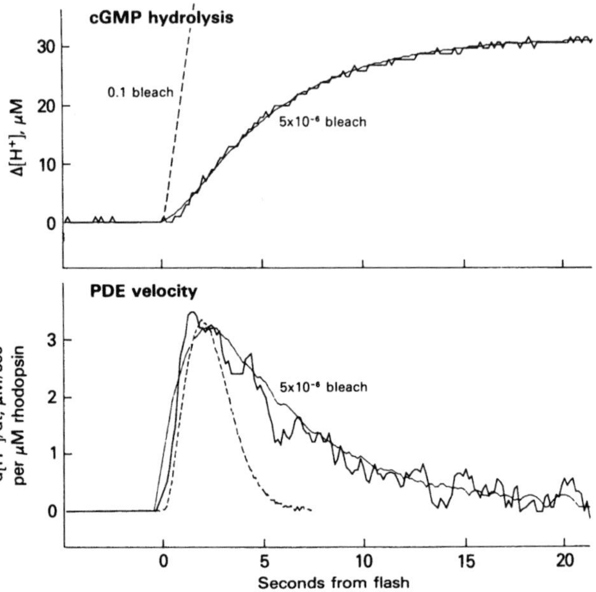

FIG. 4 - Time course of proton production by toad rod disk membrane phosphodiesterase (PDE), from Liebman et al. (51). Preparation consisted of freshly prepared, fractured Bufo rod outer segments, suspended at 3 μmol/l rhodopsin concentration in a 20 mmol/l Hepes-buffered isotonic KCl medium at pH = 8, temperature = 25°C. To this suspension was added: cGMP, 8 mmol/l; GTP, 1 mmol/l; ATP, 2 mmol/l. The PDE-catalyzed reaction was triggered at t = 0 by a 1 ms light flash, and proton production followed with sensitive pH electrometer (Liebman and Evanczuk, 1982).

amounts of added acid or base. The smooth curve fit through the data is derived from the model of phosphodiesterase activation/deactivation of Liebman and Pugh (53); differential equations governing the reaction were solved with a Runge-Kutta technique and converted to the same 12-bit integer scale used by the A/D for collecting and displaying the data. The dotted line shows the time course of the response to a flash that bleached 10% of the rhodopsin in the cuvette.

Lower panel, PDE velocity: Jagged unbroken trace shows the derivative of the experimental trace of Fig. 3A, and thus gives the velocity of the PDE-catalyzed reaction. The peak hydrolytic velocity was 3 mol/l H^+/s mol/l rhodopsin. Since only 5 x 10^{-6} was bleached, the peak velocity was 3/(5 x 10^{-6}) = 6 x 10^5/s/isomerization. The smoother unbroken trace is the derivative of the theoretical curve of Fig. 3A. The maximum velocity of the reaction (obtained in response to a 10% bleach) was 15 mol/l H^+/mol/l rhodopsin/s and was achieved within the time resolution of the electrometer. The dashed trace is the average single-photon response of the toad rod whose records are shown in Fig. 1, arbitrarily normalized to have the same peak as the PDE velocity trace. Due to the method of taking the derivatives (polynomial fitting to curve segments) and the problem of normalization, the rising phase of the velocity trace and of the single-photon photocurrent cannot be compared quantitatively in this representation. However, it is clear that the final time constant of decay of the PDE velocity is greater than that of the tail phase of the photocurrent.

Electrophysiology. There has been one report to date on the injection of activated G-protein into rods (17). These authors report that pressure injections of G-protein cause dose-dependent rod hyperpolarizations that mimic many features of the light response, supporting an intermediary role for G-protein.

Summary. Photoactivated GTP-binding protein appears to have many properties required of an intermediary excitational transmitter. It is produced with a gain of several hundred copies/isomerization, probably with exponential kinetics, on the time scale of activity of R*. Its initial production rate is expected to remain linear with light intensity over many log units of intensity, possibly up to a 10% bleach. Disturbingly, however, the lifetime of the photoactivated α-subunit is apparently much longer than the single-photon response. At present G-protein's only known function is to activate disk membrane PDE, and thus transduction schemes that incorporate G-protein as an intermediary must do so through the effects of PDE.

Protons

Protons would seem on a priori grounds to be a poor choice for an intracellular messenger. Nucleoside phosphates alone can be expected to provide several millimolar of proton buffering capacity at intracellular pH (72). In the rod outer segment, because of the disk membranes contributing ca. 140 mmol/l zwitterionic (PC, PE) and 20 mmol/l negative (PS, PI) phospholipids to the outer segment (26, 31), and because of the negatively charged protein (e.g., (3)), the buffering capacity of the outer segment is less favorable than in most cells for protons to play a messenger role. Despite this serious objection based on buffering, two facts about rods have made it reasonable to consider the hypothesis that protons act as a transduction intermediary.

The first pertinent fact is that light-activated hydrolysis of cGMP by phosphodiesterase generates in the outer segment many protons per isomerization on the time scale of the linear photocurrent (81). From the data of Fig. 4 (discussed below), we estimate that between 10^4 and 10^5 cGMP's would be hydrolyzed, and therefore protons produced, in a toad rod at 20°C in 1-2 s, the approximate peak of the single-photon response.

The second pertinent fact that led us (51, 62) to consider the hypothesis that protons act as an intermediary in transduction was the failure of several labs to observe significant decreases in the outer segment cGMP content in response to physiological light stimuli (37, 39, 43). (For an alternative view and positive results showing cGMP decrease, see (21)). Although these negative results could be taken to mean that light-activated PDE activity under intracellular conditions remains low, an alternative interpretation is that guanylate cyclase is capable of producing cGMP at nearly the rate it is hydrolyzed in response to physiological stimuli. This latter interpretation has obtained support from the work of Goldberg et al. (37), who have developed a technique based upon incorporation of ^{18}O into the α-phosphoryl upon cyclic nucleotide hydrolysis and have studied the fate of these outer segment nucleotides in light-stimulated rabbit retina. As shown in the data of Fig. 5, large light-activated turnover of cGMP is occurring in the rod in the absence of large changes in total cGMP. Given these data, a reasonable hypothesis is that the flux of hydrolyzed cGMP, or some byproduct of it (protons, 5'GMP) plays a role as an intermediary in transduction.

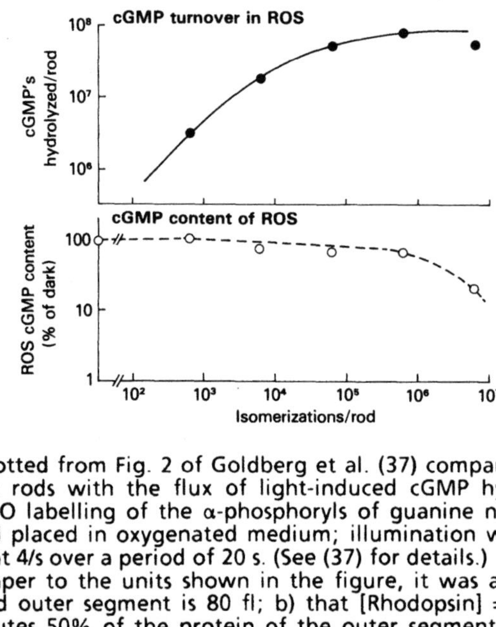

FIG. 5 - Data replotted from Fig. 2 of Goldberg et al. (37) comparing the total cGMP content of rabbit rods with the flux of light-induced cGMP hydrolysis, the latter measured with ^{18}O labelling of the α-phosphoryls of guanine nucleotides. Retinas were isolated and placed in oxygenated medium; illumination was 80 10-ms strobe flashes delivered at 4/s over a period of 20 s. (See (37) for details.) To convert the data in the original paper to the units shown in the figure, it was assumed a) that the volume of the rod outer segment is 80 fl; b) that [Rhodopsin] = 3 mmol/l; c) that rhodopsin constitutes 50% of the protein of the outer segment; d) that the water space of the outer segment is 50% of the volume; e) that a light calibrated to have a flux 0.1 photopic ft-lamberts (and assumed color temperature 3500°K) flashed for 10 ms would produce 8 isomerizations/flash in a rabbit rod illuminated end-on.

Gain. To date there have been no reports of successful measurement of proton production by light-activated PDE in living rods. Furthermore, measurement methods based upon change in pH would underestimate the production because of buffering, and possibly because the removal process could be very rapid. The ^{18}O method of Goldberg et al. (37), however, can provide one lower limit: from the data of Fig. 5 we estimate that about 4800 protons are generated per isomerization in rabbit rods, probably within

250 ms. The data of Fig. 4 can be used to compute an estimate for the toad rod, which we place conservatively at 5 x 10^4 per isomerization total at the peak, ca. 2 s, of the single-photon response (see Fig. 2, inset).

Production kinetics. PDE-catalyzed cGMP hydrolysis produces protons with three delay steps: a) time delay for laterally diffusing rhodopsin to activate the complement of G-protein that it will find before it (R*) becomes inactive; b) time decay of G-protein activity, shaped presumably by GTP hydrolysis; c) time delay for an individual, activated G-protein to find and activate a PDE. The third delay component is expected to be relatively short and probably negligible compared with the first two. The smooth curves fit to the cGMP hydrolysis data in Fig. 4 were computed with a solution to a differential equation model of the theory of Liebman and Pugh (53), which has essentially two delay steps.

Liebman and Pugh (53) have shown that cGMP hydrolysis is linear with light intensity (under the "intracellular" conditions of the data of Fig. 4) up to about 5 x 10^{-6} bleach or about five isomerizations per side of a frog or toad disk. Thus, Fig. 4 gives the impulse response for proton production for the suspension of toad disks. Extrapolating to the cell, then, one can expect that at least the initial rate of proton production should remain linear with light until about 10,000 isomerizations or slightly more.

Buffering and diffusion. The composite buffering power of the outer segment at its resting pH is likely equivalent to that of about 20-50 mmol/l of a weak acid with pK_a at that pH. This value can be deduced from the data of Hagins and Yoshikami (42) or of Emrich (28) or can be inferred from summing the weighted contributions of ionizable groups in the outer segment (23). Assuming the value 50 mmol/l, the production of ca. 5 x 10^6 protons (16 μmol/l total in the outer segment) in 1 s by a flash of 100 isomerizations in a frog or toad rod would be expected to change the pH by only about 0.0006 pH unit. This would contribute a total of about 18 extra free H^+ to an outer segment at pH = 7 containing initially 3 x 10^4 free H^+.

The diffusion coefficient of protons in water is about ten times faster than that of any other ion, ca. 10^{-4} cm^2/s. If either equilibrium binding[3] or more appropriate surface-potential theory-based calculations (59) is applied, the diffusion of protons in the outer segment is estimated to be slowed by a factor of ca. 10^{-5} or more, resulting in a diffusion coefficient of less than 10^{-9} cm^2/s. This makes it unlikely that protons released into the cytoplasm by PDE-catalzyed hydrolysis in the center of a toad interdiskal space of cGMP could make it to the plasma membrane on the time scale of the single-photon response.

Electrophysiological evidence. There have been three reports that lowered external pH causes rod a dark current to diminish (34, 51, 80). Liebman et al. (51) demonstrated that the time course of proton suppression of dark current can occur in ca. 1 s or less. They argue that protons applied internal to the plasma membrane are the effective agent of dark current suppression since CO_2 perfusion, and the aftereffect of NH_4Cl perfusion, also can cause suppression of dark current. Figure 6 shows titrations of dark current suppression by lowered external pH with and without 50 mmol/l acetate in the Ringer's: the presence of acetate (pK_a = 4.8), which acts as a neutral proton carrier in the protonated form, shifts the apparent pK_a of the dark current suppression to higher values, consistent with the idea that protons effect dark current suppression interior to the plasma membrane. Liebman et al. (51) also demonstrated that the effectiveness of protons in suppressing rod dark current depends on calcium, in much the same way that light-sensitivity does: thus proton suppression of dark current appears to require calcium. The dependence of proton suppression on calcium makes it unlikely that protons act directly on the Na conductance. Two possible interpretations of the calcium-dependence of dark current suppression are these: a) protons block plasma membrane a Na/Ca exchange, allowing Ca^{2+} to build up in the outer segment by leakage, and block the conductance; b) protons exchange for calcium sequestered in or on the disks.

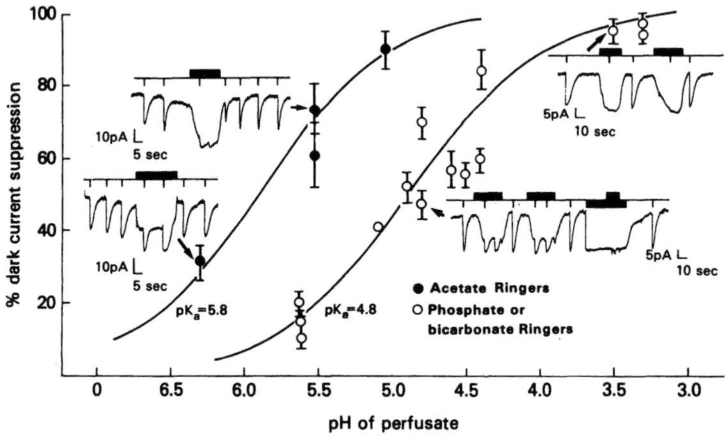

FIG. 6 - Proton suppression of frog rod dark current. Frog rods with ellipsoid were held in suction electrodes (see Fig. 1C) in "ellipsoid-in" configuration, and the exposed outer segment perfused with solutions of altered pH. Substitution of 50 mmol/l CH_3COONa (sodium acetate) for 50 mmol/l sodium chloride caused the dark current suppression titration curve to shift left - i.e., less protons in the medium were required to achieve the same degree of suppression; this is expected if the protons act internally to the plasma membrane in suppressing the dark current. The traces shown about the curve give some of the data from which the curves were derived. In some cases the perfusion causes a slight shift in the level of the saturated photocurrent response, due to junction currents. The fractional suppression is measured as (pre-perfusion photocurrent) - (photocurrent during perfusion). For details see (51).

There are two negative reports of ionophoretic injection of protons into rods with simultaneous recording. Pinto and Ostroy (69) injected 10 pC or about 6×10^7 H^+ of charge into a toad rod while recording membrane potential from a pipette containing 1 mol/l HCl and observed no effect. MacLeish, Schwartz, and Tachibana (57) continuously injected 100 pA of positive current into a salamander rod in a Ringer's containing 20 µmol/l calcium for several minutes from one barrel of a double-barrelled pipette containing 0.1 mol/l citric acid, pH = 1.8, while recording voltage-clamp current through the other barrel, and found no detectable change in the dark current. Both of these results pose serious objections to any simple scheme in which protons, released freely into the cytoplasm, play an intermediary role (although it should be noted that the low calcium Ringer's in the MacLeish et al. study would be expected, on the basis of the results of Liebman et al. (51), to diminish the effectiveness of protons more than a factor of 10).

Another serious objection to protons playing an intermediary role in transduction has been posed by Yoshikami and Hagins (82). They showed that adding extrinsic permeable weak acid (e.g., 20 mmol/l imidazole) to Ringer's could block the dark current suppression caused by the proton release aftereffect of NH_4Cl perfusion; but the same weak buffer had very little effect, if any, on the sensitivity and time course of the light response. Yoshikami and Hagins also show that protons introduced into the rod by NH_4Cl perfusion block light-triggered Ca^{2+} efflux from rods, which is almost surely carried by an Na/Ca exchanger. Thus, the likely explanation of the proton-induced suppressions of dark current reported by Liebman et al. (51) is that sufficient protons interior to the rod can suppress Na/Ca exchange, causing excess Ca^{2+} buildup in the cytoplasm and that, furthermore, the numbers of protons produced by cGMP hydrolysis are far less than the numbers introduced by perfusions that cause dark current suppression.

Summary. The findings just reviewed render it unlikely that protons are an intermediary between cGMP hydrolysis and calcium release in the outer segment. It would appear that the only way that protons could still play an intermediary role would be that the protons released by cGMP hydrolysis are in effect released extremely close to the site where they are effective, e.g., in causing Ca^{2+} release so that they escape the buffering effect of the cytoplasm. Since the proton release site is the active site of the phosphodiesterase molecule itself, what would be required for protons to play an intermediary role would be that an arbitrary PDE molecule is somehow tied to the exchange site.

5'GMP

Since PDE-catalyzed hydrolysis of cGMP produces both H^+ and 5'GMP, the latter is logically a candidate for an intermediary transmitter. The gain, production kinetics, and linearity ought all to be the same for 5'GMP as for protons, and thus the discussion above applies verbatim to these properties.

A general objection against a role for 5'GMP can be made on the basis of its resting concentration of ca. 0.5 mmol/l (72). At this concentration a toad rod with a 60 m outer segment should contain about 3 x 10^8 5'GMP molecules. If an isomerization yields 10^5 molecules of 5'GMP by the peak of the single-photon response, and even if these molecules are confined by restricted longitudinal diffusion within a 2 μm region, the change of concentration expected in this region is less than 1%. Furthermore, Robinson and Hagins (72) showed that even a 1% bleach causes relatively little change in GMP in frog, which suggests that 5'GMP may be very efficiently removed. This conclusion fits well with the more recent work of Goldberg et al. (37), who showed that the 5'GMP whose α-phosphoryl is labelled by ^{18}O upon cGMP hydrolysis rapidly turns up in the guanosine di- and triphosphate nucleotide pool of the outer segment, indicating the activity of an effective guanylate kinase. It is worth noting that microdissection data of Berger et al. (11) show that guanylate kinase is largely confined to the inner segment of rabbit and monkey.

Electrophysiology. Several laboratories have reported ionophoretic injections of 5'GMP into rods during electrophysiological recording. Nicol and Miller (63) and Miller and Nicol (61), in their original reports showing a cGMP-induced depolarizations of rods, also demonstrated that injecting comparable amounts of 5'GMP caused little effect, if any, and certainly not the hyperpolarization expected were 5'GMP a transduction intermediary. More recently, MacLeish et al. (57) reported that injections of 5'GMP (amount not given) produced no effect on the rod dark or photocurrent, measured with voltage-clamping microelectrodes. In the same report, MacLeish et al. showed that injection of cGMP caused large increases in dark current.

Cyclic Nucleotide-dependent Protein Kinase

One of the prevalent "dogmas" about cyclic nucleotides is that they produce their physiological effects through cyclic nucleotide-dependent protein kinases (40). A number of authors have hypothesized that cGMP hydrolysis in the rod might effect Na^+ conductance decrease by altering the balance between phosphorylation (catalyzed by a cGMP-dependent protein kinase) and dephosphorylation (effected by an outer segment phosphatase) of some specific target phosphoprotein. Farber, Brown, and Lolley (29) proposed that a cGMP-dependent kinase regulated the phosphorylation of a 30kD soluble protein believed to come from the outer segment and that this 30kD protein was the effector of Na^+ conductance decrease. Polans, Hermolin, and Bownds (14) discovered two small extrinsic proteins (13kD and 12kD) of frog outer segments that were phosphorylated in a cGMP-dependent manner, and Bownds (14) has proposed that these proteins could play a key role in altering the Na^+ conductance. Liebman and Pugh (53) proposed that cGMP-dependent protein kinase might directly regulate the outer segment Na^+ conductance via phosphorylation, the phosphorylated state being the more conductive one.

Space does not permit review of the details of biochemical experiments on cGMP-dependent phosphorylation in outer segments (see Bownds and Brewer, this volume), nor is anything of moment known about the gain, kinetics, linearity, etc., of these reactions, nor of any injection experiments demonstrating physiological effects of cyclic nucleotide-dependent kinases. It is, however, worth noting three points. The first is that for cGMP-dependent protein kinase to play an intermediary role a single isomerization must cause a significant change in cGMP in some compartment (however small) in which PDE, kinase, and cGMP reside. The second is that cGMP-dependent protein kinase might be a highly sensitive way to preserve the message of isomerization: if [cGMP] were to drop significantly, even in one interdiskal space, some protein kinases might be altered in their activity. This alteration could be conserved beyond the time when the cGMP level returns to its basal level (either by diffusional exchange, cyclase replenishment, or both), and thus the message of photon absorption might be preserved despite apparently little change in total [cGMP], up to levels of several hundred isomerizations/rod. The third point, made by Lamb (48), is that phosphorylation of the light-sensitive conductance as a means of control is energetically costly. To account for the rising phase of the light-response (see Fig. 1, above) each conductance must be dephosphorylated in ca. 1 ms after it is phosphorylated and opened. If the unitary conductance is 2 fA (25) and the normal dark current (of a toad rod) 30 pA, then 1.5×10^7 ATP's/s would be required by the channel phosphorylation alone to maintain the normal dark current. This value can be compared with the value of 6×10^7 ATP's/s required by the Na/K exchanger to maintain the cell's resting Na gradient in the presence of a 30 pA dark current.

Inositol-1,4,5-trisphosphate

Another substance that can be expected to receive scrutiny in the near future as an intermediary (or even terminal) transmitter candidate in rod outer segments is inositol-1,4,5-trisphosphate, or $InsP_3$ for short. Although little is known about $InsP_3$ metabolism in vertebrate rods on the time scale and at light levels relevant to transduction, several facts make $InsP_3$ an interesting candidate.

First, many types of cells that respond to hormones or neurotransmitters react (at least in part) through a mechanism that causes phosphatidylinositol (PI) hydrolysis, thereby releasing $InsP_2$ and $InsP_3$ into the cytoplasm; PI hydrolysis is effected by a specific phospholipase, phospholipase C (for a recent review, see (64)). Since the outer segment contains ca. 65 phospholipids/rhodopsin (26, 60), and 1-2% of these are PI (31), an abundance of substrate for a phospholipase C is present. Furthermore, Schmidt (74) has demonstrated a light-enhanced turnover of PI in rat rods. Although the light levels used by Schmidt were exceptionally high (the stimulus of average intensity 1200 ft-candles should cause well over 10^8 isomerizations/rod s in a rat rod oriented perpendicular to the source) and probably bleached all the rhodopsin in a few seconds, the fact remains that photolyzed rhodopsin, which has been compared with a hormonal messenger, has thus been shown to increase PI hydrolysis.

Second, it has been shown that $InsP_3$ is the earliest-appearing product of hormonally activated PI hydrolysis (12). Although $InsP_2$ is also produced, little InsP or inositol alone results from hormonal activation, indicating that the primary effect of hormones on this system is the hydrolysis of phosphatidylinositol (4,5) P_2 to yield diacylglycerol (DAG) and $InsP_3$. It is important to note that DAG, unlike the phosphorylated inositol compounds, remains membrane-bound after hydrolysis of the phosphate-ester bond and is thought to be rapidly rephosphorylated to phosphatidic acid.

Third, Streb et al. (78) have presented data that support the conclusion that $InsP_3$ functions to release calcium. Thus, if calcium were the terminal transmitter to the outer segment light-sensitive conductance, $InsP_3$ would have to be considered a strong candidate for an intermediary.

Fourth, Fein et al. (30) and Brown et al. (15) have demonstrated that injection of $InsP_3$ into Limulus photoreceptors causes discrete events that resemble light responses and likewise produces relatively local light adaptation. If, as they have argued, $InsP_3$ is part of the transduction chain in invertebrates, the possibility must be considered that GTP-binding protein might activate phospholipase C, since the effects of GTP-γ-S in Limulus photoreceptors have lent support to the hypothesis that activation of G-protein is an early stage in phototransduction in invertebrates (19, 20), as it has been hypothesized to be in vertebrate rods (32). Both invertebrate and vertebrate rhodopsin would act to catalyze the binding of GTP to G-protein. G-protein and its variants may well be protean in function, capable of activating not only adenylate cyclase and disk membrane phosphodiesterase, but also possibly phospholipase C (38) or other enzymes in signalling cascades.

Thus, one can construct a reasonable scheme leading from photolysis of rhodopsin to calcium release in vertebrate rods. The release of calcium in such a scheme would involve at least two delay steps: activation of G—protein (see above), and subsequent activation of phospholipase C; a third delay could be introduced in the mechanism by which $InsP_3$ causes calcium release. In this scheme cGMP might function only to regulate calcium storage or uptake (35, 56). Obviously, too little is known at this point to comment more on the gain, details of production kinetics, etc., of such a mechanism. Nonetheless, transduction models based on $InsP_3$ release will certainly enrich vertebrate transduction research with a new set of ideas and tools.

Acknowledgements. We thank H. Stieve for helpful comments. This work was supported by N.I.H. grants EY-02660 and EY-01583.

ENDNOTES

1 We note here that characterization of the photocurrent response as "multi-order" or "involving several delay stages" does not depend on the specifics of the mathematical models that have been fit to responses. Rather, what is critical in mathematical terminology is that $t = 0$ is a "pole of order > 4": near $t = 0$ the response behaves as tn, $n > 4$.

2 The derivation of D_{eff} can readily be made by setting up Fick's Law for fluxes: the flux of molecules in the rod in the longitudinal direction is given by

$$J_x = -DA'(\partial c/\partial x), \tag{A1}$$

where c(r, x, t) is the cytoplasmic concentration. Since A(V'/V)c(a, x, t) is the total number of molecules per unit length of the outer segment, by continuity, conservation, and the assumption of rapid radial equilibration, we must also have

$$A(V'/V)\,(\partial c/\partial t) = -\,\partial J_x/\partial x. \tag{A2}$$

Combining (A1) and (A2) gives

$$\partial c/\partial t = D_{eff}(\partial^2 c/\partial x^2). \tag{A3}$$

Note that if transmitter is also destroyed or removed uniformly at a rate k, the composite differential equation is

$$\partial c/\partial t = D_{eff}(\partial^2 c/\partial x^2) - kc. \tag{A4}$$

3 If a transmitter T is in rapid equilibrium with a population of binding sites B, according to $T + B \rightleftharpoons (TB)$, then

$$[TB] = K_A\,[T]_f\,[B]_f,$$

where K_A is the association constant of the reaction, and subscripts "f" mean "free." If $[B]_{tot} \gg [T]_{tot}$, then

$$[TB] = K_A\,[B]_{tot}\,[T]_f, \text{ and}$$

$$[T]_f/[T]_{tot} = 1/\{K_A\,[B]_{tot} + 1\}.$$

Letting $c = [T]_f$, Fick's 2nd Law in 1-dim. form becomes

$$\begin{aligned}\partial c/\partial t &= D(\partial^2 c/\partial x^2) - \partial[TB]/\partial t\\ &= D(\partial^2 c/\partial x^2) - K_A\,[B]_{tot}\,(\partial c/\partial t).\end{aligned} \tag{A5}$$

Rearranging terms in the last equation, one has

$$\partial c/\partial t = (D/(1 + K_A\,[B]_{tot}\,(\partial^2 c/\partial x^2), \tag{A6}$$

and so the effective diffusion coefficient is

$$\begin{aligned}D_{eff} &= D/\{1 + K_A\,[B]_{tot}\}\\ &= D\{[T]_f/\,[T]_{tot}\},\end{aligned}$$

the normal coefficient multiplied by the fraction of transmitter remaining free.

REFERENCES

(1) Atkins, P.W. 1978. Physical Chemistry. San Francisco: Freeman.

(2) Bader, C.R.; MacLeish, P.R.; and Schwartz, E.A. 1979. A voltage-clamp study of the light response in solitary rods of the tiger salamander. J. Physiol. **296:** 1-26.

(3) Baehr, W.; Morita, E.A.; Swanson, R.J.; and Applebury, M.L. 1982. Characterization of bovine rod outer segment G-protein. J. Biol. Chem. **257:** 6452-6460.

(4) Baylor, D.A.; Hodgkin, A.L.; and Lamb, T.D. 1974. The electrical response of turtle cones to flashes and steps of light. J. Physiol. **242:** 685-727.

(5) Baylor, D.A.; Lamb, T.D.; and Yau, K.-W. 1979a. The membrane current of single rod outer segments. J. Physiol. **288:** 589-611.

(6) Baylor, D.A.; Lamb, T.D.; and Yau, K.-W. 1979b. Responses of retinal rods to single photons. J. Physiol. **288:** 613-634.

(7) Baylor, D.A.; Matthews, G.; and Yau, K.-W. 1980. Two components of electrical dark noise in toad retinal rod outer segments. J. Physiol. **309:** 591-621.

(8) Baylor, D.A.; Matthews, G.; and Yau, K.-W. 1983. Temperature effects on the membrane current of retinal rods of the toad. J. Physiol. **337:** 723-734.

(9) Baylor, D.A., and Nunn, B.J. 1985. Electrical properties of the light-sensitive conductance of salamander rods. J. Physiol., in press.

(10) Bennett, N. 1982. Light-induced interactions between rhodopsin and the GTP-binding protein: relation with phosphodiesterase activation. Eur. J. Biochem. **123:** 133-139.

(11) Berger, S.J.; DeVries, G.W.; Carter, J.G.; Schulz, D.W.; Passonneau, P.N.; Lowry, O.H.; and Ferrendelli, J.A. 1980. The distribution of the components of the cGMP cycle in retina. J. Biol. Chem. **255:** 3128-3133.

(12) Berridge, M.J. 1983. Rapid accumulation of inositol trisphosphate reveals that agonists hydrolyse polyphosphoinositides instead of phosphatidylinositol. Biochem. J. **212:** 849-858.

(13) Bodoia, R.D., and Detwiler, P.B. 1984. Patch-clamp study of the light response of isolated frog retinal rods. Biophys. J. **45:** 337a.

(14) Bownds, M.D. 1981. Biochemical pathways regulating transduction in frog photoreceptor membranes. Curr. Top. Membr. Trans. **15:** 203-214.

(15) Brown, J.E.; Rubin, L.J.; Ghalayini, A.J.; Taver, A.P.; Irvine, R.F.; Berridge, M.J.; and Anderson, R.E. 1984. Myo-inositol polyphosphate may be a messenger for visual excitation in Limulus photoreceptors. Nature **311:** 160-162.

(16) Chabre, M.; Vuong, M.; and Stryer, L. 1982. Anisotropy of the infrared light-scattering changes induced by illumination of oriented retinal rod outer segments. Biophys. J. **37:** 247a.

(17) Clack, J.W.; Oakley, B. II.; and Stein, P.J. 1983. Injection of GTP-binding protein or cGMP phosphodiesterase hyperpolarizes retinal rods. Nature **305:** 50-52.

(18) Cohen, A.I. 1968. New evidence supporting the linkage to extracellular space of outer segment saccules of frog cones but not rods. J. Cell. Biol. **37:** 424-444.

(19) Corson, D.W., and Fein, A. 1983a. Chemical excitation of Limulus photoreceptors. I. Phosphatase inhibitors induce discrete wave production in the dark. J. Gen. Physiol. **82:** 639-657.

(20) Corson, D.W., and Fein, A. 1983b. Chemical excitation of Limulus photoreceptors. II. Vanadate, GTP-g-S and fluoride prolong excitation evoked by dim flashes of light. J. Gen. Physiol. **82:** 659-677.

(21) Cote, R.H.; Biernbaum, M.S.; Nicol, G.D.; and Bownds, M.D. 1984. Light-induced decreases in cGMP concentration precede changes in membrane permeability in frog rod photoreceptors. J. Biol. Chem. **259:** 9635-9641.

(22) Crank, J. 1975. The Mathematics of Diffusion, 2nd ed. London: Oxford Press.

(23) Dearry, A. 1981. Rod outer segment phosphodiesterase: a study on light-induced activity in whole retina using bromcresol purple. Ph.D. Dissertation, University of Pennsylvania.

(24) DeFelice, L.J. 1981. Introduction to Membrane Noise. New York: Plenum.

(25) Detwiler, P.B.; Conner, J.D.; and Bodoia, R.D. 1982. Gigaseal patch clamp recordings from outer segments of intact retinal rods. Nature **300:** 59-61.

(26) Dratz, E.A.; Miljanich, G.P.; Nemes, P.P.; Gaw, J.E.; and Schwartz, S. 1979. The structure of rhodopsin and its disposition in the rod outer segment disk membrane. Photochem. Photobiol. **29:** 661-670.

(27) Eigen, M. 1973. Diffusion control in biochemical reactions. In Quantum Statistical Mechanics in the Natural Sciences, eds. S.L. Mintz and S.M. Widmayer, pp. 37-61. New York: Plenum.

(28) Emrich, H. 1971. Optical measurements of the rapid pH change in the visual process during the metarhodopsin I-II reaction. Z. Naturforsch. **266:** 352-356.

(29) Farber, D.B.; Brown, B.M.; and Lolley, R.N. 1978. Cyclic GMP: proposed role in visual function. Vision Res. **18:** 497-500.

(30) Fein, A.; Payne, R.; Corson, D.W.; Berridge, M.J.; and Irvine, R.F. 1984. Photoreceptor excitation and adaptation by inositol-1,4,5-trisphosphate. Nature **311:** 157-160.

(31) Fliesler, S.J., and Anderson, R.E. 1983. Chemistry and metabolism of lipids in the vertebrate retina. Prog. Lipid Res. **22:** 79-131.

(32) Fung, B.K.; Hurley, J.B.; and Stryer, L. 1981. Flow of information in the light-triggered cyclic nucleotide cascade of vision. Proc. Natl. Acad. Sci. USA **78:** 152-156.

(33) Fung, B.K., and Stryer, L. 1980. Photolyzed rhodopsin catalyzes the exchange of GTP for bound GDP in retinal rod outer segments. Proc. Natl. Acad. Sci. USA **77:** 2500-2504.

(34) Gedney, C., and Ostroy, S.E. 1978. Hydrogen ion effects of the vertebrate photoreceptor: the pK's of ionizable groups affecting cell permeability. Arch. Biochem. Biophys. **188:** 105-113.

(35) George, J.S., and Hagins, W.A.H. 1983. Control of Ca2+ in rod outer segment disks by light and cGMP. Nature **303:** 344-348.

(36) Godchaux, W. III., and Zimmerman, W.F. 1979. Membrane-dependent guanine nucleotide binding and GTP-ase activities of soluble protein from bovine rod cell outer segments. J. Biol. Chem. **254:** 7874-7884.

(37) Goldberg, N.D.; Ames, A. III.; Gander, J.E.; and Walseth, T.F. 1983. Magnitude of increase in retinal cGMP metabolic flux determined by

^{18}O incorporation into nucleotide a-phosphoryls corresponds with intensity of photic stimulation. J. Biol. Chem. **258:** 9213-9219.

(38) Gomperts, B.D. 1983. Involvement of guanine nucleotide-binding protein in the gating of Ca^{2+} by receptors. Nature **306:** 64-66.

(39) Govardovskii, V.I., and Berman, A.L. 1981. Light-induced changes of cGMP content in frog retinal rod outer segments measured with rapid freezing and microdissection. Biophys. Struct. Mech. **7:** 125-130.

(40) Greengard, P. 1978. Phosphorylated proteins as physiological effectors. Science **199:** 146-152.

(41) Hagins, W.A.; Penn, R.D.; and Yoshikami, S. 1970. Dark current and photocurrent in retinal rods. Biophys. J. **10:** 380-412.

(42) Hagins, W.A., and Yoshikami, S. 1977. Intracellular transmission of visual excitation in photoreceptors: electrical effects of chelating agents introduced into rods by vesicle fusion. In Vertebrate Photoreception, eds. H.B. Barlow and P. Fatt, pp. 97-139. New York: Academic.

(43) Kilbride, P., and Ebrey, T.G. 1979. Light-initiated changes of cGMP levels in the frog retina measured with quick-freezing techniques. J. Gen. Physiol. **74:** 415-426.

(44) Kühn, H. 1980. Light- and GTP-regulated interaction of GTP-ase and other proteins with bovine photoreceptor membranes. Nature **283:** 587-589.

(45) Kühn, H. 1981. Interactions of rod cell proteins with the disk membranes: influence of light, ionic strength and nucleotides. Curr. Top. Membr. Trans. **15:** 172-199.

(46) Kühn, H.; Bennett, N.; Michel-Villaz, M.; and Chabre, M. 1981. Interactions between photoexcited rhodopsin and GTP-binding protein: kinetic and stoichiometric analyses from light-scattering changes. Proc. Natl. Acad. Sci. USA **78:** 6873-6877.

(47) Lamb, T.D. 1984. Effects of temperature on toad rod photocurrents. J. Physiol. **346:** 557-578.

(48) Lamb, T.D. 1984. Electrical responses of photoreceptors. In Recent Advances in Physiology, ed. P.F. Baker. London: Churchill Livingstone.

(49) Lamb, T.D.; McNaughton, P.A.; and Yau, K.-W. 1981. Spatial spread of activation and background desensitization in toad rod outer segments. J. Physiol. **319:** 463-496.

(50) Lewis, J.W.; Miller, J.L.; Mendel-Hartvig, J.; Schaechter, L.E.; Kliger, D.S.; and Dratz, E.A. Sensitive light-scattering probe of enzymatic processes in retinal rod photoreceptor membranes. Proc. Natl. Acad. Sci. USA **81:** 743-747.

(51) Liebman, P.A.; Mueller, P.; and Pugh, E.N., Jr. 1984. Protons suppress the dark current of frog retinal rods. J. Physiol. **347:** 85-110.

(52) Liebman, P.A., and Pugh, E.N., Jr. 1980. ATP mediates rapid reversal of cGMP phosphodiesterase activation in visual receptor membranes. Nature **287:** 734-736.

(53) Liebman, P.A., and Pugh, E.N., Jr. 1981. Control of rod disk membrane phosphodiesterase and a model for visual transduction. Curr. Top. Membr. Trans. **15:** 157-169.

(54) Liebman, P.A., and Pugh, E.N., Jr. 1982. Gain, speed and sensitivity of GTP-binding vs. PDE activation in visual excitation. Vision Res. **22:** 1475-1480.

(55) Liebman, P.A.; Weiner, H.L.; and Dryzmala, R.D. 1982. Lateral diffusion of visual pigment in rod disk membranes. Meth. Enzym. **81:** 660-668.

(56) Lipton, S.A.; Rasmussen, H.; and Dowling, J.E. 1977. Electrical and adaptive properties of rod photoreceptors in Bufo marinus. II. Effects of cyclic nucleotides and prostaglandins. J. Gen. Physiol. **70:** 771-791.

(57) MacLeish, P.R.; Schwartz, E.A.; and Tachibana, M. 1984. Control of the generator current in solitary rods of the Ambystoma tigrinum retina. J. Physiol. **348:** 645-664.

(58) McLaughlin, S. 1977. Electrostatic potentials at membrane-solution interfaces. Curr. Top. Membr. Trans. **9:** 71-144.

(59) McLaughlin, S., and Brown, J. 1981. Diffusion of calcium ions in retinal rods. J. Gen. Physiol. **77:** 475-487.

(60) Miljanivich, G.P.; Nemes, P.P.; White, D.L.; and Dratz, E.A. 1981. The asymmetric distribution of phosphotidylethanolamine, phosphatidylserine and fatty acids of the bovine retinal rod outer segment disk membrane. J. Membr. Biol. **60:** 249-255.

(61) Miller, W.H., and Nicol, G.D. 1979. Evidence that cGMP regulates membrane potential in rod photoreceptors. Nature **280:** 64-66.

(62) Mueller, P., and Pugh, E.N., Jr. 1983. Protons block the dark current of isolated retinal rods. Proc. Natl. Acad. Sci. USA **80:** 1892-1896.

(63) Nicol, G.D., and Miller, W.H. 1978. Cyclic GMP injected into retinal rod outer segments increases latency and amplitude of response to illumination. Proc. Natl. Acad. Sci. USA **75:** 5217-5220.

(64) Nishizuka, Y. 1984. Turnover of inositol phospholipids and signal transduction. Science **225:** 1365-1370.

(65) Nunn, B.J., and Baylor, D.A. 1982. Visual transduction in retinal rods of the monkey Macaca fascicularis. Nature **299:** 726-728.

(66) Olive, J. 1980. The structural organization of mammalian retinal disk membrane. Int. Rev. Cytol. **64:** 107-169.

(67) Penn, R.D., and Hagins, W.A. 1972. Kinetics of the photocurrent of retinal rods. Biophys. J. **12:** 1073-1094.

(68) Pfister, C.; Kühn, H.; and Chabre, M. 1983. Interaction between photoexcited rhodopsin and peripheral enzymes: influence of the postmetarhodopsin II decay and phosphorylation rate of rhodopsin. Eur. J. Biochem. **136:** 489-499.

(69) Pinto, L.H., and Ostroy, S.E. 1978. Ionizable groups and conductances of the rod photoreceptor membrane. J. Gen. Physiol. **71:** 329-345.

(70) Polans, A.S.; Hermolin, J.; and Bownds, M.D. 1979. Light-induced dephosphorylation of two proteins in frog rod outer segments. J. Gen. Physiol. **74:** 595-613.

(71) Pugh, E.N., Jr., and Liebman, P.A. 1980. Delays and sensitivity support lateral diffusion hypothesis of multiple PDE activation by single rhodopsin. Fed. Proc. **39:** 1815a.

(72) Robinson, W.E., and Hagins, W.A. 1979. GTP hydrolysis in intact rod outer segments and the transmitter cycle in visual excitation. Nature **280:** 398-400.

(73) Rüppel, H., and Hagins, W.A. 1973. Spatial origin of the fast photovoltage in retinal rods. In Biochemistry and Physiology of Visual Pigments, ed. H. Langer, pp. 257-262. New York: Springer.

(74) Schmidt, S.Y. 1983. Light enhances the turnover of phosphatidylinositol in rat retinas. J. Neurochem. **40:** 1630-1638.

(75) Schnapf, J. 1983. Dependence of the single photon response on longitudinal position of absorption in toad rod outer segments. J. Physiol. **343:** 147-159.

(76) Sitaramayya, A., and Liebman, P.A. 1983a. Mechanism of ATP-dependent quench of phosphodiesterase activation in rod disk membranes. J. Biol. Chem. **258:** 1205-1209.

(77) Sitaramayya, A., and Liebman, P.A. 1983b. Phosphorylation of rhodopsin and quenching of cGMP phosphodiesterase activation by ATP at weak bleaches. J. Biol. Chem. **258:** 12106-12109.

(78) Streb, H.; Irvine, R.F.; Berridge, M.J.; and Schulz, I. 1983. Release of Ca^{2+} from a non-mitochondrial store in pancreatic acinar cells by inositol-1,4,5-trisphosphate. Nature **306:** 67-69.

(79) Szabo, A.; Schulten, K.; and Schulten, Z. 1980. First passage time approach to diffusion-controlled reactions. J. Chem. Phys. **72:** 4350-4357.

(80) Wormington, C.M., and Cone, R.A. 1978. Ionic blockage of the light-regulated sodium channels in isolated rod outer segments. J. Gen. Physiol. **71:** 657-681.

(81) Yee, R., and Liebman, P.A. 1978. Light-activated phosphodiesterase of the rod outer segment: kinetic parameters of activation and deactivation. J. Biol. Chem. **253:** 8902-8909.

(82) Yoshikami, S., and Hagins, W.A. 1984. Phototransduction in rods does not require a change in cytoplasmic pH. Biophys. J. **45:** 339a.

(83) Yoshikami, S.; Robinson, W.E.; and Hagins, W.A. 1974. Topology of the outer segment membranes of retinal rods and cones revealed by a fluorescent probe. Science **185:** 1176-1179.

The Molecular Mechanism of Photoreception, ed. H. Stieve, pp. 159-169. Dahlem Konferenzen 1986. Berlin, Heidelberg, New York, Tokyo: Springer-Verlag.

Changes in Protein Phosphorylation and Nucleoside Triphosphates during Phototransduction - Physiological Correlates

M.D. Bownds and E. Brewer
Laboratory of Molecular Biology and Dept. of Zoology
University of Wisconsin
Madison, WI 53706, USA

Abstract. Illumination of vertebrate rods causes the phosphorylation of the visual pigment rhodopsin, the dephosphorylation of two small proteins, and can also change nucleoside triphosphate levels. Light-induced phosphorylation of invertebrate rhodopsin has also been observed. The physiological roles of these reactions are not understood. Vertebrate rhodopsin is phosphorylated at multiple sites nears its C-terminal end located at the surface of the disc membrane, and it seems likely that this influences its interaction with other proteins regulating the transduction pathway. Both the protein phosphorylation reactions and nucleoside triphosphate changes induced by illumination are slower than conductance changes, and they occur mainly at light levels higher than those required to saturate the conductance. This suggests that they may be more important in regulating adaptation and recovery processes than in triggering the initial conductance changes. After a bright flash rhodopsin phosphorylation occurs over the same time period as the recovery of conductance in the dark, and a GTP decrease is reversed as light sensitivity returns. The correspondence between the measured chemistry and physiology is not obligatory, however, because lowering calcium concentration in the medium accelerates the return of both conductance and sensitivity without influencing the kinetics of either rhodopsin phosphorylation or the GTP return.

INTRODUCTION

The assignment of this topic by the workshop organizing committee was based on the thought that experiments then in progress attempting to relate light-induced nucleoside triphosphate or protein phosphorylation changes to photoreceptor current, sensitivity, or noise might have proved fruitful by the time of the meeting. Unfortunately, none of the data obtained thus far suggests a unique physiological role for these reactions.

LIGHT-SENSITIVE PROTEIN PHOSPHORYLATION

Protein phosphorylation is known to regulate the activity of many enzymes and typically is studied in disrupted cells or purified extracts because the reactions can be easily assayed. The expression of the reactions in living cells is harder to appreciate. It is important that any model for the role of a protein phosphorylation change be tested by measurement of the stoichiometry and time course of the reaction either in vivo or in a physiologically viable system because regulatory cofactors and compartmentation may be present that are not measured in vitro. Stoichiometry, kinetics, and phosphorylation sites in the intact system may be different (cf. (14, 27, 33)). This article reviews the few studies on light-induced changes in the level of protein phosphorylation in intact visual cells.

Rhodopsin Phosphorylation

Light-induced phosphorylation of visual pigment has been shown to occur both in vertebrates (7, 22) and in invertebrates (35, 44). Photoexcitation of vertebrate rhodopsin makes several sites near the C-terminal of rhodopsin accessible to phosphorylation by a kinase that is not light-sensitive. While it is clear that bleached rhodopsin becomes the substrate for the kinase, several reports suggest further that at low light levels interaction between a bleached rhodopsin molecule and several of its unbleached neighbors can cause phosphorylation of the unbleached rhodopsin (7, 30, 42). This would represent a small amplification step.

Kühn and co-workers have measured the time course of rhodopsin phosphorylation in whole frogs (19) and in isolated retinas (21), finding that phosphorylation was slow ($t_{1/2} \sim 2$ min) compared with conductance changes, but that the slow rate of dephosphorylation ($t_{1/2} \sim 20$ min) was compatible with a role for the reaction in dark adaptation.

This sort of experiment has been refined by simultaneously monitoring rhodopsin phosphorylation and the physiology in purified suspensions of frog rod outer segments still attached to their mitochondria-rich inner segments (o.s.-i.s., cf. (5)). In the experiment shown in Fig. 1a, the current flowing into the outer segment, as measured by the suction electrode technique of Baylor et al. (2), is rapidly suppressed (in <200 ms) after a flash bleaching 2% of the rhodopsin present. The current half recovers in two minutes, and recovery of light sensitivity is slower ($t_{1/2} \sim 5$-6 min). (In the figure, light sensitivity is measured as the amplitude of the response to a flash of light that causes a half maximal response in the dark-adapted receptor. It is this response, taken as a percentage of the dark-adapted response, that is shown recovering with a half-time of 5-6 min.)

Rhodopsin phosphorylation is monitored by incubating the o.s.-i.s. suspension in $^{32}P_i$. The isotope is taken up and incorporated into the ATP used to phosphorylate rhodopsin. Portions of the suspension are quenched at varying times after flash illumination, and phosphorylation is measured by autoradiography of electrophoretic gels (15). The time course of rhodopsin phosphorylation does not closely match changes in either conductance or sensitivity (Fig. 1b). The possible correlation between phosphorylation and return of current observed in Ringer's solution containing 1 mmol/l calcium is not maintained when the experiment is repeated in a Ringer's solution containing 10^{-8} mmol/l calcium - this accelerates return of both current and sensitivity but appears to have little effect on the time course of

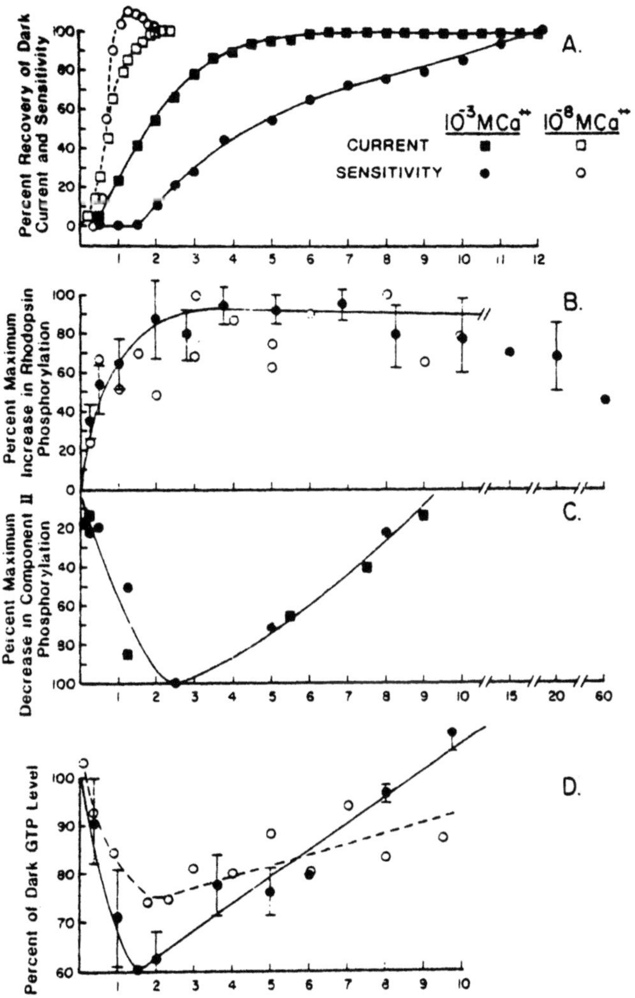

FIG. 1 - Light-initiated reactions in purified suspensions of frog outer segments still attached to inner segments (o.s.-i.s.). In each part of this figure the reaction plotted on the ordinate is expressed as a percentage of its dark-adapted value. A. Recovery of current (squares) and light sensitivity (circles) after a 2% rhodopsin bleach. Solid symbols: cells in 1 mmol/l calcium Ringer's solution (from (5)); open symbols: cells in 10^{-8}mol/l calcium (G.D. Nicol and M.D. Bownds, unpublished data). B. Time course of rhodopsin phosphorylation after a 2% bleach (Brewer et al., in preparation). Solid symbols: 1 mmol/l calcium; open symbols: 10^{-8}mol/l calcium. C. Dephosphorylation and rephosphorylation of components I and II after a 6% bleach in o.s.-i.s. suspensions subjected to slight hypoosmotic shock in 1 mmol/l calcium (Brewer et al., in preparation). D. Time course of GTP changes following a 2% rhodopsin bleach. Solid symbols: I mmol/l calcium (6); open symbols: 10^{-8}mol/l calcium (Biernbaum and Bownds, unpublished data).

rhodopsin phosphorylation. The data argue against a simple relationship between rhodopsin phosphorylation and sensitivity or conductance control. Even if rhodopsin phosphorylation plays a role at more elevated calcium levels, its function apparently can be bypassed by lowering calcium. It will be important eventually to consider possible correlation of rhodopsin phosphorylation or of the other reactions mentioned below not only with dark adaptation of the sort shown in Fig. 1, but also with adaptation to dim continuous background light (background adaptation).

One drawback to this approach of correlating changes in extent of rhodopsin phosphorylation with conductance or sensitivity changes is that we may not be able to measure the relevant phosphorylation/dephosphorylation events. For example, if only a few rhodopsin molecules need to be phosphorylated for this reaction to have its maximum effect, or if the rhodopsin phosphorylation occurring in a minor subcellular compartment is most important, then only a small fraction of the total reaction normally measured (possibly occurring rapidly and near or below limits of detectability) is the part we might expect to correlate with physiology. We also know that rhodopsin can be phosphorylated at up to nine separate sites (47), perhaps by several different kinases. It is possible that only certain of these sites are relevant, or that different sites have different regulatory effects. This is known to be the case with phosphorylated enzymes regulating glycogen metabolism (8).

Suggested roles for rhodopsin phosphorylation can be placed into two categories: those that assume that the reaction is slow (many seconds) and most likely involved in adaptation, and those that suggest that it is rapid and involved in shaping the waveform of the response to light flashes (<5 s). As candidates for regulation by rhodopsin phosphorylation we have the two known pathways that proceed directly from rhodopsin bleaching: interconversions of the retinaldehyde chromophore and cyclic GMP cascade. Also, there has been the suggestion of a link between rhodopsin phosphorylation and ion binding or flux (46). Miller and Paulsen (29) found no relationship between rhodopsin regeneration and phosphorylation-dephosphorylation in frog outer segments, but Paulsen and Bentrop (34) have recently correlated dephosphorylation of the invertebrate blowfly visual pigment with photoregeneration of rhodopsin from a metarhodopsin.

Interest has focused recently on potential roles for rhodopsin phosphorylation in the cyclic nucleotide pathway. Pfister et al. (36) have shown that binding of the G-protein to rhodopsin can inhibit the access of rhodopsin kinase to its substrate. One possibility is that phosphorylation of rhodopsin reduces its ability to bind to and activate G-protein on illumination. This in turn might result in reduced PDE activation and a smaller cyclic GMP decrease. If this could be observed, it would be consistent with rhodopsin phosphorylation contributing to the desensitization that occurs during adaptation to background illumination. Experiments by Aton and Litman (1), Shichi et al. (40), and Kühn (personal communication) found that phosphorylation of rhodopsin does not completely inhibit its ability to activate PDE. The experiments are difficult to interpret clearly, because the PDE activation that is observed might be due to trace amounts of unphosphorylated rhodopsin.

A similar idea derives from the observation of Liebman and Pugh (25) that addition of ATP causes a rapid (<1 s) quenching of PDE activation. They suggest that this effect derives from an ATP-dependent phosphorylation of opsin. This hypothesis requires that rhodopsin phosphorylation be more light-sensitive and more rapid than previously reported, because the ATP quench of PDE can be observed after only 10^2-10^3 rhodopsin molecules are bleached per outer segment. Sitaramayya and Liebman (42) have suggested that this may be the case, but experiments carefully matching the kinetics of rhodopsin phosphorylation and the ATP quench of PDE have not yet been done. Another potential mediator of the ATP-dependent quench of PDE has emerged recently, a 48-50 kD protein that has ATP-binding activity and whose binding to the disk membrane is light-dependent (20, 48).

A further idea is that rhodopsin phosphorylation has a role in the regulation of "dark noise," the increased frequency of quantum bump-like events that is observed in the dark after strong illumination (16, 23, 26, 43). One explicit model, suggested by J. Lisman (26), is that phosphorylated rhodopsin intermediates are less likely to revert to the excitatory intermediate that triggers the transduction chain. Successive phosphorylations of the same rhodopsin molecule would serve to make this shutoff reaction even more reliable, since the back reaction would now have to involve multiple dephosphorylation steps. With such a scheme, one might expect that lowering rhodopsin phosphorylation by depleting ATP levels might cause an increase in membrane noise following illumination. Stern et al. (43) have observed that internal dialysis of Limulus ventral photoreceptors with solutions lacking nucleoside triphosphates causes an elevation of quantum bump rate, but rhodopsin phosphorylation has not yet been measured under these conditions.

Cyclic Nucleotide-dependent Protein Phosphorylations

Because rhodopsin and the enzymes that it activates in the cyclic GMP cascade make up at least 90% of the protein of the vertebrate rod outer segment, it is commonly assumed that the pathway must be important in conductance or sensitivity regulation. In apparent accord with the dogma that cyclic nucleotides act as regulators of protein kinases that in turn control protein phosphorylation (31), another protein phosphorylation in search of a function has been found: a light-induced dephosphorylation of two small proteins (named components I and I, m.w. 13 and 23 kD, respectively (37)). The phosphorylation of the proteins in disrupted outer segments can be regulated by cyclic GMP (15), even though no direct evidence for a cyclic GMP-dependent kinase has been found (45). One speculation has been that the light-induced decrease in cyclic GMP that occurs in intact outer segments causes the observed dephosphorylation of these proteins, by either lowering kinase or stimulating phosphatase activity. This dephosphorylation, if it were as rapid as the cyclic GMP decrease, might be regulating a conductance decrease.

We have attempted to determine whether the time course of the light-induced dephosphorylation of the 12K protein correlates with any electrophysiological responses to illumination. The time course of the reaction in a suspension of o.s.-i.s. is shown in Fig. 1c. The dephosphorylation is much slower than the light-induced decrease in conductance and cyclic GMP levels. It is also much less sensitive to light, because at least 1% of the rhodopsin must be bleached to obtain a reliable signal. Thus, there is no obvious link between the light-induced dephosphorylation of components I and II and the measured physiology.

Why is relatively high illumination required to observe component I and II dephosphorylation? Perhaps higher rhodopsin bleaches cause some structural changes required for the cyclic GMP to have access to the kinase/ phosphatase enzymes regulating the phosphorylation or dephosphorylation of components I and II. Support for the idea of several compartments comes from the observation that component I and II phosphorylation is enhanced when cyclic GMP levels are elevated by the phosphodiesterase inhibitor IBMX (37). If, on the other hand, cyclic GMP levels are elevated by lowering external calcium concentration, no corresponding increase in component I and II phosphorylation is observed (Brewer et al., in preparation). Evidence for compartmentalization of cyclic GMP effects in heart muscle has recently been summarized by Walter (45).

The situation which has emerged with these reactions highlights a general problem in studying physiologically relevant protein phosphorylations. In other membrane systems that have been studied (27), protein phosphorylations that are cryptic in vivo are found upon disruption. Conversely, some reactions observed in the living cell cannot be reproduced in vitro (33). In the first case, one supposes that reactants that normally do not have access to each other are mixed, whreas in the second, either dilution or the loss of a structural requirement may have occurred. We have found that the phosphorylation pattern of proteins in the molecular weight range of components I and II is different in disrupted versus intact o.s.-i.s. (Brewer et al., in preparation).

Recent work on components I and II has shown that their cyclic GMP-dependent phosphorylation is inhibited by a monoclonal antibody that blocks the activity of the G-protein (13). This suggests that these proteins may be regulatory components in the cyclic GMP pathway.

Several further light-sensitive phosphorylations have been reported. Light causes the dephosphorylation of a 33 kD protein (24), which both Schuster and Farber (41) and Devries and Ferrendelli (10) report to be in the inner segment or more proximal portion of the retina and thus not involved in the primary events in phototransduction. Kapoor and Chader (17) suggest that protein kinase C may regulate several protein phosphorylations in cattle rod outer segments. Matsumoto and Pak (28) have recently shown light-induced phosphorylation of retina-specific polypeptides in Drosophila, and Paulsen (personal communication) has noted several light-sensitive phosphorylations in blowfly retinas.

LIGHT-SENSITIVE NUCLEOSIDE TRIPHOSPHATE CHANGES

Illumination of isolated outer segments causes a decrease in their GTP content (4, 38), and in o.s.-i.s. suspensions both GTP and ATP decreases can be observed (5, 6). Corresponding observations have been made in measurements on freeze-sectioned retinas (3, 9).

The GTP and ATP decreases, measured in o.s.-i.s. suspensions, are slower than the light-induced conductance decrease and also less sensitive to light. Figure 1d, taken from Biernbaum and Bownds (6), shows that the time course of GTP recovery after a 2% rhodopsin bleach approximately follows the recovery of light sensitivity in Ringer's solution containing 1 mmol/l calcium. This apparent correlation, however, is not maintained in a low calcium buffer. Incubation of the o.s.-i.s. suspensions in 10^{-8} mol/l calcium Ringer's solution enhances recovery of sensitivity but delays GTP recovery. It is difficult to give a mechanistic interpretation to such data, because GTP participates in numerous outer segment reactions, and also because actual flux of GTP - its synthesis and degradation - is not being measured. If the main use of GTP is as a substrate for guanylate cyclase and the GTPase activity of the G-protein (cf. (11, 12)), then GTP levels may reflect in a complex way the activity of the cyclic GMP pathway. Another interpretation of the ATP and GTP decreases is that they reflect utilization of high energy phosphate by ion transport process (cf. (4, 38)).

CONCLUSIONS

In spite of tantalizing speculation about the involvement of protein phosphorylation in signal amplification (32, 39) modulating receptor activation and adaptation (18), and definitive statements in review articles about the function of rhodopsin phosphorylation and other protein phosphorylation in photoreceptors (31), we do not know what these reactions are doing. Experiments measuring light-initiated physiology and biochemistry in chemically defined living photoreceptor cells are still at a primitive stage. Even given the problems outlined above, they still must be pursued if one is critically to test models suggested by studies on dissociated and reconstituted protein components. One of the most pressing needs is for techniques that would permit definition of relevant intracellular compartments and allow measurements of the chemical changes occurring in each.

Acknowledgements. This work was supported by NIH grants EY 00463 and ST-32-GM -07507.

REFERENCES

(1) Aton, B., and Litman, B.J. 1984. Activation of rod outer segment phosphodiesterase by enzymatically altered rhodopsin: A regulatory role for the carboxyl terminus of rhodopsin. Exp. Eye Res. **38:** 547-559.

(2) Baylor, D.A.; Lamb, T.D.; and Yau, K.-W. 1979. The membrane current of single rod outer segments. J. Physiol. **288:** 589-611.

(3) Berger, S.J.; Devries, G.W.; Carter, J.G.; Schulz, D.W.; Passonneau, P.N.; Lowry, O.H.; and Ferrendelli, J.A. 1980. The distribution of the components of the cyclic GMP cycle in retina. J. Biol. Chem. **255:** 3128-3133.

(4) Biernbaum, M.S., and Bownds, M.D. 1979. Influence of light and calcium on guanosine 5'-triphosphate in isolated frog rod outer segments. J. Gen. Physiol. **74:** 649-669.

(5) Biernbaum, M.S., and Bownds, M.D. 1985. Frog rod outer segments with attached inner segment ellipsoids as an in vitro model for photoreceptors on the retina. J. Gen. Physiol. **85:** 83-105.

(6) Biernbaum, M.D., and Bownds, M.S. 1985. Light-induced changes in GTP and ATP in frog rod photoreceptors: Comparison with recovery of dark current and light sensitivity during dark adaptation. J. Gen. Physiol. **85:** 107-121.

(7) Bownds, M.D.; Dawes, J.; Miller, J.; and Stahlman, M. 1972. Phosphorylation of frog photoreceptor membranes induced by light. Nature **237:** 125-127.

(8) Cohen, P. 1982. The role of protein phosphorylation in neural and hormonal control of cellular activity. Nature **296:** 613-620.

(9) de Azeredo, F.A.M.; Lust, W.D.; and Passonneau, J.V. 1981. Light-induced changes in energy metabolites, guanine nucleotides, and guanylate cyclase within frog retinal layers. J. Biol. Chem. **256:** 2731-2735.

(10) De Vries, G.W., and Ferrendelli, J.A. 1983. Localization of an endogenous substrate for cyclic AMP-stimulated protein phosphorylation in retina. Exp. Eye Res. **36:** 505-516.

(11) Fung, B.K.-K.; Hurley, J.B.; and Stryer, L. 1981. Flow of information in the light-triggered cyclic nucleotide cascade of vision. Proc. Natl. Acad. Sci. USA **78:** 152-156.

(12) Goldberg, N.D.; Ames, A. III; Gander, J.E.; and Walseth, T.F. 1983. Magnitude of increase in retinal cGMP metabolic flux determined by 18O incorporation into nucleotide alpha-phosphoryls corresponds with intensity of photic stimulation. J. Biol. Chem. **258:** 9213-9219.

(13) Hamm, H.E., and Bownds, M.D. 1984. A monoclonal antibody to guanine nucleotide binding protein inhibits the light-activated cyclic GMP pathway in frog rod outer segments. J. Gen. Physiol. **84:** 265-280.

(14) Harris, H.W.; Levin, N.; and Lux, S.E. 1980. Comparison of the phosphorylation of human erythrocyte spectrin in the intact red cell and in various cell-free systems. J. Biol. Chem. **255:** 11521-11525.

(15) Hermolin, J.H.; Karell, M.A.; Hamm, H.E.; and Bownds, M.D. 1982. Calcium and cyclic GMP regulation of light-sensitive protein phosphorylation in frog photoreceptor membranes. J. Gen. Physiol. **79:** 633-655.

(16) Hillman, P.; Hochstein, S.; and Minke, B. 1983. Transduction in invertebrate photoreceptors: Role of pigment bistability. Physiol. Rev. **63:** 668-772.

(17) Kapoor, C.L., and Chader, G.J. 1984. Endogenous phosphorylation of retinal photoreceptor outer segment proteins by calcium phospholid-dependent protein kinase. Biochem. Biophys. Res. Comm. **3:** 1397-1403.

(18) Koshland, D.E.; Goldbeter, A.; and Stock, J.B. 1982. Amplification and adaptation in regulatory and sensory systems. Science **217:** 220-225.

(19) Kühn, H. 1974. Light-dependent phosphorylation of rhodopsin in living frogs. Nature **230:** 588-590.

(20) Kühn, H. 1981. Interactions of rod cell proteins with the disk membrane: Influences of light, ionic strength, and nucleotides. In Molecular Mechanisms of Photoreceptor Transduction, Current Topics in Membranes and Transport, ed. W.H. Miller, vol. 15, pp. 172-201. New York: Academic Press.

(21) Kühn, H., and Bader, S. 1976. The rate of rhodopsin phosphorylation in isolated retinas of frog and cattle. Biochim. Biophys. Acta **428:** 13-18.

(22) Kühn, H., and Dreyer, W.J. 1972. Light dependent phosphorylation of rhodopsin by ATP. FEBS Lett. **20:** 1-6.

(23) Lamb, T.D. 1980. Spontaneous quantal events induced in toad rods by pigment bleaching. Nature **287:** 349-351.

(24) Lee, R.H.; Brown, B.M.; and Lolley, R.N. 1984. Light-induced dephosphorylation of a 33K protein in rod outer segments of rat retina. Biochemistry **23:** 1972-1977.

(25) Liebman, P.A., and Pugh, E.N. 1980. ATP mediates rapid reversal of cyclic GMP phosphodiesterase activation in visual receptor membranes. Nature **287:** 734-736.

(26) Lisman, J. 1985. The role of metarhodopsin in the generation of quantum bumps in UV-receptors in Limulus media eye: Evidence for reverse-reactions into an active state. J. Gen. Physiol., in press.

(27) Manning, D.R.; Disalvo, J.; and Stull, J.T. 1980. Protein phosphorylation: quantitative analysis in vivo and in intact cell systems. Molec. Cell Endocrin. **19:** 1-19.

(28) Matsumoto, H., and Pak, W.L. 1984. Light-induced phosphorylation of retina-specific polypeptides of Drosophila in vivo. Science **223:** 184-186.

(29) Miller, J.A., and Paulsen, R. 1975. Phosphorylation and dephosphorylation of frog rod outer segment membranes as part of the visual process. J. Biol. Chem. **250:** 4427-4432.

(30) Miller, J.A.; Paulsen, R.; and Bownds, M.D. 1977. Control of light-activated phosphorylation in frog photoreceptor membranes. Biochemistry **16:** 2633-2639.

(31) Nestler, E.J., and Greengard, P. 1983. Protein phosphorylation in the brain. Nature **305:** 583-588.

(32) Newsholme, E.A.; Challis, R.A.J.; and Crabtree, B. 1984. Substrate cycles: their role in improving sensitivity in metabolic control. Trends Biochem. Sci. **9:** 277-280.

(33) Owens, C.O., and Ohad, I. 1982. Phosphorylation of Chlamydomonas reinhardi chloroplast membrane proteins in vivo and in vitro. J. Cell Biol. **93:** 712-718.

(34) Paulsen, R., and Bentrop, J. 1984. Reversible phosphorylation of opsin induced by irradiation of blowfly retinae. J. Comp. Physiol. **155:** 39-46.

(35) Paulsen, R., and Hoppe, I. 1978. Light-activated phosphorylation of cephalopod rhodopsin. FEBS Lett. **96:** 55-58.

(36) Pfister, C.; Kühn, H.; and Chabre, M. 1983. Interaction between photoexcited rhodopsin and peripheral enzymes in frog retinal rods. Influences on the postmetarhodopsin II decay and phosphorylation rate of rhodopsin. Eur. J. Biochem. **15:** 489-499.

(37) Polans, A.S.; Hermolin, J.; and Bownds, M.D. 1979. Light-induced dephosphorylation of two proteins in frog rod outer segments. J. Gen. Physiol. **74:** 595-613.

(38) Robinson, W.E., and Hagins, W.A. 1979. GTP hydrolysis in intact rod outer segments and the transmitter cycle in visual excitation. Nature **280:** 398-400.

(39) Shacter-Norman, E.; Chock, P.B.; and Stadtman, E.R. 1983. Protein phosphorylation as a regulatory device. Phil. Trans. Roy. Soc. Lond. B **302:** 157-166.

(40) Shichi, H.; Yamamoto, K.; and Somers, R.L. 1984. GTP binding protein: Properties and lack of activation by phosphorylated rhodopsin. Vision Res. **24:** 1523-1531.

(41) Shuster, T.A., and Farber, D.B. 1984. Phosphorylation in sealed rod outer segments: effects of cyclic nucleotides. Biochemistry **23:** 515-521.

(42) Sitaramayya, A., and Liebman, P.A. 1983. Phosphorylation of rhodopsin and quenching of cyclic GMP phosphodiesterase. Activation by ATP and weak bleaches. J. Biol. Chem. **258:** 12106-12109.

(43) Stern, J.; Chinn, K.; Robinson, P.; and Lisman, J. 1985. The effect of nucleotides on the rate of spontaneous quantum bumps in Limulus ventral photoreceptors. J. Gen. Physiol., in press.

(44) Vandenberg, C.A., and Montal, M. 1984. Light-regulated events in invertebrate photoreceptors. 2. Light-regulated phosphorylation of rhodopsin and phosphoinositides in squid photoreceptor membranes. Biochemistry **23:** 2347-2352.

(45) Walter, U. 1984. Cyclic-GMP-regulated enzymes and their possible physiological functions. **In** Advances in Cyclic Nucleotide and Protein Phosphorylation Research, ed. P. Greengard et al., vol. 17, pp. 249-257. New York: Raven Press.

(46) Weller, M.; Virmaux, N.; and Mandel, P. 1975. Role of light and rhodopsin phosphorylation in control of permeability of retinal rod outer segment discs to Ca^{2+}. Nature **256:** 68-70.

(47) Wilden, U., and Kühn, H. 1982. Light-dependent phosphorylation of rhodopsin: number of phosphorylation sites. Biochemistry **21:** 3014-3022.

(48) Zuckerman, R.; Buzdygon, B.; and Liebman, P. 1984. Characterization of the 48 kilodalton protein of retinal rod outer segments as a light-dependent ATP binding protein. Inv. Opthalmol. Vis. Sci. **25:** 112.

The Molecular Mechanism of Photoreception, ed. H. Stieve, pp. 171-187. Dahlem Konferenzen 1986. Berlin, Heidelberg, New York, Tokyo: Springer-Verlag.

The Light-induced Conductance Change in the Vertebrate Rod

W.G. Owen
Dept. of Biophysics and Medical Physics
University of California, Berkeley, CA 94720, USA

Abstract. The molecular structure of the light-modulated channel in the vertebrate rod is unknown and we remain far from being able even to postulate what it might be. By discussing what is presently known about the electrical properties and ion selectivity of the channel and what they might imply for channel structure, this paper attempts to narrow the range of possibilities.

Several factors suggest that the structure of the light-modulated channel must be quite different from that of commonly studied ionic pores such as the Na^+ channels of nerve membranes. Thus, although its ion selectivity falls in the same sequence as that of the Na^+ channel in nerve, it is only poorly selective for Na^+ over K^+. Electrical measurements have revealed little evidence of any intrinsic, voltage-dependent channel gating. Moreover, the single-channel conductance, estimated on the basis of patch-clamp measurements of membrane noise, appears to be two orders of magnitude lower than that of the nerve Na^+ channel, and estimated single-channel ion fluxes are similarly very small.

Of course, ions can also cross biological membranes via membrane-bound carrier molecules. The properties of such molecules are discussed in some detail and compared with those of the light-modulated channel. It is concluded that if the light-modulated channel is an ion carrier, it is probably not a mobile carrier like Valinomycin. A hybrid structure, having some of the features of a pore, seems more likely.

INTRODUCTION

Intracellular recordings of light-evoked voltage responses of vertebrate rods were first achieved twenty years ago (7, 38). The response proved to be a graded hyperpolarization of the rod's plasma membrane which was later shown to be accompanied by a reduction in input conductance (40). A study of the aspartate-isolated late receptor potential of the frog retina (37) established that the voltage response is highly sensitive to the external concentrations of Na^+ and K^+. Because of this, it was argued that the response must be initiated by a reduction in the membrane's conductance to Na^+.

The elegant work of Hagins et al. (23) added greatly to this picture. Analysis of the radial distribution of current flowing through the extracellular space between rods in the rat retina revealed the existence of a standing current which, in darkness, enters the rod across the plasma membrane of the outer segment and is actively extruded from the inner segment by a metabolically driven pump. This "dark current" appeared to be carried by Na ions. The effect of light was to reduce the dark current by changing the permeability of the outer segment membrane. Thus, it became clear that the rod cell was not only morphologically differentiated but functionally differentiated as well, both the absorption of photons and the primary light-induced conductance change taking place in the outer segment.

The central role of Na ions in generating the light response was further supported by experiments of Brown and Pinto (10), who monitored the transmembrane potential of rods in the isolated, perfused retina of the toad. Replacing external Na^+ with Li^+, choline, or sucrose rapidly and reversibly hyperpolarized the rod and abolished the light response as would be expected if the dark current were carried by sodium. Indeed, osmotic experiments (28) seemed to indicate that the light-modulated channel preferred Na^+ over K^+ by a ratio of 100:1 and was strongly rectified in an inward direction.

During the subsequent decade, considerable research has been directed towards understanding the nature of this light-induced conductance change. Of particular importance was the finding that the voltage response is in fact shaped by two classes of conductance change: the primary change which modulates the dark current and secondary changes which result from the change in membrane potential. As will be discussed later, it is now known that the primary light-modulated conductance is itself voltage-dependent, but for simplicity I will draw a distinction between this conductance and secondary conductances which are not initiated by the direct action of light. A recent observation by Baylor and Lamb (3) that intense illumination reduces the conductance of the outer segment plasma membrane to a value expected for a lipid membrane of the same surface area, indicates that the only significant outer segment conductance is that which is modulated by light. The further implication is that all secondary, voltage-modulated conductance changes occur in the membrane of the inner segment of the cell. This is convenient since it permits the investigator to study these two classes of conductances independently by restricting measurements to the appropriate part of the cell.

In this paper I will take a narrow view of light-induced conductance changes and focus upon the primary conductance change that occurs in the outer segment of the vertebrate rod. Secondary changes due to the voltage-dependence of ionic conductances in the inner segment will not be discussed. Recent reviews which describe these voltage-dependent conductance changes include those Fain and Lisman (18) and Owen and Torre (32). Instead, my intention is to review the most recent data concerning the properties of the light-modulated conductance and to speculate upon the physical nature of the mechanism by which ions cross the outer segment plasma membrane.

The term "channel" will be used throughout the paper when referring to the pathway by which the light-modulated current crosses the ROS membrane. It should not be regarded as implying any specific molecular structure since we do not yet know whether the current crosses through an aqueous pore or via a specialized carrier molecule.

ION SELECTIVITY OF THE LIGHT-MODULATED CHANNEL

The idea that the dark current is carried by sodium ions through an inwardly rectifying channel was not seriously challenged until recently, though there were results in the literature that were not fully consistent with it. For example, Cavaggioni et al. (12), by preloading isolated rod outer segments with radioactive tracers, demonstrated a light-modulated efflux of K^+ and Rb^+ suggesting that a part of the dark current might normally be carried by K^+. This possibility later gained support from the results of Bader et al. (2), who voltage-clamped rods that had been enzymatically dissociated from the salamander retina. They measured the reversal potential of the light-modulated channel (see following section) and found it to be between 0 mV and +10 mV. The internal concentration of K^+ in toad rods is about 94 mmol/l (32), while that of Na^+ is no more than 30 mmol/l (39). Assuming similar values in salamander rods and that the concentrations and/or permeabilities of other ions are low, the relative permeability of Na^+ and K^+ through the light-modulated channel can be estimated from the reversal potential. Given the external concentrations used by Bader et al. (2), a reversal potential of +5 mV implies that P_K/P_{Na} should be at least 0.5. At a resting potential of -40 mV, therefore, a dark current of 30 pA might be expected to consist of an inward Na^+ component of ~47 pA, offset by an outward K^+ component of ~17 pA.

The finding that, by reducing external Ca^{2+} to micromolar concentrations or less, a large and reversible increase in the dark current could be produced (10) paved the way for a new approach to studies of ion selectivity since it seemed clear that this would offset the effect of reducing the external Na^+ concentration in ion substitution experiments. Of course, the extent to which this might affect the selectivity of the channel remained unclear, and investigators have been at pains to restrict the interpretation of their results to the specific conditions under which they were obtained. Given this, however, there is substantial agreement amongst the results obtained in low Ca^{2+} solutions using various techniques of measurement.

With the aid of radioactive tracers, Woodruff et al. (42) found that in the presence of 10^{-8} mol/l Ca^{2+} there was a significant light-dependent influx of Na^+, K^+, Rb^+, Cs^+, and Tl^+.

Yau et al. (44), monitoring transmembrane current with a suction pipette, found that in 0 Ca^{2+} + EGTA the channel was permeable to Na^+, Li^+, K^+, Rb^+, and Cs^+ and the divalent cations Ca^{2+} and Mg^{2+}.

Conventional intracellular voltage recordings made by Capovilla et al. (11) revealed that in a medium containing nanomolar quantities of Ca^{2+} and Mg^{2+}, the channel was permeable to Na^+, Li^+, Rb^+, and K^+. Upon raising the external concentration of Mg^{2+} they found it to be permeable to Mg^{2+} as well. Adding 0.5 mmol/l IBMX (or higher concentrations of other phosphodiesterase inhibitors) to a medium containing nanomolar concentrations of Na^+, Mg^+, and Ca^{2+}, they were able to demonstrate that other divalent cations were permeable in the sequence $Mn^{2+} > Ba^{2+} > (Ca^{2+}, Co^{2+}, Mg^{2+}$, and $Sr^{2+})$ which, they pointed out, is the activation sequence for guanylate cyclase in vertebrate rods.

It might appear on this evidence, therefore, that the light-modulated channel has several different conductance states; selective primarily for Na^+ under normal conditions, permeable to Li^+, Na^+, Rb^+, Cs^+, Tl^+, Ca^{2+}, and Mg^{2+} in submicromolar concentrations of external calcium and also to Mn^{2+}, Ba^{2+}, Co^{2+}, and Sr^{2+} in the presence of PDE inhibitors. Multiple conductance states were proposed by Capovilla et al. (11) to explain their findings. More recently, however, it has become clear that the interpretation of ion substitution experiments is complicated by the fact that they may affect a mechanism in the ROS plasma membrane distinct from the light-modulated conductance; an Na^+/Ca^{2+} exchanger similar to that described in epithelial cells, muscle cells, and nerve cells (see DiPolo and Beauge (16) for a recent review of the latter).

Na^+/Ca^{2+} exchange across a photoreceptor membrane was first implicated by Lisman and Brown (30) in their study of the Limulus lateral eye and later confirmed by Waloga et al. (41), using the metallochromic dye, arsenazo III. More recently, Gold and Korenbrot (21, 22) showed that following a light flash there is a significant efflux of Ca from the ROS and that when external Na was reduced from 27 mmol/l to 7 mmol/l, the rate of this efflux was reduced. Schnetkamp (35) demonstrated an efflux of $^{45}Ca^{2+}$ from preloaded rod outer segments which was stimulated by external Na^+, Ca^{2+}, and Sr^{2+} but not by other cations, suggesting the existence in the ROS plasma membrane of Na^+/Ca^{2+} exchange and possibly Ca^{2+}/Ca^{2+} or Ca^{2+}/Sr^{2+} exchange as well. Fain and Lisman (18) pointed out that Na^+/Ca^{2+} exchange is likely to be highly selective for external Na^+. In the squid axon, for example, Li^+ will not drive Na^+/Ca^{2+} exchange, although Na^+ and Li^+ are about equally permeable through the Na^+ channel in that preparation. Bastian and Fain (42), noting that the effect of substituting Li^+ for Na^+ on the membrane potential of toad rods was highly dependent upon the external concentration of Ca^{2+}, argued that the light-modulated conductance is normally permeable to Li^+ but, because it substitutes poorly in Na^+/Ca^{2+} exchange, replacing Na with Li in the external medium reduces the rate of extrusion of Ca^{2+} from the cytosol. The resulting accumulation of Ca^{2+} inside the ROS causes a reduction in the light-modulated conductance, accounting both for the observed hyperpolarization of the rod and for the virtual elimination of the light response. Fain and Lisman (18) pointed out that this mechanism could account for the apparent high Na^+ selectivity and inward rectification of the light-modulated conductance revealed by the experiments of Korenbrot and Cone (28).

In a recent study Hodgkin et al. (25), again monitoring transmembrane current with a suction pipette, showed that many of the effects of altering external concentrations of Na^+ and Ca^{2+} on both the dark current and the light response of the rod could be interpreted in terms of an Na^+/Ca^{2+} exchange in which two $[Na]_o$ ions are exchanged for each $[Ca]_i$ ion and Ca ions block the light-modulated channel at an internal site. They used a protocol called "priming" to activate the light-modulated conductance. The rod was first exposed to 0 Na, 1 µmol/l Ca, 1.6 mmol/l Mg which abolished the dark current. A brief exposure to 20 mmol/l Na produced a light-sensitive inward current which rapidly disappeared on removal of Na. When this exposure was repeated in the presence of a bright light, no light-sensitive current developed, but on removal of the Na, light evoked an inward current carried by Mg^{2+}. This observation was consistent with the idea that external Na activates an Na^+/Ca^{2+} exchange and that Ca normally leaks into the outer segment through the light-modulated channel. By using light to block the channel while activating the exchanger, the internal Ca concentration was sufficiently reduced that channels remained open upon return to darkness and 0 Na, revealing the Mg-permeability of the channel. Brief exposures to Li did not "prime" the light-modulated conductance, confirming that Li is a poor substitute for Na in Na^+/Ca^{2+} exchange.

In the same study, a rapid perfusion technique was used to show that whereas removing external Na produces a near instantaneous reduction in the dark current, recovery of the dark current occurs only after a delay of one to three seconds upon restoring normal Na, this being the time required to reactivate Na^+/Ca^{2+} exchange and deplete internal Ca. Thus, if the effects of ion substitution on the Na^+/Ca^{2+} exchange are not to interfere with determination of the ion selectivity of the light-modulated channel, measurements must be accomplished within rather less than one second of the solution change.

This has now been accomplished by Yau and Nakatani (45) and by Hodgkin et al. (in preparation), with generally similar results. In the experiments of Yau and Nakatani, the bathing medium around the ROS could be completely exchanged within 200 ms while transmembrane current was monitored using a suction pipette. Upon substituting all Na^+ with Li^+, they observed no immediate change in the dark current. Within ~3 s, however, the dark current fell to a very low level as Na^+/Ca^{2+} exchange was inhibited. They were also able to show that under normal physiological conditions the light-modulated channel is permeable to monovalent cations in the sequence Li > Na > K > Rb > Cs, the relative permeabilities being estimated to be ~1 : 1 : 0.7 : ~0.4 : ~0.2. Note that the high relative permeability to K^+ agrees well with the estimate made earlier on the basis of the reversal potential of the dark current. The channel was also shown to be permeable to the divalent cations Ca, Sr, and Ba. Significantly, similar results were found in low Ca^{2+} solutions.

It thus appears that the light-modulated channel has the same selectivity sequence as the Na channel in nerve but is much less selective. Moreover, it seems likely that reducing external Ca^{2+} may have little or no effect on selectivity. The simplest explanation of the effect of reducing external Ca^{2+} is that, by eliminating the concentration gradient carrying Ca^{2+} into the cell through the light-modulated channel, the rise in internal free Ca^{2+} that would normally occur upon inhibition of the Na^+/Ca^{2+} exchange is eliminated and the light-modulated conductance thereby remains unchanged. The question of whether the channel can exist in more than one conductance state must remain an open one, however, since many of the effects of phosphodiesterase inhibitors do not conform with this simple picture (11).

ELECTRICAL PROPERTIES OF THE LIGHT-MODULATED CHANNEL

By voltage-clamping rods that had been enzymatically dissociated from the salamander retina, Bader, MacLeish, and Schwartz (2) showed that the light-modulated channel has a reversal potential near +7 mV and behaves as an outward rectifier (see Fig. 1).

Depolarizing the rod from -50 mV to +30 mV produced a 30-fold increase in slope conductance. Thus, the rectification was not only significant but in the opposite direction from the expected constant field rectification. There are two possible interpretations of this finding. First, the light-modulated conductance might be nonlinearly dependent upon voltage. This possibility is discussed further in the section on CHANNEL STRUCTURE. Second, the channels themselves might be Ohmic while their mean lifetime is a nonlinear function of voltage, there being a precedent for this in the light-modulated channels of the photoreceptors in the barnacle, Balanus (9). In that case, the instantaneous I-V curve was linear but relaxed, with a time constant of about 15 ms, to a nonlinear form not unlike that measured by Bader et al. (2). This presumably reflects the time dependence of the channel's voltage-controlled gating machinery. In the rod outer segment, Bader et al. (2) found the rectification to be instantaneous within the time resolution of their voltage clamp (6 ms). This does not rule out the possibility that channel lifetime is voltage-dependent, but it sets a requirement that the time constant of the relaxation of the I-V relation be no more than 2 ms or so (however, see the section on NOISE ASSOCIATED WITH THE LIGHT-MODULATED CHANNEL).

MacLeish, Schwartz, and Tachibana (31) also measured the reversal potential of the light-modulated channel. They found in all cases that it lay between -10 mV and 0 mV, the difference between this value and that measured by Bader et al. (2) being due to a lower external concentration of Na^+ (75 mmol/l) in these experiments. Of particular interest was their finding that the reversal potential was unaffected by changes in the

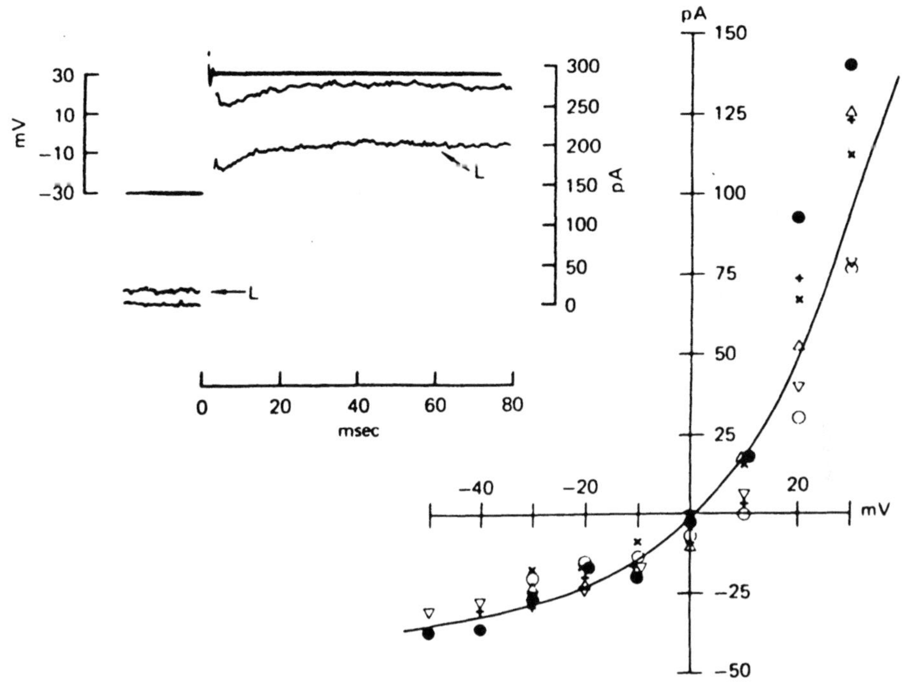

FIG. 1 - Voltage- and time-dependence of the I-V relation of the light- modulated channel (from (2)). The inset shows typical voltage-clamp records. The voltage was stepped from a holding potential of -30 mV to a more positive value, first in darkness and then during exposure to a steady bright light (trace L). The current obtained by subtracting these two measured currents, 6 ms after the beginning of the voltage step, is plotted against clamp potential to obtain the I-V relation of the light-modulated channel. No time-dependent behavior was evident in these measurements (data from six cells). (Reprinted from (2) with permission of The Physiological Society.)

external concentration of Ca^{2+}. This implies that the dark current is carried by a single type of channel whose selectivity does not depend upon external Ca^{2+}, in agreement with the conclusions drawn from the most recent studies of the ion selectivity of the light-modulated channel (see preceding section).

The experiments of MacLeish et al. (31) differed from those of Bader et al. (2) in that for the most part they were carried out in the presence of a cocktail of drugs, which included 3 mmol/l Co^{2+} to block voltage-dependent conductance changes in the inner segment of the rod. In the earlier study, D-600 had been used in place of Co^{2+}. MacLeish et al. (31) found that Co^{2+} changed the shape of the I-V relation, reducing inward currents but not outward currents. The size of this reduction increased with increasing negativity of the transmembrane voltage, producing a negative slope at voltages more negative than -40 mV. In addition, during depolarizing voltage steps a slow decline in the current with a time constant of more than a second was observed. Reducing external Ca^{2+} or injecting EGTA increased this time-varying behavior, while injecting Ca^{2+} reduced it. In the earlier experiments of Bader et al. (2), all experimental solutions contained 3 mmol/l Ca^{2+} and no such time-varying behavior was seen.

In view of a pronounced variability in its time course and because of its slowness, MacLeish et al. argued that this apparent inactivation of the light-modulated channel is best explained not by a direct effect of voltage on channel structure but by an indirect effect mediated by some cytoplasmic factor. Specifically they propose that Ca enters the ROS through the light-modulated channel, internal Ca blocks these channels, and the ability of Ca to block is regulated by cGMP, perhaps by causing a change in channel structure. The inactivation, in this case, would reflect the time course of a change in the internal calcium concentration that follows depolarization. Such a change would occur if the Na^{+}/Ca^{2+} exchange is voltage-dependent as in cardiac muscle (33).

The effect of Co^{2+}, they propose, is due to an ability of that ion to block the light-modulated channel from the outside, the blocking efficiency increasing as membrane hyperpolarization increases. Thus it may be possible to block the light-modulated channel at its inner end with Ca^{2+} and its outer end with Co^{2+}.

It is well-known that fixed charges on the surface of the membrane can profoundly affect the measured current-voltage relation, changing its shape and/or position on the voltage axis. Such charges on the outer face of the membrane can be screened by externally applied divalent ions such as calcium (5, 19). In rod outer segments, however, reducing external Ca^{2+} had little effect upon the measured current-voltage relation. MacLeish, Schwartz, and Tachibana (31) found that even with large reductions in $[Ca^{2+}]_o$, a lateral shift of the I-V curve of the kind found by Frankenhäuser and Hodgkin (19) in squid axon and shown to be due to a reduction in the screening by Ca^{2+} of surface charge did not occur. A minor change in the shape of the I-V curve was confined to potentials negative with respect to the reversal potential and was interpreted in terms of an indirect effect on channel structure mediated by cGMP.

NOISE ASSOCIATED WITH THE LIGHT-MODULATED CHANNEL

Baylor, Matthews, and Yau (4), in their study of current noise recorded with a suction pipette, found it to consist of two components: a discrete component which could be ascribed to the random thermal bleaching of rhodopsin molecules and a continuous component due to the random opening and closing of the light-modulated channels in darkness. The power spectrum of the continuous component was fitted by the product of two Lorentzians. In view of this, they suggested that the continuous component of the noise might result from impulses that were propagated through the last two stages of the four-stage transduction mechanism. The double

Lorentzian power spectrum could then be explained if each impulse gives rise to multiple transmitter particles which act subsequently to block the channels.

This interpretation is consistent with the view that the primary gating of the light-modulated channel is effected not by intrinsic gating machinery like that possessed by excitable Na channels but by the action of a transmitter.

In a recent study Detwiler, Connor, and Bodoia (15) succeeded in patch-clamping the plasma membrane at the tip of the ROS in the lizard, Gekko gekko. Their pipette was filled with a solution that contained only 0.1 mmol/l Ca^{2+} but was otherwise identical to Ringer's solution. Despite the small size of the membrane patch and the high resistance seal of pipette to membrane (1-50 Gigaohm), they were unable to resolve single-channel currents. The current crossing the patch was noisy in darkness and its variance decreased during steady illumination. By taking the ratio of the light-induced change in current variance to the change in mean current, they estimated that the unit event underlying the noise was only 12 fA in amplitude. On the assumption that a reduction in external Ca^{2+} affects the size of the single-channel current rather than mean channel lifetime (see below), they further estimated that if the patch pipette had contained a normal Ca^{2+} concentration the unitary event would have been 2 fA, corresponding to a unit conductance of 50 fS. The net flux through a single channel would thus be 1.25×10^4 ions per s. (For comparison, a single Na^+ channel in nerve has a conductance of 1-10 pS and transports about 10^8 Na^+ per second. Valinomycin, a fast, mobile, neutral carrier can transport on the order of 10^4 ions per second.) A single-photon absorption which elicits a photocurrent of ~1 pA peak amplitude must therefore cause the closure of ~500 channels. The normal dark current of ~30 pA must flow through about 15,000 channels, implying an open-channel density of about 15-μm^2 in darkness.

More recent measurements have shown ((6), and P.B. Detwiler, personal communication) that the power spectrum of the current noise, measured under patch-clamp in the presence of 0.1 mmol/l external Ca^{2+}, consists of two components. The low-frequency component is well fitted by a product of two Lorentzians with a corner frequency of about 0.6 Hz and resembles closely the continuous component of current noise measured by Baylor et al. (4) in the presence of normal external Ca^{2+}. This lends support to the idea that the large increase in dark current seen upon reducing external calcium is due either to an increase in the number of channels or to an increase in the single-channel current but not to an increase in channel lifetime, since in that case a significant reduction in the corner frequency would have been expected.

The high-frequency component is well fitted by a single Lorentzian with a corner frequency of at least 60 Hz and in some cases as high as 400 Hz (6). In other analyses of noise measured in doped artificial membranes and in patch-clamped biological membranes, noise that extends to these high frequencies has been identified as transport noise, i.e., the noise associated with the movement of ions through an open pore. Transport noise has a power spectrum that falls off at high frequencies with a slope proportional to 1/f, though if short sampling times are used (<30 s) a Lorentzian spectrum can result ((14), p. 284). Thus it is tempting to identify the high-frequency component as transport noise by analogy. The failure to detect single-channel events in these experiments, however, must make us cautious in pressing such an argument.

CHANNEL STRUCTURE

An important and so far unresolved question concerns the detailed structure of the light-modulated channel. It is not yet known whether it consists of an aqueous pore like the Na^+ channel in nerve, or a membrane-bound carrier molecule that shuttles ions across the lipid phase of the ROS plasma membrane. Certainly, both possibilities must be considered in any speculation on this question.

If the light-modulated channel is a pore, some aspects of its structure can be inferred from its ionic selectivity and its electrical properties. The selectivity sequence for the alkaline cations provides strong evidence that it is not simply a wide, unstructured, water-filled hole through the membrane. If that were the case, the sequence would reflect the ionic mobilities in free aqueous solution and would be the exact reverse of that which was observed by Yau and Nakatani (45). The observed sequence is the same as that of the Na^+ channel in squid nerve (13). It is identical with sequence XI predicted by Eisenman's theory of ion permeation, implying that the selectivity filter is an anionic binding site of high field strength. If we assume that the rate-limiting step for an ion passing through the channel is likely to be its association with and dissociation from this site, the outward rectification of the channel (2) might be explained if it were located asymmetrically within the channel towards the outer surface of the membrane. Sites of high field strength are generally weak acids (high pK_a), and in this respect it is worth noting that Wormington and Cone (43) found experimental evidence for a weak acid ($pK_a = 5.8$) close to the outer surface of the ROS plasma membrane. Protons binding to this site apparently blocked the light-modulated channel.

A lower limit on the dimensions of the pore at the selectivity filter can be obtained from the crystalline radii of the permeant ions. The largest of the permeant ions is Cs^+ with a radius of 1.69 Å. Since ions may not be fully dehydrated when they pass through the filter, the aperture of the filter must be at least 3.4 Å in diameter.

These properties are reasonably similar to those of the Na^+ channel in nerve. There are several dissimilarities, however, which suggest that the light-modulated channel may be substantially different in structure. It has already been noted that the voltage-dependent, time-varying inactivation seen by MacLeish et al. (31) is probably not due to a voltage-activated change in channel structure. Thus the light-modulated channel may lack some or all of the complicated voltage-controlled gating machinery of the Na^+ channel. Instead, at its inner margin there must be binding sites for the internal transmitter particles that do gate the channel. Second, the light-modulated channel is much less selective than the Na^+ channel. The relatively high permeability of K^+, Rb^+, and Cs^+ is surprising and suggests that factors other than the field strength of the selectivity filter may help determine the permeabilities of the ions. Third, the temperature-dependence of the dark current, and hence of the light-modulated conductance, is not well described by a straight line on an Arrhenius plot (29). Moreover, over the physiological range of temperatures from 15°C to 25°C, the slope of the line of best fit yields an activation energy of ~12 kcal/mol and a Q_{10} of ~2.0. The Q_{10} of the conductance of the Na channel in nerve is about 1.3 (20), only slightly higher than that of free diffusion in Ringer's solution, and the activation energy is ~4.6 kcal/mol. Thus the (rate-limiting) energy barrier within the light-modulated channel may be significantly higher than that within nerve Na^+ channels, and in consequence, ions are likely to pass through the light-modulated channel with rather less ease. In this regard it is worth noting that the current passing through the light-modulated channel is lower by almost three orders of magnitude (15) than that passing through Na^+ channels in nerve (24). Fourth, the light-modulated channel, unlike the Na channel in nerve, is not blocked by externally applied tetrodotoxin (17).

Carrier mechanisms are not well understood. Much of what is known derives from the study of natural and artificial ionophores in artificial bilayers. Defining the structure and properties of putative carrier mechanisms in biological systems is a difficult and uncertain process. In consequence, the following discussion must be considered tentative and not a little speculative.

Models of ion transport in which ions cross the membrane in combination with mobile carrier molecules were reviewed by Adrian (1). The rate-limiting step for an ion crossing the membrane was assumed to be the formation of the ion-carrier complex, (rate constant α, while the translocation step was assumed sufficiently rapid for the concentrations of the complex on each side of the membrane to be described by the Boltzmann relation. Thus,

$$\frac{[A]_o}{[A]_i} = exp(z_A \cdot EF/RT) \text{ and } \frac{[SA]_o}{[SA]_i} = exp(z_A + z_S)(EF/RT) \qquad (1,a\ b)$$

where $[A]_i$ and $[A]_o$ are the concentrations of free carrier, $[SA]_i$ and $[SA]_o$ the concentrations of the complex at the inner and outer surfaces of the membrane, respectively, z_A is the charge on the carrier, and the other terms have their usual meaning. The free carrier concentration was assumed to be buffered at either the inner surface or the outer surface of the membrane, an assumption which implies a virtually unlimited supply of carriers. When the outer surface concentration, $[A]_o$, was assumed to be buffered, the current carried by a given species of monovalent cation could be described by the equation

$$I_S = \alpha[A]_o F \frac{C_{in} - C_{out} \cdot exp(-EF/RT)}{exp(z_a \cdot EF/RT) + exp(-EF/RT)}, \tag{2}$$

where C_{in} and C_{out} are the internal and external concentrations of the transported ion, respectively ((1), Eq. 13.1). In the case of the light-modulated channel, the relatively high permeability of K requires that we assume the carrier to be capable of transporting either Na or K and that the net current crossing the membrane be the sum of the Na and K currents. With this modification, the above equation can be used to predict the shape of the current-voltage relation for charged and uncharged carriers. Using the known concentrations of these ions, this model predicts that a neutral, mobile carrier would generate a relation which is roughly symmetrical about the reversal potential; i.e., it would not be expected to give rise to significant outward rectification. A carrier possessing a single negative charge, on the other hand, would yield a relation highly rectified in the outward direction. If the carrier charge were -2, and even more marked outward rectification would be expected with a region of negative slope at potentials more negative than about -30 mV. Positively charged carriers would yield inward rectification.

If the carrier concentration were assumed to be buffered at the inner surface of the membrane, outward rectification would require that the carrier be positively charged which would effectively preclude its behaving as a cation transporter.

To summarize, if the light-modulated channel is a mobile carrier that behaves according to this class of model, the outward rectification measured by Bader et al. (2) would require that, in its "free" form, it carry a net negative charge and that its concentration be effectively buffered at the outer surface of the membrane.

Naturally occurring mobile ion carriers can be divided into two principal classes, neutral ionophores and carboxylic ionophores (34). Valinomycin is perhaps the best characterized of the neutral carriers, while X-537A and A23187 are probably the most familiar of the carboxylic carriers. Neutral carriers, when complexed with a cation, cross the membrane as a positively charged species, and hence the diffusional reaction is driven electrophoretically by any potential difference across the membrane. For a single complexing ion species, the equilibrium condition is defined by the Nernst potential. If gradients of more than one complexing species exist across the membrane, the equilibrium condition obeys the Goldman equation. Thus a net current and a true reversal potential can be generated by neutral carriers. For the reasons given above, however, neutral carriers are not expected to give rise to significant electrical rectification.

Carboxylic carriers possess a net negative charge, but this charge is normally protonated. Thus in its free form the carrier's charge is internally compensated and it diffuses as a neutral complex. A cation complexing with this type of carrier displaces the proton, and the complex then crosses the membrane as an electrically neutral zwitterion. Carboxylic carriers are all capable, therefore, of transporting protons. Because in both their protonated and their zwitterionic forms they are electrically neutral, the equilibrium condition depends neither on transmembrane voltage nor on the selectivity of the carrier, though it is pH-dependent. This can be seen by noting that in both Eqs. 1a and 1b (above) the effective value of the charge term will be zero and the concentrations of free carriers and of ion-carrier complexes must be the same on either surface of the membrane. The only way that this can be true is if these carriers mediate exchange diffusion, which is indeed what carboxylic carriers are known to do.

The conclusion to be drawn from this is that none of the known types of naturally occurring, mobile ion carrier can account for the measured electrical properties of the light-modulated channel.

There is another class of carrier which should be considered, however, one that combines some of the features of a pore with some aspects of a mobile carrier. Anion carriers in red blood cells, for example, are believed to consist of a protein that spans the entire membrane. The "carrier" subcomponent of the protein can be reached from either surface of the membrane via pore-like aqueous cavities. In order to cross the membrane, an ion must pass into the cavity and bind to a charged site on that subcomponent. Binding then triggers a conformational change which causes the subcomponent to rotate so that the ion, now facing the aqueous cavity on the opposite side of the membrane, can rehydrolyze and diffuse away (see (26) for a review). In the case of these anion carriers, the translocating conformational change requires the binding of an ion, and in consequence they behave as tightly coupled ion exchangers. It is now thought that a mechanism having features in common with this type of carrier, but differing in that the translocating conformational change requires the hydrolysis of ATP, might form the basis of ATP-driven pumps such as the Na^+-K^+ ATPase and the Ca^{2+}-ATPase of sarcoplasmic reticulum. These are known to generate a net transmembrane current, and we might speculate that a mechanism of this type could carry the dark current across the ROS membrane. It would be of interest in this regard to know whether or not the current crossing the ROS membrane in darkness is ATP-dependent.

IS THE LIGHT-MODULATED CHANNEL A PORE OR A CARRIER?

This remains one of the more important questions yet to be answered. For several reasons, a "hybrid" carrier of the type described above must be considered a possibility. First, the binding/translocation step will be rate-limiting, and since the binding site can be located asymmetrically within the membrane, it should be capable of conferring electrical rectification on the channel. Second, the rate at which ions can be transported by such a system will be lower than that of an aqueous pore but higher than that of a mobile carrier system which must diffuse across the whole membrane in order to translocate an ion. Third, a conformational change of the protein involves some interaction with the surrounding lipid molecules and, in consequence, the transport mediated by such mechanisms has a significantly higher temperature-dependence than that of current flow through aqueous pores. Moreover, that temperature-dependence is frequently found to be non-Arrhenius, the Q_{10} being significantly higher at low temperatures than at higher temperatures. The activation energy for chloride exchange in the erythrocyte, for example, changes from 30 kcal/mol below 15°C to 20 kcal/mol above 15°C (8). Similarly, the activation energy of hexose transport in the erythrocyte, known to be mediated by a hybrid carrier, drops from 20 kcal/mol at 20°C to only 7-10 kcal/mol at 37°C (36). Thus, while one can think of other possible explanations for the non-Arrhenius temperature-dependence of the ROS dark current reported by Lamb, (29), it is not inconsistent with what would be expected of a hybrid carrier.

The spectrum of the current noise induced by a mobile carrier such as valinomycin is white at low frequencies and increases in the kHz range (27). By contrast, in their study of ROS current noise under patch-clamp, Detwiler and his co-workers found the power spectrum of the high-frequency component to resemble a single Lorentzian (P. Detwiler, personal communication). This does not necessarily rule out carrier-mediated transport of the dark current, however. In the case of a hybrid carrier mechanism, we might conjecture that transport noise is generated in the pore-like cavities on either side of the carrier subcomponent and that this predominates at high frequencies, yielding (for short sampling times) the observed Lorentzian spectrum.

SUMMARY AND CONCLUSION

Perhaps the one thing that can be said with any certainty is that the light-modulated channel has properties which define it as being quite unlike the excitable Na^+ channel of nerve membranes. First, although its ionic selectivity falls in the same sequence as that of the Na^+ channel in nerve, it is poorly selective, preferring Na^+ over K^+ by a factor of only 1.4. Second, there is little evidence of any intrinsic, voltage-dependent gating mechanism; gating appears to be due primarily to the action of blocking particles in the ROS cytoplasm. Third, ions pass through the channel at a rate that is three orders of magnitude slower than that of Na^+ ions through Na^+ channels in nerve but faster than that of a fast mobile carrier such as valinomycin. Fourth, its electrical conductance is more than two orders of magnitude smaller than that of an Na^+ channel in nerve.

If the channel is unlike the Na^+ channel, what can we say about its structure? There is some evidence suggesting that it may possess two sites at which divalent ions can block it: a site near the external face of the membrane with low affinity for Co^{2+} and a high-affinity site for Ca^{2+} at the internal face. The rest of the structure remains mysterious. If it were a pore, the marked outward rectification would suggest that the rate-limiting energy barrier for ions passing through it is located towards the outer surface of the membrane while the selectivity sequence is suggestive of a high-strength anionic binding site. If, on the other hand, a carrier mechanism were involved, it seems more likely that it would be of the hybrid type than a mobile carrier such as valinomycin.

With the recent application of powerful new techniques for studying the light-modulated channel, rapid progress has been made towards an understanding of its functional properties. Although many substantive questions remain, there seems little doubt, in view of the present intense interest in the phototransduction process, that many of these questions will be answered in the near future.

Acknowledgements. Supported by grant number EY03785 from the National Institutes of Health (USPHS).

REFERENCES

(1) Adrian, R.H. 1969. Rectification in muscle membrane. Prog. Biophys. Molec. Biol. **19:** 341-369.

(2) Bader, C.R.; MacLeish, P.R.; and Schwartz, E.A. 1979. A voltage-clamp study of the light response in solitary rods of the tiger salamander. J. Physiol. **296:** 1-26.

(3) Baylor, D.A., and Lamb, T.D. 1982. Local effects of bleaching in retinal rods of the toad. J. Physiol. **328:** 49-71.

(4) Baylor, D.A.; Matthews, G.; and Yau, K.-W. 1980. Two components of electrical dark noise in retinal rod outer segments. J. Physiol. **309:** 591-621.

(5) Begenisch, T. 1975. Magnitude and location of surface charges in Myxicola giant axons. J. Gen. Physiol. **66:** 47-65.

(6) Bodoia, R.D., and Detwiler, P.B. 1984. Patch-clamp study of the light response of isolated frog retinal rods. Biophys. J. **45:** 337a.

(7) Bortoff, A. 1964. Localization of slow potential responses in the Necturus retina. Vision Res. **4:** 627-636.

(8) Brahm, J. 1977. Temperature-dependent changes of chloride transport kinetics in human red cells. J. Gen. Physiol. **70:** 283-306.

(9) Brown, H.M.; Hagiwara, S.; Koike, H.; and Meech, R.W. 1970. Membrane properties of a barnacle photoreceptor examined by the voltage-clamp technique. J. Physiol. **208:** 385.

(10) Brown, J.E., and Pinto, L.H. 1974. Ionic mechanism for the photoreceptor potential of the retina of Bufo marinus. J. Physiol. **236:** 575-591.

(11) Capovilla, M.; Caretta, A.; Cervetto, L.; and Torre, V. 1983. Ionic movements through light-sensitive channels of toad rods. J. Physiol. **343:** 295-310.

(12) Cavaggioni, A.; Sorbi, R.T.; and Turini, S. 1973. Efflux of potassium from isolated rod outer segments: a photic effect. J. Physiol. **232:** 609-620.

(13) Chandler, W.K., and Meves, H. 1965. Voltage clamp experiments on internally perfused giant axons. J. Physiol. **180:** 788.

(14) DeFelice, L.J. 1981. Introduction to Membrane Noise. Plenum: New York.

(15) Detwiler, P.B.; Connor, J.D.; and Bodoia, R.D. 1982. Gigaseal patch clamp recordings from outer segments of intact retinal rods. Nature **300:** 59-61.

(16) DiPolo, R., and Beauge, L. 1983. The calcium pump and sodium-calcium exchange in squid axons. Ann. Rev. Physiol. **45:** 313-324.

(17) Fain, G.L.; Gerschenfeld, H.M.; and Quandt, F.N. 1980. Ca^{2+} spikes in rods. J. Physiol. **303:** 495-513.

(18) Fain, G.L., and Lisman, J.E. 1981. Membrane conductances of photoreceptors. Prog. Biophys. Molec. Biol. **37:** 91-147.

(19) Frankenhäuser, B., and Hodgkin, A.L. 1957. The action of calcium on the electrical properties of squid axons. J. Physiol. **137:** 218-244.

(20) Frankenhäuser, B., and Moore, L.E. 1963. The effect of temperature on the sodium and potassium permeability changes in myelinated fibres of Xenopus laevis. J. Physiol. **169:** 431-437.

(21) Gold, G.H., and Korenbrot, J.I. 1980. Light-induced Ca release by intact retinal rods. Proc. Natl. Acad. Sci. USA **77:** 5557-5561.

(22) Gold, G.H., and Korenbrot, J.I. 1981. The regulation of calcium in the intact retinal rod: a study of light-induced calcium release by the outer segment. Curr. Top. Membr. Trans. **15:** 307-330.

(23) Hagins, W.A.; Penn, R.D.; and Yoshikami, S. 1970. Dark current and photocurrent in retinal rods. Biophys. J. **10:** 380-412.

(24) Hille, B. 1970. Ionic channels in nerve membranes. Prog. Biophys. Molec. Biol. **21:** 1-32.

(25) Hodgkin, A.L.; McNaughton, P.A.; Nunn, B.J.; and Yau, K.-W. 1984. Effect of ions on retinal rods from Bufo marinus. J. Physiol. **350:** 649-680.

(26) Knauf, P.A. 1979. Erythrocyte anion exchange and the band 3 protein: Transport kinetics and molecular structure. Curr. Top. Membr. Trans. **12:** 249-363.

(27) Kolb, H.-A., and Lauger, P. 1978. Spectral analysis of current noise generated by carrier-mediated ion transport. J. Membr. Biol. **41:** 167-187.

(28) Korenbrot, J.I., and Cone, R.A. 1972. Dark ionic flux and the effects of light in isolated rod outer segments. J. Gen. Physiol. **60:** 20-45.

(29) Lamb, T.D. 1984. Effects of temperature changes on toad rod photocurrents. J. Physiol. **346:** 557-578.

(30) Lisman, J.E., and Brown, J.E. 1972. The effects of intracellular ionophoretic injection of calcium and sodium ions on the light response of Limulus photoreceptors. J. Gen. Physiol. **59:** 701.

(31) MacLeish, P.R.; Schwartz, E.A.; and Tachibana, M. 1984. Control of the generator current in solitary rods of the Ambystoma tigrinum retina. J. Physiol. **348:** 645-664.

(32) Owen, W.G., and Torre, V. 1981. Ionic studies of vertebrate rods. Curr. Top. Membr. Trans. **15:** 33-57.

(33) Pitts, B.J.R. 1979. Stoichiometry of sodium-calcium exchange in cardiac sarcolemmal vesicles. J. Biol. Chem. **254:** 6232-6235.

(34) Pressman, B.C. 1976. Biological applications of ionophores. Ann. Rev. Biochem. **45:** 501-530.

(35) Schnetkamp, P.P.M. 1980. Ion selectivity of the cation transport system of isolated intact cattle rod outer segments: evidence for a direct communication between the rod plasma membrane and the rod disk membranes. Biochim. Biophys. Acta **598:** 66-90.

(36) Sen, A.K., and Widdas, W.F. 1962. Determination of the temperature and pH dependence of glucose transfer across the human erythrocyte membrane measured by glucose exit. J. Physiol. **160:** 392-403.

(37) Sillman, A.J.; Ito, H.; and Tomita, T. 1969. Studies on the mass receptor potential of the isolated frog retina. II. On the basis of the ionic mechanism. Vision Res. **9:** 1443-1451.

(38) Tomita, T. 1965. Electrophysiological study of the mechanisms subserving colour coding in the fish retina. Cold S.H. Symp. Quant. Biol. **30:** 559-566.

(39) Torre, V. 1982. The contribution of the electrogenic sodium-potassium pump to the electrical activity of toad rods. J. Physiol. **333:** 315-341.

(40) Toyoda, J.; Nosaki, H.; and Tomita, T. 1969. Light-induced resistance changes in single photoreceptors of Necturus and Gekko. Vision Res. **9:** 453-463.

(41) Waloga, G.; Brown, J.E.; and Pinto, L.H. 1975. Detection of changes in Cain from Limulus photoreceptors using arsenazo III. Biol. Bull. **149:** 449.

(42) Woodruff, M.L.; Fain, G.L.; and Bastian, B. 1982. Light-dependent ion influx into toad photoreceptors. J. Gen. Physiol. **80:** 517-536.

(43) Wormington, C.M., and Cone, R.A. 1978. Ionic blockage of the light-regulated channel in isolated rod outer segments. J. Gen. Physiol. **71:** 657-681.

(44) Yau, K.-W.; McNaughton, P.A.; and Hodgkin, A.L. 1981. Effect of ions on the light-sensitive current in retinal rods. Nature **292:** 502-505.

(45) Yau, K.-W., and Nakatani, K. 1984. Cation selectivity of the light-sensitive conductance in retinal rods. Nature **309:** 352-354.

The Molecular Mechanism of Photoreception, ed. H. Stieve, pp. 189-198. Dahlem Konferenzen 1986. Berlin, Heidelberg, New York, Tokyo: Springer-Verlag.

Properties of the Light-dependent Membrane Conductances in Invertebrate Photoreceptors

J.E. Lisman
Dept. of Biology
Brandeis University
Waltham, MA 02254, USA

Abstract. Substantial progress has been made in understanding the ionic basis of the receptor potential in invertebrate photoreceptors. In receptors that depolarize in response to light, the receptor potential is generated primarily by an increase in conductance to Na^+. The permeability of this conductance for K^+ is approximately half that of Na^+. In Limulus, light also modulates a voltage-dependent K^+ conductance. This modulation appears to be mediated by a light-induced rise in intracellular Ca^{2+}. The second messenger involved in activating the Na^+ conductance is not known, though there has been much recent work in this area. The channels that carry the light-activated Na current have been studied using the patch-clamp technique. Results have been obtained in both cell-attached and excised patches.

INTRODUCTION

This review is divided into four sections. The first section deals with the somewhat surprising finding that light can activate more than one type of ionic conductance in invertebrate photoreceptors. The second section describes the macroscopic properties of these conductances and the cytoplasmic factors that may be involved in the control of these conductances. The third section describes recent patch-clamp studies of the light-activated Na^+ conductance in Limulus ventral photoreceptors. The final section deals with preliminary work on excised patches.

EVIDENCE FOR MULTIPLE LIGHT-DEPENDENT CONDUCTANCES

Limulus ventral photoreceptors have provided a useful model system for the study of light-dependent membrane currents because the cells can be easily impaled with two microelectrodes and voltage-clamped. Optical measurements using voltage-sensitive dyes indicate that the photoreceptor cell body is isopotential under voltage-clamp (6). If only one ionic conductance is affected by light, there should be a unique reversal potential for the light-dependent current measured under voltage-clamp. Measurements show, however, that there is no voltage at which there is no light-induced current (22).

Fortunately, the membrane currents at different holding voltages are readily explainable in terms of two light-dependent conductances, one that is rapidly activated by light and has a unique reversal potential just positive of zero, and a second that responds much more slowly to changes in light intensity and which produces an inward current at all voltages more positive than -50 mV. The first conductance mechanism, g_{Na}, has been studied by Millecchia and Mauro and shown to be a conductance increase primarily to sodium (24); the second, slower process has been identified as a light-dependent modulation of the delayed rectifier potassium channel, g_K (21). This voltage-dependent K^+ conductance is large in the dark. Light reduces this conductance, thereby leading to an apparent (voltage-dependent) inward current at voltages more positive than E_K. It should be emphasized that the control of both these conductances by light does not require a change in membrane voltage. In this sense, modulation of the K^+ channel is "direct" and can be distinguished from the indirect effects of light on a variety of voltage-dependent channels that follow as a secondary result of the light-induced change in voltage (4, 12).

Evidence for multiple light-activated conductances has also been obtained in barnacle and Hermissenda photoreceptors. Barnacle photoreceptors contain a Ca-activated K^+ conductance similar to that seen in many types of neurons (2). Illumination of barnacle photoreceptors leads to a rise in the intracellular free Ca^{2+} concentration (5), which then appears to activate a K^+ conductance (20). Hermissenda photoreceptors (15) appear to contain a Na^+ conductance activated by light, a K^+ conductance activated by light, and a K^+ conductance inactivated by light. Why photoreceptors should contain multiple light-activated conductances remains quite unclear. In Limulus, the slow reduction in g_K produced by a constant light occurs in parallel with a slow decrease in g_{Na}. As a result, the voltage remains more constant than if g_K were fixed (21). This may be of functional value to the system.

MACROSCOPIC PROPERTIES OF LIGHT-DEPENDENT CONDUCTANCES

The most detailed study of the ion selectivity of the light-activated Na^+ conductance in Limulus was done by Brown and Mote (7). They showed that when Na^+ was replaced by sucrose, the reversal potential became ~ 50 mV more negative and the receptor potential became hyperpolarizing. Preliminary experiments showed that K^+ could go through the channel quite well, and the estimate of the Na^+ to K^+ permeability ratio was about 2. Substitution of Na^+ by Li^+ caused no change in reversal potential, indicating that Li^+ is as permeable as Na^+. This was somewhat surprising, given that substituting Li^+ for Na^+ leads to a dramatic reduction in the amplitude of the receptor potential. The likely explanation is that by making this substitution, Na^+/Ca^{2+} exchange is reduced; since the leakage of Ca^{2+} into the cell is not counteracted, there is a rise in the concentration of intracellular free Ca^{2+}; this, in turn, reduces the number of channels activated by light (see section by Brown (this volume) regarding the effects of intracellular Ca^{2+}).

In Brown and Mote's ion substitution experiments, no organic cation was found that shifted the reversal potential as much as sucrose did. It is thus possible that organic cations such as choline can pass through the channel. A more detailed study of the permeability of inorganic and organic cations through the light-activated Na^+ conductance would seem desirable. A recent study of the ion selectivity of the light-activated K^+ channel in the hyperpolarizing photoreceptors of the scallop (13) is exemplary in this regard.

The dependence of the receptor potential on Na^+ has been demonstrated in a wide variety of depolarizing photoreceptors, but it seems unlikely that many of these preparations will be useful for doing quantitative ion selectivity measurements. One problem is that intracellular concentrations can undergo secondary changes in response to extracellular ion substitutions. Coles and Orkand (11) found that substituting sucrose for extracellular Na^+ in the drone caused a large and fairly rapid reduction of the intracellular Na^+ concentration. This may explain in part why depolarizing receptor potentials can persist during such substitutions (19). A second problem is the control of the concentrations in the extracellular space around the photoreceptor by glia. The ommatidial structures of compound eyes are tightly wrapped in glia, and the extracellular space surrounding the microvilli is not in equilibrium with the bath (12). The influence of the glial cells that surround Limulus ventral photoreceptors (which are not compound eyes) is unclear. The fact that the reversal potential can be dramatically shifted by substituting sucrose for Na^+ suggests that the Na^+ concentration around the microvilli must be greatly reduced under these conditions. Whether the Na^+ concentration around the villi is quanti tatively controlled is not known. If the subject of ion selectivity of the light-activated Na^+ conductance in Limulus is to be reexamined, it would seem desirable to strip off the glial cells, a procedure that is now routinely possible (28).

Because of the importance of intracellular Ca^{2+} in invertebrate photoreceptors, it is of interest to know whether the light-activated Na^+ channel is permeable to Ca^{2+}. If so, the channel would provide one of several pathways by which intracellular Ca^{2+} might be influenced by light (the others being released from intracellular stores (5) and entry through voltage-dependent Ca^{2+} channels (25)). In barnacle photoreceptors the answer to this question is quite clear (4); in zero Na^+ solutions the reversal potential of the light-activated inward current varies with the extracellular Ca^{2+} concentration, indicating that Ca^{2+} can go through the channel.

Although the conductance that is activated by light cannot be activated by changing the voltage in the dark, the magnitude of the light-activated conductance depends on the voltage at which the measurements are made. This was first shown by Millecchia and Mauro (24) and has been recently reexamined (9). Similar results have also been obtained in barnacle photoreceptors (4). In both cell types, the conductance is nearly constant in the physiological range (-70 to 0 mV) but increases dramatically at more positive voltages. Chinn, Bacigalupo, and Lisman (9) found that the time integral of the light-induced current evoked by a brief flash is a more appropriate quantitative measure of the light-dependent conductance than the peak current because the kinetics of the response to a brief flash vary with voltage. Using this measure of voltage-dependence, they found that the conductance at positive voltages could be an order of magnitude larger than at negative voltages.

The voltage-dependence of the light-activated channel serves as a signature of the light-activated conductance and can be used as a basis for identifying the channel in excised patches or reconstituted systems. For similar purposes, it would be extremely useful to have pharmacological agents that specifically block the channel. Unfortunately, no one has yet reported any channel blocker that seems to reduce the light-activated conductance.

The instantaneous current-voltage curve of the light-activated conductance has been measured in barnacle photoreceptors and was found to be linear (4). The voltage-dependence of the conductance developed with a time constant of 14 ms. Thus, whatever is responsible for the voltage-dependence of the light-activated conductance is a time-dependent property of the channel rather than a property of the single-channel conductance per se, a conclusion supported by recent single-channel measurements discussed below. It is intriguing that the light-dependent conductance in vertebrate rods (3) shows a voltage-dependence somewhat similar to that found in barnacle and Limulus photoreceptors. Whether this is coincidental or important is unclear.

The possibility that the voltage-dependence of the light-activated conductance in Limulus is dependent on extracellular divalent ions has been recently examined (9). The voltage-dependence was not removed by reductions in the extracellular divalent cation concentration (Ca_o = 0.3 mmol/l; Mg_o = 3 mmol/l). It was not possible completely to remove divalents since this leads to a huge increase in the membrane conductance (32). Whether this increase is due to an activation of the light-dependent conductance, as suggested by Stieve and Bruns (31), or to some other conductance is an important and controversial (17) question that needs to be pursued.

There are many factors that affect the response to light in Limulus, but the second messenger responsible for controlling the channel that is activated by light remains elusive. Cyclic nucleotides at reasonable concentration levels do not have a dramatic effect on the macroscopic currents (30).

Recent work from Fein's and Brown's laboratories with inositol 1,4,5-triphosphate ($InsP_3$) indicates that $InsP_3$ can activate the light-activated conductance (8, 18). However, $InsP_3$ is unlikely to be directly responsible for the activation of channels by light because the effects of $InsP_3$ are completely blocked by EGTA, whereas the response to light becomes larger after EGTA injection (26). Measurements with aequorin indicate that $InsP_3$ causes a rapid rise in intracellular Ca^{2+} (14). Although previous work with intracellular Ca^{2+} ionophoresis showed that Ca^{2+} reduces the response to light (23), it now appears that Ca^{2+} can have additional effects if introduced rapidly (26). Pressure injection of 1 mmol/l Ca causes a brief inward current having a reversal potential close to that of the light-activated current. It is thus possible that a rise in Ca^{2+} may be the mechanism by which light-activated channels are opened by $InsP_3$.

In an attempt to identify the second messenger involved in the control of the light-activated channel in invertebrates, Stern and Lisman (29) prepared partially purified extracts of small molecules from either squid photoreceptors or Limulus hepatopancreas. The extracts were assayed for activity using an intracellular bioassay: introduction of the extract into functioning Limulus ventral photoreceptors using the internal perfusion technique. The extract produced a large and reversible decrease in the amplitude of the response. Similar reversible effects were observed after pressure injection of the extract into intact cells, indicating that the active factor is metabolized by the cell. The effect of the extract was much greater in dark-adapted Limulus photoreceptors than in light-adapted ones, possibly because the metabolism of the factor is enhanced by light. Although the effect of the extract superficially resembles the adapting effect of intracellular Ca^{2+}, the active factor is not calcium: the extract has a low Ca^{2+} concentration; moreover, the effect of elevated intracellular Ca^{2+} is to speed up the response to a test-flash, whereas the extract slowed the kinetics. One radical but not implausible idea is that the extract contains the second messenger for excitation. It must be supposed that the second messenger is present in the dark and blocks the channels; channels open during light because the blocking molecules are destroyed during light. Flooding a cell with a high concentration of blocker would lead to the observed reduction in response amplitude. While this line of reasoning is highly speculative, it gains some support from recent work showing high channel activity in excised patches (see below).

In contrast to the slow progress in establishing how the light-activated Na^+ channels are controlled, considerable progress has been made in understanding the mechanism by which the voltage-dependent K^+ conductance in Limulus is reduced by light. Ca^{2+} meets all the qualitative criteria for the second messenger: Ca increases during illumination (5); Ca injection mimics the effect of light by producing a decrease in K^+ conductance (10); the effect of light on K^+ conductance can be greatly reduced by intracellular injection of EGTA (10). What remains unclear is whether Ca^{2+} interacts directly with the K^+ channel. Furthermore, the mechanism by which Ca^{2+} is released from intracellular stores is uncertain, though the possibility that $InsP_3$ is involved seems plausible.

PATCH-CLAMP RECORDINGS OF LIGHT-ACTIVATED CHANNELS IN LIMULUS

Bacigalupo and Lisman (1, 2) have obtained gigohm seals on the light-transducing lobe of ventral photoreceptors and observed channels activated by light. These channels are not activated as a secondary result of the light-induced membrane depolarization since depolarization in the dark fails to produce a similar activation. The probability of the channel opening is graded with light intensity and depends on the state of adaptation of the cell. In several experiments, the activation of channels by a light of fixed intensity was examined before and after intracellular injection of EGTA. Such injections dramatically increased the macroscopic light-activated current and greatly increased the probability of channel opening. The current-voltage curve for the single-channel current is linear in the range of -80 to +60 mV and has a slope of ~40 pS. The reversal potential is the same as that of the macroscopic currents. Although the conductance of the open channel is independent of voltage, the probability of the channel being in the open state is highly voltage-dependent at voltages more positive than the reversal voltage (25). The probability of being open increases primarily because the lifetime of the open channel increases at positive voltages, but also because the probability of a channel opening increases. Taken together, these gating processes can account for the voltage-dependence of the macroscopic conductance. In addition to the 40 pS light-activated channel described above, all patches show light-activated channel openings having about half that conductance. In most patches these smaller events are rare, but in some they can be quite frequent. Whether this smaller channel represents a truly distinct, light-activated channel or a different state of the 40 pS channel is an interesting question that will have to be pursued.

The data outlined above show a close similarity between the macroscopic and microscopic properties of light-activated current. In contrast, the kinetics of channel activation in some patches was grossly different from that of the macroscopic current: in extreme cases, channels were activated by light with a latency of several seconds as compared to the 100 ms latency of the macroscopic currents. In other patches this discrepancy was much less; recently, Bacigalupo and Lisman have obtained patches in which the discrepancy is very small (less than 100 ms). Why this discrepancy should exist and be so variable is unclear. It may be related to the difficulty of obtaining seals on the microvillar surface of the plasma membrane. Since seals are obtained in only a small fraction of the attempts, it is possible that successful seals form only on structurally atypical regions of the plasma membrane. Because these regions may be distant from the microvilli, the channels open with longer than average latency. This discrepancy cannot be attributed to the method used to prepare the cells, since the discrepancy is between the macroscopic and microscopic currents measured simultaneously in the same cell. The method of preparation is, however, a cause for concern. Bacigalupo and Lisman found that to obtain good seals on the microvillar lobe it was necessary to sonicate the cells mildly in a bath sonicator. While this sounds horrendous, the sonication was so brief that it had no effect on cell morphology visible in the light-microscope, nor did it obviously alter the macroscopic response to light. Recently, Bacigalupo and Lisman tried again to obtain seals without sonication but had no success. Obtaining seals on the transducing lobe has been difficult in other invertebrate photoreceptors (Nasi, personal communication) under conditions where seals could be easily obtained on the non-transducing (non-microvillar) region of the cell.

The single-channel lifetime measured in Limulus appears to be very short (3 ms). This is much shorter than the time constant of the declining phase of the quantum bump. It is therefore likely that the duration of a quantum bump is determined by the time-dependence of the second messenger rather than being rate-limited by the closing of channels. This, however, is an inference since single-channel events were not measured during actual quantum bumps. Recently, Dirnberger et al. (16) put forward the interesting suggestion that the channel lifetime may be shortened by the light-adaptation process and that the channel lifetime measured by Bacigalupo and Lisman under light-adapted conditions may not be representative of the dark-adapted values that would obtain during quantum bumps. In support of this idea, they present evidence that the falling phase of quantum bumps has a time constant that becomes smaller as the cell is light-adapted. Similar kinds of argument based on measurements of the falling phase of the receptor potential have been proposed by Raggenbass (27). This idea now needs to be rigorously tested using single-channel recording. The principal difficulty with measuring single-channel events during single-photon events stems from the highly localized conductance change associated with quantum bumps.

Bacigalupo, Stern, and Lisman (unpublished observations) placed two suction pipettes (diameters >4 microns) within 5 microns of each other on the transducing lobe of a stripped ventral photoreceptor. Large inward currents with the kinetics of quantum bumps could be measured by each of the electrodes during a small fraction of the quantum bumps. In no case was inward current measured simultaneously in both electrodes. This indicates that the inward membrane current during quantum bumps is quite restricted. The implication is that inward single-channel currents measured by a small patch pipette will occur during only a small fraction of the quantum bumps (less than 1%).

PROPERTIES OF EXCISED PATCHES CONTAINING LIGHT-ACTIVATED CHANNELS

Recently, Bacigalupo and Lisman (unpublished observations) obtained recordings from excised patches placed in a solution resembling intracellular medium (high potassium, low Na, 10^{-8} mol/l Ca^{2+}). In each case (n = 5), recordings of the cell-attached mode indicated that the patch contained light-activated channels. There was no channel activity in the dark. After excision, channel activity was extremely high: the activity was greater than or equivalent to the highest activity observed during bright illumination of the cell-attached patch. Illumination of the excised patch had no effect. The reversal potential of the currents was close to that seen in the cell-attached patches. At negative voltages, the currents were inward and had a lifetime in the millisecond range, similar to that in the cell-attached patch. The probability of the channel being in the open state was voltage-dependent, increasing dramatically at voltages more positive than reversal voltage. This again is characteristic of the light-activated channel in cell-attached patches. The single-channel conductance of the excised channel was 15-20 pS. This is smaller than the light-activated channel seen most often in cell-attached patches (40 pS) but comparable to the lower conductance, light-activated channels. One interesting possibility is that the 20 pS channel is a different state of the 40 pS channel and that excision favors the low-conductance state for unknown reasons. An alternative possibility is that the light-activated inward current is carried by two different types of channels that respond differently to excision. The observation that excision activates a channel is surprising and unprecedented. One interpretation of this result is that the cell contains a blocking particle which keeps the channel closed and that removal of the patch from the cell is equivalent to the reduction in blocker concentration that occurs during light. An alternative hypothesis is that activation of the channel by excision reflects the loss of some regulatory factor that is not involved in the normal light-dependent gating of the channel.

While it is not yet certain that the channels seen in excised patches are the light-activated Na^+ channels, it is nevertheless useful to consider the type of information that can potentially be learned by applying substances to excised patches. If applied substances affect channel activity, it is safe to assume that the substance affects either the channel itself or a closely associated molecule. This is a much more definitive conclusion than can be reached from experiments on intact cells. For instance, injection of Ca^{2+} into Limulus photoreceptors reduces the light-activated conductance, but it is not clear whether this is due to an effect on the channel or whether it follows as a secondary result of an effect of Ca^{2+} on the metabolism of the second messenger. In excised patches the second messenger is absent, so effects can be attributed to interactions with either the channel or a closely associated membrane-bound molecule. Thus, excised patches may be a useful preparation for examining the interaction of the final soluble messenger with the photoreceptor membrane.

REFERENCES

(1) Bacigalupo, J., and Lisman, J.E. 1983. Single channel currents activated by light in Limulus ventral photoreceptors. Nature **304:** 268-270.

(2) Bacigalupo, J., and Lisman, J.E. 1984. Light activated channels in Limulus ventral photoreceptors. Biophys. J. **45:** 3-5.

(3) Bader, C.R.; MacLeish, P.R.; and Schwartz, E.A. 1979. A voltage-clamp study of the light response in solitary rods of the tiger salamander. J. Physiol. **296:** 1.

(4) Brown, H.M.; Hagiwara, S.; Koike, H.; and Meech, R.W. 1970. Membrane properties of a barnacle photoreceptor examined by the voltage-clamp technique. J. Physiol. **208:** 385.

(5) Brown, J.E., and Blinks, J.R. 1974. Changes in intracellular free calcium concentration during illumination of invertebrate photoreceptors: detection with aequorin. J. Gen. Physiol. **64:** 643-665.

(6) Brown, J.E.; Harary, H.H.; and Waggoner, A. 1979. Isopotentiality and an optical determination of series resistance in Limulus ventral photoreceptors. J. Physiol. **296:** 357-372.

(7) Brown, J.E., and Mote, M.I. 1974. Ionic dependence of reversal voltage of the light response in Limulus ventral photoreceptors. J. Gen. Physiol. **63:** 337.

(8) Brown, J.E.; Rubin, L.J.; Ghalayini, A.J.; Tarver, A.P.; Irvine, R.F.; Berridge, M.J.; and Anderson, R.E. 1984. Evidence that myo-inositol polyphosphate may be a messenger for visual excitation in Limulus photoreceptors. Nature **311:** 160-163.

(9) Chinn, K.; Bacigalupo, J.; and Lisman, J.E. 1984. Rectifying properties of the light-activated conductance observed at the microscopic and macroscopic current level in Limulus ventral photoreceptors. Soc. Neurosci. Abstr. **10:** 620.

(10) Chinn, K., and Lisman, J. 1984. Calcium mediates the light-induced decrease in maintained K^+ current in Limulus ventral photoreceptors. J. Gen. Physiol. **84:** 447-462.

(11) Coles, J.A., and Orkand, R.K. 1982. Sodium activity in drone photoreceptors. J. Physiol. **322:** 16-17.

(12) Coles, J.A., and Tsacopoulos, M. 1979. K^+ activity in photoreceptors, glial cells and extracellular space in the drone retina: changes during photostimulation. J. Physiol. **290:** 525-549.

(13) Cornwall, M.C., and Gorman, A.L.F. 1983. The cation selectivity and voltage dependence of the light-activated potassium conductance in scallop distal photoreceptor. J. Physiol. **340:** 287-305.

(14) Corson, D.W.; Fein, A.; and Payne, R. 1984. Detection of an inositol 1,4,5-triphosphate-induced rise in intracellular free Ca with aequorin in Limulus ventral photoreceptors. Biol. Bull. **167:** 524-525.

(15) Detwiller, P.B. 1976. Multiple light-evoked conductance changes in the photoreceptors of Hermissenda crassicornis. J. Physiol. **256:** 691-708.

(16) Dirnberger, G.; Keiper, W.; Schnakenberg, J.; and Steive, H. 1985. Comparison of time constants of single channel patches, quantum bumps and noise analysis in Limulus ventral photoreceptors.

(17) Fain, G.L., and Lisman, J.E. 1981. Membrane conductances of photoreceptors. Prog. Biophys. Molec. Biol. **37:** 91-147.

(18) Fein, A.; Payne, R.; Corson, D.W.; Berridge, M.J.; and Irvine, R.F. 1984. Photoreceptor excitation and adaptation by inositol 1,4,5 triphosphate. Nature **311:** 157-160.

(19) Fulpius, B., and Baumann, F. 1969. Effects of sodium, potassium and calcium ions on slow and spike potentials in single photoreceptor cells. J. Gen. Physiol. **53:** 541-561.

(20) Hanani, M., and Shaw, C. 1977. A potassium contribution to the response of the barnacle photoreceptor. J. Physiol. **270:** 151-163.

(21) Leonard, R.J., and Lisman, J.E. 1981. Light modulates voltage-dependent potassium channels in Limulus ventral photoreceptors. Science **212:** 1273-1275.

(22) Lisman, J.E., and Brown, J.E. 1971. Two light-induced processes in the photoreceptor cells of Limulus ventral eye. J. Gen. Physiol. **59:** 701-719.

(23) Lisman, J.E., and Brown, J.E. 1972. The effects of intracellular sodium and calcium ions on the light response of Limulus ventral photoreceptors. J. Gen. Physiol. **59:** 701-719.

(24) Millecchia, R., and Mauro, A. 1969. The ventral photoreceptor cells of Limulus. III. A voltage-clamp study. J. Gen. Physiol. **54:** 331-351.

(25) O'Day, P.M.; Lisman, J.E.; and Goldring, M. 1982. Functional significance of voltage-dependent conductances in Limulus ventral photoreceptors. J. Gen. Physiol. **79:** 211-232.

(26) Payne, R.; Fein, A.; and Corson, D.W. 1984. A rise in intracellular Ca^{2+} is necessary and perhaps sufficient for photoreceptor excitation and adaptation by inositol 1,4,5-triphosphate. Biol. Bull. **167:** 531.

(27) Raggenbass, M. 1983. Effects of extracellular calcium and of light adaptation on the response to dim light in honey bee drone photoreceptors. J. Physiol. **344:** 525-548.

(28) Stern, J.; Chinn, K.; Bacigalupo, J.; and Lisman, J. 1982. Distinct lobes of Limulus ventral photoreceptors. I. Functional and anatomical properties of lobes revealed by removal of glial cells. J. Gen. Physiol. **80:** 825-837.

(29) Stern, J., and Lisman, J. 1984. Isolation of a substance with internal transmitter-like activity in Limulus ventral photoreceptors. Invest. Ophthalmol. Vis. **25:** 113.

(30) Stern, J., and Lisman, J.E. 1982. Internal dialysis of Limulus ventral photoreceptors. Proc. Natl. Acad. Sci. USA **79:** 7580-7589.

(31) Stieve, H., and Bruns, M. 1978. Extracellular calcium, magnesium, and sodium ion competition in the conductance control of the photosensory membrane of Limulus nerve photoreceptor. Z. Naturforsch. **33c:** 574- 579.

(32) Stieve, H., and Pflaum, M. 1978. Lowering the ratio of extracellular calcium to sodium mimics the effects of light on the photosensory membrane of Limulus ventral nerve photoreceptor. Vision Res. **18:** 883-885.

The Molecular Mechanism of Photoreception, ed. H. Stieve, pp. 199-230. Dahlem Konferenzen 1985. Berlin, Heidelberg, New York, Tokyo: Springer-Verlag.

Bumps, the Elementary Excitatory Responses of Invertebrates

H. Stieve
Institut für Biologie II
Verfügungszentrum, RWTH Aachen
5100 Aachen, F.R. Germany

Abstract. Due to the successful absorption of a photon, a rhodopsin molecule is light-activated and starts a sequence of causal steps which lead in the Limulus photoreceptor to the generation of a relatively large elementary excitatory response, the "bump." A bump is a transient increase of the cation conductance of the visual cell membrane and follows photon absorption after a long, greatly variable delay (on the average, latency ca. 200 ms). The bump size also varies greatly (average amplitude $\bar{A}$ ca. 1 nA, or average current-time integral $\bar{F}$ ca. 50 pAs), indicating the variation of the degree of amplification in the transduction process. A bump is based on the transient opening of a large number (up to 10^3-10^4) of ion-specific channels through the cell membrane.

The visual cell changes its sensitivity according to the ambient illumination. A major part of this adaptation is accomplished by a control process which mainly regulates the degree of amplification that determines the size of the bump (bump adaptation, according to the adapting bump model). There are at least two mechanisms responsible for light adaptation: a faster one, a feedback loop, which regulates the sensitivity of the photoreceptor cell by variation of the intracellular level of free calcium ions, and a slower one which is not - or is much less - calcium-dependent.

A plausible description of the mechanism of bump generation includes the enzymatic production of transmitter and transmitter diffusion to the light-controlled ion channels which are distributed over a large area of the photosensory membrane. A time-consuming process which activates an enzyme could determine the latency. Transmitter diffusion over the "bump-speck" could be responsible for the bump rise, and either the stochastic closure of ion channels or the time course of transmitter decay could determine the exponential bump decay.

BUMPS, THE ELEMENTARY EXCITATORY EVENTS*

If the dark-adapted photoreceptor cell of Limulus is stimulated by light flashes which are so weak that not every flash evokes a light response, one observes "bumps" (Fig. 1, left), responses of the photoreceptor to the successful absorption of single photons. The conclusion that light-evoked bumps are single photon-evoked events follows from the statistical analysis of the number of bumps evoked by dim flashes (10, 89, 90) and of the bump interval distribution (42). The quantum efficiency for bump generation is, according to Lillywhite (44), greater than 0.5 in the locust retinular cell. This number indicates that almost every light activation of a rhodopsin molecule triggers the generation of a bump, since the quantum efficiency of rhodopsin photoisomerization is 0.6-0.7.

A bump is based on considerable transient increase in cation (mainly sodium) conductance of the photosensory membrane, which follows photon absorption after a considerable delay. Bumps can also be generated in the cell spontaneously after a long stay in the dark.

The current bump amplitude is variable; it can be as large as 6 nA. We do not know the smallest bump size, since we can only detect bumps when they are larger than the noise level of our registrations, which means bumps with amplitudes larger than about 20 pA. There are only very few bumps which have amplitudes just above the noise level (Figs. 3 and 8), however, one cannot exclude the possibility of many smaller bumps hidden in the noise. In unclamped Limulus photoreceptors one often finds amplitude distributions with many small voltage bumps which are separated by an intermediate frequency minimum from a maximum of many bumps of medium amplitudes (73, 91).

A large bump of a dark-adapted Limulus photoreceptor is based upon a cation conductance increase of about 10-20 nS in bump maximum. The plausible assumption that a bump is based on the superposition of many incompletely synchronized, i.e., concerted, opened ion channels is supported by the registration of single-channel events (3). One can estimate that in the maximum of a large bump up to 10^3-10^4 ion channels with a single-channel conductance of 10-20 pS (3, 84) are opened simultaneously. Bump generation thus involves a considerable amplification (as defined by the number of ion channels opened due to the light activation of a single rhodopsin molecule).

---

*Bumps can be recorded as membrane current signals (current bumps) or as membrane voltage signals (voltage bumps). Voltage bumps are secondarily modified due to voltage-sensitive conductances (e.g., regenerative components) and saturation phenomena. The phenomenology of the bumps of invertebrate photoreceptors is treated here mainly through the example of the current bumps (recorded under voltage-clamp conditions) of the photoreceptor of Limulus, the best investigated example. Comparison will be made with bumps of other species (such as locust and fly).

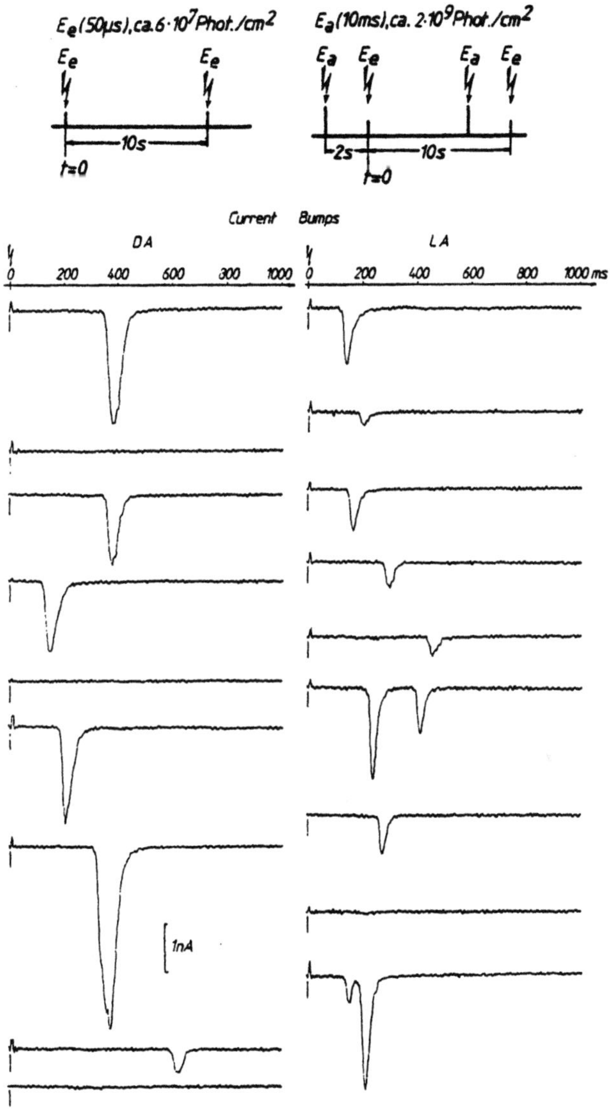

FIG. 1 - Current bumps measured from a Limulus ventral nerve photoreceptor under voltage-clamp conditions. Left column: dark-adapted photoreceptor; right column: photoreceptor weakly light-adapted by a conditioning, light-adapting flash, two seconds prior to the bump-evoking flash. Bump-evoking flash: ca. 6 x 10^7 photons/cm^2, duration 50 µs, repetition time 10 s; conditioning, light-adapting flash:ca. 2 x 10^9 photons/cm^2; both flashes 450 nm, membrane potential clamped to -40 mV; 15°C. Top: stimulus regime.

Variability of Bump Parameters

In order to characterize the size and shape of bumps, a number of parameters have been measured: the latency TLAT, the bump size characterized by bump amplitude A or by the current-time integral F, the bump width TB, etc. Keiper, Schnakenberg, and Stieve (43) have demonstrated a useful parameterization of the bump (Fig. 2), including the bump latency TLAT (ca. 200 ms), the duration of the bump rise TR, the slope $\underline{m}$ of the almost linear phase of the bump rise, and the time constant λ of the exponential phase of the bump decay. We call these four parameters "primary" parameters because they are not correlated with one another as shown by Keiper (42), whereas the bump amplitude and current-time integral are "secondary" parameters, depending mainly on slope and duration of the bump rise (cf. Schnakenberg and Keiper, this volume).

Under identical stimulus conditions (identical photons, identical state of adaptation) bumps vary greatly in size and shape, which is expressed in the variation of the described parameters. Most parameters have bell-shaped, more or less asymmetric frequency distributions (Fig. 3 and 8). Grzywacz and Hillman (32) and Goldring and Lisman (personal communication) have investigated the frequency distribution of the bump current-time integrals and found this always to be exponentially falling distributions. This does not quite agree with our observations. We have found in many cases monotonically falling distributions but also a substantial percentage of cells with asymmetric, more or less bell-shaped curves (see Figs. 3 and 8; Stieve and Klomfaß, unpublished; and (42)).

It has been shown, e.g., that the variations of size and latency of light-evoked bumps (Fig. 1, left, Fig. 3, DA) are not correlated with each other ((38, 43, 74, 87), and Goldring and Lisman, personal communication). The great independent variation of latency and bump size demonstrates that

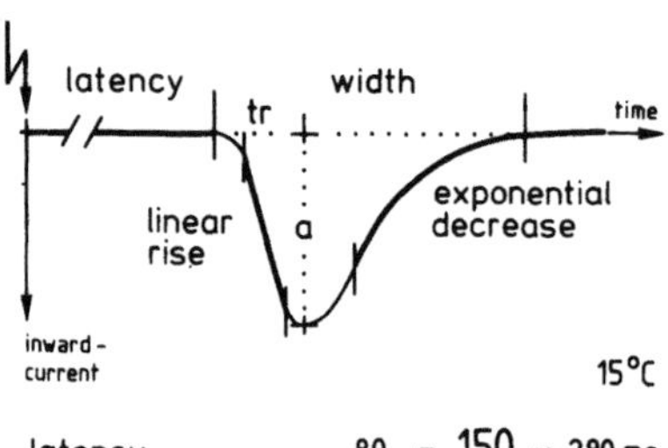

latency	80 - 150 - 280 ms
amplitude	0.2 - 0.8 - 3.0 nA
width	40 - 80 - 140ms
t-rise : t-decrease	ca.1 : 2

FIG. 2 - Schematic representation of the time course of a current bump with specification of the variation of some shape parameters.

delay and size of the bumps are determined by different processes, each depending on reactions of only a small number of molecules. The great variation of latency and of time course of the bumps of invertebrates described here is in striking contrast to the good synchronization of the single photon-evoked events of vertebrates which show almost no variation in latency and time course and vary in current only by a scaling factor (6).

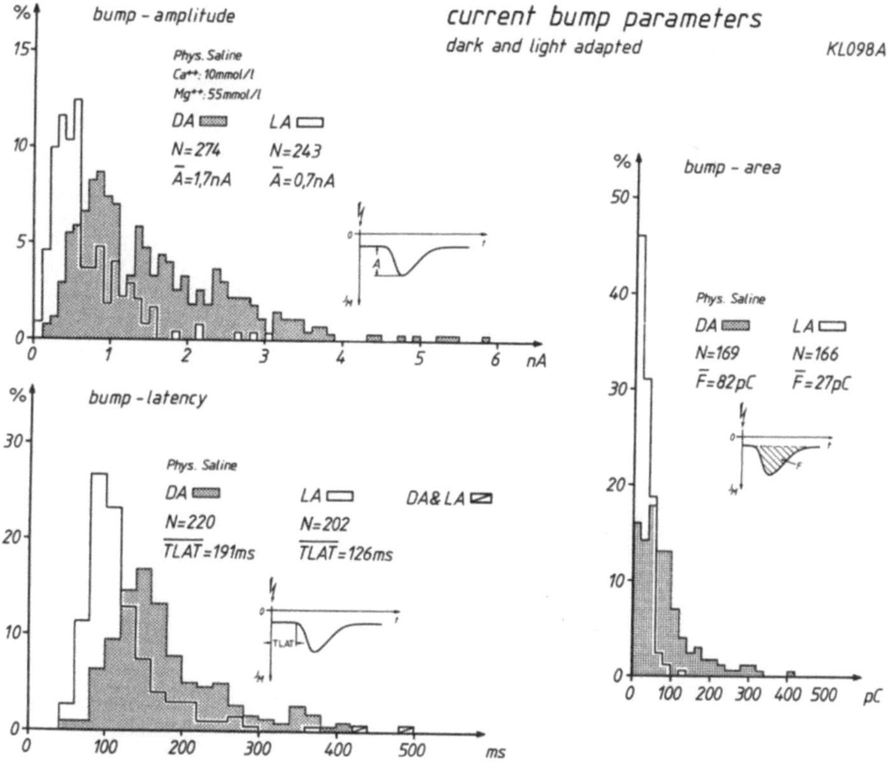

FIG. 3 - Frequency distributions of bump amplitudes (above), bump current-time integrals (area, right), and bump latencies (below) of a Limulus ventral nerve photoreceptor cell in dark-adapted (DA) and weakly light-adapted (LA) conditions. The stimulus program is described in Fig. 1. Only latencies of first bumps following the bump-evoking flash are plotted; amplitudes plotted from all single and first bumps, current-time integrals of all single bumps. N is the number of the bumps accounted for, A, F, TLAT are the arithmetical averages of the bump amplitudes, bump current-time integrals, and bump latencies, respectively. Bump-evoking flash: E_e ca. 9 x 10^7 photons/cm^2; conditioning, light-adapting flash: E_c ca. 8 x 10^9 photons/cm^2; otherwise as in Fig. 1.

Bump size and latency are influenced differently by the mutation norp A in Drosophila (59). The bump parameters depend on temperature to different degrees. The latency shows a much stronger temperature-dependence (temperature coefficient Q_{10} ca. 4) as compared to bump amplitude and bump width (Q_{10} around 2.5) (2, 66, 87). This indicates that the different processes, which are rate-determining for latency and other bump parameters, differ distinctly in their activation energy.

The bump parameters also differ in their dependence on the extracellular calcium concentration. Lowering the extracellular calcium concentration from 10 mmol/l to 250 µmol/l (the Mg^{2+} concentration increased from 55 to 100 mmol/l; see Table 1) causes an increase of the average bump amplitude by a factor of 1.2 and that of the current-time integral by a factor of 2.1; the latency is prolonged by a factor of 1.4 (Table 1, and Stieve and Klomfaß, unpublished).

TABLE 1 - Average current bump parameters dependent on adaptation and external calcium concentration. DA, dark-adapted; LA, weakly light- adapted. A: Amplitude; TLAT: latency; T1, T2: half-time of rise and decline; TB: width; F: current-time integral. Latencies from all first bumps; amplitudes and T1 from all single and first bumps; T2, TB, and F for all single bumps. About 80 bumps per sample of the individual experiment. First column: Mean ± S.E. of the mean. Errors for values of other columns have similar magnitude.

	Physiological saline 10 mmol/l Ca^{2+} 55 mmol/l Mg^{2+} (eleven experiments)		250 µmol/l Ca^{2+} 55 mmol/l Mg^{2+} (six experiments)		250 µmol/l Ca^{2+} 100 mmol/l Mg^{2+} (five experiments)	
	DA	LA	DA	LA	DA	LA
A (nA)	0.93 ± 0.13	0.78	1.66	0.79	1.14	0.79
TLAT (ms)	311.00 ± 26.00	249.00	432.00	330.00	647.00	417.00
T1 (ms)	16.00 ± 1.00	14.00	20.00	15.00	19.00	16.00
T2 (ms)	27.00 ± 2.00	23.00	39.00	29.00	38.00	27.00
TB (ms)	103.00 ± 10.00	84.00	149.00	100.00	144.00	87.00
F (pAs)	46.00 ± 8.00	32.00	111.00	40.00	95.00	38.00

BUMP SUMMATION

The number of bumps evoked by a light stimulus increases linearly with the intensity of the light stimulus, that is to say, with the number of absorbed photons (65, 90). With stronger light stimuli, bumps superimpose and fuse to a relatively smooth signal, the macroscopic receptor current (1, 25). Figure 4 shows a comparison of a macroscopic receptor current elicited by a strong stimulus with the artificial summation of bump registrations, the "bump sum." The artificial bump sum shows the general appearance of a macroscopic receptor current but also significant differences in shape to the macroscopic receptor current. Most conspicuously, the bump sum has a much slower time course than the macroscopic receptor current (especially, the latency and time-to-peak are much longer. These differences are only to a small part due to the different experimental conditions (different cells, different light intensities, and different light adaptation; for details see legend of Fig. 4). This means that the time course of the macroscopic receptor current cannot be sufficiently described by linear bump summation alone (see below).

Changes in the external calcium concentration, in some respects, cause similar changes in the artificial bump sum to those observed in the macroscopic receptor current: decreasing the calcium concentration from 10 mmol/l to 250 µmol/l prolongs in both cases the latency, raises the peak amplitude, and slows down the declining phase ((74, 76), and Stieve et al., unpublished).

Since the number of light-evoked bumps is proportional to the number of absorbed photons (90), it was to be expected that the amplitude of the

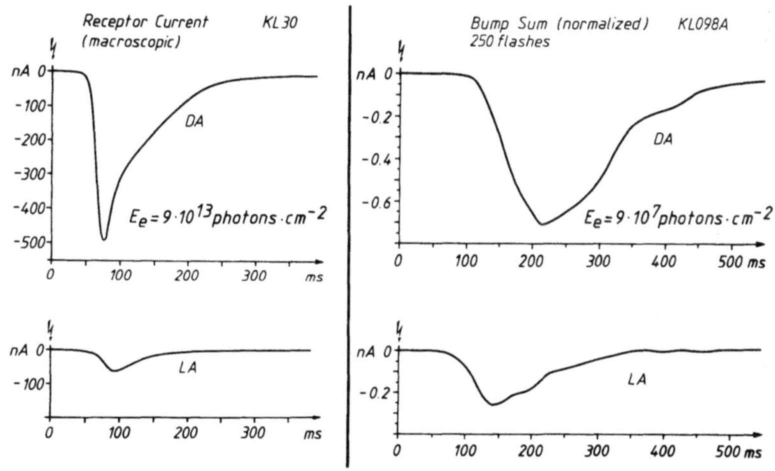

FIG. 4 - Comparison of the macroscopic receptor currents evoked by strong stimuli (left) of a dark-adapted or light-adapted photoreceptor cell with the artificial summation of current bump responses to very weak flashes (right) of another cell in the dark- or light-adapted state. Experimental conditions - left: receptor current-evoking flash ca. 9×10^{13} photons/cm^2, duration 10 ms, membrane potential clamped to -50 mV; record of the light-adapted state: 18 s prior to the receptor current-evoking flash as above, the cell was illuminated for 2 s by a conditioning illumination (intensity ca. 2.3×10^{15} photons/cm^2); right: bump-evoking flash 9×10^7 photons/cm^2, flash duration 50 µs, flash repetition time 10 s, membrane potential clamped to -40 mV; records of the light-adapted state: 2 s prior to the bump-evoking flash the cell was illuminated by a 10 ms conditioning flash (8×10^9 photons/cm^2). Wavelength of stimulating and conditioning light, 540 nm; temperature, 15°C. The bump sum was formed by linear summation, millisecond for millisecond, starting from the light stimulus. The current of the bump sum is scaled for a single bump response. The current-time integrals of the signals are left: DA 38 µAs, LA 4.2 µAs; right: DA 140 pAs, LA 32 pAs.

receptor current would rise linearly with increasing stimulus intensity. Such a linear rise has been observed only for the low-intensity range in Limulus ventral nerve photoreceptor by Lisman and Brown (47). In a medium range of stimulus intensities the rise is even steeper, "supralinear," and at still

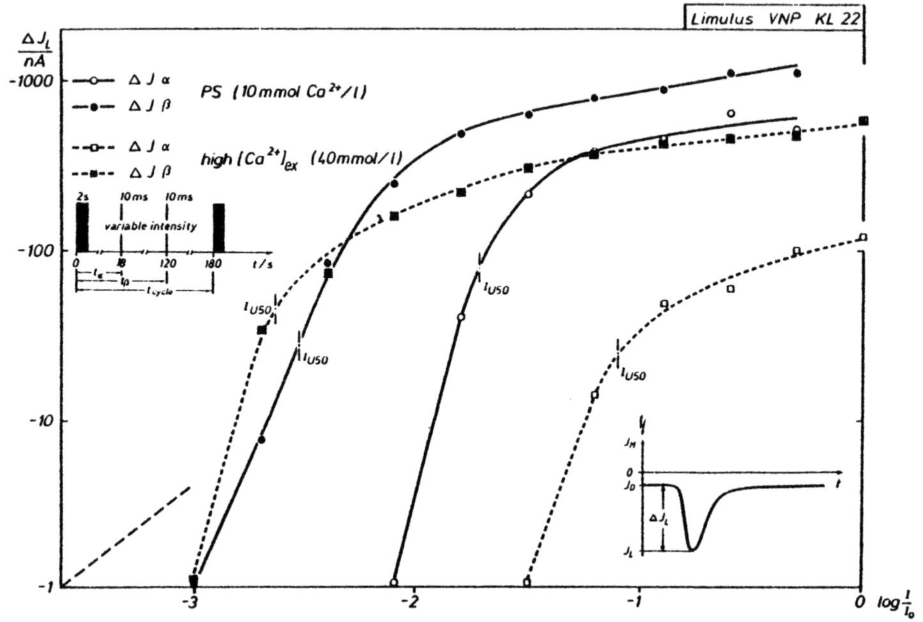

FIG. 5 - Amplitude of receptor current as function of stimulus intensity of a Limulus ventral nerve photoreceptor cell in two defined states of adaptation and two external Ca^{2+} concentrations. The upper left inset explains the stimulus program, which was repeated every 3 min (t_{cycle}). A strong, constant, light-adapting illumination (duration 2 s, intensity ca. 4 x 10^{16} photons/cm^{2}s; 540 nm) was followed after 13 s (t_α) and 120 s (t_β) by 10 ms test flashes, the intensity of which was varied in order to obtain two stimulus response curves. I_0 ca. 3.7 x 10^{16} photons/cm^2, 540 nm, 15°C; membrane potential clamped to -50 mV. The first test stimulus (α) stimulates the relatively light-adapted photoreceptor, the second test flash (β) the fairly dark-adapted photoreceptor. The amplitude ΔJ_L of the light-induced receptor current (inset right, below) is plotted versus the test light intensity with a double logarithmic scale. The extent of the shift of the light-adapted curve with respect to that of the dark-adapted depends, ceteris paribus, upon the degree of light adaptation and becomes larger the higher the external Ca^{2+} concentration is. The two slopes of the curve of the dark-adapted photoreceptor in physiological Ca^{2+} concentration are 3.1 and 0.23, respectively. The dashed line in the corner indicates slope 1. The intensity I_{U50} marks the half-saturation intensity of the membrane voltage response (77).

higher intensities the slope becomes much less steep ((13, 71), and Grzywacz and Hillman, personal communication). In a double logarithmic plot (such as Fig. 5) the amplitude of the receptor current rises with the intensity of the stimulating light flash with a slope of about one (not seen in Fig. 5), and then with a steeper slope up to a certain intensity I_K. Above this knee the slope of the curve is much less steep. According to Brown and Coles (13), the stimulus energy at the knee of the curve corresponds in the dark-adapted Limulus ventral nerve photoreceptor cell to about 350 rhodopsin photoisomerizations.

The supralinear dependence of the light response upon the stimulus intensity has been observed at low and medium stimulus intensities by Lisman and Brown (47), Fein and Charlton (26), Coles and Brown (18), and Stieve and Pflaum (79). In the supralinear range the response amplitude versus stimulus intensity rises in a linear plot with increasing slope and in a double logarithmic plot with a slope greater than one (up to 3 to 5, Fig. 5). This strong supralinearity is not due to increasing bump synchronization with rising flash intensity since it is also seen when the current-time integral and not the amplitude of the receptor current is plotted (Grzywacz and Hillman, unpublished; Stieve, unpublished). A hypothetical explanation for this supralinearity is given below.

The macroscopic receptor current is composed of many bumps; however, as shown in Fig. 4, its time course is not simply the result of the linear summation of bumps which are independent of each other. If we assume that the bump latency distribution is (to a first approximation) not much altered during a response to a stronger light flash, we can try to mimic the macroscopic response (as in Fig. 4, left) by the adjusted, time-dependent attenuation only of the bump sizes in the course of the light response (Fig. 6). One can simulate the shape of the macroscopic response by convoluting the bump sum (or the bump latency distribution of all bumps) with an "attenuation function" $\underline{a}(t)$, the value of which increases with time. The early bumps in a dark-adapted photoreceptor are not or very little attenuated, whereas consecutively occurring bumps are more attenuated the later they occur. The attenuation function $\underline{a}(t)$, which is used by the cell to control the size of the bumps constituting the macroscopic receptor current, could be the current-time integral of the receptor current (as assumed in Fig. 6) or the instantaneous values of the intracellular Ca^{2+} concentration as it has been measured in the Arsenazo response (39, 50, 57). By such convolutions we (Stieve et al., unpublished) obtained naturalistic shapes of simulated macroscopic receptor currents which have several properties like experimentally recorded receptor currents and their changes with stimulus intensity and with light adaptation (see below). The increasing influence of this attenuation function $\underline{a}(t)$ with increasing intensity of the stimulating light flash results in the shortening of the time-to-peak. Moreover, this convolution (as demonstrated in Fig. 6) gave unexpectedly an explanation of the two components C1 and C2 of the receptor current characterized by Maaz, Nagy, Stieve, and Klomfaß (49).

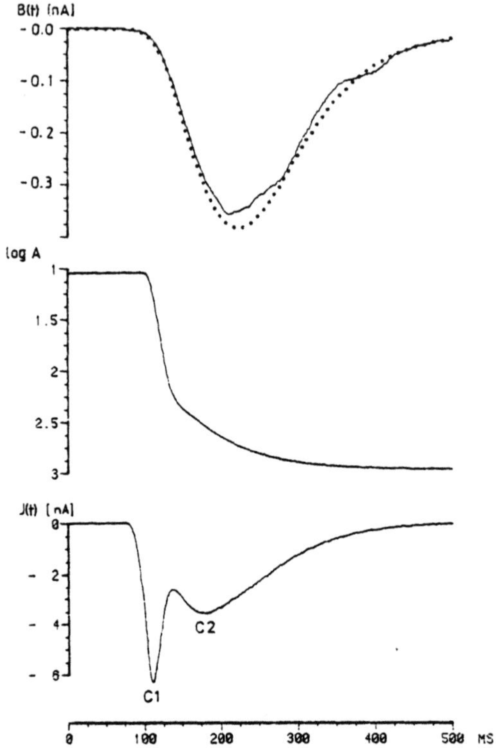

FIG. 6 - Simulation of macroscopic receptor current signal by convoluting the bump sum $\underline{b}(t)$ with an attenuation function $\underline{a}(t)$. Top: Bump sum achieved by linear summation of 250 bump responses to very dim light flashes (normalized, same as in Fig. 4). Dotted line: function $\underline{b}(t)$ approximating the shape of the bump sum. Middle: Attenuation function $\underline{a}(t)$ plotted logarithmically. Example chosen: modified current-time integral of receptor current.

$$\underline{a}(t) = 1 + a_o + s \cdot \int_0^{t-\Delta t} J(t')\,dt' .$$

Bottom: simulated receptor current

$$J(t) = \frac{N \cdot \underline{b}(t)}{\underline{a}(t)} ,$$

where $\underline{b}(t)$ is normalized and N represents the number of bumps evoked by the light flash. Different intensities of the light stimulus can be represented by different values of N, different levels of preadaptation by different values of a_o. In agreement with our experimental observations, the components of C1 and C2 of the receptor current depend strongly on the state of adaptation. In order to demonstrate the two components distinctly in the example shown, a somewhat light-adapted state (a_o = 10) is assumed (Stieve et al , unpublished).

Besides the time-to-peak, also the latency of the macroscopic receptor current is drastically shortened with increasing stimulus intensity (e.g., (75)). The end of the latency is marked by the first occurring bump constituting the macroscopic receptor current, that is, the bump with the shortest latency following the stimulating flash. If we assume that the latency of the first bump is independent of the other bumps triggered by the same flash, then the drastic shortening of the latency of the macroscopic receptor current with increasing stimulus intensity would be entirely a consequence of the frequency distribution of bump latencies (Fig. 3): the more photons are absorbed from a flash, the more bumps with shorter, less probable latencies will occur. That is to say, the more photons, the higher the probability for rare, short latencies. In the fly eye, however, an additional process as demonstrated by Hamdorf and Kirschfeld (34) further shortens the latency: this will be discussed below.

ADAPTATION

The photoreceptor cell adjusts its sensitivity within a large range according to the mean ambient brightness. In the foregoing we have considered the visual cell in the state of maximal sensitivity, dark-adapted after a long stay in the dark. Figure 4, left, shows that after light adaptation the same light stimulus causes a much smaller macroscopic receptor current than that of the dark-adapted cell. In Fig. 5 it can be seen that light adaptation causes the receptor current amplitude versus stimulus intensity curve to shift to the right, to higher stimulus intensities; the amount of shift depends upon the degree of light adaptation. The intensity I_K which corresponds to the location of the knee between the higher steepness and the lower steepness of the curve is also shifted to the right. According to the "adapting bump model" as formulated by Dodge, Knight, and Toyoda (25), the macroscopic receptor current of the light-adapted cell is smaller, since the size of the bumps constituting the receptor current is diminished by light adaptation (bump adaptation). This was shown by noise analysis (25, 84, 88) and could also be shown by directly observed bumps (73, 74).

Figure 1, right column, shows bumps evoked by identical flashes as in the left (dark-adapted) column, but here the bump-evoking flashes were delivered two seconds after a light-adapting, desensitizing flash had been administered. The light adaptation due to the conditioning flash was weak enough that individual bumps could still be observed. On the average, these bumps of the weakly light-adapted photoreceptor are smaller than those of the dark-adapted one. More frequently, two bumps are observed to follow the bump-evoking flash (see below). In contrast to observations of Srebro and Behbehani (66), we observed that light adaptation influences not only the bump size but also the latency and time course of the bump. Figure 3 shows that the frequency distributions of bump amplitudes, current-time integral, and bump latencies are narrowed and shifted to smaller values due to light adaptation. The bump width is shortened by light adaptation, and the bump rise as well as the bump decline are both accelerated (Table 1, and (78)).

Artificially summated bump responses under those light-adapted conditions (Fig. 4) result in a "light-adapted" sum curve which shows some characteristic features of light adaptation: diminution of the response signal size and shortening of the latency. Comparison to the light-adapted response of the macroscopic receptor current again reveals similarities and significant differences: again, the differences are only to a small part due to the different stimulus conditions indicated in the legend of Fig. 4 (see above), and a similar shape for the macroscopic receptor current signal of the light-adapted photoreceptor is obtained by convolution of the bump sum (or the latency distribution of all bumps) of the light-adapted photoreceptor with the above-described attenuation function a(t) (see Fig. 6; and Stieve, Schnakenberg, Kühn, and Reuss, unpublished). This may, however, not be sufficient to model the prolongation of the latency of the macroscopic receptor current due to stronger light adaptation as described by Stieve, Bruns, and Gaube (75).

Raising the extracellular Ca^{2+} concentration causes changes in bump parameters similar to those due to light adaptation; lowering of the extracellular Ca^{2+} concentration causes changes in the opposite direction, resembling increasing dark adaptation ((74), and Table 1).

There are at least two different processes which cause bump adaptation, i.e., bump size diminution in light adaptation (see also (68)).

1. A feedback mechanism regulating the sensitivity of the photoreceptor cell by controlling the intracellular calcium ions was demonstrated by Lisman and Brown (46, 48) and Fein and Charlton (27). Light-adapting illumination causes an increase in the intracellular concentration of free calcium ions which causes desensitization of the cell ((11, 12, 39, 50, 57), and Brown, this volume). The feedback loop consists of two elements: a) the larger the increase in intracellular Ca^{2+} concentration, the greater the electrical light response or light stimulus; and b) the higher the increase in intracellular Ca^{2+} concentration, the stronger the desensitization (response diminution).

The most probable explanation for the reduction in receptor current is that the raised intracellular calcium concentration attenuates the bump size; this has, however, not yet been demonstrated by noise analysis nor by directly observed bumps.

The calcium-dependent feedback loop for sensitivity control can also be demonstrated in the shift of the response versus stimulus intensity curve due to light adaptation in different extracellular calcium concentrations (Fig. 5), although changes in external Ca^{2+} concentration are not simply reflected in changes in the intracellular free calcium ion concentration. Whereas the dark-adapted curve is almost uninfluenced by external calcium concentration, the shift due to light adaptation becomes greater, the higher the external calcium concentration is (Fig. 5, and (76)). These results indicate that the dark-adapted cell can maintain more or less its normal intracellular Ca^{2+} level when the extracellular calcium level is changed, whereas in response to the light-adapting illumination the intracellular Ca^{2+} concentration is transiently raised. A substantial amount of this calcium enters the cell from extracellular sources driven by the electrochemical calcium gradient (39).

2. Weak light adaptation, weak enough that bumps can still be observed (as demonstrated in Figs. 1 and 3) depends only slightly on the extracellular calcium concentration between ca. 1 nmol/l to 100 mmol/l (74), as opposed to the strong calcium dependence of strong light adaptation. Even in an external Ca^{2+} concentration as low as ca. 1 nmol/l, a considerable bump size reduction by weak light adaptation can still be observed. The mechanism of this calcium-independent (or only slightly dependent) light adaptation has not yet been elucidated. Table 1 shows that in external Ca^{2+} concentration lowered to 250 µmol/l, bump parameters which have been enlarged for the dark-adapted photoreceptor are reduced by light adaptation to about the same absolute values as in physiological Ca^{2+} concentration, except for the latency which is less reduced.

FACILITATION

Facilitation was first described by Hanani and Hillman (36) for the macroscopic light response of the barnacle. It has also been observed on the bump level in the Limulus photoreceptor (73). Figures 7 and 8 show that the same stimulus program can cause either light adaptation or facilitation depending on the strength of the conditioning illumination. Very weak conditioning flashes (10^6-10^7 (540 nm) photons/cm^2) two seconds prior to the bump-evoking illumination cause facilitation, namely, an increase in number and size of the observed light-evoked bumps; the distribution of bump latencies, however, is already shifted to shorter values as in light adaptation under these conditions. Stronger conditioning flashes cause a diminution of bump size (Figs. 7 and 8).

The increase in the number of light-evoked bumps means that the quantum efficiency is raised, that is to say, the probability that a photoisomerization causes a bump is increased. Facilitation is smaller in 40 mmol/l Ca^{2+} concentration than in 10 mmol/l (by about the same amount by which bump size and frequency are greater in 40 mmol/l calcium than in 10 mmol/l calcium (73)). However, studied over a wider range of external Ca^{2+} concentrations, the extent of facilitation did not show a systematic dependence on the extracellular calcium concentration which was varied between 1 nmol/l and 100 mmol/l (74). The quantum efficiency of the dark-adapted photoreceptor is greater in a calcium concentration raised to 40 mmol/l than in physiological 10 mmol/l calcium. This had been interpreted by us as a calcium dependence of the coupling between the light-activated rhodopsin and the trigger for bump generation (73), which has been recently confirmed directly by Ca injection into Limulus photoreceptors (A. Fein, personal communication, and see below).

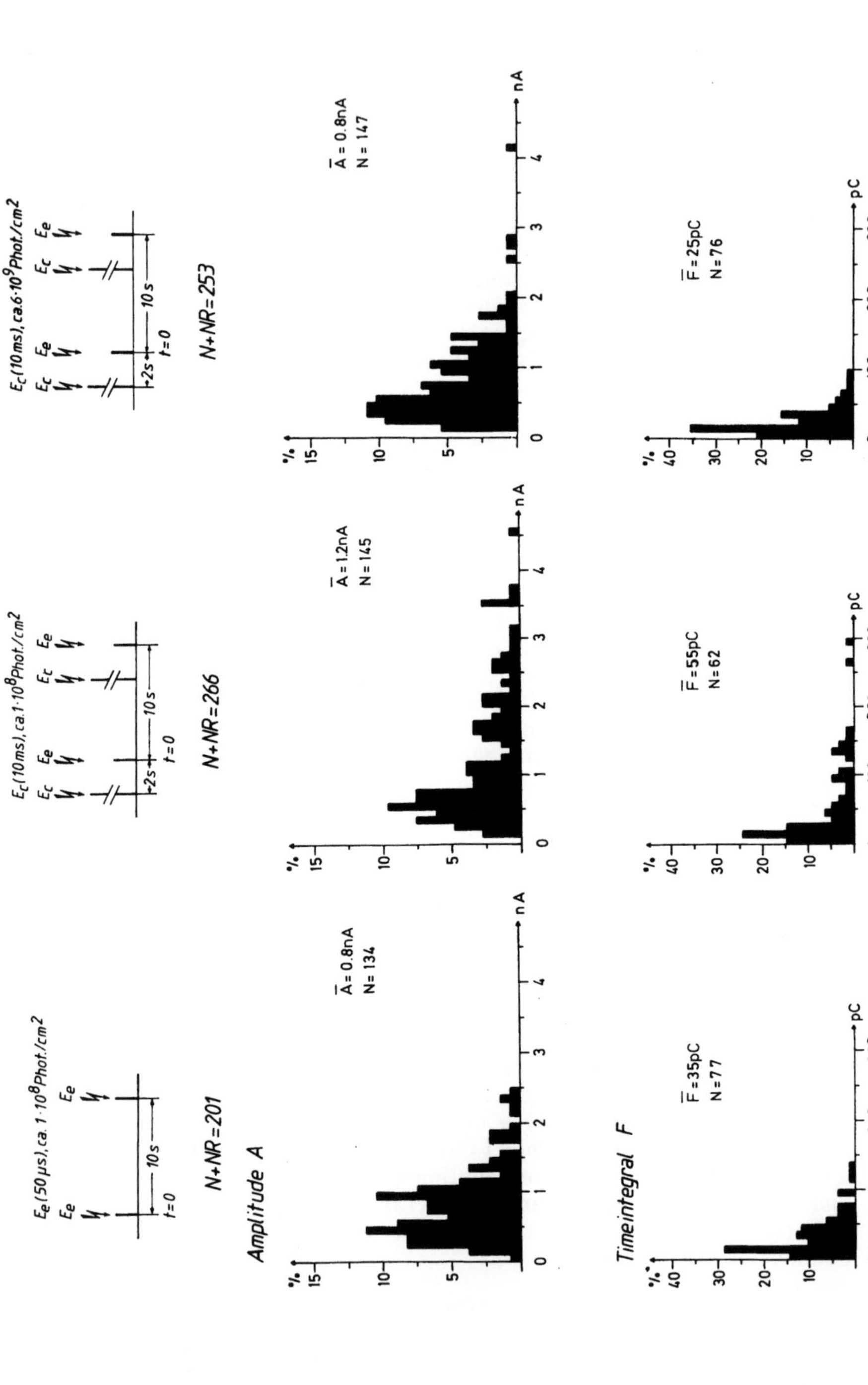

Frequency Distributions of Bump Parameters in Dependence of Strength of Light Adaptation

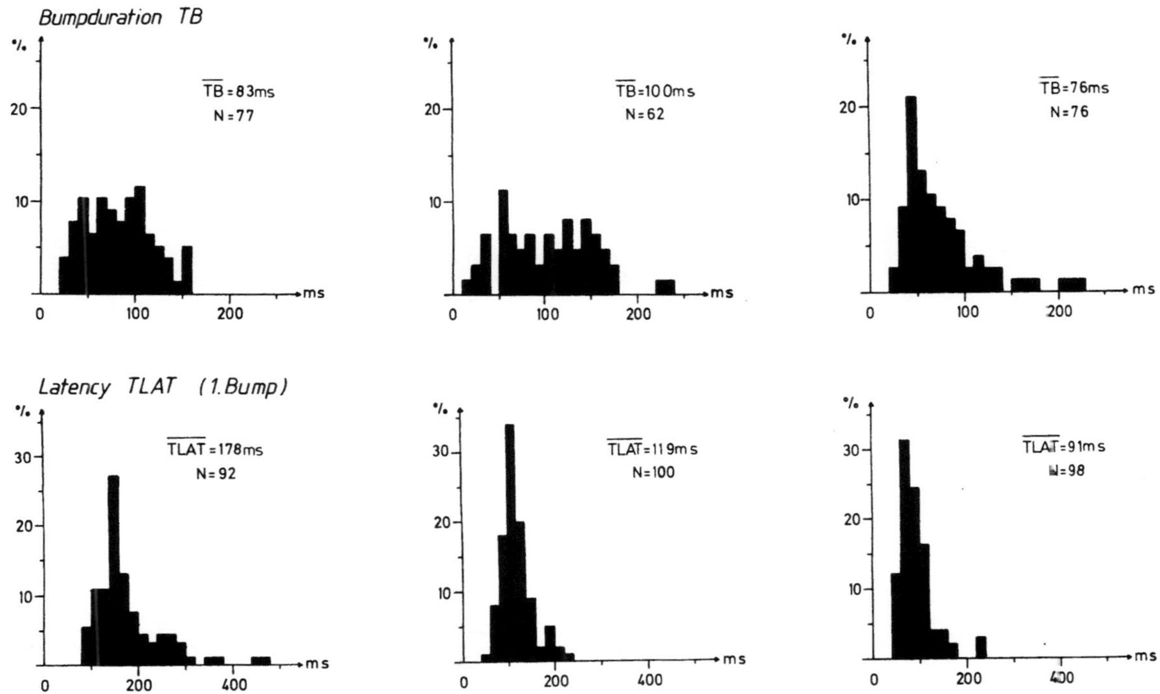

FIG. 7 - Frequency distribution of bump parameters (amplitude, current-time integral, duration, and latency) of Limulus ventral photoreceptors in three different states of adaptation. First column: dark-adapted; second column: very weakly light-adapted; third column: weakly light-adapted. Same experiment as in Fig. 7. Distributions of amplitudes for all single and first bumps, current-time integrals and width for all single bumps, latencies for all first bumps.

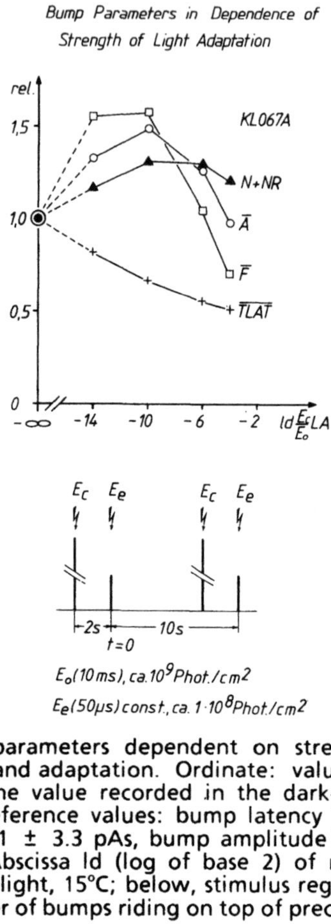

FIG. 8 - Average bump parameters dependent on strength of light adaptation demonstrating facilitation and adaptation. Ordinate: value of bump parameter normalized with respect to the value recorded in the dark-adapted state of Limulus ventral photoreceptor. Reference values: bump latency $\overline{TLAT}$ 178 ± 8 ms, bump current-time integral $\overline{F}$ 35.1 ± 3.3 pAs, bump amplitude $\overline{A}$ 0.83 ± 0.05 nA, bump frequency N + NR 201. Abscissa ld (log of base 2) of normalized energy of the conditioning flash; 540 nm light, 15°C; below, stimulus regime. N: number of single and first bumps; NR: number of bumps riding on top of preceding bumps.

SPONTANEOUS BUMP

Spontaneous bumps can be recorded selectively after a long stay (more than several seconds) of the photoreceptor cell in the dark. The frequency of the spontaneous bumps varies greatly. It is on the order of up to 0.2 bumps/s and is reduced by about half when the extracellular calcium concentration is raised from 10 to 40 mmol/l (73, 74). Adolph (2) and Srebro and Behbehani (67) measured a temperature coefficient Q_{10} of around 4 for the spontaneous bump frequency, demonstrating a high activation energy for the rate-limiting step. The average amplitude is smaller for spontaneous bumps than for light-evoked bumps (e.g., $\overline{A}$ 0.51 nA and 0.98 nA), but both amplitude distributions overlap (74, 91). Spontaneous bumps are diminished in size by light adaptation, like light-evoked bumps (74). Fein and Hanani (29) have demonstrated that the rate and size of spontaneous bumps can be greatly increased for several minutes after a dark-adapted photoreceptor has been stimulated by conditioning illumination; these bumps may not, however, be of the same kind as other spontaneous bumps.

Kaplan and Barlow (40) have shown that the frequency of spontaneous bumps in the lateral eye of Limulus is strongly influenced by diurnal rhythmics; the spontaneous bump frequency is larger in the daytime than during the night. The same spontaneous bump level as occurs at night can be recorded when the efferent nerve fibers projecting to the retinular cells are electrically stimulated or when the neurotransmitter octopamine is added to the saline bathing the photoreceptors (41). These treatments suppress only spontaneous bumps, not light-evoked bumps.

Goldring (31) and Lisman (45) have investigated the nature of the trigger of spontaneous bumps. According to their working hypothesis, spontaneous bumps are triggered by an activated state of the metarhodopsin molecule which can be reached from the inactivated metarhodopsin state of the molecule (Fig. 9 illustrates a rhodopsin reaction scheme proposed by these authors). By experiments with the UV-sensitive median eye of Limulus, Lisman (45) could show that the spontaneous bump rate is higher if a large amount of rhodopsin has been shifted to the metarhodopsin state than if only a small fraction of rhodopsin was in the meta-state. This makes this plausible explanation more probable, if it can be ruled out that these bumps are not caused by a long-lasting PDA (prolonged depolarizing afterpotential)

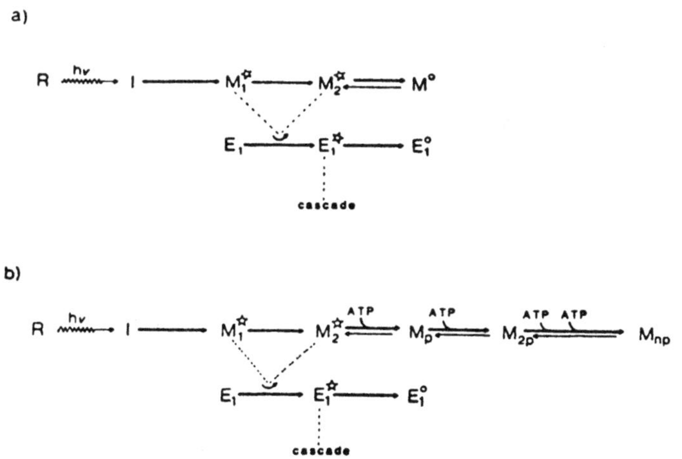

FIG. 9 - Reaction scheme that could generate spontaneous bumps, proposed by Lisman (45), including two sequential active states of rhodopsin M^*_1 and M^*_2. a): Reversion from M^o to M^*_2 generates a bump that is smaller than light-induced bumps which are generated by passage through both of the M^*_1 and M^*_2 states. b): Similar to a) except that the inactivation mechanism is more specific. A single phosphorylation reaction makes M^*_2 inactive; subsequent phosphorylations reduce the probability of reversion of phosphorylated photoproduct to M^*_2.

mechanism (see Minke, this volume). Lisman suggests that phosphorylated metarhodopsin from time to time becomes spontaneously dephosphorylated and thus activated. (Such an experiment could not be done in the ventral photoreceptor or lateral eye of Limulus, since here the absorption spectra of rhodopsin and metarhodopsin overlap practically completely.) According to Lisman (45), two sequential, activated states of rhodopsin M^*_1 and M^*_2 which trigger bump generation could be the basis of the observed differences in average size of spontaneous and light-evoked bumps. According to that scheme, reactions favoring metarhodopsin phosphorylation would selectively reduce the spontaneous bump rate without inhibiting light-evoked bump generation.

CHEMICALLY INDUCED BUMP

A number of pharmaca and treatments evoke bumps or increase the frequency of spontaneous bumps in the photoreceptor of Limulus:

1. phosphatase inhibitors or cyclase activators, fluoride, vanadate, tungstate, molybdate, poorly hydrolyzable GTP analogues such as GTP-γ-S (9, 21);
2. diamide, a sulfhydryl oxidizing agent (35);
3. shifting the extracellular pH from 7 to 8 (20);
4. metabolic inhibitors (60, 61, 69);
5. Recently Brown et al. (16) and Fein et al. (30) demonstrated that inositol polyphosphate injected into Limulus ventral photoreceptors evoked bumps in the dark.

Corson and Fein (21) demonstrated that chemically induced bumps have the typical bump shape and the same reversal potential as light-evoked bumps. They have, as spontaneous bumps, a smaller average bump size than light-evoked bumps. Their size is also diminished by light adaptation. Corson and Fein (21) suggested that these chemically induced bumps may arise from activation of a GTP-binding protein.

Stern, Chinn, Robinson, and Lisman (69) (see also (61, 70)) found that treatment of Limulus ventral photoreceptors with 1-deoxy-glucose or internal dialysis with solutions lacking nucleotides raised the spontaneous bump rate. The spontaneous bump rate can be reduced after intracellular injection of ATP (also observed by Fein and Corson (28)). These authors assume that phosphorylation of metarhodopsin by utilizating ATP keeps metarhodopsin in the inactive, multiphosphorylated state (Fig. 9). Bumps or "noisy" depolarizations with similar characteristics to light-induced bumps of smaller amplitudes are also evoked in insect photoreceptor cells when these are treated with fluoride (55, 62), vanadate, or poorly hydrolyzable GTP analogues. Noise analysis indicates that these noisy depolarizations are due to small bumps (55).

Chemically induced bumps (by vanadate, fluoride, or GTP-γ-S and in external calcium concentration lowered to 1 mmol/l) cluster together in time (22). Stieve and Bruns (73) have shown that the probability for bump generation under steady dim illumination is not strictly time-independent. It is greater shortly after a bump has been generated. Conditioning illumination increases the number of bumps evoked by a light flash, that is to say, the quantum efficiency (facilitation, Stieve and Klomfaß, unpublished; see Fig. 6). Corson, Fein, and Walthall (22) have shown that after treatment with fluoride in an external calcium concentration lowered to 1 mmol/l, a single light-activated rhodopsin molecule may cause the generation of more than one light-evoked bump. The probability as to whether photoisomerization leads to the generation of one (or even more than one) bump, i.e., the quantum efficiency, varies in the cell even under physiological conditions and may thus contribute to light/dark adaptation (73). According to the suggestion of Corson, Fein, and Walthall (22) and Payne (62), GTP hydrolysis normally may help to turn off the production of both light-activated and spontaneous bumps.

THE trp MUTANTS OF DROSOPHILA*

This mutation of Drosophila, of which now several alleles have been isolated (58), causes a defect in the transduction mechanism. The membrane voltage response (receptor potential) to sustained illumination consists merely of a transient but not steady depolarization (23, 56). The expression of this mutation is temperature-sensitive: the degree and rate of decline of the light response during sustained illumination is larger, the higher the temperature during development or during later life of the fly (54). Minke, Wu, and Pak (56) and Minke (53), using noise analysis, showed that the decline in response is caused by a fast decrease in quantum efficiency of bump production during sustained illumination; the bumps which are generated during the response decay are normal in size. This desensitization is not due to a rise in intracellular calcium (53). It is remarkable that the desensitization due to the PDA-producing light exposure (see Minke, this volume) has the same characteristics and about the same duration in the trp mutant as in the wild type, although in the mutant the electrical light response is only a short transient. This indicates that this PDA-related desensitization is caused by a step of the transduction chain prior to the electrical response which is not affected by the mutation (52).

Recently the trp gene has been cloned and the molecular basis of the mutation has been determined at the RNA level (Wong, this volume; (85, 86)). It appears that the trp mutant prevents the expression of the trp gene in that no mRNA is made from the trp gene. It was also observed that the trp gene may be encoding for two distinct mRNAs, perhaps by alternate splicing mechanisms. These results imply that the lack of the trp gene product (one or both of the proteins encoded by the trp gene) would lead to the observed defect in bump occurrence. The trp gene product may be a familiar protein or a novel protein. Whatever it is, its identity and exact role in transduction can be determined much more easily now that the gene

*See also Wong, this volume.

is available. Analysis of this sort should lead to ultimate understanding of the molecular basis of visual transduction.

BUMP VARIATION

Bumps recorded from photoreceptors of different individual Limuli vary considerably in spontaneous frequency and in the shape of the frequency distributions of bump parameters. These differences may be partially due to the differences in the physiological state of the photoreceptors. Bayer and Barlow (5) and Barlow and Kaplan (4) have shown that the mean amplitude of voltage bumps in the Limulus photoreceptor becomes smaller when the blood supply is interrupted; it becomes larger when organ culture medium is used, and it is influenced by octopamine treatment. Goldring and Lisman (personal communication) suggested that the ratio of different populations of bumps may vary from cell to cell. Keiper, Schnakenberg, and Stieve (43) suggest that local differences in the photoreceptors, such as channel densities, may be responsible for the differences in the distributions of "current" parameters as opposed to the more uniform "time-like" parameters of the bumps in different cells (see Schnakenberg and Keiper, this volume).

Since local photon absorption leads to local bump generation, local differences in different areas of the photosensory membrane or in its environment may lead to bump variation. Yeandle and Spiegler (92) have investigated whether bumps evoked in different regions of the same cell by light spots with diameters of ca. 5-10 μm have different average latencies; the results were negative. However, it still may be that local differences lead to different distributions of other bump parameters besides the latency, especially "current" parameters.

There are even greater differences between bumps from different species of animals. In some photoreceptors, such as those of the barnacle and the bee drone, no bumps could be observed. In the fly, Calliphora, the bump latency distribution and the bump size distribution appear to be much narrower than in Limulus (33). No spontaneous bumps have been observed in the retinular cells of the locust (38). The bumps of most species in which bumps have been observed are smaller than those of Limulus, and the largest after Limulus so far are those of the locust (38, 65). It seems plausible to assume that the degree of amplification, that is, the size of the bumps, may differ from species to species. In barnacles and drones the single-photon responses of the dark-adapted photoreceptor cell seem to be so small (but still greater than those of the light-adapted cell) that individual bumps are buried in the noise.

The bump size is of physiological significance: the larger the bumps, the better the ability of the photoreceptor to detect single photons above the noise level, but the weaker the resolution for light intensities, i.e., the discrimination between different grades of brightness.

THE BUMP-GENERATION MECHANISM

What do we know about the mechanisms which trigger and generate bumps?

Nature of Latency

The bump latency, as indicated by its great variation (see above), is determined by a very small number of molecules. The latency seems to be too long by a factor of about 100 to be caused by unhindered Fickian diffusion in a homogeneous medium (37, 42). In a locust, in which the length of the microvilli change during diurnal rhythm by more than twofold, Williams (83) did not find a change in the average bump latency accompanying these changes in microvillus length. This indicates that neither diffusion rate nor any kind of spatial velocity along the microvillus is a main factor determining the latency. The strong temperature-dependence and the dependence on extracellular calcium concentration of the frequency distribution of bump latencies may indicate a calcium-dependent chemical reaction as the latency-determining step, plausibly the process of enzyme activation (see Schnakenberg and Keiper, this volume). Results of Hamdorf and Kirschfeld ((33); see Kirschfeld, this volume) indicate that the activated state of the visual pigment must be present in the fly eye for 3 to 4 ms in order to initiate an electrical light response.

In Limulus the drastic shortening of the latency of the macroscopic receptor current with stimulus intensity may be explained solely on the basis of the statistics of the bump latency distribution (see above). This, however, seems not to be the case in Calliphora. According to Hamdorf and Kirschfeld (34), in this fly the bumps evoked by very dim flashes never have latencies as short as macroscopic receptor potentials evoked by strong stimuli. This indicates nonlinearities: shorter latencies occur when the density of absorbed photons becomes higher. Hamdorf and Kirschfeld (34) report that the latency is shortened stepwise when two photons out of a flash are absorbed in two neighboring microvilli of the fly's visual cell, and even more shortened when two photons from a flash are absorbed within the same microvillus (Hamdorf, personal communication). These authors propose a model in which the latency is spent to cumulate a substance until a certain local threshold is reached which triggers the firing of a bump (see also (63)). So far, indications for this mechanism have not been found in the Limulus photoreceptor.

The frequency distribution of bump latencies is narrowed and shortened when the photoreceptor is weakly light-adapted by a conditioning light flash two seconds prior to the bump-evoking flash (Figs. 7 and 8). This result contrasts with the fact that stronger light adaptation causes a prolongation of the latency of the macroscopic response (75, 76). The underlying mechanism responsible for the shortening of the latency due to weak light adaptation may be a calcium-dependent process. The latency distribution is narrowed and shifted to shorter values by high (40 mmol/l) external calcium concentration and is broadened and shifted to longer latencies by low (250 μmol/l) external calcium concentration (51, 74). Possibly the increase in intracellular calcium concentration due to the conditioning illumination accelerates the time-consuming processes of enzyme activation which determine the duration of the latency.

The long latency may be necessary to provide the system with a high safety factor for the information of successful photon absorption: since Rh* lives long enough, the light-activated rhodopsin molecule Rh* may sequentially induce several reaction steps which come to a successful end (e.g., an activated enzyme). Such a mechanism would warrant a single-photon response with high probability in addition to a very low probability for a "false alarm" (i.e., bump generation without activated rhodopsin) (Schnakenberg, personal communication).

"Bump-speck"

As suggested by Behbehani and Srebro (7) and shown by Brown and Coles (13), a light-activated rhodopsin molecule in the dark-adapted Limulus ventral photoreceptor causes the opening of sodium channels in regions of the cell membrane beyond the microvillus in which the photon was absorbed, but not necessarily more than about 2 μm away. Rhodopsin photoisomerization causes opening of ion channels in a surface area, the "bump-speck," which may include about 1000-2000 microvilli or 1000-2000 light-activated ion channels. A knee in the intensity dependence curve of the receptor current, e.g., Fig. 5, is found (for the dark-adapted photoreceptor) at a stimulus intensity where the stimulating flash evokes about 400 bumps (13). This intensity corresponds to a density of absorbed photons in the photosensory membrane, where the bump-specks start to touch each other and overlap. For the spreading of information over the bump-speck, the hypothesis of diffusing internal transmitter T molecules, first proposed by Cone (19), seems to be reasonable and plausible. An excitatory internal transmitter has not yet been identified.

Inositol polyphosphate evokes bumps of similar shape as light-evoked bumps when pressure-injected into Limulus photoreceptor cells (16, 30). Therefore, it does not seem to be the "terminal" transmitter T which itself opens the light-activated ion channels. Stieve and Bruns (73) reported results which indicate that the coupling between rhodopsin reactions and bump generation is calcium-dependent. This is in accordance with a new report by Fein (personal communication) that calcium injection into Limulus photoreceptor leads to inward current signals similar to light responses. Brown and Rubin (15) demonstrated that inositol triphosphate induces an intracellular calcium increase in Limulus ventral nerve photoreceptor that arises from release of calcium from intracellular stores.

The high amplification results in the opening of about 10^3 ion channels underlying one bump, which may indicate that at least as many T molecules have to be formed, released, or activated in the process of bump generation. This large amplification is most probably achieved by enzymatic reactions. In the vertebrate photoreceptor cell an enzyme cascade could be demonstrated which causes the degradation of cGMP (see Chabre and Applebury, this volume). The high amplification is responsible for the high maximal sensitivity of the Limulus ventral photoreceptor, enabling the cell to detect single-photon responses way above the noise level. The great variation in the size of bumps evoked by identical photons under identical experimental conditions indicates that reactions of only a small number (<~10) of molecules are determinative for the degree of bump size amplification. As mentioned above, latency and amplification are determined by different processes. It seems that substantial amplification does not start before the end of the latency.

Intermediate Processes and the Role of Cyclic Nucleotide

After photon absorption, rhodopsin most probably does not itself form an ion channel through the photosensory membrane but rather, as shown for the vertebrate photoreceptor, activates an enzyme. However, there is as yet no strong evidence that G-protein plays an important act in the phototransduction in Limulus, although some remarkable indications for the importance of G-protein have been found in photoreceptors of Limulus and other invertebrates. Corson and Fein (21) showed that substances which influence the level of cyclic nucleotides in the cell raise the frequency of spontaneous bumps in Limulus photoreceptors (see above). Saibil and Michel-Villaz (64) showed that light-scattering signals similar to those from vertebrate rod outer segments indicate the binding of G-protein to photosensory membranes of the squid, and Brown, Kaupp, and Malbon (14) showed that light influences the level of cyclic AMP in the Limulus photoreceptor but that the changes in the level of cyclic AMP are not rigidly correlated to the excitation of the cell. Injection of the α-subunit of bovine G-protein activated by cGMPγS into Limulus ventral nerve photoreceptors did not cause significant effects on the light response (Stieve, Kühn, and Bruns, unpublished).

Recently Blumenfeld et al. (8) demonstrated, by using membrane preparations from fly eyes, that illumination activates the GTPase activity of the GTP-binding protein in strict correlation to the light-induced formation of activated metarhodopsin as monitored by the prolonged depolarizing afterpotential (PDA) (see also (17, 81, 82, 93)).

Light-activated enzyme reactions could provide the source Q of production, activation, or release of transmitter molecules (Fig. 10). Active transmitter molecules T could start to occur at the end of the latency, and immediately as they occur they could start to open light-activated cation channels Ch (probably by binding to them).

The gating mechanism of the light-activated ion channels is not known. A calcium/sodium antagonism and a light-induced affinity change of a channel gating structure seem to play an important part in the gating process (72, 80). By analogy one might assume that the channels are opened following binding of internal transmitter T. The number of T molecules which have to be bound in order to cause the opening of an ion channel may be one or greater than one. The assumption that the binding of more than one (e.g., two) T molecule is needed to open a channel could provide a satisfactory explanation for both the phenomenon of supralinearity and that of facilitation (Stieve et al., unpublished).

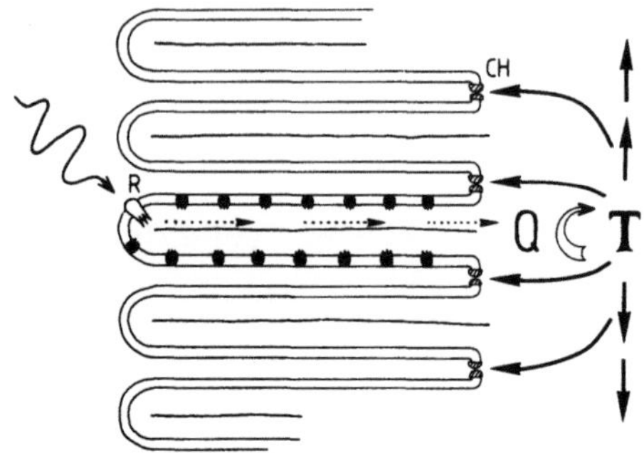

FIG. 10 - Schematical diagram of proposed mechanism of bump generation in Limulus ventral nerve photoreceptor. A light-activated rhodopsin molecule R in a microvillus starts the activation of an enzyme (cascade) which finally leads to the activation of a transmitter source Q. This transmitter source thereupon produces, activates, or releases many internal transmitter molecules which are built from precursors. The transmitter T diffuses along the bases of the microvilli and is bound by the ion channels Ch which plausibly may be situated close to the bases of the microvilli. Following transmitter binding, the ion channels are transiently opened. Thus a more or less circular "bump-speck" develops with a diameter of about 4 µm. On the average, a bump-speck of a light-adapted photoreceptor cell is smaller. The bump amplitude is proportional to the number of simultaneously opened ion channels and could therefore be more or less proportional to the area of the bump-speck.

After a very weak illumination a photosensory membrane would contain three classes of channels:

1. many Ch channels which have no transmitter molecule bound and are closed;
2. some Ch-T_1 channels which have one transmitter molecule bound and are still closed ("unlocked");
3. fewer Ch-T_2 channels which have two transmitter molecules bound and are open.

Under those conditions, the number of open channels rise with increasing stimulus energy in a certain range supralinearly. If the Ch-T_1 state had a lifetime on the order of seconds, a weak conditioning flash would generate a number of channels in the unlocked Ch-T_1 state. A following flash would evoke an enlarged (facilitated) response if it were administered during the lifetime of the Ch-T_1 state.

It seems plausible that the almost linear rising phase of the bump is caused by the diffusion of the transmitter over the bump-speck, opening more and more peripheral ion channels (Fig. 10; (43)). The longer and the stronger the production of active T molecules by source Q, the larger the bump. The bump rise ends when the transmitter production dwindles. According to this model, the transmitter-producing source is active only for a very short period (about 25 to 30 ms); compared to this the latency is about 5 to 6 times as long (see Schnakenberg and Keiper, this volume).

The bump decline is caused by two processes: decay in transmitter concentration and the frequency distribution of the lifetimes of the light-activated ion channels. There seem to be two possible explanations for the exponential phase of the bump decay: it could be due to an exponential decay of the transmitter concentration, as suggested by Bacigalupo and Lisman (3), or, if the transmitter decay is fast compared to the channel lifetime, it could be due to the stochastics of the channel closures (lifetimes) (24). The flow chart in Fig. 11 summarizes our model for bump generation.

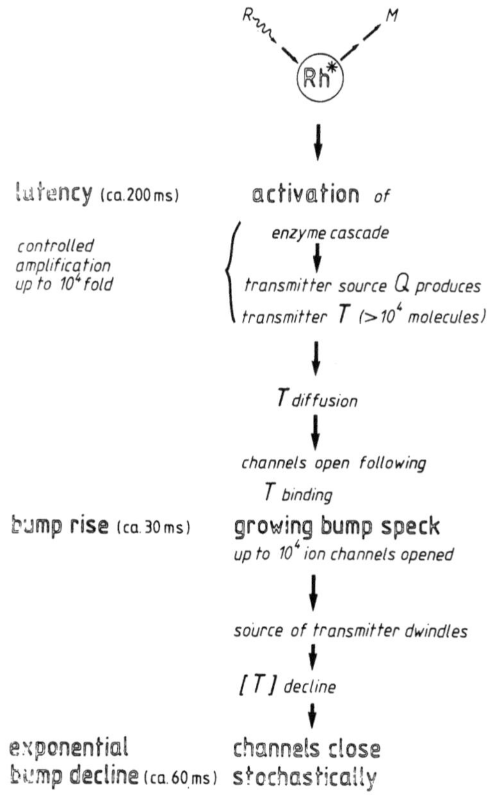

FIG. 11 - Flow chart of the proposed mechanism of bump generation (schematically) in the Limulus photoreceptor cell. R, M: inactive forms of rhodopsin; Rh*: activated state of rhodopsin; [T]: concentration of transmitter molecules.

The variation in quantum efficiency due to facilitation or to changes in Ca^{2+} concentration indicate that not every rhodopsin photoisomerization leads to the generation of a bump. It may be that the critical step (early?) in the transduction process is a Ca-dependent reaction necessary for the activation of the enzyme cascade.

In light adaptation the bumps become smaller because the amplification is lowered, i.e., the number of channels involved in a bump becomes smaller and, as indicated by the shift of the knee between the two slopes in the intensity dependence of the receptor current (Fig. 5), the bump-specks become smaller. The rising phase TR of the bump becomes shorter, possibly because the source Q produces transmitter molecules for a shorter period, and the bump decay is accelerated, either because the average lifetime of the channels is shortened or because the transmitter concentration declines faster.

The described model for bump generation which seems plausible to us is far from proven, but it is testable and will be tested.

Acknowledgements. I wish to thank J. Schnakenberg, K. Baer, A. Kuhn, and H. Reuss for stimulating discussions, S. Barash, P. Hillman, and B. Minke for valuable comments on the manuscript, M. Bruns and R. Schanzer for patient technical help, and A. Eckert for efficient help with the manuscript and for compiling the references. This work was supported by the Deutsche Forschungsgemeinschaft, SFB 160.

REFERENCES

(1) Adolph, A. 1964. Spontaneous slow potential fluctuations in the Limulus photoreceptor. J. Gen. Physiol. **48:** 297-322.

(2) Adolph, A. 1968. Thermal and spectral sensitivities of discrete slow potentials in Limulus eye. J. Gen. Physiol. **52:** 584-599.

(3) Bacigalupo, J., and Lisman, J.E. 1983. Single-channel currents activated by light in Limulus ventral photoreceptors. Nature **304:** 268-270.

(4) Barlow, R.B., and Kaplan, E. 1977. Properties of visual cells in the lateral eye of Limulus in situ. J. Gen. Physiol. **69:** 203-220.

(5) Bayer, D.S., and Barlow, R.B. 1978. Physiological properties of photoreceptor cells in an organ culture medium. J. Gen. Physiol. **72:** 539-563.

(6) Baylor, D.A.; Lamb, T.D.; and Yau, K.-W. 1979. Responses of retinal rods to single photons. J. Physiol. **288:** 613-634.

(7) Behbehani, M., and Srebro, R. 1974. Discrete waves and phototransduction in voltage-clamped ventral photoreceptors. J. Gen. Physiol. **64:** 186-200.

(8) Blumenfeld, A.; Erusalimsky, J.; Heichal, O.; Selinger, Z.; and Minke, B. 1985. Light-activated guanosintriphosphatase in Musca eye membranes mimics the prolonged depolarizing afterpotential in photoreceptor cells. Proc. Natl. Acad. Sci. USA **82:** 7116-7120.

(9) Bolsover, S.R., and Brown, J.E. 1982. Injection of guanosine and adenosine nucleotides into Limulus ventral photoreceptor cells. J. Physiol. **332:** 325-342.

(10) Borsellino, A., and Fuortes, M.G.F. 1968. Response to single photons in visual cells of Limulus. J. Physiol. **196:** 507-539.

(11) Brown, J.E., and Blinks, J.R. 1974. Changes in intracellular free calcium concentration during illumination of invertebrate photoreceptors. J. Gen. Physiol. **64:** 643-665.

(12) Brown, J.E.; Brown, P.K.; and Pinto, L.H. 1977. Detection of light induced changes of intracellular ionized calcium concentration in Limulus ventral nerve photoreceptors using Arsenazo III. J. Physiol. **267:** 299-320.

(13) Brown, J.E., and Coles, J.A. 1979. Saturation of the response to light in Limulus ventral photoreceptors. J. Physiol. **296:** 373-392.

(14) Brown, J.E.; Kaupp, U.B.; and Malbon, C.C. 1984. 3',5'-cyclic adenosine monophosphate and adenylate cyclase in phototransduction by Limulus ventral photoreceptors. J. Physiol. **353:** 523-539.

(15) Brown, J.E., and Rubin, L.J. 1984. A direct demonstration that inositol-triphosphate induces an increase in intracellular calcium in Limulus photoreceptors. Biochem. Biophys. Res. Commun. **125:** 1137-1142.

(16) Brown, J.E.; Rubin, J.L.; Traver, A.P.; Ghalayini, A.J.; Irvine, R.F.; and Anderson, R.E. 1984. Myo-inositol polyphosphate may be a messenger for visual excitation in Limulus photoreceptors. Nature **311:** 160-163.

(17) Calhoon, R.; Tsuda, M.; and Ebrey, T.G. 1980. A light-activated GTPase from octopus photoreceptors. Biochem. Biophys. Res. Commun. **94:** 1452-1457.

(18) Coles, J.A., and Brown, J.E. 1976. Effects of increased intracellular pH-buffering capacity on the light response of Limulus ventral photoreceptor. Biochim. Biophys. Acta **436:** 140-153.

(19) Cone, R.A. 1973. The internal transmitter model of visual excitation, some quantitative implications. In Biochemistry and Physiology of Visual Pigments, ed. H. Langer, pp. 275-282. Berlin: Springer-Verlag.

(20) Corson, D.W., and Fein, A. 1980. The pH dependence of discrete wave frequency in Limulus ventral photoreceptors. Brain Res. **193:** 558-561.

(21) Corson, D.W., and Fein, A. 1983. Chemical excitation of Limulus photoreceptors. J. Gen. Physiol. **82:** 639-657.

(22) Corson, D.W.; Fein, A.; and Walthall, W.W. 1983. Chemical excitation of Limulus photoreceptors II. J. Gen. Physiol. **82:** 659-667.

(23) Cosens, D.J., and Manning, A. 1969. Abnormal electrogram from a Drosophila mutant. Nature **224:** 285-287.

(24) Dirnberger, G.; Keiper, W.; Schnakenberg, J.; and Stieve, H. 1984. Comparison of time constants of single channel patches, quantum bumps, and noise analysis in Limulus ventral photoreceptors. J. Membr. Biol. **83:** 39-43.

(25) Dodge, F.A.; Knight, B.W.; and Toyoda, J.I. 1968. Voltage noise in Limulus visual cells. Science **160:** 88-90.

(26) Fein, A., and Charlton, J.S. 1977a. Enhancement and phototransduction in the ventral eye of Limulus. J. Gen. Physiol. **69:** 553-569.

(27) Fein, A., and Charlton, J.S. 1977b. A quantitative comparison of the effects of intracellular calcium injection and light adaptation on the photoresponse of Limulus ventral photoreceptors. J. Gen. Physiol. **70:** 591-600.

(28) Fein, A., and Corson, D.W. 1982. Internal injection of ATP can reduce discrete wave activity. Biol. Bull. **163:** 395.

(29) Fein, A., and Hanani, M. 1978. Light-induced increase in discrete waves in the dark in Limulus ventral photoreceptors. Brain Res. **156:** 157-161.

(30) Fein, A.; Payne, R.; Corson, D.; Berridge, M.J.; and Irvine, R.F. 1984. Photoreceptor excitation and adaptation by inositol 1,4,5-triphosphate. Nature **311:** 157-160.

(31) Goldring, M.A. 1982. Tests of models for the mechanism of the single photon response in Limulus ventral photoreceptors. Ph.D. Thesis, Brandeis University.

(32) Grzywacz, N.M., and Hillman, P. 1985. Statistical test of linearity of photoreceptor transduction process: Limulus passes, others fail. Proc. Natl. Acad. Sci. US-Biol. Sci. **82:** 232-235.

(33) Hamdorf, K., and Kirschfeld, K. 1980. "Prebumps": Evidence for double-hits at functional subunits in a rhabdomeric photoreceptor. Z. Naturforsch. **35c:** 173-174.

(34) Hamdorf, K., and Kirschfeld, K. 1980. Reversible events in the transduction process of photoreceptors. Nature **283:** 859-860.

(35) Hanani, M., and Fein, A. 1981. Diamide, a sulfhydryl reagent, modifies the light response of Limulus ventral nerve photoreceptor. Neurosci. Lett. **21:** 165-170.

(36) Hanani, M., and Hillman, P. 1976. Adaptation and facilitation in the barnacle photoreceptor. J. Gen. Physiol. **67:** 235-249.

(37) Hillman, P. 1981. The biophysics of intermediate processes in photoreceptor transduction: "Silent" stages, nonlocalities, single-photon response and models. In Proceedings of the Symposium on the Biology of Photoreceptor Cells, pp. 443-475. Cambridge: Cambridge University Press.

(38) Howard, J. 1983. Variations in the voltage response to single quanta of light in the photoreceptors of Locusta migratoria. Biophys. Struct. Mech. **9:** 341-348.

(39) Ivens, I., and Stieve, H. 1984. Influence of the membrane potential on the intracellular light induced Ca^{2+} concentration change of the Limulus ventral photoreceptor monitored by Arsenazo III under voltage clamp conditions. Z. Naturforsch. **39c:** 986-992.

(40) Kaplan, E., and Barlow, R.B. 1980. Circadian clock in Limulus brain increases response and decreases noise of retinal photoreceptors. Nature **286:** 393-395.

(41) Kass, L., and Barlow, R.B. 1984. Efferent neurotransmission of circadian rhythms in Limulus lateral eye. J. Neurosci. **4:** 908-917.

(42) Keiper, W. 1983. Zur Theorie der Photorezeption. Dissertation, RWTH Aachen, F.R. Germany.

(43) Keiper, W.; Schnakenberg, J.; and Stieve, H. 1984. Statistical analysis of quantum bump parameters in Limulus ventral photoreceptor. Z. Naturforsch. **39c:** 781-790.

(44) Lillywhite, P.G. 1977. Single photon signals and transduction in an insect eye. J. Comp. Physiol. **122:** 189-200.

(45) Lisman, J.E. 1985. The role of metarhodopsin in the generation of spontaneous quantum bumps in ultraviolet receptors of Limulus median eye. J. Gen. Physiol. **85:** 171-187.

(46) Lisman, J.E., and Brown, J.E. 1972. The effects of intracellular ionophoretic injection of calcium and sodium ions on the light response of Limulus ventral photoreceptors. J. Gen. Physiol. **59:** 701-719.

(47) Lisman, J.E., and Brown, J.E. 1975a. Light-induced changes of sensitivity in Limulus ventral photoreceptors. J. Gen. Physiol. **66:** 473- 488.

(48) Lisman, J.E., and Brown, J.E. 1975b. Effects of intracellular injection of calcium buffers on light adaptation in Limulus ventral photoreceptors. J. Gen. Physiol. **66:** 489-506.

(49) Maaz, G.; Nagy, K.; Stieve, H.; and Klomfaß, J. 1981. The electrical light response of the Limulus ventral nerve photoreceptor, a superposition of distinct components - observable by variation of the state of light adaptation. J. Comp. Physiol. **141:** 303-310.

(50) Maaz, G., and Stieve, H. 1980. The correlation of the receptor potential with the light induced transient increase in intracellular calcium-concentration measured by absorption change of Arsenazo III injected into Limulus ventral nerve photoreceptor cell. Biophys. Struct. Mech. **6:** 191-208.

(51) Martinez, J.M., and Srebro, R. 1976. Calcium and the control of discrete wave latency in the ventral photoreceptor of Limulus. J. Physiol. **261:** 535-562.

(52) Minke, B. 1979. Transduction in photoreceptors with bistable pigments: Intermediate processes. Biophys. Struct. Mech. **5:** 163-174.

(53) Minke, B. 1982. Light-induced reduction in excitation efficiency in the trp-mutant of Drosophila. J. Gen. Physiol. **79:** 361-385.

(54) Minke, B. 1983. The trp is a Drosophila mutant sensitive to developmental temperature. J. Comp. Physiol. **151:** 483-486.

(55) Minke, B., and Stephenson, R.S. 1985. The characteristics of chemically induced noise in musca photoreceptors. J. Comp. Physiol. **156:** 339-356.

(56) Minke, B.; Wu, C.F.; and Pak, W.L. 1975. Induction of photoreceptor voltage noise in the dark in Drosophila mutant. Nature **258:** 84-87.

(57) Nagy, K., and Stieve, H. 1983. Changes in intracellular calcium ion concentration, in the course of dark adaptation measured by Arsenazo III in the Limulus photoreceptor. Biophys. Struct. Mech. **9:** 207-223.

(58) Pak, W.L.; Conrad, S.K.; Kremer, N.E.; Larrivee, D.C.; Schinz, R.H.; and Wong, F. 1980. Photoreceptor function. In Development and Neurobiology of Drosophila, eds. O. Siddiqu, P. Babu, L.M. Hall, and J.C. Hall, pp. 331-346. New York: Plenum Publishing Corp.

(59) Pak, W.L.; Ostroy, S.E.; Deland, M.C.; and Wu, C.F. 1976. Photoreceptor mutant of Drosophila: Is a protein involved in intermediate steps of phototransduction? Science **194:** 956-959.

(60) Payne, R. 1980. Voltage noise accompanying chemically-induced depolarization of insect photoreceptors. Biophys. Struct. Mech. **6:** 235-251.

(61) Payne, R. 1981. Suppression of noise in a photoreceptor by oxidative metabolism. J. Comp. Physiol. **142:** 181-188.

(62) Payne, R. 1982. Fluoride blocks in inactivation step of transduction in a locust photoreceptor. J. Physiol. **325:** 261-269.

(63) Payne, R., and Howard, J. 1981. Response of an insect photoreceptor: a simple log-normal model. Nature **290:** 415-416.

(64) Saibil, H.R., and Michel-Villaz, M. 1984. Squid rhodopsin and GTP-binding protein crossreact with vertebrate photoreceptor enzymes. Proc. Natl. Acad. Sci. USA **81:** 5111-5115.

(65) Scholes, J. 1965. Discontinuity of the excitation process in locust visual cells. Cold S.H. Symp. Quant. Biol. **30:** 517-527.

(66) Srebro, R., and Behbehani, M. 1972. Light adaptation of discrete waves in the Limulus photoreceptor. J. Gen. Physiol. **60:** 86-101.

(67) Srebro, R., and Behbehani, M. 1972. The thermal origin of spontaneous activity in the Limulus photoreceptor. J. Physiol. **224:** 349-361.

(68) Srebro, R., and Behbehani, M. 1974. Light adaptation in the ventral photoreceptor of Limulus. J. Gen. Physiol. **64:** 166-185.

(69) Stern, J.; Chinn, K.; Robinson, P.; and Lisman, J.E. 1985. The effect of nucleotides on the rate of spontaneous quantum bumps in Limulus ventral photoreceptors. J. Gen. Physiol., in press.

(70) Stern, J.H., and Lisman, J.E. 1982. Internal dialysis of Limulus ventral photoreceptor. Proc. Natl. Acad. Sci. USA **79:** 7580-7584.

(71) Stieve, H. 1983. Transduction of light energy to electrical signals in photoreceptor cells. In The Biology of Photoreception, Society for Experimental Biology Symposium XXXVI, eds. D.J. Cosens and D. Vince-Prince, pp. 249-274. Great Britain: Society for Experimental Biology.

(72) Stieve, H., and Bruns, M. 1978. Extracellular calcium, magnesium, and sodium ion competition in the conductance control of the photosensory membrane of Limulus ventral nerve photoreceptor. Z. Naturforsch. 33c: 574-579.

(73) Stieve, H., and Bruns, M. 1980. Dependence of bump rate and bump size in Limulus ventral nerve photoreceptor on light adaptation and calcium concentration. Biophys. Struct. Mech. **6:** 271-285.

(74) Stieve, H., and Bruns, M. 1983. Bump latency distribution and bump adaptation of Limulus ventral nerve photoreceptor in varied extracellular calcium concentration. Biophys. Struct. Mech. **9:** 329-339.

(75) Stieve, H.; Bruns, M.; and Gaube, H. 1983. The intensity dependence of the receptor potential of the Limulus ventral nerve photoreceptor in two defined states of light- and dark adaptation. Z. Naturforsch. **38c:** 1043-1054.

(76) Stieve, H.; Bruns, M.; and Gaube, H. 1984. The sensitivity shift due to light adaptation depending on the extracellular calcium ion concentration in Limulus ventral nerve photoreceptor. Z. Naturforsch. **39c:** 662-679.

(77) Stieve, H.; and Klomfaß, J. 1981. Calcium dependence of light evoked membrane current signal and membrane voltage signal and their changes due to light adaptation in Limulus photoreceptor. Biophys. Struct. Mech. **7:** 345.

(78) Stieve, H., and Klomfaß, J. 1983. Distribution and bump latency and bump shape parameters in dependence on adaptation and external Ca^{2+} concentration in Limulus photoreceptor. Abstract. Jahrestagung, Deutsche Gesellschaft für Biophysik, Neuherberg.

(79) Stieve, H., and Pflaum, M. 1978. The response height versus stimulus intensity curve of the ventral nerve photoreceptor of Limulus depending on adaptation and external calcium concentration. Vision Res. **18:** 747-749.

(80) Stieve, H.; Pflaum, M.; Klomfaß, J.; and Gaube, H. 1985. Calcium/sodium binding competition in the gating of light-activated membrane conductance studied by voltage clamp technique in Limulus ventral nerve photoreceptor. Z. Naturforsch., in press.

(81) Vandenberg, C.A., and Montal, M. 1984a. Light-regulated biochemical events in invertebrate photoreceptors I. Biochemistry **23:** 2339-2347.

(82) Vandenberg, C.A., and Montal, M. 1984b. Light-regulated biochemical events in invertebrate photoreceptors II. Biochemistry **23:** 2347-2353.

(83) Williams, D.S. 1983. Changes of photoreceptor performance associated with the daily turnover of photoreceptor membrane in locusts. J. Comp. Physiol. **150:** 509-519.

(84) Wong, F. 1978. Nature of light-induced conductance changes in ventral photoreceptors of Limulus. Nature **276:** 76-79.

(85) Wong, F. 1985. Molecular analysis of visual mutation in Drosophila. Proc. Symp. Cont. Sens. Neurobiol., in press.

(86) Wong, F.; Hokanson, K.M.; and Chang, T.L. 1985. Molecular basis of an inherited retinal defect in Drosophila. Inv. Ophthalmol. Vis., in press.

(87) Wong, F.; Knight, B.W.; and Dodge, F.A. 1980. Dispersion of latencies in photoreceptors of Limulus and the adapting bump model. J. Gen. Physiol. **76:** 517-537.

(88) Wong, F.; Knight, B.W.; and Dodge, F.A. 1982. Adapting bump model for ventral photoreceptors of Limulus. J. Gen. Physiol. **79**: 1089-1113.

(89) Yeandle, S. 1958. Electrophysiology of the visual system - discussion. Am. J. Ophthalmol. **46**: 82.

(90) Yeandle, S., and Fuortes, M.G.F. 1964. Probability of occurrence of discrete potential waves in the eye of Limulus. J. Gen. Physiol. **47**: 443-463.

(91) Yeandle, S., and Spiegler, J.B. 1973. Light-evoked and spontaneous discrete waves in the ventral nerve photoreceptor of Limulus. J. Gen. Physiol. **61**: 552-572.

(92) Yeandle, S., and Spiegler, J.B. 1974. Independence of location of light absorption and discrete wave. J. Gen. Physiol. **64**: 494-502.

(93) Yoshika, T.; Inoue, H.; Inomata, K.; Hayashi, F.; Takagi, M.; and Takenaka, T. 1981. The parallel study of protein phosphorylation and phosphatidyl-inositol metabolism in the photoreceptor of squid retina. Proc. Jpn. Acad. **57**: 309.

The Molecular Mechanism of Photoreception, ed. H. Stieve, pp. 231-240. Dahlem Konferenzen 1986. Berlin, Heidelberg, New York, Tokyo: Springer-Verlag.

Calcium and Light Adaptation in Invertebrate Photoreceptors

J.E. Brown
Dept. of Ophthalmology
Washington University School of Medicine
St. Louis, MO 63110, USA

Abstract. A light-induced increase in the concentration of intracellular calcium ions has been proposed to be a step in the cascade of reactions mediating light adaptation in invertebrate photoreceptors. The physiological and anatomical evidence consistent with that proposal is reviewed.

Absorption of photons by rhodopsin in photoreceptors initiates a sequence of reactions that leads to the electrical signal of excitation, the receptor potential. Absorption of photons also initiates a sequence of reactions that leads to adaptation, which is a modulation of the excitatory sequence. In some invertebrate photoreceptors, an increase in the cytosolic concentration of calcium ions has been proposed to be a step in the adaptation sequence. Evidence supporting this proposal has been found in a variety of invertebrate photoreceptors. The present paper will be restricted to a few invertebrate photoreceptors, those of the ventral eyes of Limulus, the lateral ocelli of Balanus, the eyes of the honeybee drone, and the ocelli of Hirudo.

In many invertebrate photoreceptors, light induces a receptor potential that makes the inside of the cell more positive (2, 8, 26, 32, 40). Voltage-clamp analyses of Limulus (14, 42) and Balanus (8) have shown that the receptor potential is generated by a light-induced increase in the membrane conductance; the inward current is normally carried predominately by sodium ions.

That calcium ions might participate in the modulation of the light-induced current was proposed by Millecchia and Mauro (41). This initial suggestion was based on the observation that removal of extracellular calcium led to an increase in the size of receptor response rather than the decrease that would be expected if calcium ions were a principle species carrying inward current. These observations have been replicated and extended many times (33, 34, 49-51). Similarly, reduction of the concentration of extracellular calcium ions leads to an increase of the amplitude of the receptor potentials in honeybee drone retina (27, 47) and Balanus photoreceptors (8). Moreover, the amplitude of light-induced current recorded by a voltage clamp also increased when extracellular calcium concentration was reduced (in Limulus ventral photoreceptors (42); in Balanus photoreceptors (8)) - this finding implies that the membrane conductance that was changed by light was modulated by calcium ions.

There are several changes in the physiological state of a photoreceptor cell when it becomes light-adapted: prior illumination reduces both a) the sensitivity of the cell (defined as peak light-induced current per unit stimulus irradiance) and b) the time scale of the light responses (i.e., the latency, time-to-peak, and time of decline of the transient response) (see, for example, (28)). The current induced during a prolonged stimulus is initially large and then, while the process of light adaptation occurs, declines to a plateau. Another manifestation of light adaptation is that the amplitude of the plateau current is nearly a linear function of light intensity for dim illumination when there is minimal adaptation but is a much less than linear function at high-intensity illumination (35).

The proposal that calcium ions might serve as an intracellular messenger for light adaptation gained its first substantial support from the observation that intracellular ionophoretic injection of calcium ions led to a decrease in the responsiveness of Limulus ventral photoreceptor cells (34). In addition, ionophoretic injection of calcium ions reduced the sensitivity of the cells as well as shortening the time scale of the light responses (13, 23). The time course of the recovery of sensitivity after injection of calcium ions closely resembles that after bright illumination (21). Thus, experimentally induced, prolonged increases in the concentration of intracellular calcium ions mimicked three aspects of light adaptation.

Light adaptation has also been found to occur locally within invertebrate photoreceptors (20); that is, adaptation is most pronounced at the locus of prior illumination and is less evident at more remote loci. The reduction of sensitivity that follows intracellular injection of calcium ions is also most pronounced at the locus of the injection (24) and is small at more distant loci. Thus, intracellular ionophoretic calcium injection mimics an additional aspect of light adaptation.

An adequate analysis of ion injection experiments such as those described above must be controlled for the effects of changing membrane voltage. The photoreceptors have voltage-dependent calcium conductances, so that a depolarization attendant upon ionophoretic injection of a cation may cause the influx of calcium ions. Therefore, injection experiments are most easily interpretable if done either under voltage-clamp (e.g., (13)) or by pressure injection (e.g., (1)).

Intracellular injection of other common ions does not directly mimic light adaptation. Injections of potassium, lithium, chloride, or sulfate ions do little (17, 34). Injections of sodium ions do lead to reductions of sensitivity, but this action is probably mediated indirectly through a sodium-calcium exchange mechanism ((22, 34, 48); see also (43) in Musca eyes).

Another line of evidence that supports the "calcium hypothesis for light adaptation" is that procedures that tend to prevent light-induced increases of intracellular calcium ions also tend to reduce light adaptation. For example, the intracellular (pressure or ionophoretic) injection of the calcium-sequestering substance EGTA (ethylene glycol-bis-(β-amino ethyl ether)-N,N,N',N' tetra acetic acid) tends to prevent an adapting flash from reducing the sensitivity of the cell to a subsequent test flash (36). Intracellular EGTA also tends to increase the amplitude of the plateau current during a prolonged stimulus; the light response thus becomes more "square." In addition, the range of the nearly linear portion of the graph of plateau current versus light intensity becomes extended to much higher light intensities after intracellular injection of EGTA. All these findings (36) indicate that reductions of light-induced increases of intracellular Ca^{2+} tend to reduce the several manifestations of light adaptation.

Pressure injections of both calcium and EGTA ions also have been made into photoreceptors of honeybee drones (1). The findings of those experiments were similar to those described for Limulus ventral photoreceptors. Injection of calcium ions mimicked the effects of light adaptation and injections of EGTA tended to prevent the effects of light adaptation. Also, injection of calcium ions into Balanus photoreceptors has been found to reduce the amplitude of currents induced by repetitive flashes (7).

If calcium ions signal light adaptation in invertebrate photoreceptors, then there ought to be a measurable light-induced increase in intracellular calcium ion concentration. Brown and Blinks (11) demonstrated a light-induced increase in the concentration of intracellular calcium ions in both Limulus and Balanus photoreceptors using the photoprotein, aequorin. This study showed that the increase in intracellular calcium both rose more slowly and reached its peak later than the light-induced current. The aequorin also indicated that during a prolonged stimulus, the concentration of intracellular calcium ions rose transiently and fell to a level higher than that observed in the dark-adapted cell (J.E. Brown, unpublished observations).

The light-induced increase in the concentration of intracellular calcium ions also has been detected by the use of a metallochromic indicator dye, arsenazo III, in Limulus ventral photoreceptors (12, 30, 38, 44, 52). These studies confirmed that light induced a transient increase, but the concentration of intracellular calcium ions fell during prolonged illumination. At the peak of the transient, the intracellular calcium ion concentration rose to as high as 200-300 µmol/l (12, 30).

More recently, arsenazo III has been used to study the spatial distribution of changes in concentration of intracellular calcium ions (31). This study showed that the changes were local within the cell and were probably restricted to the rhabdomeric lobe (or lobes) of the cell. Moreover, the diffusion of calcium ions was remarkably restricted, as anticipated from the local effects of calcium ion injection (24).

Calcium-selective microelectrodes have also been used to detect changes in intracellular concentration of calcium ions in both Limulus ventral photoreceptors (S. Levy, personal communication) and Balanus photoreceptors (7).

These lines of evidence described above indicate that intracellular calcium ions satisfy many of the criteria necessary to identify an intracellular messenger for light adaptation in invertebrate photoreceptors: Calcium ions (i) are present in the cytosol and (ii) increase in concentration after illumination. (iii) The maximum concentration occurs while the light-adaptation process begins. (iv) Reduction of light-induced concentration increase reduces light adaptation. (v) Injection of calcium ions mimics several aspects of light adaptation. (vi) The effects of an injection of calcium ions, the naturally occurring increase in calcium ion concentration, and light adaptation are all localized within the cell.

An increase of intracellular hydrogen concentration has been proposed as an alternative hypothesis for light adaptation (39). Consistent with this "proton hypothesis" is the observation that light induces an acidification of the cytosol of both Balanus photoreceptors (7, 9, 10) and Limulus photoreceptors (6) as detected by phenol red. There are two lines of evidence inconsistent with the "proton hypothesis." First, the injection of exogenous pH buffers to intracellular concentrations of 200 mmol/l or more does not appreciably alter either the sensitivity of the cell (17) or the shape of the response elicited by prolonged flashes (4). Second, the time course of the light—induced change of intracellular pH in Limulus photoreceptors differs dramatically from the expected time course of a messenger for adaptation: the cell begins to recover sensitivity after a bright adapting light while the pH continues to become more acidic (6). For these reasons, a change in cytosolic concentration of hydrogen ions is unlikely to serve as a signal for the adaptation process.

The increase in concentration of intracellular ionized calcium that is induced by illumination might arise either from influx of calcium ions across the plasma membrane or from release of calcium ions from some intracellular stores. A light-induced influx of calcium ions has been reported for Balanus photoreceptors (8); in the absence of extracellular sodium ions, the reversal voltage of the light-induced current changes with the concentration of extracellular calcium ions in a nearly Nernstian fashion. Moreover, there is an easily measured steady-state increase in Ca_i in Balanus photoreceptors (by aequorin (11)). For Limulus ventral photoreceptors, no electrophysiologically measured calcium current has been reported; even in the absence of extracellular sodium ions, a decrease in Ca_i leads to an increase in the remanent light-induced current (14). Thus, different invertebrate photoreceptors apparently possess differing contributions of calcium current to the total light-induced current.

Release of calcium ions from intracellular stores apparently also plays a role in the light-induced changes of Ca_i. For Limulus photoreceptors, light can induce a rise of Ca_i in the prolonged absence of extracellular calcium ions plus the presence of extracellular EGTA (5, 11). Repetitive flashes (5) or combined steady (low, level) illumination and flashes (38) can lead to a reduction in the light-induced increase of Ca_i to below the limits of detectability. These studies indicate that a light-labile intracellular store participates in the control of Ca_i and that the store can be loaded and unloaded by manipulation of extracellular calcium ion concentration.

The locus of the intracellular store has been investigated by anatomical techniques. For honeybee retinular cells, Perrelet and Bader (46) suggested that calcium ions were stored in the cisternae of endoplasmic reticulum that underlie the rhabdomeric microvilli. Although those experiments might be criticized on technical grounds, more recent studies have also indicated the merit in their suggestion. Walz (53-55) has shown that the smooth endoplasmic reticulum in leech (Hirudo) photoreceptors does sequester calcium. Also, Coles and Rick (18) have found that the total calcium content is higher in the region of the subrhabdomeric cisternae than in the rest of the cytoplasm of the honeybee drone retinular cells.

Lisman and Strong (37) showed that adaptation, and presumably light-induced release of calcium ions, is governed by isomerization of rhodopsin in the plasma membrane. Thus, an intracellular diffusible messenger can be imputed to trigger the release of calcium ions from the light-labile intracellular store. Recent work (16, 25) suggests that an inositol polyphosphate may serve that function. Moreover, intracellular injection of inositol-trisphosphate has been shown directly to induce a rise of intracellular Ca^{2+} in Limulus ventral photoreceptors (by the aequorin technique (15, 19)).

As reviewed above, calcium ions apparently participate in the mechanism for light adaptation within invertebrate photoreceptors. Calcium ions have also been proposed to participate in "facilitation" (29) and within the excitation processes. Although outside the scope of this review, two indications of this involvement in excitation ought to be mentioned.

First, the sensitivity of Limulus ventral photoreceptors becomes reduced after being repetitively illuminated while bathed in seawater containing EGTA and micromolar or less free calcium ions (3, 5). In this state (called "calcium-deprived"), the light-induced rise of Ca_i (as detected by aequorin) becomes much smaller. Moreover, in the calcium-deprived state, intracellular pressure injection of EGTA ions <u>reduces</u> the sensitivity whereas intracellular ionophoretic injection of calcium ions <u>increases</u> the sensitivity. Both of these latter findings are opposite to the effects observed while the cells are bathed in normal seawater. The individual light responses recorded in the calcium-deprived state often show a delayed rise in amplitude. In addition, the light-induced current vs. light intensity relationship is more steep in the calcium-deprived state than normally (originally reported by Stieve and Pflaum (51)). Taken together, these findings indicate that calcium ions participate in the excitation mechanism. Presumably the efficiency of excitation is high in the normal, optimal level of intracellular calcium ion concentration. In the calcium-deprived state, the efficiency is reduced. Either exogenous injection of calcium or the light-induced release of calcium from the light-labile store can raise the efficiency of the excitation mechanism (3, 5).

Second, it should be noted that recently (45) calcium ions have been pressure-injected into Limulus ventral photoreceptors out of an electrode filled with 1 mmol/l Ca^{2+}. Such brief, abrupt injection of a small bolus of 1 mmol/l Ca^{2+} induces an inward current that mimics several aspects of the light-induced current (e.g., sign and reversal voltage measured in ASW). These findings indicate that calcium ions may play a direct role in the generation of light-induced current in Limulus ventral photoreceptors.

Acknowledgement. Supported by National Institutes of Health grants EY05166 and EY05168.

REFERENCES

(1) Bader, C.R.; Baumann, F.; and Bertrand, D. 1976. Role of intracellular calcium and sodium in light adaptation in the retina of the honey bee drone (Apis mellifera). J. Gen. Physiol. **67:** 475-491.

(2) Baumann, F. 1968. Slow and spike potentials recorded from retinula cells of the honeybee drone in response to light. J. Gen. Physiol. **52:** 855-875.

(3) Bolsover, S.R., and Brown, J.E. 1982. Calcium injections increase sensitivity in calcium depleted Limulus ventral photoreceptor cells. Biol. Bull. **163:** 394-395.

(4) Bolsover, S.R., and Brown, J.E. 1982. Light adaptation of invertebrate photoreceptors: influence of intracellular pH buffering capacity. J. Physiol. **330:** 297-305.

(5) Bolsover, S.R., and Brown, J.E. 1985. Calcium ion, an intracellular messenger of light adaptation, also participates in excitation of Limulus ventral photoreceptor cells. J. Physiol. **364:** 381-393.

(6) Bolsover, S.R.; Brown, J.E.; and Goldsmith, T.G. 1986. Intracellular pH of Limulus ventral photoreceptor cells: measurement with phenol red. In Optical Methods in Cell Physiology, eds. P. DeWeer and B.M. Salzberg. New York: J. Wiley and Sons.

(7) Brown, H.M. 1984. Intracellular changes of H^+ and Ca^{++} in Balanus photoreceptors. In Photoreceptors, eds. A. Borsellino and L. Cervetto. NATO ASI Series: A. Life Sciences **75:** 99-118.

(8) Brown, H.M.; Hagiwara, S.; Koike, H.; and Meech, R.W. 1970. Membrane properties of a barnacle photoreceptor examined by the voltage-clamp technique. J. Physiol. **208:** 385-413.

(9) Brown, H.M., and Meech, R.W. 1976. Intracellular pH and light adaptation in barnacle photoreceptor. J. Physiol. **263:** 218P.

(10) Brown, H.M., and Meech, R.W. 1979. Light-induced changes of internal pH in a barnacle photoreceptor and the effect of internal pH on the receptor potential. J. Physiol. **297:** 73-94.

(11) Brown, J.E., and Blinks, J.R. 1974. Changes in intracellular free calcium concentration during illumination of invertebrate photoreceptors. Detection with aequorin. J. Gen. Physiol. **64:** 643-665.

(12) Brown, J.E.; Brown, P.K.; and Pinto, L.H. 1977. Detection of light-induced changes of intracellular ionized calcium concentration in Limulus ventral photoreceptors using arsenazo III. J. Physiol. **267:** 299-320.

(13) Brown, J.E., and Lisman, J.E. 1975. Intracellular Ca modulates sensitivity and time scale in Limulus ventral photoreceptors. Nature **258:** 252-254.

(14) Brown, J.E., and Mote, M.I. 1974. Ionic dependence of reversal voltage of the light response in Limulus ventral photoreceptors. J. Gen. Physiol. **63:** 337-350.

(15) Brown, J.E., and Rubin, L.J. 1985. A direct demonstration that inositol-trisphosphate induces an increase in intracellular calcium in Limulus photoreceptors. Biochem. Biophys. Res. Comm. **125:** 1137-1142.

(16) Brown, J.E.; Rubin, L.J.; Ghalayini, A.J.; Tarver, A.P.; Irvine, R.F.; Berridge, M.J.; and Anderson, R.E. 1984. Myoinositol polyphosphate may be a messenger for visual excitation in Limulus photoreceptors. Nature **311:** 160-163.

(17) Coles, J.A., and Brown, J.E. 1976. Effects of increased intracellular pH-buffering capacity on the light response of Limulus ventral photoreceptor. Biochim. Biophys. Acta **436:** 140-153.

(18) Coles, J.A., and Rick, R. 1985. An electron microprobe analysis of photoreceptors and outer pigment cells in the retina of the honeybee drone. **156:** 213-222.

(19) Corson, D.W.; Fein, A.; and Payne, R. 1984. Detection of an inositol 1,4,5-trisphosphate-induced rise in intracellular free calcium with aequorin in Limulus ventral photoreceptors. Biol. Bull. **167:** 524-525 (Abstract).

(20) Fein, A., and Charlton, J.S. 1975. Local adaptation in the ventral photoreceptors of Limulus. J. Gen. Physiol. **66:** 823-836.

(21) Fein, A., and Charlton, J.S. 1977a. A quantitative comparison of the effects of intracellular calcium injection and light adaptation on the photoresponse of Limulus ventral photoreceptors. J. Gen. Physiol. **70:** 591-600.

(22) Fein, A., and Charlton, J.S. 1977b. Increased intracellular sodium mimics some but not all aspects of photoreceptor adaptation in the ventral eye of Limulus. J. Gen. Physiol. **70:** 601-620.

(23) Fein, A., and Charlton, J.S. 1978. A quantitative comparison of the time-course of sensitivity changes produced by calcium injection and light adaptation in Limulus ventral photoreceptors. Biophys. J. **22:** 105-113.

(24) Fein, A., and Lisman, J.E. 1975. Localized desensitization of Limulus photoreceptors produced by light or intracellular calcium ion injection. Science **187:** 1094-1096.

(25) Fein, A.; Payne, R.; Corson, D.W.; Berridge, M.J.; and Irvine, R.F. 1984. Photoreceptor excitation and adaptation by inositol 1,4,5-trisphosphate. Nature **311:** 157-160.

(26) Fioravanti, R., and Fuortes, M.G.F. 1972. Analysis of responses in visual cells of the leech. J. Physiol. **227:** 173-194.

(27) Fulpius, B., and Baumann, F. 1969. Effects of sodium, potassium, and calcium ions on slow and spike potentials in single photoreceptor cells. J. Gen. Physiol. **53:** 541-561.

(28) Fuortes, M.G.F., and Hodgkin, A.L. 1964. Changes in time scale and sensitivity in the ommatidia of Limulus. J. Physiol. **172:** 239-263.

(29) Hanani, H., and Hillman, P. 1976. Adaptation and facilitation in the barnacle photoreceptor. J. Gen. Physiol. **67:** 239-249.

(30) Harary, H.H. 1983. Optical probes of the physiology of Limulus ventral photoreceptors. Ph.D. Thesis, Harvard University.

(31) Harary, H.H., and Brown, J.E. 1984. Spatially nonuniform changes in intracellular calcium ion concentrations. Science **224:** 292-294.

(32) Lasansky, A., and Fuortes, M.G.F. 1969. The site of origin of electrical responses in the visual cells of the leech, Hirudo medicinalis. J. Cell Biol. **42:** 241-252.

(33) Lisman, J.E. 1976. Effects of removing extracellular Ca^{2+} on excitation and adaptation in Limulus ventral photoreceptors. Biophys. J. **16:** 1331-1335.

(34) Lisman, J.E., and Brown, J.E. 1972. The effects of intracellular iontophoretic injection of calcium and sodium ions on the light response of Limulus ventral photoreceptors. J. Gen. Physiol. **59:** 701-719.

(35) Lisman, J.E., and Brown, J.E. 1975a. Light-induced changes of sensitivity in Limulus ventral photoreceptors. J. Gen. Physiol. **66:** 473- 488.

(36) Lisman, J.E., and Brown, J.E. 1975b. Effects of intracellular injection of calcium buffers on light adaptation in Limulus ventral photoreceptors. J. Gen. Physiol. **66:** 489-506.

(37) Lisman, J.E., and Strong, J.A. 1979. The initiation of excitation and light adaptation in Limulus ventral photoreceptors. J. Gen. Physiol. **73:** 219-243.

(38) Maaz, G., and Stieve, H. 1980. The correlation of the receptor potential with the light induced transient increase in intracellular calcium-concentration measured by absorption changes of arsenazo III injected into Limulus ventral nerve photoreceptor cell. Biophys. Struct. Mech. **6:** 191-208.

(39) Meech, R.W., and Brown, H.M. 1976. Invertebrate photoreceptors: A survey of recent experiments on photoreceptors from Balanus and Limulus. Persp. Exp. Biol. **I:** 331-351.

(40) Millecchia, R.; Bradbury, J.; and Mauro, A. 1966. Simple photoreceptors in Limulus polyphemus. Science **154:** 1199-1201.

(41) Millecchia, R., and Mauro, A. 1969a. The ventral photoreceptor cells of Limulus II. The basic photoresponse. J. Gen. Physiol. **54:** 310-330.

(42) Millecchia, R., and Mauro, A. 1969b. The ventral photoreceptor cells of Limulus III. A voltage-clamp study. J. Gen. Physiol. **54:** 331-351.

(43) Minke, B., and Armon, E. 1984. Activation of electrogenic Na-Ca exchange by light in fly photoreceptors. Vision Res. **24:** 109-115.

(44) Nagy, K., and Stieve, H. 1983. Changes in intracellular calcium ion concentration, in the course of dark adaptation measured by arsenazo III in the Limulus photoreceptor. Biophys. Struct. Mech. **9:** 207-223.

(45) Payne, R.; Fein, A.; and Corson, D.W. 1984. A rise in intracellular calcium is necessary and perhaps sufficient for photoreceptor excitation and adaptation by inositol 1,4,5-trisphosphate. Biol. Bull. **167:** 531-532 (Abstract).

(46) Perrelet, A., and Bader, C.R. 1978. Morphological evidence for calcium stores in photoreceptors of the honeybee drone retina. J. Ultrastruct. Res. **63:** 237-243.

(47) Raggenbass, M. 1983. Effects of extracellular calcium and of light adaptation on the response to dim light in honeybee drone photoreceptors. J. Physiol. **344:** 525-548.

(48) Stieve, H. 1981. Roles of calcium in visual transduction in invertebrates. In Sense Organs, eds. M.S. Laverack and D.J. Casens, pp. 163-185. Glasgow: Blackie and Sons, Ltd.

(49) Stieve, H., and Bruns, M. 1978. Extracellular calcium, magnesium and sodium ion competition in the conductance control of the photosensory membrane of Limulus ventral nerve photoreceptor. Z. Naturforsch. **33c:** 574-579.

(50) Stieve, H., and Klomfass, J. 1981. Calcium dependence of light evoked membrane current signal and membrane voltage signal and their changes due to light adaptation in Limulus photoreceptor. Biophys. Struct. Mech. **7:** 345.

(51) Stieve, H., and Pflaum, M. 1978. The response height versus stimulus intensity curve of the ventral nerve photoreceptor of Limulus depending on adaptation and external calcium concentration. Vision Res. **18:** 747- 749.

(52) Waloga, G.; Brown, J.E.; and Pinto, L.H. 1975. Detection of changes in Ca(in)from Limulus ventral photoreceptors using arsenazo III. Biol. Bull. **149:** 449-450.

(53) Walz, B. 1979. Subcellular calcium localization and ATP-dependent Ca^{2+} uptake by smooth endoplasmic reticulum in an invertebrate photo receptor cell. An ultrastructural, cytochemical and X-ray microanalytical study. Eur. J. Cell Biol. **20:** 83-91.

(54) Walz, B. 1982a. Ca^{2+}-sequestering smooth endoplasmic reticulum in an invertebrate photoreceptor. I. Intracellular topography as revealed by OsFeCN staining and in situ Ca accumulation. J. Cell Biol. **93:** 839- 848.

(55) Walz, B. 1982b. Ca^{2+}-sequestering smooth endoplasmic reticulum in an invertebrate photoreceptor II. Its properties as revealed by microphotometric measurements. J. Cell Biol. **93:** 849-859.

The Molecular Mechanism of Photoreception, ed. H. Stieve, pp. 241-265. Dahlem Konferenzen 1986. Berlin, Heidelberg, New York, Tokyo: Springer-Verlag.

Photopigment-dependent Adaptation in Invertebrates - Implications for Vertebrates

B. Minke
Dept. of Physiology
The Hebrew University-Hadassah Medical School
Jerusalem 91010, Israel

Abstract. Bleaching light in vertebrate rods and strong colored light which converts a substantial fraction of rhodopsin (R) to metarhodopsin (M) in invertebrate photoreceptors both induce a prolonged excitation in the dark which desensitizes both kinds of photoreceptors for a long time in the dark. There are remarkable similarities between the characteristics of the prolonged dark excitation in both vertebrates and invertebrates: a) both are induced by pigment conversion from R. b) Both can be cut short by regeneration of R molecules. c) Both are composed of discrete events similar to photon-induced events. d) Both are accompanied by a large reduction in sensitivity to light. e) Both are localized to the illuminated area and do not spread all over the cell. A molecular mechanism is suggested to account for the prolonged dark excitation. The model suggests that when large amounts of R molecules are converted to the M state, many activated M molecules continue to be active in the dark due to lack of phosphorylation giving rise to the prolonged dark excitation. The decline of the prolonged excitation is due to phosphorylation's turnoff of the active M molecules. The mechanisms by which excitation desensitizes the photoreceptor are still unknown, but in invertebrates part of it is mediated via an increase in intracellular free Ca^{2+}.

INTRODUCTION

The large and long-term reduction in sensitivity of the vertebrate rods in the isolated retina, as a consequence of bleaching rhodopsin, is known as bleaching adaptation or pigment adaptation (see, for example, (19, 20, 49-51, 63)). The mechanism underlying bleaching adaptation is still not clear.

It is a common notion that vertebrates and invertebrates are completely different with respect to bleaching adaptation. In the vertebrate rods (for example, in the skate and the axolotl) the desensitization following bleaching light is roughly logarithmically related to the concentration of pigment remaining in the bleached state (14, 20, 49). The sensitivity (i.e., the amplitude of the receptor potential in response to a weak test light, or the light intensity needed to induce a criterion response) of invertebrate photoreceptors (4, 22) a long time after a light stimulus (which reduces the rhodopsin concentration) seems not to be affected by the reduction in rhodopsin concentration, except for a small (linear) reduction in sensitivity due to the reduction in the probability to absorb photons. However, there seems to be a problem with the above comparison between vertebrates and invertebrates since it was based on data which were obtained under different conditions. In vertebrate rods bleaching light induces a prolonged excitation in the dark (i.e., a prolonged voltage change in the dark (21) which is due to summation of unitary events (30)). This prolonged dark excitation was most pronounced when photocurrent and not voltage was measured from the outer segment (1, 7). Obviously, measurements of sensitivity in the vertebrate rods were influenced by this prolonged dark excitation (30) which lasted many minutes, depending on the amount of bleaching light (7). In invertebrates, measurements of sensitivity were done either when no such prolonged dark excitation was found (4) or when it was already completed (22). Since a prolonged dark excitation, which strongly affects the sensitivity, can be induced in many invertebrate species in a well controlled manner, it seems desirable to examine the effects of this phenomenon on the sensitivity of the photoreceptor to light and to compare the characteristics of the prolonged dark excitation in both vertebrates and invertebrates. The large body of information which already exists about the prolonged dark excitation in invertebrates (known as prolonged depolarizing afterpotential, PDA) may be used as a guideline for investigating this process together with the reduced sensitivity level in vertebrate rods and to suggest a molecular mechanism for these phenomena.

THE CHARACTERISTICS OF THE PROLONGED DEPOLARIZING AFTERPOTENTIAL (PDA) IN INVERTEBRATES

Dependence on Pigment Conversions

The PDA is observed only when a considerable amount of pigment (>10%) is converted from rhodopsin (R) to its dark stable photoproduct acid metarhodopsin (M). Fig. 1A, B shows that the larger the amount of R to M conversion, the longer the PDA. The duration of the PDA depends in a supralinear manner on the amount of light (i.e., the duration of the stimulus multiplied by the light intensity) (Fig. 1B). The PDA does not depend on either stimulus duration or light intensity alone but only on their product. The PDA also shows large facilitation (for reviews see (22, 25)).

The PDA can be depressed at any time by pigment conversion from M to R (M to R conversion). The degree of PDA depression depends on the amount of M to R conversion (26). After the depression of maximal PDA, additional PDA can be induced immediately by R to M conversion (Fig. 2). The situation is more complex when the PDA-depressing light is given following the decline of the PDA (see below). When R to M conversion is maximal, a maximal PDA is induced. Additional strong light stimuli, which do not change the distribution of the pigment between R and M, do not affect the duration of the PDA but only induce light-coincident receptor potentials (LRP) which are superimposed on the PDA (22).

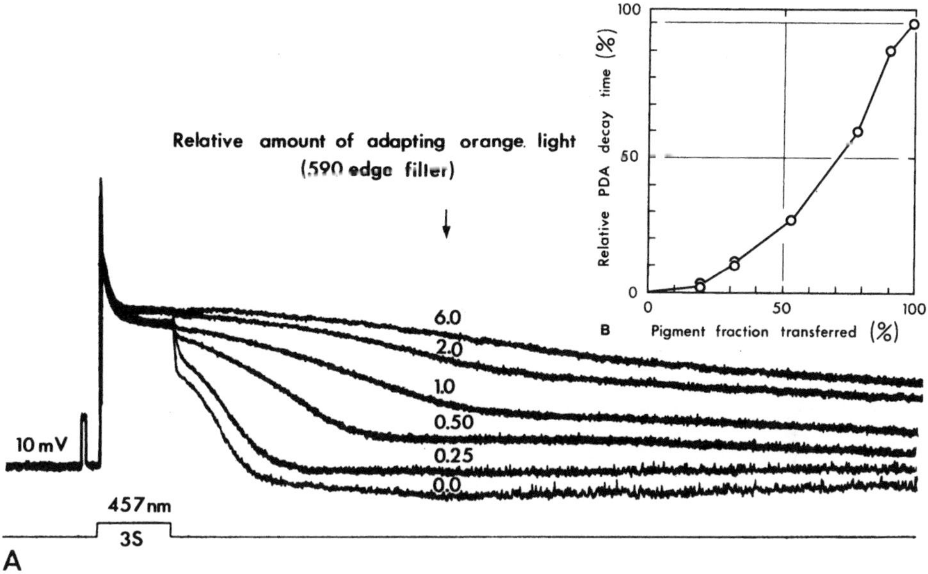

FIG. 1 - The dependence of the prolonged depolarizing afterpotential (PDA) on the amount of pigment conversion from rhodopsin (R) to metarhodopsin (M). The PDA was recorded intracellularly from a single photoreceptor of the white-eyed fly Calliphora. A: Superimposed traces which show the receptor potential (LRP) during 3 s blue light and the following PDA. Before each trace the cell received saturating blue light (457nm) which put maximal (i.e., ~80%) pigment in the M state. Then, also before each trace, an orange light (590 edge filter) with variable amounts of photons was given. The relative amount of orange light is indicated above the corresponding trace. After 4 min in the dark the saturating 3 s blue light was given again. This blue light put the pigment system back into photoequilibrium in which ~80% of the pigment molecules were in the M state. The responses to this constant blue light are presented in Fig. 1A. The differences among the various traces are only due to increasing amount of pigment conversion from R to M due to the adapting orange lights which put increasing amounts of pigment in the R state. B: The dependence of the relative decay time (of the various PDA's of Fig. 1A) to 1/e of the steady-state amplitude of the receptor potential during light on the relative amount of pigment conversion by the orange adaptation. The relative decay time was obtained by normalizing the time needed for the various PDA's to decay from the steady-state amplitude of the LRP to 1/e of the steady-state amplitude. This is a rough measure of PDA duration without any correction. The 100% point corresponds to the maximal PDA (upper trace in Fig. 1A). The decay time to 1/e of this trace is outside the figure. The scale of pigment conversion was determined by measuring the Early Receptor Potential (ERP) which is a linear manifestation of the changes in the visual pigment (from (41)).

The duration of a maximal PDA varies greatly among different species. Even in closely related species of the fly such as Calliphora and Drosophila, the duration of maximal PDA can be very different. For example, the duration of a maximal PDA is only a few minutes in Calliphora (22) as

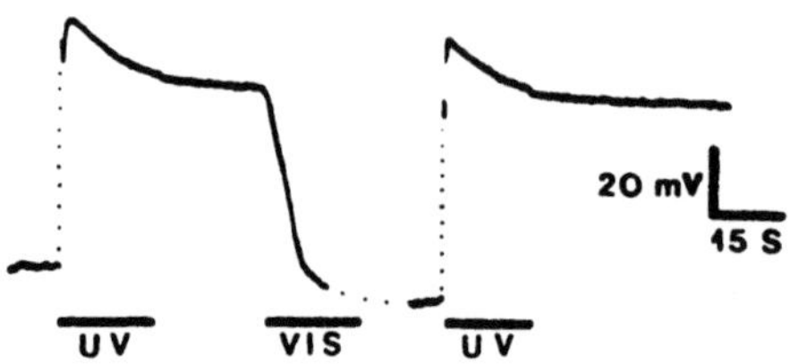

FIG. 2 - A demonstration of PDA depression by pigment conversion from M to R. The PDA was recorded intracellularly from a single UV photoreceptor of the median eye of the Limulus. The cell was initially adapted with saturating green light (555nm) which put nearly all the pigment in the R state. After 3 min in the dark the cell was illuminated with saturating UV (360nm) light which put maximal pigment in the M state. The LRP and the PDA which resulted from this UV light are presented in the figure. The resulting PDA, which could last longer than an hour, was depressed by the following saturating visible (555nm) light, which put the pigment back into the R state. Additional UV light which was given shortly later (15 s) induced a maximal PDA again. In this experiment PDA induction and depression could be repeated many times (from (39)).

compared to a few hours in Drosophila (43). The duration of a maximal PDA in a specific photoreceptor is tightly linked to the total amount of its visual pigment. When the amount of the visual pigment is reduced by vitamin A deprivation (22, 40, 56) or by mutation (32), the duration of the maximal PDA is reduced in correlation with the amount of pigment present until the PDA (but not the LRP) is completely abolished.

Since invertebrates have a long-lived M, the pigment distribution between R and M does not change (in most species) for hours. Following the decline of maximal PDA a regenerating light, which converts the maximal amount of pigment back from M to R, results in a period in which induction of additional PDA is strongly inhibited. A certain dark time (in the range of seconds to minutes) has to pass before a maximal PDA can be induced again. This dark time is called the anti-PDA period (26, 37). Its presence in Calliphora is demonstrated in Fig. 3.

Superposition of Discrete Events

There is evidence that the later receptor potential (LRP) is composed of discrete potentials (bumps) (13, 64). In those invertebrate species where discrete potentials are observed, either in the dark or during dim light, the PDA is also accompanied by voltage fluctuations (noise). The PDA noise can be used to compare quantitatively between the LRP and the PDA. However, this noise is not easily observed during a maximal PDA (Fig. 5, below). The amplitude of the light-induced noise is strongly attentuated during intense light ((13, 64), and see Fig. 4). A similar attenuation is observed during maximal PDA (see below).

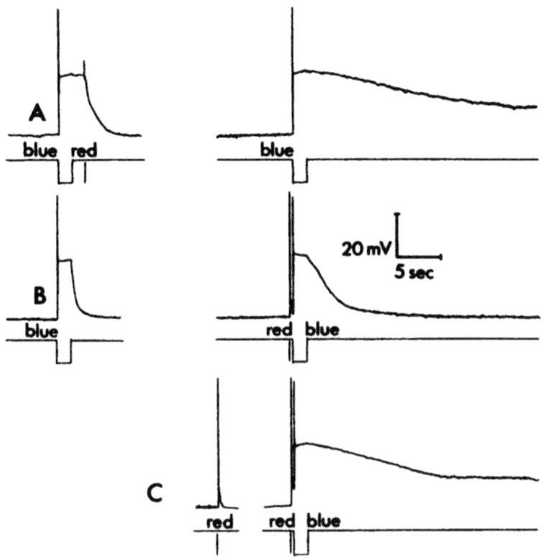

FIG. 3 - A demonstration of the anti-PDA phenomenon, a period in which induction of PDA is strongly inhibited by pigment conversion from M to R. All traces were recorded intracellularly from a single photoreceptor of white-eyed Calliphora. A: Left - A PDA was induced by saturating blue (457nm) light after red-orange (590 edge filter) adaptation and depressed by short intense red-orange flash. Right - A presentation of maximal PDA which was induced in similar manner to the PDA presented on the left but without depression with the red light. This PDA declined within a few minutes. B: Left - After the decline of the maximal PDA of trace A (right), additional saturating blue light induced LRP without a PDA. This blue light was given to the blue-adapted cell and therefore did not cause any net pigment conversion. Such a stimulus did not affect the PDA in any way. The fast decline of the response indicates that strong light without pigment conversion did not induce an afterpotential. Right - A pair of lights, red followed by blue, was given to the blue-adapted cell 1 min after trace B (left) was recorded. The red flash was strong enough to shift all the pigment from M to R within 2 ms (see Fig. 3A, left). The following saturating blue light which was given 0.4 s later did not induce a pronounced PDA in spite of the fact that it converted maximal (80%) pigment from R to M in similar manner to the blue light of line A. C: The paradigm of lines A and B was repeated except that a longer dark time (50 s) was given between the red light (M to R conversion) and the blue light (R to M conversion). The additional red flash which was given to the red-adapted cell (C) before the blue light did not change the pigment distribution between R and M and therefore had no effect on the PDA. It was given as a control to ensure that the state of light adaptation would be similar in lines B and C (from (37)).

In the Drosophila mutant trp, where light causes a reduction in excitation efficiency (38), a low-amplitude PDA with enhanced noise level could be observed even after maximal pigment conversion. This noise could be completely abolished by M to R conversion ((42), and see Fig. 5E). The autocorrelation function calculated from the noise of the PDA closely resembled the same function calculated from the noise of the receptor

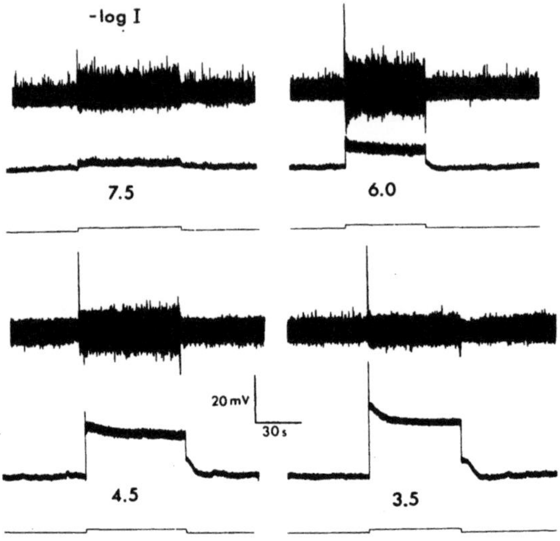

FIG. 4 - The LRP is accompanied by an elevated noise level which is reduced when the intensity of the light stimulus is relatively strong. The figure shows the intracellularly recorded LRP from white-eyed Musca photoreceptor in response to blue (<480nm) light stimuli of various intensities (as indicated). First and fourth rows - AC-coupled recordings at high gain (0.4-1000 Hz). Second and fifth rows - DC-coupled recordings at lower gain (as indicated). Even the stronger blue light (-log I = 3.5) was not strong enough to induce a PDA. The unattenuated light intensity in Figs. 4-6 is $4.7 \cdot 10^{16}$ photons$\cdot$cm^{-2}s^{-1} (from Barash and Minke, unpublished).

potential (42). A noise which could be resolved into bumps was also found in other species during the PDA or during an afterpotential induced by a strong light (18, 23, 24, 60). A comparison between the noise during the LRP, in response to increasing intensities of light, and during the PDA, at various phases of its decay, was carried out by Hamdorf and Razmjoo (23).

A detailed comparison, using shot-noise analysis, between the noise during the PDA and during the receptor potential (LRP) in the same cell is now in progress (Barash and Minke, unpublished). Preliminary results are presented in Figs. 4-6. Figures 4 and 5 show voltage noise recorded from Musca photoreceptors during LRP (LRP noise, Fig. 4) and PDA (PDA noise, Fig. 5). Figure 6 shows the power spectra calculated from the noise which accompanied the LRP of Musca in response to various light intensities as compared to power spectra calculated from the PDA of the same cell. Power spectra which were calculated from the LRP noise changed shape with increasing light intensity. Similarly, power spectra calculated from the PDA noise also changed shape when calculated at different times during the decline of the PDA, or during short PDA's induced by increasing amounts of R to M conversions, or during the anti-PDA period. Among the various shapes of the power spectra calculated from the LRP and the PDA

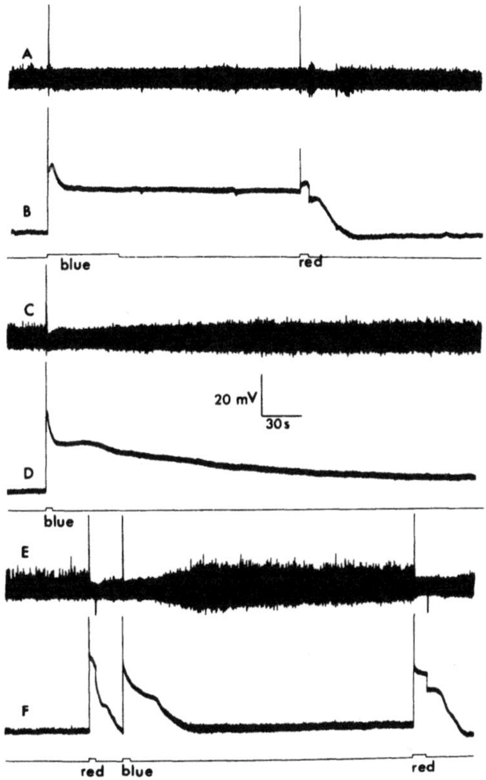

FIG. 5 - The PDA is accompanied by an elevated noise level which is very small when the PDA is maximal. All traces are from the same cell of Fig. 4. Traces A, C, and E are AC-coupled recordings (0.4-1000 Hz), and traces B, D, and F are the corresponding DC-coupled recordings. Trace B shows the maximal PDA (and its noise, trace A) which was induced by maximal intensity blue (<480nm) and depressed by maximal intensity red (>580nm) light. Trace D shows a short-duration PDA (and its noise, trace C) in response to maximal intensity short blue light which was attentuated by 2 log units. Trace F (middle response) shows a short duration noisy PDA in response to blue light identical to that of trace D. This blue light was given during the anti-PDA period. The anti-PDA period was induced by a maximal intensity red light (initial response, trace F) which was given after the decline of maximal PDA (not shown). The elevated noise level of the PDA was reduced by additional red light (trace F, right) (from Barash and Minke, unpublished).

of the same cell, one could find power spectra of certain LRP and PDA which were remarkably similar (Fig. 6). We do not know yet if for every power spectrum calculated from the PDA we can find a correspondingly

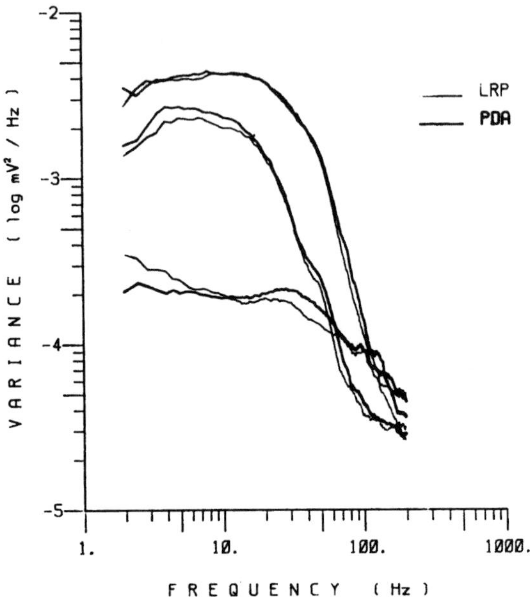

FIG. 6 - Power spectra calculated from the voltage fluctuations (noise) obtained during various PDA's and during LRP's of various amplitudes which were recorded from the same cell seen in Figs. 4 and 5. The bottom curves were calculated from the PDA and LRP resulting from blue light of the maximal intensity (Fig. 5A, B) and the middle curves when they had minimal amplitudes. The upper curves were calculated from LRP with maximal noise level and from PDA during the anti-PDA period (Fig. 5E). The power spectra of the LRP were calculated from the steady-state response to blue lights of maximal intensity (lower curve) and to lights which were attenuated by 8.0 (middle curve) and 6.5 (upper curve) log units. Each power spectrum was calculated from 1024 points, sampled at 500 Hz. The voltage responses were filtered by 0.4 Hz high-pass filter and 1000 Hz low-pass filter. The power spectra were calculated by fast Fourier transform with an 11/23 DEC computer. The power spectra presented in the figure are the average of fifteen similar spectra calculated from nonoverlapping segments (from Barash and Minke, unpublished).

similar power spectrum determined from the LRP. Nevertheless, as a first approximation one can conclude from noise analysis and from qualitative observations (23) that the PDA is composed of bumps with characteristics roughly similar (under specific conditions) to the bumps that sum to produce the LRP.

The PDA Is a Local Phenomenon

As mentioned above, the PDA duration depends on the amount of pigment conversion from R to M. When the total amount of pigment in the cell is greatly reduced by vitamin A deprivation, a detectable PDA cannot be induced (22, 56). Recent experiments conducted in the barnacle lateral eye indicate that the duration of the PDA depends on the concentration of the pigment converted (2). When a localized strong laser light converted a small amount (~4%) of pigment from R to M in a restricted area of the cell, a low-amplitude but long-duration PDA was induced (Fig. 7a). When the same amount of pigment was converted from R to M by a light which was diffuse all over the cell, no detectable PDA was observed. A much larger amount of pigment (i.e., >30%) had to be converted by a diffuse light in order to induce a detectable PDA (Fig. 7b). These experiments indicated that the range of interaction of the mechanisms underlying PDA induction has an upper limit on the order of 10 μm. These observations together with the effects of low pigment concentration and the supralinear dependence of the PDA on pigment conversion are consistent with the hypothesis that the PDA is induced when neighboring pigment molecules are converted from R to M. In additional experiments conducted in the barnacle, a PDA which was induced in one half of the cell by R to M conversion could not be depressed by M to R conversion in the other half of the cell (3). The above experiments indicated that the processes of PDA induction and depression and of the anti-PDA are local phenomena which do not spread all over the cell as suggested by Hamdorf (22).

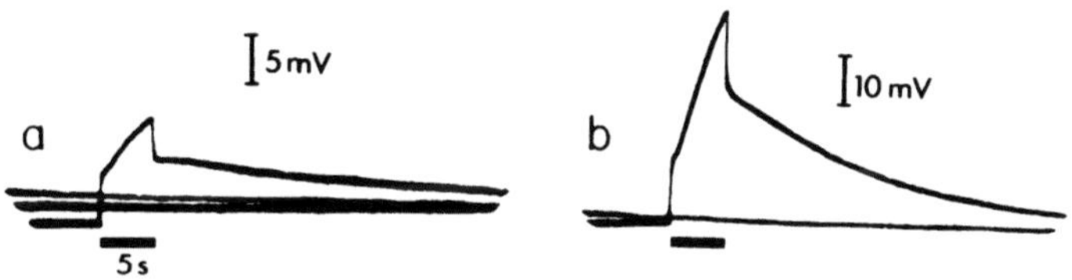

FIG. 7 - A comparison between the PDA resulting from localized illumination in a restricted region of the cell (a) with the PDA resulting from diffuse illumination all over the cell which converted 10 times more pigment from R to M (b). The PDA was recorded intracellularly from a single photoreceptor of the barnacle lateral eye (B. eburneus). The cell was initially blue- adapted (495nm). This illumination put maximal pigment (80%) in the R state. The cell was then dark-adapted for 5 min. The PDA's were induced by strong red laser (He-Ne) light (632.8nm). The red spotlight (a) which covered about 13% of the rhabdomere area shifted ~4% of the pigment while the diffuse light shifted ~40% of the pigment molecules from R to M. The amount of pigment conversion was determined by the ERP. Four successive traces were photographed on the same scale to show the relatively slow decay of the PDA resulted from the spot illumination relative to diffuse illumination (where two successive traces were superimposed). LRP and the initial PDA amplitudes are considerably larger following the diffuse illumination relative to spot illumination. However, the time in which the PDA in (b) declined to baseline is much shorter, although 10 times more pigment was converted from R to M by the diffuse light relative to the spotlight (from Almagor, Hillman, and Minke (2)).

Reduction in Sensitivity to Light During the PDA

Although a detailed quantitative study of the reduction in sensitivity to light during a PDA has not been done in invertebrates (but see (23)), qualitative observations indicated that the PDA is accompanied by a large

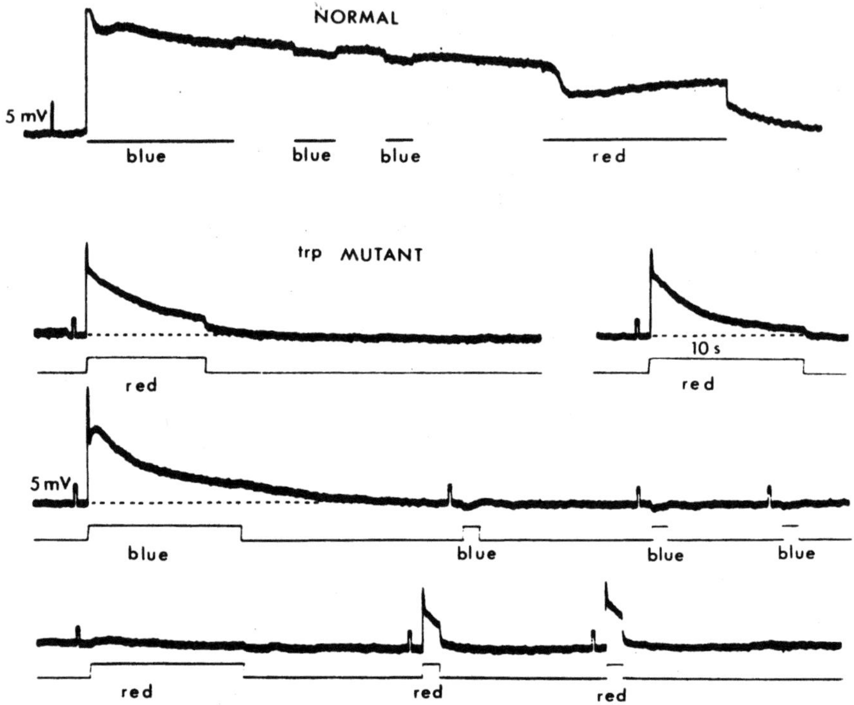

FIG. 8 - The prolonged depolarization and desensitization which are observed during the PDA can be separated from each other by a mutation in Drosophila. The upper row shows the PDA of white-eyed Drosophila which was induced by saturating blue (480nm) light following red adaptation. Additional blue lights gave only small hyperpolarizing responses for a few minutes, thus indicating that the cell was desensitized. Red light (600nm) which converted the pigment from M to R resulted in PDA depression and restoration of the sensitivity to light. The second row shows the receptor potential of the white-eyed trp mutant of Drosophila which is characterized by the decline of the response during illumination. When the light stimulus did not cause a net pigment conversion (i.e., red light after red adaptation - second row) the responsiveness recovered within 60 s after illumination. The dark interval between the first and second red lights was 50 s. However, when a PDA-inducing blue stimulus was given, the receptor potential still decayed to baseline but the cell remained desensitized for many minutes, as demonstrated by the lack of responses to blue and red stimuli after various periods of dark time (third and fourth rows). A PDA-depressing red light (fourth row), which converted the pigment from M to R, reestablished sensitivity to light. (Upper trace from (43); lower traces from (37).)

reduction in sensitivity to light. In Drosophila, where the PDA is very long, even an intense light did not result in further depolarization (Fig. 8) as long as the PDA was maximal. The reduced sensitivity level persisted during the entire duration of the PDA. The depolarization and the desensitization which accompanied the PDA could be separated from each other by using specific Drosophila mutations (34, 37). In these mutants the receptor potential was abolished during prolonged light (Fig. 8, second row) but it recovered in the dark in less than 1 min. Following "PDA-inducing light" (Fig. 8, third row) the receptor potential did not recover for many minutes (even though no PDA was observed), unless a regenerating ("PDA-depressing") light was first given (Fig. 8, fourth row). Thus the desensitization which accompanied the PDA persists even in the absence of voltage response (due to a mutation) and it is apparently related to the mechanism underlying the PDA. Similar observations were reported by Wong et al. (65) in normal Drosophila: the desensitization which accompanied the PDA persisted in the absence of voltage response due to anoxia.

The characteristics of the PDA discussed above are very similar to the characteristics of the prolonged aftercurrent (or after hyperpolarization) found in vertebrate rods.

THE CHARACTERISTICS OF THE PROLONGED AFTERCURRENT IN VERTEBRATE RODS

Dependence on Pigment Conversion

In vertebrate rods, where metarhodopsin is unstable in the dark, every intense light in the visible range causes a reduction in rhodopsin concentration. Bright illumination in the range which induces a saturated light response (>1% bleach of the visual pigment) induces a current change that continues for minutes in the dark (prolonged aftercurrent). The duration of the aftercurrent increases with increase in the amount of bleaching light up to many minutes (Fig. 9). A systematic study which correlates the duration of this aftercurrent to the amount of bleached rhodopsin has not been done. The prolonged dark response was observed in several species, for example in toad rods (7), in the rods of the axolotl (20, 21), in the rods of the tiger salamander (12), and in rat rods (1). In rat rods the prolonged aftercurrent could be depressed to baseline when rhodopsin was regenerated by adding 11-cis retinal (but not all-trans retinal) to a totally bleached retina ((1), and Fig. 10).

Superposition of Discrete Events

The prolonged aftercurrent in toad rods following small bleaches of rhodopsin was accompanied by current fluctuations (Fig. 11; (7, 30)). Shot-noise analysis of these current fluctuations indicated that they are very similar to light-induced current fluctuations (8, 30). It was concluded by Lamb (30) that spontaneous discrete events similar to photon-induced events occur in the dark in the rod outer segments at a much higher rate after a bleach than after full dark adaptation.

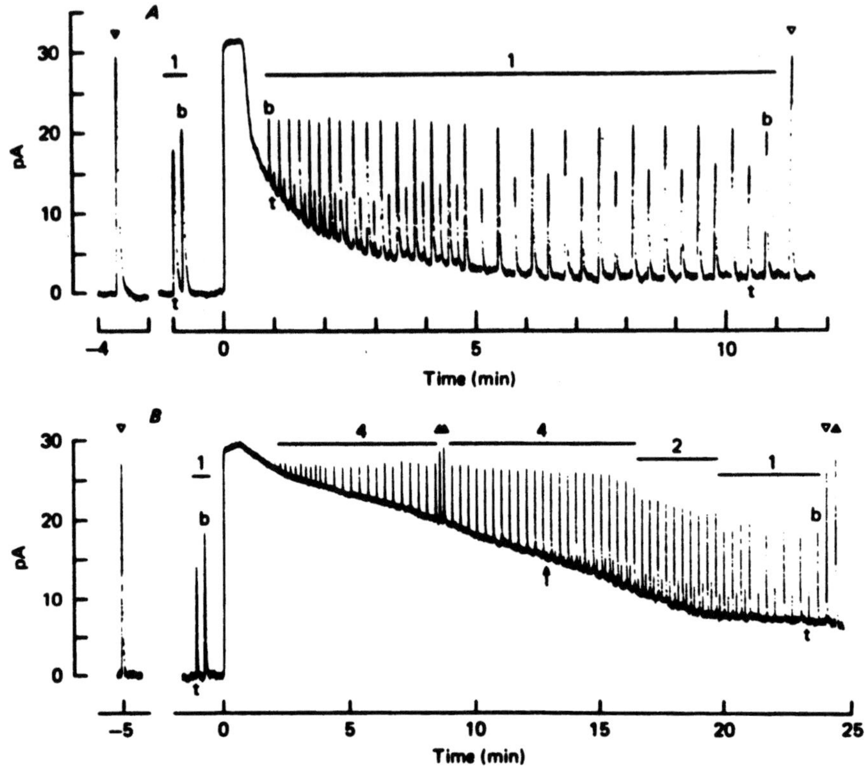

FIG. 9 - A demonstration of the prolonged aftercurrent which was recorded with a suction electrode from a toad rod following a small bleach (A) and larger bleach (B) of rhodopsin. The figure also shows the recovery of responsiveness to bright tip and base test flashes following two uniform bleaches. The short responses which are superimposed on the aftercurrent were induced by exposing the base (b-larger responses) and tip (t-smaller response) of the outer segment to the same test flash: A - bleach of about 0.8%; B - bleach of about 3.3%. In each case t and b indicate responses to flashes at tip and base, respectively (25 µm diam.), and triangles indicate responses to bright diffuse test flashes. Bars above each trace show the period during which local tip and base flashes were alternated; numbers 1,2, and 4 give flash strength relative to 42.5 photons μm^{-2} (about 740 isomerizations per rod) per flash. Relative to this same intensity, the diffuse test flashes have strengths of 1▽, 4Δ, 8Δ; Arrow in B (↑) indicates first detectable response to a tip flash. Bleach in A - 1.1×10^5 photons $\mu m^{-2}\,s^{-1}$ for 10 s, isomerizing approximately 2×10^7 rhodopsins; B - 4.7×10^5 photons $\mu m^{-2} s^{-1}$ for 10 s, bleaching approximately 8.2×10^7 rhodopsins. The figure shows that the duration and amplitude of the prolonged aftercurrent increased with increasing the amount of pigment bleached. The figure also shows that the aftercurrent is accompanied by a reduction in sensitivity to light. The sensitivity was much smaller at the tip of the outer segment relative to its base (from (7)).

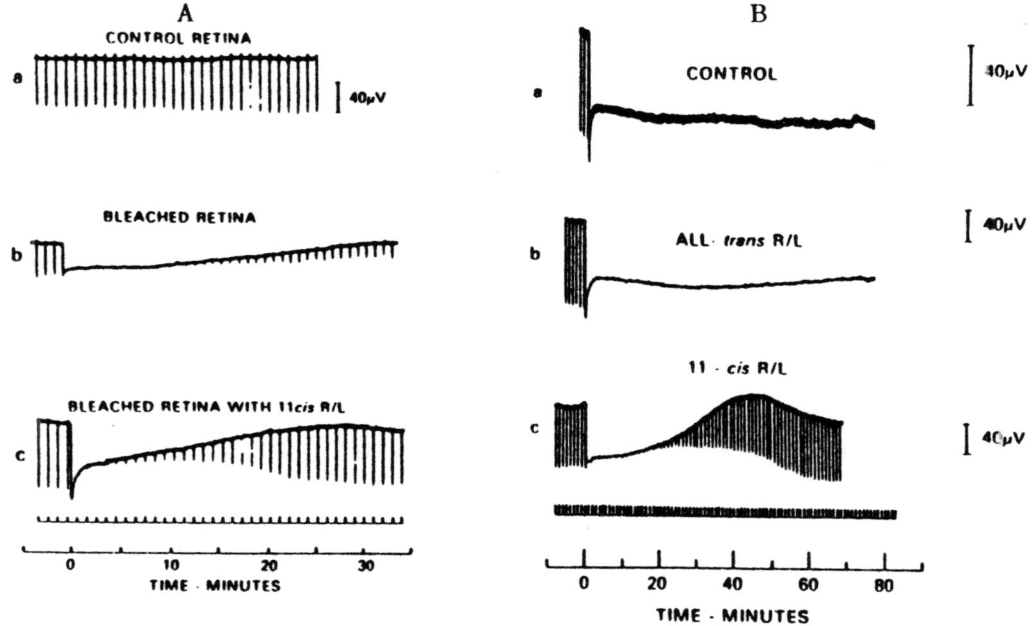

FIG. 10 - An induction of prolonged aftercurrents in the rods of albino rat and depression of the prolonged aftercurrent by adding 11-cis retinal to the retina. The dark current of the rods was measured extracellularly by two microelectrodes (1). A: (a) The stability of dark current and its suppression by light flashes of approximately 100 photons absorbed per rod. (b) Effects of bleaching (8%) of the retinal rhodopsin on the dark current. The bleaching light was applied at t = 0. Even though the current has returned fully at 30 min, it still took more light than 1000 photons absorbed per rod to shut the dark current completely (not shown). (c) Less than 5 s after an 8% rhodopsin bleach, liposomes bearing 11-cis retinalydehyde were added to the Ringer's perfusing the retina. The dark current recovered earlier and the 100-photon flashes once again suppressed the dark current completely. The train of short vertical bars above the time line indicates the stimulus light flashes. B: Recovery of the dark current from a 100% bleach. (a) The response of a retina to a 2 min exposure of orange light which bleached all the rhodopsin. There is a quick but small recovery of the dark current which was insensitive to light. The dark current remained suppressed at this level for the duration of the experiment. (b) Retina treated exactly as in (a), but all-trans retinaldehyde bound to lipids was added to the Ringer's perfusing the bleached retina. (c) Retina treated exactly as in (b), but 11-cis retinaldehyde was added before the bleach (from (1)).

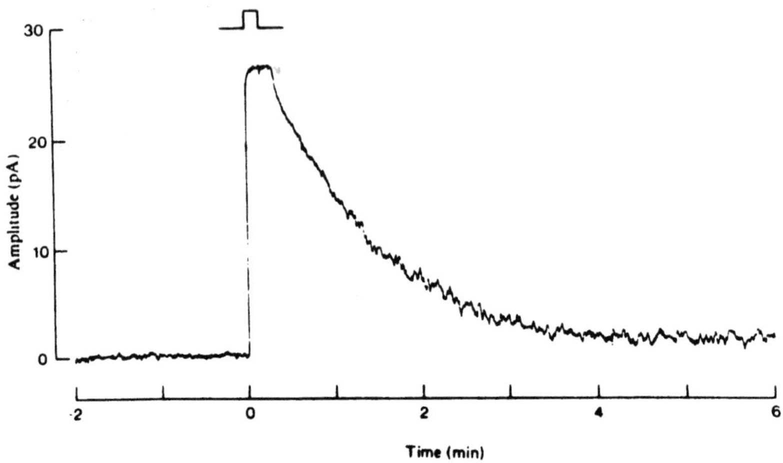

FIG. 11 - The prolonged aftercurrent is accompanied by quantal events, similar to photon-induced events in toad rod. The recording technique is similar to that of Fig. 9. The figure shows a response of a single rod to a bleach of about 0.7% (7.2 x 10^5 photons μm^{-2} calculated to isomerize 1.8 x 10^7 rhodopsins). Quantal events before the bleach are not visible, as the cell was relatively insensitive. The response was filtered by a low-pass filter at 5 Hz (from (30)).

The Reduction in Sensitivity to Light is Related to the Prolonged Aftercurrent

The prolonged aftercurrent is accompanied by a reduction in sensitivity. The reduced sensitivity was found to be highly localized to the bleached area (7, 12). Like the PDA of the invertebrate (22, 23), the prolonged aftercurrent of the vertebrate rod seems to be related to the reduced sensitivity. When the amount of bleaching is small and the aftercurrent is short, the sensitivity recovers faster (Figs. 9, 10) (1, 7). The reduced sensitivity remains as long as the aftercurrent is observed. In the toad retina, following a bleach of 3.3%, the aftercurrent remains longer than 25 min (7). In the rat rods, after total bleach, no sign of recovery is observed even after 80 min. However, when 11-cis retinal is added to the rat retina, depression of the aftercurrent and recovery of sensitivity occurs roughly in parallel ((1), and Fig. 10).

In toad rods when sensitivity is measured locally in a retricted area of the rod, a uniform bleach of the rod outer segment desensitizes the tip of the outer segment considerably more than its base (Fig. 9). This difference between tip and base desensitization increases with a larger bleach (Fig. 9). The difference in desensitization between tip and base may be correlated with differences in the magnitude (duration) of the aftercurrent in these two regions. The data of Baylor and Lamb (7) suggest that a longer aftercurrent is induced at the tip relative to the base of the outer segment with equal bleach in the two regions.

The coincidence between desensitization and the presence of aftercurrent was not observed in the experiments of Cornwall et al. (12) in the rods of the tiger salamander. In their experiments the aftercurrent returned to baseline 20 min following the bleaching light (which bleached >80% pigment) while the reduced sensitivity in the bleached area remained low for more than 100 min.

The time course of rod dark adaptation (i.e., the recovery from desensitization) in isolated frog retina, during a phase which was not distorted by the response of the cones, was found to depend on the amount of pigment bleached. When a 5 μs bright laser flash and a 10 s weaker light bleached a similar amount of pigment, a similar time course of dark adaptation was observed (16). Thus both the duration of rod desensitization and the PDA depend on the amount of the adapting light (i.e., on the product of stimulus intensity and its duration).

SIMILARITIES BETWEEN THE PDA OF INVERTEBRATES AND THE PROLONGED AFTERCURRENT OF VERTEBRATE RODS - A SUMMARY

1) Both the PDA and the prolonged aftercurrent are induced by pigment conversion (bleach) from the rhodopsin state. The larger the pigment conversion, the longer the effect (Figs. 1, 9, 10). In vertebrate rods smaller amounts of pigment conversion are needed to induce the effect.
2) Both phenomena can be cut short at any time by regeneration of R: by light in invertebrates (Fig. 2), by adding 11-cis retinal in vertebrates (Fig. 10).
3) Both phenomena are accompanied by appearance of discrete events similar to photon-induced events which are manifested as elevated noise in the dark (Figs. 5, 11).
4) Both phenomena are accompanied by a large reduction in sensitivity to light which is much larger than the reduction in the probability to absorb photons (Figs. 8-10).
5) Both phenomena are localized in the illuminated area and do not spread all over the photoreceptors (Figs. 7, 9).

Due to the similarities between the PDA and the prolonged aftercurrent, a mechanism which explains the PDA may also explain the prolonged aftercurrent.

A MOLECULAR MODEL FOR THE PDA

Several models have been suggested to account for the PDA phenomenology (for reviews see (22, 25)). A model which explains the PDA at the level of the photopigment, called the clusters model, can best explain the experimental data. This model is based on principles similar to those of the model of Hamdorf and Razmjoo (which is described in Fig. 12) and on the ideas and data of Paulsen and Bentrop (Fig. 13, below; (45)).

The clusters model is based on two kinds of experiments: a) Experiments of the first kind suggest that the PDA depends on the absolute concentration of pigment which is converted from R to M. As suggested above, a PDA is induced only when a sufficient amount of neighboring pigment molecules are converted from R to M. Thus no PDA can be induced by light which converts less than 10% of the pigment from R to M. Also, photoreceptors whose pigment concentration has been reduced by vitamin A deprivation or by mutation show little or no PDA. b) The other type of observations is related to the phosphorylation of bleached rhodopsin. In vertebrate rods and invertebrate photoreceptors bleached rhodopsin is phosphorylated at multiple locations (27, 29, 45-47, 53, 61, 62). This phosphorylation is probably required for the inactivation of the light-dependent phosphodiesterase activity in the rods (33, 55) and may constitute a mechanism for turning off excitation.

The model suggests that when neighboring pigment molecules in sufficient amounts are converted from R to M, phosphorylation-resistant clusters of M molecules are formed. These clusters are equivalent to the ⟁M state in the

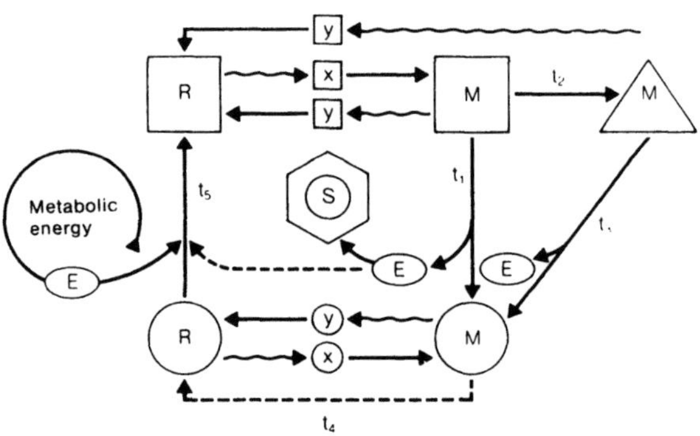

FIG. 12 - Hypothetical reaction scheme of visual pigments in the microvillus membrane of invertebrate photoreceptors according to Hamdorf. The successful absorption of a photon by dark-adapted rhodopsin [R] transforms the latter to "energy-rich" metarhodopsin [M], which can activate a site Ⓢ of the membrane with the release of energy. In this process [M] is deactivated to Ⓜ. Absorption of another photon can transform Ⓜ to Ⓡ, which can be restored to the active form [R] with the help of metabolic energy. This cycle is dominant under normal, physiological light intensities. Non-physiologically intense illumination increases the transitions [R]⇝[M]. Since the rate t_1 is limited by the rate at which the membrane can use energy, some [M] molecules go over into a storage state ⟁M. This state is transformed in a much slower deactivating reaction ($t_3 << t_1$) into Ⓜ, producing a prolonged depolarization afterpotential (PDA). The intermediate states of the reactions in the forward and backward directions are indicated by [x] and [y]. The dashed arrows indicate the possibility of energy transfer from [M] or ⟁M directly to Ⓡ. The rates t_1 to t_5 of the dark reactions are thermally controlled, t_4 indicates the possibility of a very slow metabolic dark regeneration of the visual pigment (from (22)).

model of Hamdorf and Razmjoo (see Fig. 12). Phosphorylation of the pigment in these clusters by protein kinase takes place at a much slower rate than during and after light which does not induce a PDA. As a result, the active M molecules during a PDA continue to excite the photoreceptor in the dark for a long time until each M molecule is finally phosphorylated at multiple locations. In principle, one active M molecule can evoke more than one bump during the PDA. However, no experimental evidence is available to support such a hypothesis. The slow decline of PDA is therefore due to the inability of the kinase to phosphorylate all the active M molecules at multiple locations fast enough after the cessation of intense light which converts large amounts of R to M. As long as the PDA does not decline appreciably (minimal phosphorylation), a regenerating stimulus which converts the pigment from M to R will depress the PDA and a following R to M pigment conversion will result in a new PDA (Fig. 2). However, following the decline of the PDA (maximal phosphorylation), a regenerating stimulus (M to R conversion) will result in phosphorylated R molecules which will have to be dephosphorylated during a certain period before additional PDA can be induced (the anti-PDA period).

An alternative general mechanism for extending the lifetime of an active substance, when it is present in high concentration, is the depletion of an additional material necessary for its deactivation and which is used up by that deactivation. However, the observation that a long PDA is induced by intense localized illumination (Fig. 7) means that this additional substance must be unable to diffuse more than a few μm in the cell, and so presumably is not ATP or similar materials. Thus a shortage in kinase molecules when a large amount of pigment is converted from R to M may account for long-lived non-phosphorylated M molecules.

Experiments which have recently been performed in fly retina by Selinger, Minke, and co-workers (9) strongly support the hypothesis that the mechanisms underlying the PDA take place at the photopigment level. These experiments show a strict quantitative correlation between induction and suppression of the PDA (in intact eye) and light-induced increase and decrease of GTPase activity in cell-free membrane preparation. The light-dependent GTPase of fly eye membrane preparation thus faithfully mimics the induction and suppression of PDA induced by pigment conversion in intact photoreceptors of the fly. GTPase activity is believed to be due to activation of GTP-binding protein (see Chabre and Applebury, this volume). The GTP-binding protein of both vertebrates (28, 59) and invertebrates (52) is known to be activated directly by the photopigment. The reflection of the PDA phenomenology in GTPase activity indicates that the mechanisms underlying the PDA operate at the photopigment level and not at a later stage of the transduction process.

The models which assume long-lived "active" non-phosphorylated M molecules as a mechanism for the PDA are strongly supported by recent experiments which were conducted in Calliphora by Paulsen and Bentrop

(45). Figure 13 shows measurement of phosphorylation of M as a function of time during (13Aa) or following (13Ab) saturating blue light which converted maximal pigment from R to M. The figure shows that the phosphorylation level is minimal in conditions when the PDA in intact cells is expected to be maximal (i.e., 2 min. after the beginning of continuous light (a); or at the cessation of the shorter blue light (b)). The level of phosphorylation increased over a time which is roughly similar to the time needed for a maximal PDA in Calliphora to decline (i.e., several min; see Fig. 3). Figure 13B shows that in the blue-adapted retina (following the decline of a maximal PDA, in intact cells) when the phosphorylation level was maximal, a regenerating red light resulted in dephosphorylation of rhodopsin in less than a minute. This time of dephosphorylation fits nicely with the duration of the anti-PDA in intact cells of Calliphora (Fig. 3).

The hypothesis suggesting that non-phosphorylated M molecules can give rise to excitation (bump production) in the dark is in agreement with recent experiments reported by Lisman (see Bownds and Brewer and Stieve, this volume) who used the UV cells of the median eye of Limulus. Unlike the model which suggests the existence of long-lived non-phosphorylated M molecules, according to the model suggested by Lisman all the M molecules are quickly phosphorylated, first at single locations. The PDA and accompanying noise are due to dephosphorylation of the phosphorylated M molecules. When the M molecules are phosphorylated at multiple locations the probability of back reactions is highly reduced and the frequency of dark bumps is reduced accordingly.

A recent experiment of pigment phosphorylation in the frog's rods is consistent with the above models of the PDA. Shichi and Williams (54) and Paulsen and Rudolphi ((47); see also (53)) found a longitudinal gradient of pigment phosphorylation in the disks along the outer segment of the frog. Interestingly, the phosphorylation level at the base of the outer segment was larger than the phosphorylation level of the tip. When one compares their results to those of Baylor and Lamb (Fig. 9), one notices that the base of the toad rod outer segment was desensitized <u>less</u> than the tip by the same bleaching light. Also, at the base a smaller aftercurrent seemed to be induced relative to the tip following equal bleach.

Thus phosphorylation level seems to be reduced in regions of the photoreceptor where the prolonged aftercurrent is relatively long. Recent experiments on membrane preparation of bovine rod outer segments are also consistent with the result obtained in the frog. Sitaramayya and Liebman (55) found that at very weak bleaches, cGMP phophodiesterase quenching by ATP-dependent phosphorylation is completed very quickly in less than 2 s. On the other hand, for very strong bleaches it takes many minutes for the phosphorylation to be completed. Furthermore, the ratio of phosphates incorporated per bleached rhodopsin is reduced 270 times when the fraction of rhodopsin bleached is increased from $1.4 \cdot 10^{-4}$ to $2.2 \cdot 10^{-1}$. These results fit previous measurements in frog reported by Bownds et al. (10), Shichi and

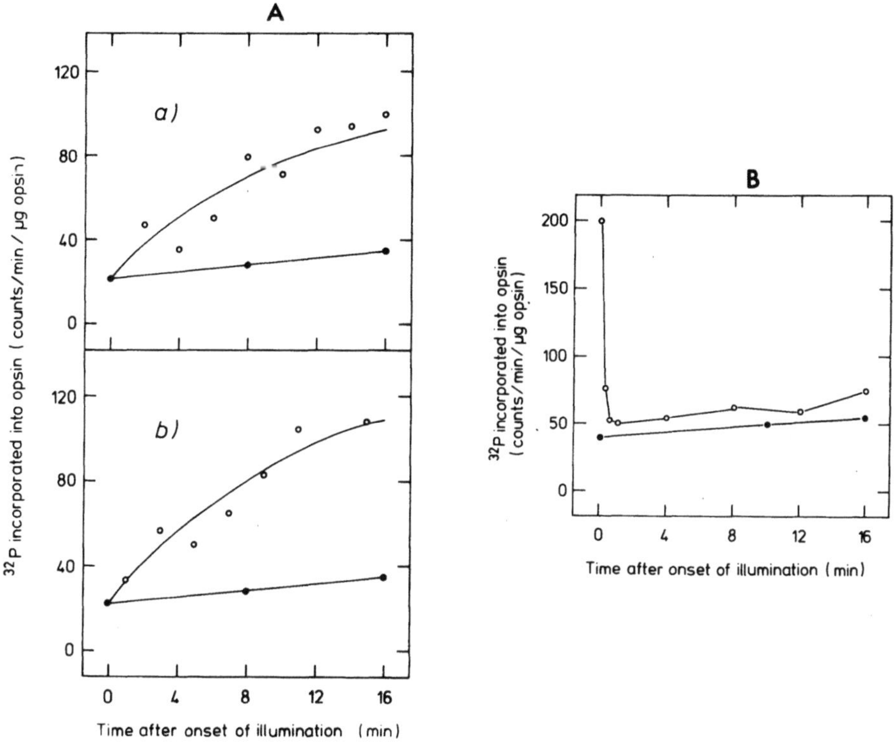

FIG. 13 - A: Time course of light-activated phosphorylation of opsin in isolated blowfly (Calliphora) retinae. Samples of 14 retinae in Hepes- Ringer's containing p^{32}-orthophosphate were incubated for 15 min. in the dark at 25°C. The retinae were then a) continuously irradiated with blue light (λ = 400-460mm) (o) or red light (610 edge filter) (●); b) irradiated for 1 min with blue light (o) or red light (●). The amounts of blue and red lights were sufficient to put the pigment system into photoequilibrium within 1 min. The amount of phosphate bound to opsin was measured after separation of the membrane proteins by SDS-PAGE. B: Time course of opsin dephosphorylation following the conversion of M into R. Phosphorylation performed as described in A except that it was initiated by irradiating each sample for 2 min with blue light (o). The retinae were then kept in the dark for 13 min at 25°C. Dephosphorylation was initiated by exposing retinae subsequently for 2 min to red light. The amount of phosphate bound to opsin was measured after isolating rhabdoms from the retinae and separating the membrane proteins by SDS-PAGE. Phosphorylation of opsin in control samples exposed only to red light (●). The maximum level of phosphate incorporation is the mean of measurements taken from three samples, each containing 14 retinae (from (45)).

Somers (53), and Wilden and Kühn (62). All the above data fit very well with the notion that a strong bleach, with a high probability of activating neighboring pigment molecules, is accompanied by a large increase in the fraction of bleached rhodopsin molecules which are not phosphorylated.

The relevance of photopigment concentration to bleaching adaptation was demonstrated by the experiments of Engbretson and Witkovsky (15): bleaching light induces bleaching adaptation in normal Xenopus photoreceptors. However, reduction in pigment concentration by vitamin A deprivation depressed the sensitivity only proportionally to the loss of pigment. The similarity between the effects of vitamin A deprivation on sensitivity to light in both vertebrates and invertebrates calls for experiments on Xenopus in which the duration of the dark aftercurrent and the sensitivity of the rods will be measured in vitamin A-deprived Xenopus following bleaching lights.

In summary, there is evidence in both invertebrate and vertebrate photoreceptors suggesting that activation of neighboring pigment molecules results in production of long-lived non-phosphorylated metarhodopsin molecules which excite the photoreceptor for a long time after the cessation of the light. This prolonged dark excitation "light adapts" the cell in a manner which may be at least partially similar to the effect of continuous background light (31). The experiments with the trp mutant (Fig. 8) indicate, however, that other factors not directly related to the voltage response are also involved (see also (48)).

One mechanism by which excitation of the invertebrate photoreceptor desensitizes the photoreceptor is via an increase in intracellular free Ca^{2+} concentration ($[Ca^{+}]_i$) (see Brown, this volume). However, there is evidence indicating that an increase in $[Ca^{+}]_i$ cannot account for all the phases of light adaptation in invertebrates (see below).

LIGHT ADAPTATION IN INVERTEBRATE PHOTORECEPTORS WHICH IS NOT Ca^{2+}-MEDIATED

Dark adaptation of invertebrate photoreceptors shows two separate phases of adaptation, a fast one (with a time constant of seconds) and a slower decay time (with a time constant of minutes): lobster (4), Limulus ventral eye (11, 17). The two phases of dark adapatation in the Limulus result from different rates of recovery of two components of the LRP (11, 35). Direct measurements of the increase in $[Ca^{2+}]_i$ using Arsenazo III indicated that a Ca-influx accompanies mainly the first but not the second phase (44). The Ca^{2+}-independent (or weakly dependent) light adaptation is not restricted to time of dark adaptation but rather to degree of adaptation. It can also be demonstrated with very weak conditioning flashes where individual bumps can still be observed. Bump adaptation, i.e., diminution of bump size due to pre-illumination, is even observable when the external Ca concentration is lowered to the nanomolar range (57) under conditions which abolish the measurable light-induced Ca^{2+} increase (as observed by Maaz and Stieve (36)) and abolish the (desensitizing) sensitivity shift due to stronger light adaptation (58). As suggested by the above authors, perhaps Ca-dependent regulation of sensitivity is a threshold phenomenon. If $[Ca^{2+}]_i$ is higher than a certain threshold level, it regulates the sensitivity of the photoreceptor; below threshold another, yet unknown, process determines photoreceptor sensitivity. It is still not clear how these two mechanisms, the Ca^{2+}-

dependent and the Ca^{2+}-independent (or less dependent), are related to the reduction in sensitivity during the PDA in invertebrates.

CONCLUSIONS

Bleaching adaptation is an unsolved problem in vision research. Barlow (5, 6), followed by Lamb (30), suggested that the elevation of "equivalent background" during dark adaptation, following bleaching light, is caused by events similar to those evoked by photons in the rod outer segment. Similar phenomena are observed in invertebrate photoreceptors during the prolonged depolarizing afterpotential.

A molecular mechanism for the induction of this prolonged dark excitation has been suggested. This mechanism is based on the observation that bleaching light in the vertebrate rods and pigment conversion from rhodopsin to metarhodopsin in invertebrate photoreceptors both induce a prolonged excitation in the dark which is very similar to the excitation during light. This prolonged excitation desensitizes the photoreceptors like the excitation during light (LRP) via still unknown mechanisms. However, the mechanisms by which these two excitations desensitize the photoreceptor may not be the same (48). In the trp mutant, following R to M conversion the desensitization occurs even without membrane conductance change. The mechanism which is proposed to account for the prolonged dark excitation suggests that a large R to M conversion results in many activated M molecules which continue to be non-phosphorylated for a long time in the dark, giving rise to the PDA and the accompanying voltage fluctuations.

Acknowledgements. I thank S. Barash for his permission to use unpublished observations. I also thank H. Stieve, P. Hillman, T.D. Lamb, and A. Lewis for critical reading of the manuscript. Supported by National Institutes of Health grant EY 03529.

REFERENCES

(1) Albani, C.; Nöll, G.N.; and Yoshikami, S. 1980. Rhodopsin regeneration, calcium and the control of the dark current in vertebrate rods. Photochem. Photobiol. **32:** 515-520.

(2) Almagor, E.; Hillman, P.; and Minke, B. 1986. Spatial properties of the prolonged depolarizing afterpotential (PDA) in barnacle photoreceptors: I. The induction process. J. Gen. Physiol., in press.

(3) Almagor, E.; Hillman, P.; and Minke, B. 1986. Spatial properties of the prolonged depolarizing afterpotential: II. Antagonistic interactions. J. Gen. Physiol., in press.

(4) Barnes, S.N., and Goldsmith, T.H. 1977. Dark adaptation, sensitivity, and rhodopsin level in the eye of the lobster, Homarus. J. Comp. Physiol. **120:** 143-159.

(5) Barlow, H.B. 1956. Retinal noise and absolute threshold. J. Opt. Soc. Am. **46:** 634-639.

(6) Barlow, H.B. 1964. Dark-adaptation: a new hypothesis. Vision Res. **4:** 47-58.

(7) Baylor, D.A., and Lamb, T.D. 1982. Local effects of bleaching in retinal rods of the toad. J. Physiol. **328**: 49-71.

(8) Baylor, D.A.; Lamb, T.D.; and Yau, K.-W. 1979. Responses of retinal rods to single photons. J. Physiol. **288**: 613-634.

(9) Blumenfeld, A.; Erusalimsky, J.; Heichal, O.; Selinger, Z.; and Minke, B. 1985. Light-activated guanosinetriphosphatase in Musca eye membranes resembles the prolonged depolarizing afterpotential in photoreceptor cells. Proc. Natl. Acad. Sci. USA **82**: 7116-7120.

(10) Bownds, D.; Dawes, J.; Miller, J.; and Stahlman, M. 1972. Phosphorylation of frog photoreceptor membranes induced by light. Nature New Biol. **237**: 125-127.

(11) Classen-Linke, I.., and Stieve, H. 1981. Time course of dark adaptation in the Limulus ventral nerve photoreceptor - measured as constant response amplitude curve - , and its dependence upon extracellular calcium. Biophys. Struct. Mech. **7**: 336-337.

(12) Cornwall, M.C.; Fein, A.; and MacNichol, E.F., Jr. 1983. Spatial localization of bleaching adaptation in isolated vertebrate rod photoreceptors. Proc. Natl. Acad. Sci. USA **80**: 2785-2788.

(13) Dodge, F.A.; Knight, B.W.; and Toyoda, J. 1968. Voltage noise in Limulus visual cells. Science **160**: 88-90.

(14) Dowling, J.E., and Ripps, H. 1970. Visual adaptation in the retina of the skate. J. Gen. Physiol. **56**: 491-520.

(15) Engbretson, G.A., and Witkovsky, P. 1978. Rod sensitivity and visual pigment concentration in Xenopus. J. Gen. Physiol. **72**: 801-819.

(16) Ernst, W., and Kemp, C.M. 1979. Reversal of photoreceptor bleaching and adaptation by microsecond flashes. Vision Res. **19**: 363-365.

(17) Fein, A., and DeVoe, R.D. 1973. Adaptation in the ventral eye of Limulus is functionally independent of photochemical cycle, membrane potential and membrane resistance. J. Gen. Physiol. **61**: 273-289.

(18) Fein, A., and Hanani, M. 1978. Light-induced increase in discrete waves in the dark in Limulus ventral photoreceptors. Brain Res. **156**: 157-161.

(19) Frank, R.N. 1971. Properties of "neural" adaptation in components of the frog electroretinogram. Vision Res. **11**: 1113-1123.

(20) Grabowski, S.R., and Pak, W.L. 1975. Intracellular recordings of rod responses during dark adaptation. J. Physiol. **247**: 363-391.

(21) Grabowski, S.R.; Pinto, L.H.; and Pak, W.L. 1972. Adaptation in retinal rods of axolotl: intracellular recordings. Science **176**: 1240- 1243.

(22) Hamdorf, K. 1979. The physiology of invertebrate visual pigments. In Handbook of Sensory Physiology. Comparative Physiology and Evolution of Vision in Invertebrates. A. Invertebrate Photoreceptors, ed. H. Autrum, vol. 7, pt. 6A, pp. 145-224. Berlin: Springer-Verlag.

(23) Hamdorf, K., and Razmjoo, S. 1979. Photoconvertible pigment states and excitation in Calliphora; the induction and properties of the prolonged depolarizing afterpotential. Biophys. Struct. Mech. **5**: 137- 161.

(24) Hardie, R.C.; Franceschini, N.; and McIntyre, P.D. 1979. Electrophysiological analysis of fly retina. II. Spectral and polarisation sensitivity in R7 and R8. J. Comp. Physiol. **133:** 23-39.

(25) Hillman, P.; Hochstein, S.; and Minke, B. 1983. Transduction in invertebrate photoreceptors. Role of pigment bistability. Physiol. Rev. **63:** 668-772.

(26) Hochstein, S.; Minke, B.; and Hillman, P. 1973. Antagonistic components of the late receptor potential in the barnacle photoreceptor arising from different stages of the pigment process. J. Gen. Physiol. **62:** 105-128.

(27) Kühn, H. 1978. Light-regulated binding of rhodopsin kinase and other proteins to cattle photoreceptor membranes. Biochemistry **17:** 4389-4395.

(28) Kühn, H.; Bennett, N.; Michel-Villaz, M.; and Chabre, M. 1981. Interactions between photoexcited rhodopsin and GTP-binding protein: kinetic and stoichiometric analysis from light-scattering changes. Proc. Natl. Acad. Sci. USA **78:** 6873-6877.

(29) Kühn, H., and Dryer, W.J. 1972. Light-dependent phosphorylation of rhodopsin by ATP. FEBS Lett. **20:** 1-6.

(30) Lamb, T.D. 1980. Spontaneous quantal events induced in toad rods by pigment bleaching. Nature **287:** 349-351.

(31) Lamb, T.D. 1981. The involvement of rod photoreceptors in dark adaptation. Vision Res. **21:** 1773-1782.

(32) Larrivee, D.C.; Conrad, S.K.; Stephenson, R.S.; and Pak, W.L. 1981. Mutation that selectively affects rhodopsin concentration in the peripheral photoreceptors of Drosophila melanogaster. J. Gen. Physiol. **78:** 521-545.

(33) Liebman, P.A., and Pugh, E.N. 1980. ATP mediates rapid reversal of cyclic GMP phosphodiesterase activation in visual receptor membrane. Nature **287:** 734-736.

(34) Lo, M.-V.C., and Pak, W.L. 1978. Desensitization of peripheral photoreceptors shown by blue-induced decrease in transmittance of Drosophila rhabdomeres. Nature **273:** 722-774.

(35) Maaz, G.; Nagy, K.; Stieve, H.; and Klomfass, J. 1981. The electrical light response of the Limulus ventral nerve photoreceptor, a superposition of distinct components - observable by variation of the state of light adaptation. J. Comp. Physiol. **141:** 303-310.

(36) Maaz, G., and Stieve, H. 1980. The correlation of the receptor potential with the light induced transient increase in intracellular calcium-concentration measured by absorption change of Arsenazo III injection into Limulus ventral nerve photoreceptor cell. Biophys. Struct. Mech. **6:** 191-208.

(37) Minke, B. 1979. Transduction in photoreceptors with bistable pigments: intermediate processes. Biophys. Struct. Mech. **5:** 163-174.

(38) Minke, B. 1982. Light-induced reduction in excitation efficiency in the <u>trp</u> mutant of Drosophila. J. Gen. Physiol. **79:** 361-385.

(39) Minke, B.; Hochstein, S.; and Hillman, P. 1973. Antagonistic process as source of visible-light suppression of afterpotential in Limulus UV photoreceptors. J. Gen. Physiol. **62:** 787-791.

(40) Minke, B., and Kirschfeld, K. 1979. The contribution of a sensitizing pigment to the photosensitivity spectra of fly rhodopsin and metarhodopsin. J. Gen. Physiol. **73:** 517-540.

(41) Minke, B., and Kirschfeld, K. 1984. Non-local interactions between light induced processes in Calliphora photoreceptors. J. Comp. Physiol. A **154:** 175-187.

(42) Minke, B.; Wu, C.-F.; and Pak, W.L. 1975. Induction of photoreceptor voltage noise in the dark in Drosophila mutant. Nature **258:** 84-87.

(43) Minke, B.; Wu, C.-F.; and Pak, W.L. 1975. Isolation of light-induced response of central retinular cells from electroretinogram of Drosophila. J. Comp. Physiol. **98:** 345-355.

(44) Nagy, K., and Stieve, H. 1983. Changes in intracellular calcium ion concentration, in the course of dark adaptation measured by arsenazo III in the Limulus photoreceptor. Biophys. Struct. Mech. **9:** 207-223.

(45) Paulsen, R., and Bentrop, J. 1984. Reversible phosphorylation of opsin induced by irradiation of blowfly retinae. J. Comp. Physiol. A **155:** 39-45.

(46) Paulsen, R., and Hopp, I. 1978. Light-activated phosphorylation of cephalopod rhodopsin. FEBS Lett. **96:** 55-58.

(47) Paulsen, R., and Rudophi, P. 1980. Rhodopsin phosphorylation in the frog retina: Analysis by autoradiography. Neurochemistry **1:** 287-297.

(48) Pepperberg, D.R. 1984. Rhodopsin and visual adaptation: Analysis of photoreceptor thresholds in the isolated skate retina. Vision Res. **24:** 357-366.

(49) Pepperberg, D.R.; Lurie, M.; Brown, P.K.; and Dowling, J.E. 1976. Visual adaptation: effects of externally applied retinal on the light-adapted, isolated skate retina. Science **191:** 394-396.

(50) Pugh, E.N., Jr. 1975. Rushton's paradox: rod dark adaptation after flash photolysis. J. Physiol. **248:** 413-431.

(51) Rushton, W.A.H. 1965. Bleached rhodopsin and visual adaptation. J. Physiol. **181:** 645-655.

(52) Saibil, H., and Michel-Villaz, M. 1984. Squid rhodopsin and GTP-binding protein crossreact with vertebrate photoreceptor enzymes. Proc. Natl. Acad. Sci. USA **81:** 5111-5115.

(53) Shichi, H., and Somers, R.L. 1978. Light-dependent phosphorylation of rhodopsin. J. Biol. Chem. **253:** 7040-7046.

(54) Shichi, H., and Williams, T.C. 1979. Rhodopsin phosphorylation suggests biochemical heterogeneities of retinal rod disks. J. Supramol. Struct. **12:** 419-424.

(55) Sitaramayya, A., and Liebman, P.A. 1983. Phosphorylation of rhodopsin and quenching of cyclic GMP phosphodiesterase activation by ATP at weak bleaches. J. Biol. Chem. **258:** 12106-12109.

(56) Stark, W.S.; Ivanyshyn, A.M.; and Hu, K.G. 1976. Spectral sensitivities and photopigments in adaptation of fly visual receptors. Naturwiss. **63:** 513-518.

(57) Stieve, H., and Bruns, M. 1983. Bump latency distribution and bump adaptation of Limulus ventral nerve photoreceptor in varied extracellular calcium concentration. Biophys. Struct. Mech. **9:** 329-339.

(58) Stieve, H.; Bruns, M; and Gaube, H. 1984. The sensitivity shift due to light adaptation depending on the extracellular calcium ion concentration in Limulus ventral nerve photoreceptor. Z. Naturforsch. **39c:** 662-679.

(59) Stryer, L. 1983. Transducin and the cyclic GMP phosphodiesterase: Amplifier proteins in vision. Cold S.H. Symp. Quant. Biol. **48:** 841-852.

(60) Tsukahara, Y., and Horridge, G.A. 1978. The distribution of bumps in the tail of the locust photoreceptor afterpotential. J. Exp. Biol. **73:** 1-14.

(61) Vandenberg, C.A., and Montal, M. 1984. Light-regulated biochemical events in invertebrate photoreceptors. 2. Light regulated phosphorylation of rhodopsin and phosphoinositides in squid photoreceptor membranes. Biochemistry **23:** 2347-2352.

(62) Wilden, U., and Kühn, H. 1982. Light-dependent phosphorylation of rhodopsin: Number of phosphorylation sites. Biochemistry **21:** 3014- 3022.

(63) Witkovsky, P.; Gallin, E.; Hollyfield, J.G.; Ripps, H.; and Bridges, C.D.B. 1976. Photoreceptor thresholds and visual pigment levels in normal and vitamin A-deprived Xenopus tadpoles. J. Neurophysiol. **39:** 1272-1287.

(64) Wong, F. 1977. Nature of light-induced conductance changes in ventral photoreceptors of Limulus. Nature **276:** 76-79.

(65) Wong, F.; Wu, C.-F.; Mauro, A.; and Pak, W.L. 1976. Persistence of prolonged light-induced conductance change in arthropod photoreceptors on recovery from anoxia. Nature **264:** 661-664.

The Molecular Mechanism of Photoreception, ed. H. Stieve, pp. 267-286. Dahlem Konferenzen 1986. Berlin, Heidelberg, New York, Tokyo: Springer-Verlag.

Photoreceptor Adaptation - Vertebrates

T.D. Lamb
Physiological Laboratory, University of Cambridge
Cambridge CB2 3EG, England

Abstract. It seems that much of the adaptational behavior of the overall visual system may be explicable in terms of properties of the rod and cone photoreceptors. This paper reviews the properties of vertebrate photoreceptors within the framework of the performance of the visual system and attempts to explore those aspects of photoreceptor adaptation which may be important at the behavioral level. For convenience it is useful to consider separately the phenomena of background adaptation and bleaching adaptation, as quite different mechanisms appear to be involved. Background adaptation involves desensitization and acceleration of the response to light, probably as a result of modifications to the transduction process at the level of biochemical reactions. During bleaching adaptation, which follows exposure to extremely intense light, the rod photoreceptors appear to experience something equivalent to the arrival of a steam of photons, and it is conceivable that the adaptational effects are essentially the same as if a real light had been absorbed. The equivalent light may originate from reverse reactions in the chain of steps involved in the removal of activated rhodopsin.

INTRODUCTION AND PSYCHOPHYSICS

The ability of the visual system to "adapt" to a given ambient level of illumination is of fundamental importance to vision. It is this property which enables us to see adequately over a 10 log-unit range intensities, from a dim starlit night to a bright sunny beach. Much of the adaptational behavior of the overall visual system can be accounted for in terms of properties of the rod and cone photoreceptors. The purpose of this paper is to review the adaptational properties of the photoreceptors and to attempt to relate them to our visual performance.

In order to simplify the study of adaptational properties it is convenient to distinguish two quite separate situations, termed "background adaptation" and "bleaching adaptation." These terms, which are defined below, cover many aspects of adaptation but it should be borne in mind that in real life more complicated situations, including combinations of the two, may occur.

Under most conditions normally encountered the eye adapts very rapidly to changes in ambient light levels. Typically, following a change in illumination of, say, 1 log unit (either an increase or a decrease), a new steady state of visual performance is established within seconds (17). This behavior will be referred to as background adaptation. It is synonymous with Rushton's (40) term "field adaptation" and is often referred to as "light adaptation," although this term may be confusing because the concept also encompasses reductions in light intensity.

The second term, bleaching adaptation (synonymous with "dark adaptation"), refers not simply to the behavior following reduction in light level, but is instead reserved for the special case of recovery from extremely intense lights which "bleach" a substantial fraction of the photopigment in the receptors. Following exposure to intense illumination which bleaches more than, say, 90% of the rhodopsin, one's visual sensitivity is initially enormously reduced and recovers very slowly. Even after 20 minutes of dark adaptation, visual threshold is elevated about 2 log units above the final dark-adapted level which is eventually reached after 40-50 minutes.

An Analogy

Although it is sometimes thought that the reduction in visual sensitivity which accompanies background adaptation (to higher intensities) is detrimental to visual performance, this is not the case. Craik (16) realized this and likened it to changing the range-setting switch on a multi-range meter. Rose (38) more accurately compared adaptation to the automatic gain control of a radio. In a radio receiver the automatic gain control (AGC) automatically adjusts the receiver sensitivity fairly rapidly in approximately inverse proportion to the signal strength, so that a nearly constant sound output is obtained for stations of very different signal strengths. In this way the output of the radio is proportional to the modulation of the transmitted signal rather than to the mean signal strength. For a weak station the gain is high and the output may be very noisy, while for a strong station the gain is greatly reduced and relatively noise-free reception is obtained.

In much the same way background adaptation acts fairly rapidly automatically to adjust visual sensitivity approximately in inverse proportion to the mean light intensity. This implies that early stages in visual processing detect contrast (or modulation) in the visual scene. As has recently been pointed out by Shapley and Enroth-Cugell (41), such contrast coding may be of immense importance to the visual system. Most of the visual scenes which we normally encounter involve reflecting objects, and in a reflected scene the contrast is independent of the mean level of illumination. Extraction of contrast information means that for reflected scenes the signals sent from the retina to the brain are invariant with the ambient level of illumination. In this way the brain is presented primarily with information about the scene rather than with information about the light level.

In terms of the radio receiver analogy, bleaching adaptation would correspond to the situation of tuning from an extremely powerful, local station to a very weak, distant station and finding (to one's surprise) that it was necessary to wait a long time for the receiver to resensitize so that the station could be heard. Obviously, the manufacturer of a radio set with this kind of performance would have some trouble selling it, and one is left wondering why such an undesirable property should be built into the visual system. It seems attractive to suppose that the slowness of visual dark adaptation represents an unavoidable consequence of some other important property of visual transduction - perhaps a consequence of the biochemistry needed to obtain the enormously high sensitivity of the rod photoreceptors.

Psychophysics

Typical plots of background adaptation and bleaching adaptation in human observers are plotted in Figs. 1 and 2, respectively.

Backgrounds. Figure 1 is often referred to as a threshold versus intensity (TVI) curve or increment threshold curve and plots the threshold intensity for the detection of some particular visual stimulus versus the intensity of the background against which the stimulus is being detected. For the

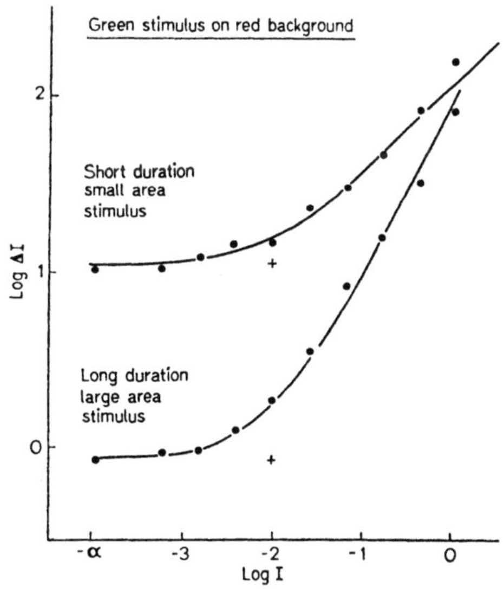

FIG. 1 - Threshold versus intensity curve (or increment threshold curve) showing Weber law and square root law behavior. Ordinate plots the threshold intensity ,ΔI, required for detection of a stimulus against a background of intensity I on the abscissa, in double logarithmic coordinates. Lower curve: large area (1° diam) long duration (1 s) stimulus gives Weber's law (unity slope). Upper curve (displaced downwards 1 log unit for convenience): small area (0.01° diam) short duration (7 ms) stimulus gives square root law (slope of 0.5). + indicates intersection of sloping and horizontal asymptotes. Photopic foveal vision. Reproduced with permission from (7).

present purposes we will restrict consideration to backgrounds which are of very large area (say, 10° diameter), uniform, steady, and monochromatic.

The observed behavior then depends on the form of the stimulus being detected, for which we will restrict consideration to patches of monochromatic light of the same wavelength as the background, presented near the center of the background. For stimuli of relatively large area ($\geq 1^\circ$) and long duration (≥ 0.1 s) the observed behavior is typically that shown below in Fig. 1. Over a wide range of backgrounds the threshold intensity is directly proportional to background intensity, i.e., the curve has unity slope in double log coordinates (1). In contrast to this, as the stimulus is made either small in area or short in duration (or both small and short), the results obtained tend towards the upper curve. Here the threshold intensity is approximately proportional to the square root of the background intensity, i.e., the slope in double log coordinates approaches a value of around 0.5. Barlow (4) showed that such square root law behavior could apply over several log units of intensity.

The case of unity slope in double log coordinates is called Weber law behavior and is described by the equation

$$\Delta I = \Delta I_D + k\, I_B, \qquad \text{(1A)}$$

or

$$\Delta I/\Delta I_D = 1 + I_B/I_D, \qquad \text{(1B)}$$

where ΔI is threshold intensity, ΔI_D is threshold in darkness, and I_B is the background intensity. The constants k and I_D are referred to, respectively, as the "Weber fraction" and the "dark light" and are related by

$$k = \Delta I_D/I_D. \qquad \text{(2)}$$

When plotted in linear coordinates the Weber fraction is the slope of the threshold versus intensity relation, and the dark light I_D is the magnitude of the intercept on the intensity scale (see Fig. 4). When plotted in logarithmic coordinates (e.g., Fig. 1) the dark light I_D is the intensity at the intersection of the horizontal and sloping straight-line regions (symbols "+" in Fig. 1). The term "dark light" originated because the visual system can be described as behaving as if it were experiencing an apparent light, I_D, even under conditions of total darkness.

For the case in Fig. 1 with a small, brief target and with log-log slope approaching 0.5, the modified form of Eq. 1,

$$\Delta I/\Delta I_D = 1 + (I_B/I_D)^n, \qquad \text{(3)}$$

applies with exponent n being 0.5. Note that it is an experimentally determined fact that the dark light I_D is the same in the two cases, as indeed it is for any other stimulus configuration (see, e.g., (7)). For human scotopic vision the dark light corresponds to about 0.01 isomerizations s^{-1} per rod, while for photopic vision it is about 1000 isomerizations s^{-1} per cone.

Bleaching. Bleaching adaptation behavior is shown in Figs. 2A and B, which plot the time course of threshold recovery following exposure to a series of different intensities which bleached from as little as 10% to more than 98% of the rhodopsin. Note that Fig. 2B was obtained from a rod monochromat (32) and therefore lacks the classical cone-rod transition shown in Fig. 2A. For each of the traces in Fig. 2B there is a considerable period at intermediate times during which log threshold elevation declines linearly with time, i.e., straight-line falling behavior.

For fuller coverage of the psychophysics of visual adaption the reader is referred to the reviews by Barlow (7), MacLeod (31), and Shapley and Enroth-Cugell (41). In the following sections some aspects of the role of vertebrate photoreceptors in overall visual adaptation will be examined separately for background adaptation and for bleaching adaptation. Because of limitations of space, the treatment is by no means exhaustive and many important references have inevitably been omitted. The case of invertebrates is dealt with by Brown and by Minke (both this volume).

BACKGROUND ADAPTATION

Weber law background adaptation behavior is observed at the level of the photoreceptors in cones and rods of lower vertebrates, but not in the rods of primates. (At the time of writing, the adaptational behavior of primate cones has not been properly examined.)

Cones

The effect of background illumination on the sensitivity of a turtle cone to light flashes is shown in Fig. 3 (9). The ordinate plots response amplitude divided by flash intensity and is in units of sensitivity (S_F, µV/photoisomerization); the uppermost trace was determined from flashes presented in darkness, and the lower traces were determined from flashes presented on backgrounds of increasing intensity. Quite clearly the flash sensitivity drops very significantly with increasing backgrounds, and in addition, the time course of the response accelerates. In Fig. 3 the time to the peak of the dim flash response is about 110 ms in darkness but falls to about 50 ms in the presence of bright background illumination.

These responses can be compared with psychophysics by taking the reciprocal of sensitivity at the peak of the response ($1/S_F$). This represents the intensity required to elicit a criterion response amplitude and is therefore analogous to threshold ΔI (although it must be stressed that vertebrate photoreceptors do not have a threshold - their response amplitude is continuously graded with flash intensity). Reciprocal sensitivity is plotted as a function of background intensity in linear coordinates for a turtle cone in Fig. 4A, and Weber law behavior is observed, i.e.,

$$1/S_F = 1/S_F{}^D + c\, I_B, \qquad (4A)$$

or

$$S_F{}^D/S_F = 1 + I_B/I_D, \qquad (4B)$$

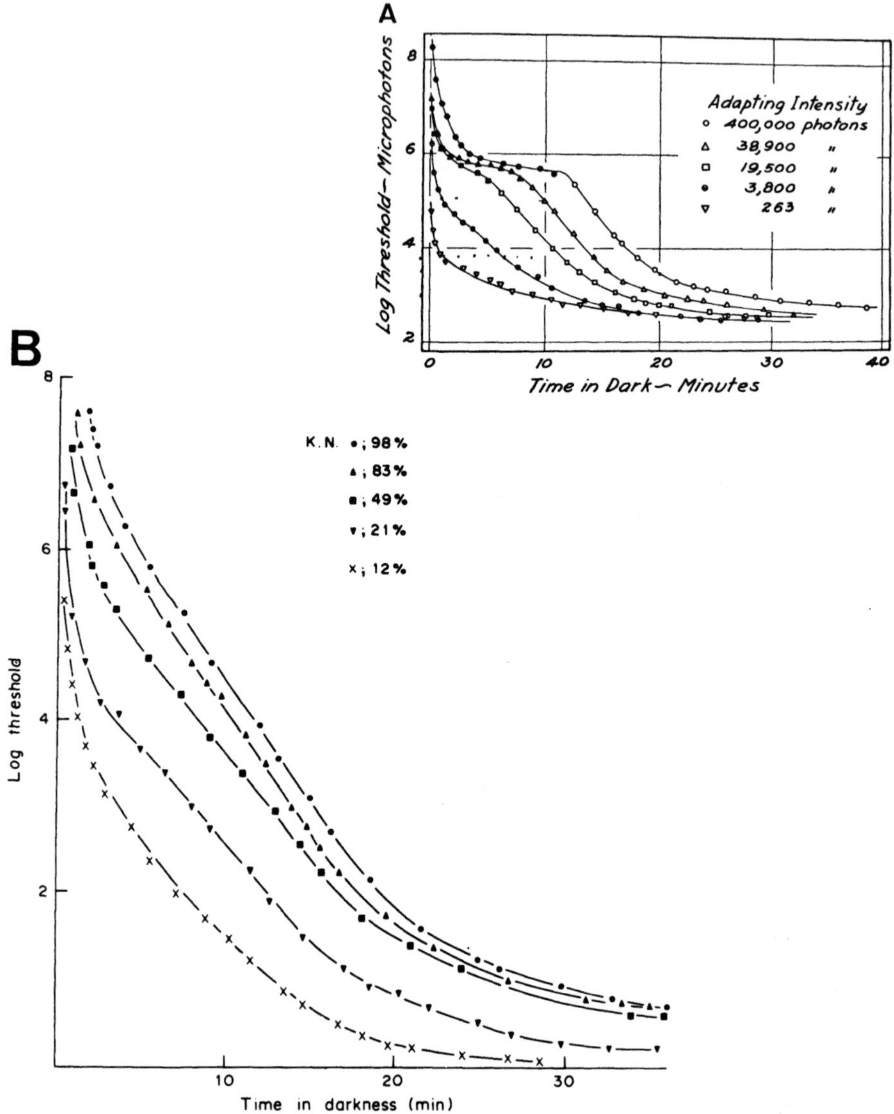

FIG. 2 - Time course of dark adaptation following various bleaching exposures. A: classical data reproduced with permission from Hecht, Haig, and Chase (24) for a normal observer. B: recent data reproduced with permission from Nordby, Stabell, and Stabell (32) for a rod monochromat. The uppermost trace in both parts corresponds to an almost total bleach. Note that for the rod monochromat, desensitization of the rod system can be followed far above the level corresponding to cone threshold in the normal.

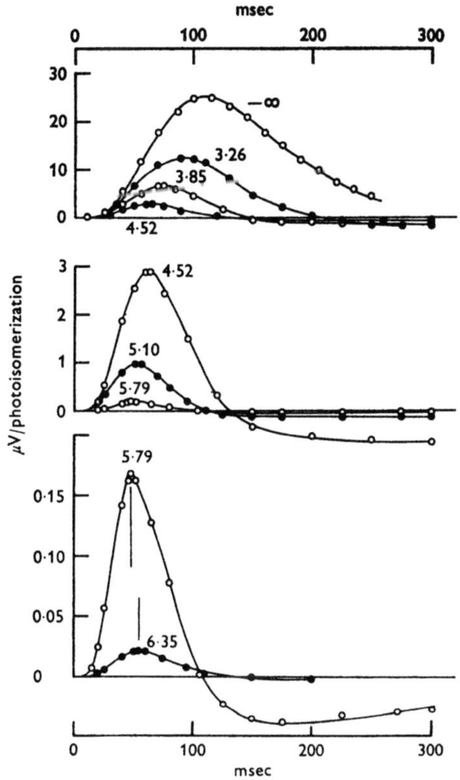

FIG. 3 - Desensitization of turtle cone by steady lights. Each trace is for a different background intensity and represents the voltage response amplitude (as a function of time after the flash) divided by the test flash intensity (i.e., it is in units of sensitivity). The uppermost trace was obtained with no background, and for the lower traces backgrounds of increasing intensity were applied. Numbers near traces represent $\log_{10}$ of background intensity expressed in photoisomerizations s^{-1} $cone^{-1}$. Test flashes were delivered 1 s after onset of the background. Reproduced with permission from (9).

exactly analogous to Eq. 1, with the slope in Eq. 4A given by $c = 1/(S_F{}^D\ I_D)$.

To test this relation over a wider range of intensities, $1/S_F$ - $1/S_F{}^D$ was plotted as a function of intensity in double log coordinates (Fig. 4B), and the straight line with unity slope shows Weber's law. So in turtle cones steady background illumination desensitizes the cone and improves its time resolution, just as occurs in the overall visual system.

One difficulty in comparing the data of Figs. 3-4 with psychophysics is that they were obtained for flashes presented only 1 s after the onset of background illumination, and at 20°C this is insufficient time for a steady state to be reached (33). The time course of onset of desensitization has not been

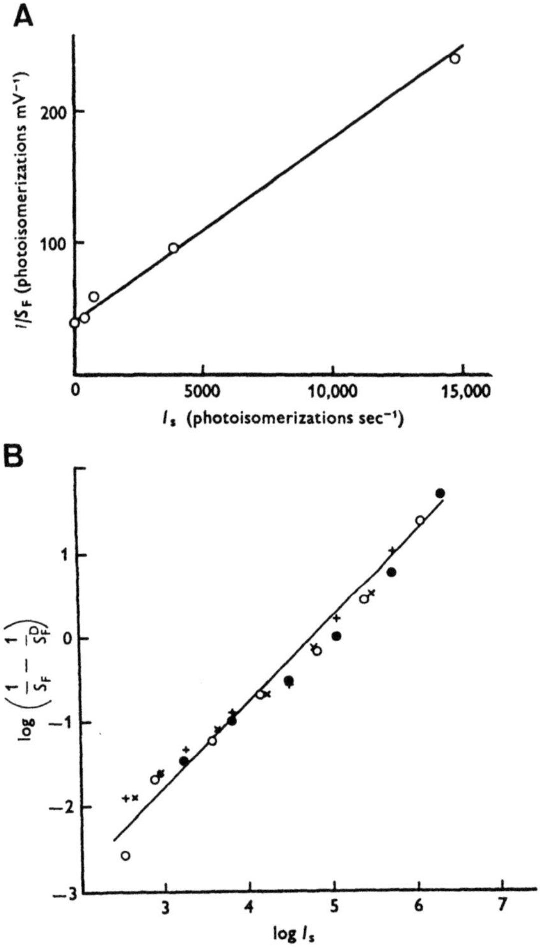

FIG. 4 - Relation between reciprocal of flash sensitivity and steady light intensity. A: with both coordinates linear. B: $1/S_F - 1/S_F^D$ against I_B in double log coordinates. Same cell as Fig. 3. Reproduced with permission from (9).

studied in great detail in vertebrate photoreceptors but appears to comprise both a rapid component and a slower component. Initially a very substantial desensitization is found, particularly with very bright illumination, but over a time course of seconds to tens of seconds this desensitization may relax somewhat. It seems probable that at longer times the cone responses would obey Weber's law even more closely and over a wider range of intensities.

By comparing the Weber fractions c and k in Eqs. 1A and 4A for cone electrophysiological data and for behavioral work on turtles, Fain, Granda, and Maxwell (21) estimated that a membrane voltage signal of about 10 μV in the red-sensitive cones corresponds to behavioral threshold with large area stimuli. A value of 10 μV in the receptors seems reasonable for the limit of detection, given the large number of cones over which the visual system is able to average. (A summing area 1° in diameter in the turtle would contain several hundred cones.)

In this species (the turtle, Pseudemys scripta) the dark light expressed in terms of Eq. 4B corresponds to 2000 isomerizations s^{-1} per red-sensitive cone (9), i.e., this is the intensity of real light required to halve the flash sensitivity of the cones.

There is, however, another phenomenon in photoreceptors which can be referred to as dark light, and this involves noise in the receptors. It has been found that even in darkness photoreceptors are surprisingly noisy, and it is possible to convert this noise to an equivalent light intensity by calculating the intensity of random arrival of photons required in an identical silent photoreceptor to give the same level of noise. In turtle cones the calculation indicates that the dark noise is equivalent to a light intensity of around 2500 isomerization s^{-1} (30), quite close to the intensity required to halve flash sensitivity. As discussed below, this equivalent light in cones is many orders of magnitude higher than that found in rods, but we have at present little understanding of the physical basis for the enormous difference.

These results are consistent with the idea that turtle cones desensitize in direct proportion to the sum of real light plus dark light, where the dark light is a spontaneous phenomenon giving rise to both noise and desensitization indistinguishable from the effects of real light. Furthermore, comparison with behavioral results suggests that the behavioral photopic Weber law relation may simply be a direct result of the Weber law desensitization of the cones.

Temporal Effects

The effects of background adaptation on the time course of flash responses was shown in Fig. 3 and is examined in the frequency domain in Fig. 5A (43). These results, although obtained in turtle horizontal cells, presumably reflect mainly the activity of cones. The effect of the adaptation is to reduce the gain at low temporal frequencies in a Weber law manner, but to leave unchanged the high frequency behavior. This property is also shown in Fig. 3 by the fact that at early times the responses begin rising along a common curve.

For comparison, the psychophysical behavior of the overall photopic visual system is shown in Fig. 5B (25). Here the modulation sensitivity for the perception of flicker is plotted as a function of flicker frequency at a range of mean luminances for a human observer. The similarity between Figs. 5A and 5B is impressive and suggests that much of the temporal response of the

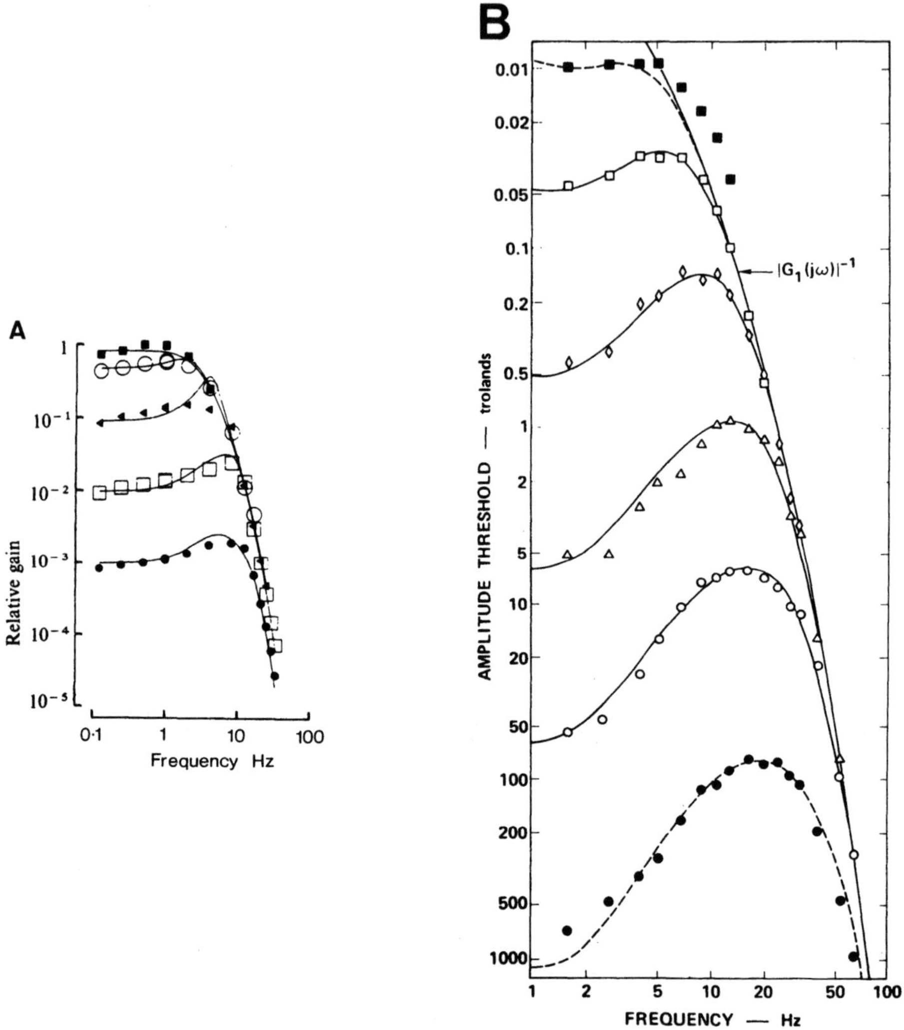

FIG. 5 - Temporal properties of responses during background adaptation, plotted in terms of sensitivity as a function of frequency of modulation. A: modulation transfer function for turtle horizontal cells (presumed to reflect cone behavior) at five background intensities. Reproduced with permission from Tranchina, Gordon, and Shapley (43). B: Comparable plots for the perception of flicker by a human observer. Large, uniformly flickering field, at luminances from 0.06 to 9300 trolands. Reproduced with permission from (25).

human observer in the photopic range may be explained simply in terms of the kinetics of the cone responses.

Rods

The voltage sensitivity of rod photoreceptors of the toad has been shown to obey Weber's law over a 5-log-unit range of intensities (20), and this has been confirmed in recordings of toad rod currents over a smaller intensity range (11, 29). The intensity required here to halve the sensitivity of the rods themselves corresponds to about 5 isomerizations s^{-1} per toad rod, nearly three orders of magnitude lower than the value for turtle cones. In terms of psychophysics, however, this would represent an enormously high value of dark light for the overall rod system. For comparison, the human psychophysical dark light in the rod system corresponds to about 0.01 isomerization s^{-1} per human rod.

Toad rods also exhibit noise in darkness. Spontaneous photon-like events (which are believed to be caused by thermal isomerizations) occur on average about once every 50 s per rod at 20°C, i.e., at a mean rate of about 0.02 s^{-1} (11). To compare this value with that in the human rod it is necessary to allow for the different size of toad and human rods and for the different temperatures. The combined effects of these two factors extrapolate from the toad to a predicted value of about 0.008 s^{-1} per human rod, extremely close to the psychophysical value of 0.01 s^{-1}.

These results would suggest that in the rod system the behavioral "dark light" is set by spontaneous thermal isomerizations (see (6, 7)) rather than by rod desensitization which sets in at intensities several log units higher. Similarly, Weber law desensitization at the ganglion cell level sets in at a much lower intensity than that required to begin desensitizing the rods themselves (19, 23).

Recently, data has been obtained from monkey rods which provides further support for this idea. Baylor, Nunn, and Schnapf (13) have found a spontaneous rate of photon-like events of about 0.006 isomerization s^{-1} at 37°C, close to the psychophysical dark light in humans. In marked contrast to lower rods, however, the monkey rod flash sensitivity does not desensitize in a Weber law manner, but instead saturates abruptly at around 150 isomerizations s^{-1} (34). This intensity is close to the intensity of psychophysical saturation of the rod system, which Nunn and Baylor (34) attribute to the sensitivity saturation they observe in the rod outer segment.

One might then reasonably ask why it is that the rods of lower vertebrates should desensitize in a Weber law manner beginning at moderately low intensities, when primate rods show no such desensitization and instead retain their full flash sensitivity until saturation.

The answer, I think, lies simply in the prevention of saturation in the larger and slower rods of the lower vertebrates, as was proposed by Fain (20). Because of their much larger size (ca. 20x) and much slower responses (ca.

10x), a given light intensity will lead in lower rods to a state of activation several hundred times greater than in primate rods. That is to say, at some arbitrary intensity the number of simultaneous photon-effects will be vastly greater in a toad rod than in a human rod. In order to permit detection at extremely low light levels, all rods appear to have evolved a single-photon response amplitude of a few per cent of their dark current. At higher scotopic intensities, where primate rods could still function normally, the rods of lower vertebrates would in the absence of the automatic gain control be totally saturated. In this way desensitization (the automatic gain control) serves to prevent saturation, and hence to improve the sensitivity of the visual system.

Site of Background Adaptation

In photoreceptors, adaptation to background illumination occurs primarily in the light-sensitive outer segment, but additional changes may also occur in the inner segment. From consideration of the very low intensities at which rod desensitization occurs, Donner and Hemila (18) and Bastian and Fain (8) concluded that the desensitization resulting from a single-photon absorption spreads to affect many disks, and therefore that it presumably involves a diffusible messenger (see Fain, this volume).

The extent of the spread of desensitization within the outer segment is illustrated in Fig. 6 (29). The inset shows the positions of the test stimuli (A-E) spaced at 7 μm intervals and of the adapting stimulus (C). The panels below illustrate responses to test flashes at corresponding positions, in the presence (heavy traces) and absence (lighter traces) of steady adapting light at position C. These responses were averaged over several minutes after turning the light on or off. For superimposed test and adapting stimuli (C), the adapting light reduced the test reponse to about one quarter of its dark-adapted value, but for displacements of 14 μm (A and E) there was negligible desensitization. Further experiments showed that the desensitizing effect decayed approximately exponentially with distance from the site of light absorption, with a length constant of about 6 μm (i.e., about 200 disks).

These observations were accounted for in terms of the following model (29). In toad rods, light absorption alters the concentration of a desensitizing substance which spreads within the outer segment cytoplasm by simple aqueous diffusion. Longitudinal diffusion (along the axis of the rod) is, however, very substantially hindered by the "baffling" action of the closely spaced stack of disks, and this accounts for the relatively short degree of longitudinal spread.

Nature of the Desensitization

The desensitization by background lights must involve modification of one or more steps (presumably biochemical) in the phototransduction process itself. A model was proposed by Baylor and Hodgkin (9) for the case of the turtle cone, in which the presence of steady light led to acceleration of one of the rate constants in the transduction chain. The specific scheme chosen

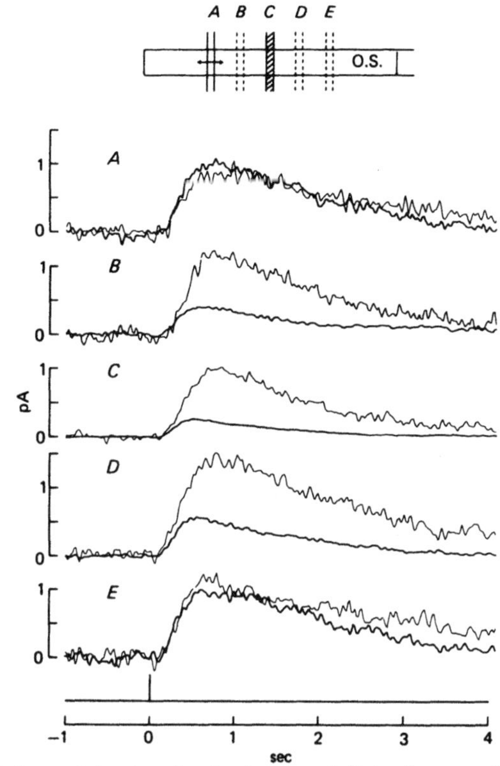

FIG. 6 - Spread of desensitization by background light in a toad rod outer segment. A-E: responses to stimuli at the positions shown in the inset (spacing 7 µm). Lighter traces: responses to control stimuli in darkness; heavier traces: responses in the presence of a steady adapting stimulus at position C, averaged over periods of several minutes. Reproduced with permission from (29).

involved acceleration of the step which removed "internal transmitter" from the cytoplasm. This step was assumed to be the slowest of the several reactions giving rise to the light response. On this model it was possible to describe to a first approximation the speeding up and desensitization shown in Fig. 3. It was additionally possible, without any further manipulation of parameters, to provide a very good description of the relationship between membrane voltage and light intensity in the "steady state" applying 1 s after the beginning of steps of illumination.

One of the reasons for the choice of this class of model was that it necessarily predicts the observed initial rise along a common curve in Fig. 3, equivalent to the common high frequency decline in Fig. 5.

The model has a number of quantitative shortcomings, some of which were elaborated by Baylor, Hodgkin, and Lamb (10) in the original treatment, and others which are described by Shapley and Enroth-Cugell (41) and Tranchina,

Gordon, and Shapley (43). It nevertheless serves as a useful starting point for the consideration of models of photoreceptor desensitization by steady background illumination.

"BLEACHING ADAPTATION"

The following is a personal and no doubt controversial account of certain aspects of bleaching adaptation.

The "classical" theory and description of bleaching adaptation (or dark adaptation) was given by Rushton in his Ferrier Lecture to the Royal Society (40). The fundamental equation in this work is the "Dowling-Rushton equation" relating threshold elevation to bleached pigment

$$\Delta I/\Delta I_D = 10^{aB}, \tag{5}$$

where B is the fraction of pigment bleached at any time and a is a constant. This equation was reputed to describe both the elevation of human psychophysical threshold following bleaches and the bleach-induced desensitization of rod voltage responses in various species.

The equation, however, no longer appears to provide a reasonable fit to either the psychophysical or photoreceptor data on bleaching, and it is without theoretical foundation. Its widespread use has, in my opinion, been one of the main factors retarding our understanding of dark adaptation. A more complicated form was used to describe the additional effects of backgrounds by Alpern, Rushton, and Torii (3), and like them, I feel that "the formula is not easy to interpret, and can only be regarded as a compact parcel of trouble."

It has been well established that following an intense bleaching light a) visual threshold is enormously elevated and slowly recovers, and b) bleached pigment slowly regenerates. In the work of Rushton and others it has been postulated that there is a direct causal link between the fraction B of bleached pigment remaining and the size of the threshold elevation, at least in the later stages of dark adaptation. It seems important to establish whether any such causal link in fact exists.

The regeneration of rhodopsin following a bleach can be followed by densitometry - that is, spectrophotometry. In many cases the regeneration has been reported to follow an exponential time course (e.g., see (2)). In some studies, however, the regeneration has been very nearly linear in time until most of the pigment has been regenerated (e.g., in the cat (14)).

It has been shown in many studies that the sensitivity of the rod photoreceptors themselves is greatly reduced following a bleach, and that sensitivity slowly recovers over a period of many minutes (e.g., (22, 26)). In the isolated retina preparation, where little or no rod pigment regeneration occurs, bleaching leads to a permanent desensitization which is graded with the magnitude of the bleach and which is much greater than expected from reduced quantal catch (e.g., (18, 35)). Addition of the chromophore 11-*cis* retinal during this permanent desensitization leads to a rapid recovery of sensitivity, as shown in Fig. 7 (35, 36).

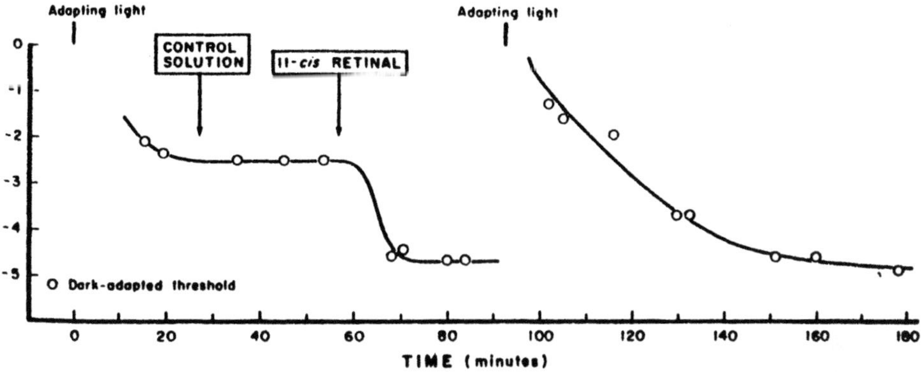

FIG. 7 - The sensitization of skate photoreceptors by treatment with 11-cis retinal following a bleach of about 40%; the second exposure bleached about 80%. After the first bleach the threshold recovered only to a plateau level well above the original dark-adapted level. Application of 11-cis retinal caused the threshold to recover to near its original value. Following a subsequent bleaching exposure full recovery occurred, presumably because of the continued presence of 11-cis retinal. Reproduced with permission from (35).

A question of some theoretical and practical interest is whether a change in rod flash sensitivity would be expected to cause a change in sensitivity of the visual system. In the fully dark-adapted state the rod photoreceptors are extremely sensitive and respond to individual photoisomerizations with a signal of up to about 5% of their maximal response. Psychophysical evidence indicates that in this region visual threshold may be limited by signal-to-noise considerations, where the "noise" is introduced by photon fluctuations in the background (plus dark light) as well as by photon fluctuations in the stimulus (5, 39). In this region the rod system appears to behave approximately as a photon-counting system of fixed quantum efficiency. If this were the case, then moderate changes in rod sensitivity might be expected to have little effect on photon-counting ability, so that visual threshold might barely be affected.

Equivalent Background Intensity

One of the most important observations on the psychophysical recovery of threshold following bleaching was made by Stiles and Crawford in 1932 (42). They showed, and it has since repeatedly been confirmed, that during bleaching adaptation the elevation of threshold (and other changes) can be quite accurately explained in terms of an "equivalent background intensity." That is, the visual system appears to be seeing the world through a veiling light which slowly fades away during the progress of dark adaptation. It is, I think, of considerable importance to obtain an understanding of the physical basis for this equivalence, particularly insofar as it involves the rod photoreceptors themselves. Relevant to this, Clack and Pepperberg (15) found in the skate retina that the rods desensitized in a qualitatively similar way with bleaches as with backgrounds, although there were large quantitative differences.

Using the equivalent background hypothesis, and analyzing Pugh's psychophysical results (38), I formulated a model (28) which provides an alternative view to Rushton's. In this model the aftereffects of extremely intense illumination give rise to events in the rod photoreceptors which are indistinguishable from the effects of photoisomerizations. These "photon-like" events are assumed to occur as a result of inactivated photopigment molecules spontaneously reverting to the activated form as follows:

$$\mathrm{Rh} \xrightarrow{h\nu} \mathrm{Rh}^* \rightleftarrows S_1 \rightleftarrows S_2 \rightleftarrows S_3 \longrightarrow \cdot \qquad (6)$$

Here Rh* is the activated or excited form of rhodopsin and S_1, S_2, and S_3 are various inactivated intermediates (presumably somehow related to the metarhodopsins). In this scheme Rh* may be formed either by real light or by the reverse reactions ("dark light"), and the magnitude of the dark light will be directly proportional to the concentration of S_1, and hence to the amount of pigment bleached. Such photon-like events are indeed observed in toad rods (27) following bleaches on the order of 1%, as illustrated in Fig. 8.

SUMMARY

There is reasonable circumstantial evidence that much of the adaptational behavior of the visual system results directly from properties of the rods and cones. In the cone system, photopic desensitization by background light seems to correspond to desensitization of the cones, and the photopic dark light seems to arise from a relatively high level of residual activity in the cones in darkness. In the rod system, on the other hand, desensitization of the overall system does not appear to result directly from desensitization of the rods in either lower vertebrates or primates, but is a more central phenomenon. It could conceivably be a consequence of "optimization" of noise performance but, unfortunately, we do not know what "trade-offs" are involved. The scotopic dark light is orders of magnitude smaller than the photopic dark light and seems to arise from an extremely low rate of spontaneous thermal isomerization of rhodopsin molecules. The molecular mechanisms by which background light desensitizes vertebrate photoreceptors are at present not understood. It is known that the amplification of the transduction process itself becomes decreased and that the speed of response is accelerated, but the means by which this occurs is unclear.

Bleaching adaptation is a very undesirable property of vision. It can be explained approximately in terms of an elevated dark light, and it is attractive to think that it might arise from an elevated rate of photon-like events as an inevitable consequence of the activation of vast amounts of rhodopsin, caused perhaps by reverse chemical reactions in the chain of inactivation reactions.

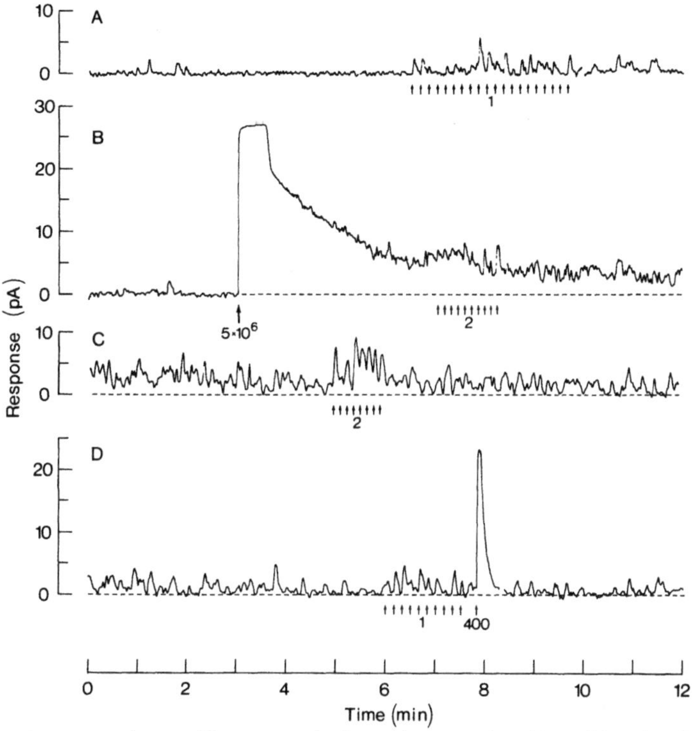

FIG. 8 - Spontaneous photon-like events induced in a toad rod by a bleach of about 1%. Single continuous record of 48 min. In the fully dark-adapted state prior to the bleach there was a low rate of occurrence of events. Following the bleach, the photocurrent was initially saturated, but slowly recovered and showed pronounced fluctuations. At later times the fluctuations appear to have separated into resolvable occurrences of the discrete events. Arrows mark timing of flashes, and numbers give their relative intensity.

Acknowledgements. This work was supported by a grant from the M.R.C. and by a Locke Research Fellowship from the Royal Society.

REFERENCES

(1) Aguilar, M., and Stiles, W.S. 1954. Saturation of the rod mechanism at high levels of stimulation. Optica Acta **1:** 59-65.

(2) Alpern, M. 1971. Rhodopsin kinetics in the human eye. J. Physiol. **217:** 447-471.

(3) Alpern, M.; Rushton, W.A.H.; and Torii, S. 1970. The attenuation of rod signals by bleachings. J. Physiol. **207:** 449-461.

(4) Barlow, H.B. 1957. Increment thresholds at low intensities considered as signal noise discriminations. J. Physiol. **136:** 469-488.

(5) Barlow, H.B. 1962. Measurements of the quantum efficiency of discrimination in human scotopic vision. J. Physiol. **160:** 169-188.

(6) Barlow, H.B. 1964. Dark-adaptation: a new hypothesis. Vision Res. **4:** 47-58.

(7) Barlow, H.B. 1972. Dark and light adaptation: Psychophysics. In Handbook of Sensory Physiology, eds. D. Jameson and L.M. Hurvich, vol. VII-4, pp. 1-28. Berlin: Springer-Verlag.

(8) Bastian, B.L., and Fain, G.L. 1979. Light-adaptation in toad rods: requirement for an internal messenger which is not calcium. J. Physiol. **242:** 729-758.

(9) Baylor, D.A., and Hodgkin, A.L. 1974. Changes in time scale and sensitivity in turtle photoreceptors. J. Physiol. **242:** 729-758.

(10) Baylor, D.A.; Hodgkin, A.L.; and Lamb, T.D. 1974. Reconstruction of the electrical responses of turtle cones to flashes and steps of light. J. Physiol. **242:** 759-791.

(11) Baylor, D.A.; Matthews, G.; and Yau, K.-W. 1980. Two components of electrical dark noise in retinal rod outer segments. J. Physiol. **309:** 591-621.

(12) Baylor, D.A.; Matthews, G.; and Yau, K.-W. 1983. Temperature effects on the membrane current of retinal rods of the toad. J. Physiol. **337:** 723-734.

(13) Baylor, D.A.; Nunn, B.J.; and Schnapf, J. 1984. The photocurrent, noise and spectral sensitivity of rods of the monkey Macaca fascicularis. J. Physiol. **357:** 575-607.

(14) Bonds, A.B., and MacLeod, D.I.A. 1974. The bleaching and regeneration of rhodopsin in the cat. J. Physiol. **242:** 237-253.

(15) Clack, J.W., and Pepperberg, D.R. 1982. Desensitization of skate photoreceptors by bleaching and background light. J. Gen. Physiol. **80:** 863-883.

(16) Craik, K.J.W. 1938. The effect of adaptation on differential brightness discrimination. J. Physiol. **92:** 406-421.

(17) Crawford, B.H. 1947. Visual adaptation in relation to brief conditioning stimuli. Proc. Roy. Soc. Lond. B **134:** 238-302.

(18) Donner, K.O., and Hemila, S. 1978. Excitation and adaptation in the vertebrate rod photoreceptor. Med. Biol. **56:** 52-63.

(19) Enroth-Cugell, C., and Shapley, R.M. 1973. Flux, not retinal illumination, is what cat retinal ganglion cells really care about. J. Physiol. **233:** 311-326.

(20) Fain, G.L. 1976. Sensitivity of toad rods: dependence on wave-length and background illumination. J. Physiol. **261:** 71-101.

(21) Fain, G.L.; Granda, A.M.; and Maxwell, J.H. 1977. The voltage signal of photoreceptors at visual threshold. Nature **265:** 181-183.

(22) Grabowski, S.R.; Pinto, L.H.; and Pak, W.L. 1972. Adaptation in retinal rods of axolotl: intracellular recordings. Science **176:** 1240- 1243.

(23) Green, D.G.; Dowling, J.E.; Siegel, I.M.; and Ripps, H. 1975. Retinal mechanisms of visual adaptation in the skate. J. Gen. Physiol. **65:** 483-502.

(24) Hecht, S.; Haig, C.; and Chase, A.M. 1937. The influence of light-adaptation on subsequent dark adaptation of the eye. J. Gen. Physiol. **20:** 831-850.

(25) Kelly, D.H. 1971. Theory of flicker and transient responses. I. Uniform fields. J. Opt. Soc. Am. **61:** 537-546.

(26) Kleinschmidt, J., and Dowling, J.E. 1975. Intracellular recordings from gecko photoreceptors during light and dark adaptation. J. Gen. Physiol. **66:** 617-648.

(27) Lamb, T.D. 1980. Spontaneous quantal events induced in toad rods by pigment bleaching. Nature **287:** 349-351.

(28) Lamb, T.D. 1981. The involvement of rod photoreceptors in dark adaptation. Vision Res. **21:** 1773-1782.

(29) Lamb, T.D.; McNaughton, P.A.; and Yau, K.-W. 1981. Spatial spread of activation and background desensitization in toad rod outer segments. J. Physiol. **319:** 463-496.

(30) Lamb, T.D., and Simon, E.J. 1977. Analysis of electrical noise in turtle cones. J. Physiol. **272:** 435-468.

(31) MacLeod, D.I.A. 1978. Visual sensitivity. Ann. Rev. Psychol. **29:** 613-645.

(32) Nordby, K.; Stabell, B.; and Stabell, V. 1984. Dark-adaptation of the human rod system. Vision Res. **24:** 841-849.

(33) Normann, R.A., and Perlman, I. 1979. The effects of background illumination on the photoresponses of red and green cones. J. Physiol. **286:** 509-524.

(34) Nunn, B.J., and Baylor, D.A. 1983. Visual transduction in single photoreceptors of the monkey Macaca fascicularis. **In** Colour Vision, Physiology and Psychophysics, eds. J.D. Mollon and L.T. Sharpe, pp. 1-11. London: Academic.

(35) Pepperberg, D.R.; Brown, P.K.; Lurie, M.; and Dowling, J.E. 1978. Visual pigment and photoreceptor sensitivity in the isolated skate retina. J. Gen. Physiol. **71:** 369-396.

(36) Perlman, J.I.; Nodes, B.R.; and Pepperberg, D.R. 1982. Utilization of retinoids in the bullfrog retina. J. Gen. Physiol. **80:** 885-913.

(37) Paulsen, R., and Bentrop, J. 1983. Activation of rhodopsin phosphorylation is triggered by the lumirhodopsin-metarhodopsin I transition. Nature **302:** 417-419.

(38) Pugh, E.N., Jr. 1975. Rushton's paradox: rod adaptation after flash photolysis. J. Physiol. **248:** 413-431.

(39) Rose, A. 1948. The sensitivity performance of the human eye on an absolute scale. J. Opt. Soc. Am. **38:** 196-208.

(40) Rushton, W.A.H. 1965. The Ferrier Lecture 1962. Visual adaptation. Proc. Roy. Soc. Lond. B **162:** 20-46.

(41) Shapley, R., and Enroth-Cugell, C. 1984. Visual adaptation and retinal gain controls. In Progress in Retinal Research, eds. N.N. Osborne and G.J. Chader, vol. 3, pp. 263-346. Oxford: Pergamon.

(42) Stiles, W.S., and Crawford, B.H. 1932. Equivalent adaptation levels in localized retinal areas. In Report of a Joint Discussion on Vision, pp. 194-211. Physical Society of London. Cambridge: Cambridge University Press. (Reprinted in Stiles, W.S. 1978. Mechanisms of Colour Vision. London: Academic.)

(43) Tranchina, D.; Gordon, J.; and Shapley, R.M. 1984. Retinal light adaptation - evidence for a feedback mechanism. Nature **310:** 314-316.

The Molecular Mechanism of Photoreception, ed. H. Stieve, pp. 287-302. Dahlem Konferenzen 1986. Berlin, Heidelberg, New York, Tokyo: Springer-Verlag.

Turnover of Vertebrate Photoreceptor Membranes

D.J. Roof
Berman-Gund Laboratory
Harvard Medical School
Boston, MA 02114, USA

Abstract. Photoreceptor membrane turnover in vertebrate retinas is measured in two ways: by monitoring incorporation of labelled precursors into new membranes or by quantitating the amount of discarded membrane which appears within the large phagosomes of the overlying pigment epithelium. Using these assays, several key factors in the membrane turnover cycle have been identified. All outer segments add membranes at their bases and shed at their tips. Rods differ from cones in segregating newly synthesized disc membranes into discrete packets which move up the rod length until shed. Newly synthesized cone membranes are quickly mixed with old membranes and thus cannot be distinguished along the length of the cone outer segment. On average for both rods and cones, the amount of shed membrane is equivalent to the amount of newly added membrane so that in the adult, rods and cones do not change length. However, not every outer segment sheds every day and the shedding vs. addition time course is not precisely matched over a single light/dark period. The timing of membrane shedding differs for rods and cones and varies widely among cones of different species. The contribution of circadian and light/dark-dependent components also varies among species. For species with circadian components of shedding, the "clock" resides within the eye. The detailed kinetics of membrane addition are less well-known. Disc addition in rods is initiated by light in some species but not in others and is probably not circadian. Finally, it remains to be discovered both how and why photoreceptor membranes are renewed. Although many extracellularly applied substances (cyclic nucleotides, calcium, depolarizing and hyperpolarizing agents, and cytoskeleton-perturbing drugs) can affect the appearance of disc-derived phagosomes in the retinal pigment epithelium, the normal role for these factors in in vivo shedding has not been shown.

INTRODUCTION

The process of membrane turnover in vertebrate photoreceptors is at present very poorly understood both in rods and cones. Controversy surrounds the molecular mechanisms by which membrane is added to and eliminated from the outer segment and extends to a description of which and how many steps are involved. In many cases these controversies arise from the nature of the assays which have been developed to quantitate membrane turnover. As discussed in detail below, there are very few types of assays available to measure either addition or shedding, none can be applied to isolated single cells, and the initiation of turnover must often be inferred from an indirect measurement. However, there is now a gathering consensus that in both rods and cones of many species not only circadian but light- and dark-dependent processes are involved in disc addition and/or shedding. This has led very recently to the study of the effects of light- and dark-related parameters on the membrane turnover process. These variables include calcium, cyclic nucleotides, and electrical activity, and all have been more intensively studied in relation to transduction in the rod outer segment (ROS). This review will discuss these parameters as related to membrane turnover with the idea that all light/dark-regulated processes in the photoreceptor may ultimately share some part of their molecular organization. In addition, filamentous and cytoskeletal elements are now being intensively studied primarily for their role in membrane turnover. A discussion of these "structural" components is also included with the expectation that these may also be profitably studied for their contribution to transduction.

MEASUREMENT OF MEMBRANE TURNOVER

Membrane Addition

In rod photoreceptors, membrane is added at the base of the outer segment by a process which includes at least three steps: protein and lipid synthesis (and glycosylation), insertion of these components into membranes, and assembly of newly formed membranes into discs. Until recently, only one assay was available to monitor all three steps (see Fig. 1). Radioactively labelled precursors for lipid, protein, or sugar are either injected into whole eyes or included in the incubation medium of eyecups or isolated retinas. Synthetic products are then monitored by autoradiography after a short pulse of radioactive precursors. Within two hours after labelling, products begin appearing in the ROS disc membranes. Label is confined to a small "packet" of basal discs which traverses the length of the ROS as a discrete unit over the next several days. Useful variations on this assay have recently been developed. Papermaster and others have combined autoradiography with immunocytochemistry (47) to follow the fate of newly synthesized rhodopsin as it progresses from rod inner segment (RIS) to rod outer segment. Alternatively, Hollyfield et al. (33) and Matsumoto and Bok (45) have combined radioactive labelling with cell fractionation and biochemical techniques to assess the specific activity of opsin and rhodopsin within each cellular compartment in relation to active disc assembly. In the latter case, the rate of biosynthesis is measured and fluctuation in rhodopsin pool size within each cellular compartment is inferred. The implications of these more sophisticated treatments with regard to the mechanism of disc formation are discussed below.

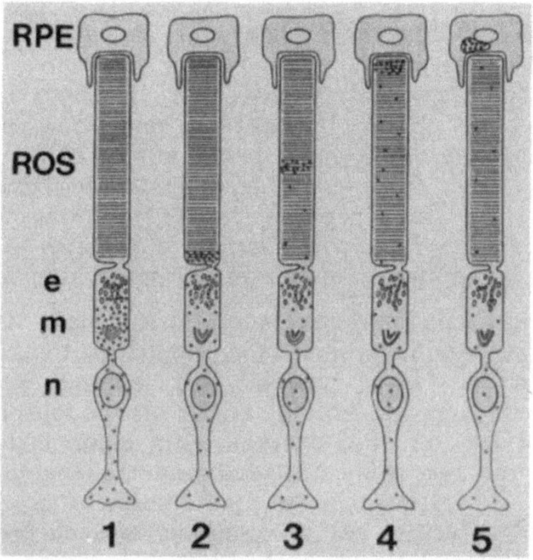

FIG. 1 - Membrane turnover in a rod photoreceptor. RPE, retinal pigment epithelium; ROS, rod outer segment; e, ellipsoid which contains mitochondria; M, myoid which contains the Golgi apparatus. Within minutes after extracellular injection of ^{3}H-leucine (small dots) label is associated with Golgi-like profiles in the rod myoid [1]. Over the next two hours, the label progresses to the base of the ROS where it is incorporated into newly formed discs [2]. The labelled discs move as a discrete band along the ROS over the next several days [3, 4]. After 5-80 days, depending on the species, the labelled disc stack disappears from the tip of the ROS and is phagocytosed by the overlying RPE [5]. (From Alberts et al., 1983. Molecular Biology of the Cell. New York: Garland.)

Recently, two entirely new types of assays have been developed to study specific aspects of disc addition in the rod. The first, developed by Hollyfield and Rayborn (31), uses thin section electron microscopy to count the number of newly formed basal "open discs." This number can be correlated with disc membrane turnover rate as measured by more conventional EM autoradiography at several different temperatures. The basis for this correlation has not been conclusively shown. However, under conditions of rapid membrane addition like those used by Hollyfield and Rayborn, disc assembly presumably becomes rate-limiting and open discs accumulate. At present, open disc analysis is the only assay which can monitor the process of disc assembly as separate and distinct from protein synthesis and membrane formation.

The second newly developed assay is that used by Kaplan which measures disc packet movement directly using polarized light microscopy (34, 35, 37). Previously, Besharse et al. monitored radioactive disc packet movement after ^{3}H-leucine injection into Xenopus eyes. They reported that several days after injection the rate of packet movement slows down although the synthesis rate is unchanged (13). This implies that disc movement along the length of the ROS is probably a separate process from protein synthesis and disc addition. However, the autoradiographic assay is so cumbersome and time-consuming that the mechanism by which discs move toward the tip of the ROS has remained completely unstudied. Kaplan's recent observation that discs formed at different times of day show different physical properties (birefringence bands) has allowed him to distinguish specific disc packets and follow their subsequent movement along the ROS. This is a particularly attractive assay because the disc-marking treatment is non-invasive and the assay is relatively simple. It remains to be seen whether this technique can be applied to membrane turnover in cones.

Very little is known about membrane addition in cones. Assays using pulses of radioactive precursors have been used, similar to those in rods, but with very different results. Cone proteins are renewed but are apparently quickly randomized over the entire length of the outer segment. This putative rapid mixing of new protein with older cone outer segment membranes has made the study of membrane movement within the outer segment impossible. Neither are detailed studies of protein synthesis and membrane assembly possible yet in cones because no proteins specific to cones have been isolated or characterized. Nevertheless, Anderson (D. Anderson, personal communication) has recently attempted to study a specific cone protein by monitoring fucose incorporation into cone membranes. Prior to this study, an unidentified fucose-containing moiety was described as a major component of cones but not of rods in some species (17). Anderson finds that this fucose-labelled moiety first appears at the base of the cone outer segment and only gradually disperses over the outer segment during the following 12-24 hours. It is not yet clear whether this labelling pattern is typical of other cone proteins.

Membrane Elimination

Membrane is eliminated from rods and cones in a complex process involving both photoreceptor outer segments and the overlying retinal pigment epithelial cells (RPE). The only assay for membrane shedding currently in use is the simple method of counting bundles of shed disc membranes (large phagosomes) as these appear in the RPE (see Fig. 1). Autoradiographic pulse/chase experiments conclusively show that such large phagosomes originate from the tips of rod outer segments (62). Morphological observations in a number of species (human, monkey, cat, and tree squirrel ((55, 3, 20, 2), respectively)), including one species with an all-cone retina (lizard (7)) suggest that these membranous bundles also originate from the tips of cone outer segments. Furthermore, since the length of outer segments within the retina remains constant, the rate of disc addition should be exactly equal to the rate at which disc membranes appear within the RPE. In practice, it has been difficult to match the independently measured disc addition values with the total amount of membrane recovered in phagosome material (20, 30). This is probably the result of technical difficulties in summing the phagosome membrane properly but points out one of the major difficulties with this assay: it is physically and temporally rather far removed from the primary events of disc shedding.

Attempts have been made to separate RPE-specific and ROS-specific processes by assays specific for RPE effects. One example is the use of cultured RPE cells which are "fed" pieces of ROS membrane in vitro (28). The rate at which ROS membranes are taken up provides a measure of RPE phagocytic activity. In a very indirect way, variables which influence RPE phagocytosis directly both in vivo and in this in vitro assay may be inferred to affect events which originate primarily in the RPE and not the ROS. Because ROS from isolated retinas will not shed at all without the influence of the RPE, the converse assays cannot be done. Furthermore, it is likely that some processes may take place in the RPE but require ROS "priming" or vice versa (see (28) for one suggestion about priming).

Another approach for defining the molecular mechanisms of shedding is the use of mutants which are defective in some aspect of the shedding process. Mutants with inherited photoreceptor degenerations may fall into this category (see (53) for a review). Although some results with regard to cyclic nucleotide metabolism will be mentioned below, in general the mutant phenotypes are too poorly defined to be useful in such studies. It is simply not clear whether any inherited photoreceptor degenerations are a result of an aberration in the normal turnover mechanism or whether they represent, for instance, developmental defects unrelated to turnover. It seems much more likely that a deeper understanding of normal photoreceptor membrane turnover will subsequently lead to a sorting out of the molecular lesions in these mutants, rather than the reverse.

General Results

The assays described above have been applied to photoreceptors from several species, at different stages in the life cycle, under a wide variety of lighting conditions, using whole eyes, eyecups, and isolated retinas. This has resulted in some difficulty in comparing results from different species (Rana pipiens and Xenopus laevis) and from different life stages (Xenopus tadpoles vs. adults). However, a consensus has emerged on several aspects of turnover in both rods and cones. The areas of general agreement include the pathway of protein synthesis and insertion into the membrane for two specific ROS proteins, the rate of membrane turnover, the effect of light and dark on both membrane addition and shedding, and the partial localization of the circadian clock in those species where some aspects of shedding are independent of ambient light.

Protein synthesis and insertion of proteins into membranes have been studied in detail for only two rod photoreceptor integral membrane proteins, rhodopsin and the large intrinsic "rim" protein (14, 47-49) and only in amphibian species (Rana pipiens and Xenopus laevis). The early steps in disc membrane formation seem to fit the classical models very well: newly synthesized rhodopsin progresses vectorially from rough endoplasmic reticulum (RER) to Golgi-like profiles, through small "transport" vesicles to the

ROS plasma membrane. The exact site and mechanism of disc membrane assembly is still controversial. However, the pathway apparently prevents rhodopsin from entering inner segment plasma membranes during the entire synthesis and assembly cycle (46). The situation may be different in mammals. In preliminary work on rat rod photoreceptors, Andrews (3) was unable to demonstrate the existence of "transport" vesicles. Thus the generality of the model (RER to Golgi to transport vesicles to plasma membrane) remains to be established for other types of rod proteins (peripheral and soluble), cone proteins, and for diverse species.

Turnover rate. The rate at which new membrane is added to the outer segment has now been measured for both rods and cones of several species. Since average outer segment length does not change with time in the adult, the rate of membrane addition is assumed equal to membrane shedding. Turnover rate can thus be determined in three different ways for rod outer segments, using either the rate of appearance of newly labelled "basal" discs, the rate of radioactive band displacement, or the total amount of membrane which appears as large phagosomes in the RPE. Band displacement is the most accurate (see (20) for a discussion) and is the method of choice for rods but, unfortunately, cannot be used for cones. Phagosome quantitation is therefore used to measure turnover of cone outer segments, giving more variable results. Literature values have been expressed in two ways: as the average length of outer segment added per photoreceptor per day, or the total time for an entire outer segment to be renewed (see Table 1). The extremes in length added per day range from slow in adult Rana rods at 0.9 μm/day (24°C) to rapid in lizard cones at 2.7 μm/day (21°C). When expressed as total turnover time, the extremes range from complete renewal in 55-80 days in Rana rods to renewal in 5-7 days in cat, squirrel, and lizard cones and cat rods. Perhaps a better basis for comparing rods and cones of different lengths and diameters is the total volume of outer segment membrane which must be turned over per day. Even when the data are expressed in this way, the values are not similar (see Table 1). Cone membranes, in general, seem to cycle more slowly than rods, and membranes in small mammalian rods are added at a slower rate than those in large amphibian rods. The mechanistic significance of these numbers is completely unknown. However, it is interesting to note that if membrane age is in all cases a critical trigger for the shedding process, all outer segments might be expected to show similar total turnover times. This is not the case. Alternatively, if membrane position within the ROS somehow initiates shedding, then the length of outer segment added per day should be similar for all outer segments. As discussed above, this value seems to show the least variability (maximum of threefold) among rods and cones from the several species studied. This is an agreement with recent work by Kaplan (36) which demonstrates directly that for Xenopus ROS, position rather than age is a critical variable for disc shedding.

Light effects. The rate at which outer segment membrane turns over is relatively steady if measured over a time span of several days in intact

TABLE 1 - Outer segment renewal rates.

	Species	Temperature	OS Length/day (μm)	Total Renewal Time (days)	Volume (μm^3) of OS/day	Reference
Cones	Lizard	24°C	2.7	5.0	3.0	(7)
	Cat	37°C	1.3	5.4	2.3	(20)
	Squirrel	37°C	1.8	5.0	5.6	(56)
Rods	Rana adult	24°C	0.9	55-80	25	(63)
	Rana tadpole	33°C	1.8	27	52	(30)
	Rana tadpole	23°C	0.9	55	25	(30)
	Xenopus					
	early larvae	22°C	1.6	31	61	(39)
	late larvae	21°C	2.1	25	60	(31)
		22°C	1.6	31	61	(39)
	adult	22°C	1.9	26	73	(38)
	Cat	37°C	1.9	6.5	3.3	(20)
	Mouse	37°C	2.1	10.4	51	(40)
	Squirrel	37°C	1.6	7.0	5.0	(1)

animals under 12 light:12 dark lighting conditions. However, it has been known for almost a decade (41) that the daily cycle of disk addition and shedding in rods depends on the light/dark pattern. Disc addition is light-dependent in both Xenopus and Rana (13) and is completed within eight hours after light onset. In pigmented mice, membrane addition occurs at a constant rate (12) and is probably neither light-stimulated nor circadian. In contrast, disc shedding in some species is circadian (rat (24)), in others shedding depends on the ambient light cycle (Rana (6)), and in still others, shedding has some components which are light-dependent and some which are circadian (Xenopus (9, 21, 22, 30)). More recently, membrane shedding in cones has also been studied and is circadian in gray squirrels (56) but shows both circadian and light-dependent components in lizards (7). The pattern of ROS disc shedding under a normal (12 light:12 dark) light cycle is the same for all species studied, showing a low level in the dark and a peak within two hours after light onset (42). In contrast, patterns of cone shedding are highly variable from species to species. All show a peak of shedding, but for some species the peak is 1-2 hours following light onset (cat, monkey, tree shrew), in others the peak comes shortly after light offset (lizard, chicken), and in some species the peak of shedding occurs at about the middle of the dark period (goldfish, gray squirrel) (see (7, 20) for a summary of all species). This is contrary to the naive expectation that the bulk of photoreceptor turnover might occur during each cell's "off" period: light for rods and dark for cones.

This idea, that membrane shedding might perturb normal photoreceptor function, has yet to be rigorously tested. Recently, Birch et al. (16) attempted to quantitate the effects of photoreceptor turnover on both amplitude and threshold of photoresponses in human retinas entrained to a daily light/dark cycle. They found a significant (about 10%) drop in both ERG b-wave amplitude and threshold at one and one-half hours after light onset. This timing coincides with the predicted peak of rod outer segment disc shedding in humans. However, the observed drop in threshold is larger than that predicted solely from the loss of photopigment through shedding, which leads Birch et al. to suggest that shedding may perturb photoreceptor function in more profound ways.

What is the role of light in noncircadian addition and shedding processes? Is it, for example, a simple trigger for shedding? Surprisingly, light itself is neither necessary nor sufficient to support disc shedding in Rana pipiens rods. When kept in constant darkness, rods persistently shed at background rates of 3-5% of normal levels. In addition, rods kept in constant darkness can be induced to shed in a normal burst following shift to a higher temperature. Thus, light is not strictly necessary for shedding to take place. On the other hand, rods kept in constant light do not shed at high levels continuously. These constant light rods shed massively only when returned to darkness (dark-primed shedding (6, 9)). Thus light is not sufficient for shedding either. Together these results imply that both light- and dark-dependent processes are required to maintain normal shedding in rods.

Much recent work has been directed toward elucidating the mechanisms by which light and dark influence disc addition and shedding (see (8) for a review). Yet, some basic parameters have been rather neglected in this regard. A complete intensity series, action spectrum, and detailed kinetics (especially rapid kinetics) have not been reported for either disc addition or shedding. Basinger and Hollyfield (6) estimated that, in Rana, a 5-10% rhodopsin bleach is necessary to initiate shedding and greater than 20% bleach is required to elicit a maximum shedding response. The full shedding response is achieved at 60-90 minutes after light onset, but the effect of light intensity on this time course was not reported. Gordon et al. (25) report preliminary studies on wavelength dependence of shedding and suggest that green-sensitive rods shed in response to green light but receive no shedding-stimulus input from other photoreceptors. This is, at present, the extent of the data on intensity, wavelength dependence, and kinetics of disc shedding processes.

Much less is known about the specific effects of light on disc addition. Refinements in the basic assays (see above) have recently allowed a distinction between the effect of light on rhodopsin biosynthesis vs. disc membrane formation in rods (33, 45). Although results differ somewhat in Rana and Xenopus, it appears that, at least in Xenopus, light-stimulated disc addition is not correlated with a burst of rhodopsin biosynthesis. This implies the existence of large pools of disc membrane precursors within the inner segment. These pools may vary with the lighting cycle while the rate of protein synthesis remains relatively constant. This clearly focusses attention on the disc assembly process as the critical light-regulated step in membrane addition.

Circadian oscillators. Those species which shed photoreceptor membranes in a circadian manner have been used to localize the "clock" mechanism. In rats, removal of the pineal, superior cervical ganglion, hypophysis, thyroid, and parathyroid glands does not affect rhythmic shedding of ROS disc membranes. Rhythmic shedding persists in constant darkness after optic nerve transection in rats (see (57) for a discussion). In Xenopus eyes kept in vitro, rhythmic shedding also persists in constant darkness. These results all point to a circadian oscillator which controls disc shedding and is located within the eye (see (20) for a discussion).

CONTROVERSY

The major outstanding questions concerning membrane turnover in vertebrate photoreceptors are: Why? and How? Both of these may be related to aspects of transduction.

Why?

One rationale for turnover has always been that ROS components somehow age and because these "worn out" parts are not efficient in transduction any longer, they must be replaced. Recent data make this idea unlikely. Kaplan (36) has shown that in Xenopus rods it is probably not membrane age but rather membrane position which is the critical variable for eliciting disc shedding. In agreement with this idea, Williams (61) has examined six properties of rhodopsin in old vs. new disc membranes and found five of these (absorption spectrum, concentration, dichroic ratio, lateral diffusion rates, and photoconversion rates) to be completely unaffected by disc age. Regeneration rates in vivo do seem to be slower in ROS tips in the toad. However, rat rods show faster regeneration in the older tip membranes. Williams goes on to suggest that a variable turnover rate could be one way that photoreceptors adjust to their lighting environment. In this view, turnover would become primarily an "adaptation" mechanism. This seems unlikely. As estimated by Birch et al. (16), a maximal 10% loss of photopigment per day would raise the threshold of the rod photoresponse by less than .05 log units.

How?

The molecular mechanisms for both disc addition and shedding are presently unknown. Many variables have been tested for their effects on shedding, fewer for their effects on addition. Of these, a large number can be considered "transduction-related" variables such as: light (discussed above), cyclic nucleotides, calcium, and electrical activity. These will be discussed below. Another approach has been to study "structure-related" variables, most of which have not yet been considered for their effects on transduction. These are also discussed. A third set of variables includes those which might be considered "non-transduction" related such as melatonin and its analogs (10), amino acids (27), and those factors which effect RPE-photoreceptor interactions, such as lectins, surface carbohydrates, and soluble sugars (29). These will not be discussed here.

Cyclic Nucleotides

There is not so much controversy as confusion over the effect of cyclic nucleotides on membrane turnover. When applied directly to Xenopus eyecups in vitro, dibutyryl cAMP (dbcAMP) inhibits one "noncircadian" component of shedding. The burst of shedding normally induced by darkness after constant light treatment (dark-primed shedding) is diminished 50% by dbcAMP and/or phosphodiesterase inhibitors. dbcGMP, 5'AMP, and adenosine had no effect (11). This clearly implicates cAMP but not cGMP in the shedding process. In contrast, three mutants in which photoreceptors degenerate extensively all show defects in cGMP but not cAMP metabolism (see (43) for a discussion). As mentioned above, the sites of the genetic lesions in these mutants are unknown and they may all be poor indicators of normal turnover processes. In addition, although cGMP changes do precede photoreceptor degeneration in all cases, the cGMP levels go up in some mutants and down in others and no cause and effect relationship can be conclusively demonstrated. However, the fact that chronic application of dbcGMP and phosphodiesterase inhibitors to normally developing Xenopus retinas (32, 44) can both prevent differentiation of photoreceptors and cause the disruption and degeneration of rudimentary outer segments may indicate a role for cGMP in both normal development and photoreceptor maintenance. Whether the site of the cyclic nucleotide effects is primarily photoreceptor or RPE has also not been established. Edwards and Bakshian (19) have reported that in cultured RPE cells which are "fed" pieces of ROS, cAMP can block phagocytosis by 50%. However, dbcGMP has only slightly smaller effects and other adenosine derivatives are equivalent to cAMP. In summary, cyclic nucleotides may contribute to rod photoreceptor shedding, but which nucleotide (cAMP vs. cGMP), in which cell (photoreceptors vs. RPE), and by what mechanism are still unknown.

Calcium

Only one study (26) has directly looked at the effect of divalent cations on photoreceptor shedding in vertebrates. Shedding is significantly reduced at external calcium concentrations less than 10 micromolar in Xenopus eyecups. This is true for both light-evoked shedding and for dark-primed shedding. The effect is reversible and specific in that it cannot be mimicked by lowering magnesium concentration. No gross morphological changes are observed in the RPE processes or junctions in low calcium. However, the primary locus of the effect has not been defined and may involve photoreceptors, RPE, and/or other retinal layers. The molecular mechanism is also not known and may include indirect effects through altered cyclic nucleotide levels and/or effects on cytoskeletal elements (see below). Specific changes in photoreceptor and/or RPE membrane potential are probably not directly involved (see below).

Electrical Activity

One of the most severe limitations on currently available shedding assays is the inability to follow the shedding response of a specific preselected photoreceptor. This makes it impossible, for example, to alter the electrical properties of a photoreceptor and assess the effects of the treatment on subsequent shedding events in that cell. However, some agents which are known to depolarize rods such as phosphodiesterase inhibitors, low calcium, and low bicarbonate concentrations (15) also suppress disc shedding. Only very recently have Williams et al., using Xenopus eyecups in vitro, applied pharmacological agents designed more specifically to hyperpolarize and depolarize photoreceptors (and other retinal cells) and assessed the gross effects on photoreceptor shedding. In preliminary work (59, 60), both hyperpolarizing (choline or lithium substitution for sodium) and depolarizing (ouabain or strophanthidin) agents induced massive shedding even in darkness. Williams and Fisher (59) favor the interpretation that, since agents which affect membrane potential in opposite ways have similar effects on shedding, light-evoked shedding cannot be triggered by hyperpolarization of the rod or RPE cells.

Cytoskeletal/Structural Components

The molecular basis for the "motility" of disc membranes as they are formed, move along the disc stack, and are ultimately pinched off and discarded was, until recently, largely ignored in considerations of membrane turnover. The pioneering work of Burnside and others first focussed attention on cytoskeletal elements within rod and cone inner segments and assessed their role in retinomotor movements. More recently, the discovery of unidentified filamentous arrays with the rod outer segment (52, 58) has suggested that even the transducing organelle itself possesses an extensive "cytoskeleton." This has led to a search for more conventional cytoskeletal elements within the ROS and as these are found and characterized, to a consideration of the functional roles for such cytoskeletal elements in the context of photoreceptor physiology. The unexpected location of some of these elements immediately suggests a role in disc membrane addition while pharmacological studies imply a contribution to disc shedding processes. However, too little is yet known about the ROS "cytoskeleton" to dampen speculation that "structural" elements may also be involved in transduction.

The only clue that cytoskeletal elements might be involved in disc addition is the preferential clustering of actin, microtubules, fodrin, and, surprisingly, calmodulin just at the site where new discs are being assembled (18, 51). Although the complete localization of actin within the ROS is still controversial, Chaitin (18) has shown that actin is one component of a "periciliary ridge" complex which is composed of ridges and grooves and is found at the very base of the ROS in lower vertebrates (4, 50). It has been suggested that the function of the complex is to segregate newly forming disc membrane from inner segment membranes and/or to direct opsin-containing vesicles to the proper site for disc assembly (18, 50).

Evidence that cytoskeletal elements might participate in disc shedding comes from pharmacological studies using actin- and microtubule-disrupting drugs. Besharse and Dunis (9) have reported that actin filament-disrupting drugs inhibit both light-activated and dark-primed disc shedding in Xenopus eyecups in vitro. This implies that microfilaments are somehow required for shedding in vivo. A role for microtubules is less clear. Microtubule-disrupting drugs at high concentrations (1 mmol/l) can activate disc shedding, but under rather special culture conditions (low bicarbonate). The physiological role of microtubules in disc shedding is thus doubtful. The site of action of all these drugs is also undefined and may involve photoreceptors, RPE, or both.

Evidence that structural components are involved in transduction is practically nonexistent. A connection between calcium and both microfilaments and microtubules is well-known in other systems and has already been invoked to explain the effect of low calcium on disc shedding (26). It may be significant that a high local concentration of calmodulin is also associated with the cytoskeleton-rich region of the ROS (51). Perhaps calcium provides a connection between the light/dark history of the photoreceptor and disc cycling processes. There is, in addition, a connection between cyclic nucleotide metabolism and the cytoskeleton. Fleischman (23) has shown by cell fractionation that the entire ROS complement of guanylate cyclase is associated with the ROS "axoneme." This is precisely the cytoskeleton-rich part of the ROS. However, the significance of this is unknown. Other structural oddities of the ROS may also be related to the transduction process. Kaplan has described a gradient in birefringence from tip to base within the ROS which may reflect some underlying structural gradient (34). A similar tip-to-base gradient in physiological parameters has also been described. Schnapf (54) has shown that in Bufo ROS, illumination of the ROS tip gives single-photon responses which are smaller and slower than those from the base. In addition, background light reduces sensitivity at the tip more than at the base. The basis for either the birefringence or "transduction" gradient has not been found and a relation between the two has not been established.

Acknowledgements. This work was supported by grants to D.J. Roof from the Muscular Dystrophy Association and the U.S. National Institutes of Health (EY05257).

REFERENCES

(1) Anderson, D.H., and Fisher, S.K. 1975. Disc shedding in rodlike and conelike photoreceptors of tree squirrels. Science **187:** 953-955.

(2) Anderson, D.H., and Fisher, S.K. 1976. The photoreceptors of diurnal squirrels: outer segment structure, disc shedding, and protein renewal. J. Ultrastruct. Res. **55:** 119-141.

(3) Anderson, D.H.; Fisher, S.K.; Erikson, P.A.; and Tabor, G.A. 1980. Rod and cone shedding in the Rhesus monkey retina: a quantitative study. Exp. Eye Res. **30:** 559-574.

(4) Andrews, L.D. 1982. Freeze-fracture studies of vertebrate photoreceptor membranes. In The Structure of the Eye, ed. J.G. Hollyfield, pp. 11-23. Amsterdam: Elsevier.

(5) Andrews, L.D. 1984. Are transport vesicles involved in rod outer segment renewal in the rat? Inv. Opthalmol. Vis. Sci. **25a:** 65.

(6) Basinger, S.F., and Hollyfield, J.G. 1980. Control of rod shedding in the frog retina. In Neurochemistry of the Retina, eds. N.G. Bazan and R.N. Lolley, pp. 81-92. Oxford: Pergamon.

(7) Bernstein, S.A.; Breding, D.J.; and Fisher, S.K. 1984. The influence of light on cone disk shedding in the lizard, Sceloporus occidentalis. J. Cell Biol. **99:** 379-389.

(8) Besharse, J.C. 1982. The daily light-dark cycle and rhythmic metabolism in the photoreceptor-pigment epithelial complex. Prog. Retinal Res. **1:** 81-124.

(9) Besharse, J.C., and Dunis, D.A. 1982. Rod photoreceptor disc shedding in vitro: inhibition by cytochalasins and activation by colchicine. In The Structure of the Eye, ed. J.G. Hollyfield, pp. 85-96. Amsterdam: Elsevier.

(10) Besharse, J.C., and Dunis, D.A. 1983. Methoxyindoles and photoreceptor metabolism: activation of rod shedding. Science **219:** 1341-1343.

(11) Besharse, J.C.; Dunis, D.A.; and Burnside, B. 1982. Effects of cyclic adenosine 3',5'-monophosphate on photoreceptor disc shedding and retinomotor movement. J. Gen. Physiol. **79:** 775-790.

(12) Besharse, J.C., and Hollyfield, J.G. 1979. Turnover of mouse photoreceptor outer segments in constant light and darkness. Inv. Ophthalmol. Sci. **18:** 1019-1024.

(13) Besharse, J.C.; Hollyfield, J.G.; and Rayborn, M.E. 1977. Turnover of rod photoreceptor outer segments. II. Membrane addition and loss in relationship to light. J. Cell Biol. **75:** 507-527.

(14) Besharse, J.C., and Pfenninger, K.H. 1980. Membrane assembly in retinal photoreceptors. I. Freeze-fracture analysis of cytoplasmic vesicles in relationship to disc assembly. J. Cell Biol. **87:** 451-463.

(15) Besharse, J.C.; Terrill, R.O.; and Dunis, D.A. 1980. Light-evoked disc shedding by rod photoreceptors in vitro: relationship to medium bicarbonate concentration. Inv. Ophthalmol. Vis. Sci. **19:** 1512-1517.

(16) Birch, D.G.; Berson, E.L.; and Sandberg, M.A. 1984. Diurnal rhythm in the human rod ERG. Inv. Ophthalmol. Vis. Sci. **25:** 236-238.

(17) Bunt, A.H., and Klock, I.B. 1980. Comparative study of ^{3}H-fucose incorporation into vertebrate photoreceptor outer segments. Vision Res. **20:** 739-747.

(18) Chaitin, M.H.; Schneider, B.G.; Hall, M.O.; and Papermaster, D.S. 1984. Actin in the photoreceptor connecting cilium: immunocytochemical localization to the site of outer segment disk formation. J. Cell Biol. **99:** 239-247.

(19) Edwards, R.B., and Bakshian, S. 1980. Phagocytosis of outer segments by cultured rat pigment epithelium. Reduction by cyclic AMP and phosphodiesterase inhibitors. Inv. Ophthalmol. Vis. Sci. **19:** 1184-1188.

(20) Fisher, S.K.; Pfeffer, B.A.; and Anderson, D.H. 1983. Both rod and cone disc shedding are related to light onset in the cat. Inv. Ophthalmol. Vis. Sci. **24:** 844-856.

(21) Flannery, J.G., and Fisher, S.K. 1979. Light-triggered rod disc shedding in Xenopus retina in vitro. Inv. Ophthalmol. Vis. Sci. **18:** 638-642.

(22) Flannery, J.G., and Fisher, S.K. 1984. Circadian disc shedding in Xenopus retina in vitro. Inv. Ophthalmol. Vis. Sci. **25:** 229-232.

(23) Fleischman, D. 1981. Rod guanylate cyclase located in axonemes. Curr. Top. Membr. Transp. **15:** 109-119.

(24) Goldman, A.I.; Tierstein, P.S.; and O'Brien, P.J. 1980. The role of ambient lighting in circadian disc shedding in the rod outer segment of the rat retina. Inv. Ophthalmol. Vis. Sci. **19:** 1257-1267.

(25) Gordon, W.C.; Sherry, D.M.; and Fauchier, T.T. 1984. The effects of intensity and wavelength on frog outer segment shedding. Inv. Ophthalmol. Vis. Sci. **25a:** 241-244.

(26) Greenberger, L.M., and Besharse, J.C. 1983. Photoreceptor disc shedding in eyecups. Inhibition by deletion of extracellular divalent cations. Inv. Ophthalmol. Vis. Sci. **24:** 1456-1464.

(27) Greenberger, L.M., and Besharse, J.C. 1984. Rod photoreceptor disc shedding is massively stimulated by aspartate and other amino acids. Inv. Ophthalmol. Vis. Sci. **25a:** 242.

(28) Hall, M.O. 1978. Phagocytosis of light- and dark-adapted rod outer segments by cultured pigment epithelium. Science **202:** 526-528.

(29) Heath, A.R., and Basinger, S.F. 1983. Simple sugars inhibit rod outer segment disc shedding by the frog retina. Vision Res. **23:** 1371-1377.

(30) Hollyfield, J.G.; Besharse, J.C.; and Rayborn, M.E. 1977. Turnover of rod photoreceptor outer segments I. Membrane addition and loss in relationship to temperature. J. Cell Biol. **75:** 490-506.

(31) Hollyfield, J.G., and Rayborn, M.E. 1982. Membrane assembly in photoreceptor outer segments: progressive increase in "open" basal discs with increased temperature. Exp. Eye Res. **34:** 115-119.

(32) Hollyfield, J.G.; Rayborn, M.E.; Farber, D.B.; and Lolley, R.N. 1982. Selective photoreceptor degeneration during retinal development: the role of altered phosphodiesterase activity and increased levels of cyclic nucleotides. **In** The Structure of the Eye, ed. J.G. Hollyfield, pp. 97-114. Amsterdam: Elsevier.

(33) Hollyfield, J.G.; Rayborn, M.E.; Verner, G.E.; Maude, M.B.; and Anderson, R.E. 1982. Membrane addition to rod photoreceptor outer segments: light stimulates membrane assembly in the absence of increased membrane biosynthesis. Inv. Ophthalmol. Vis. Sci. **22:** 417-427.

(34) Kaplan, M.W. 1981. Light cycle-dependent axial variations in frog rod outer segment structure. Inv. Ophthalmol. Vis. Sci. **21:** 395-402.

(35) Kaplan, M.W. 1982. Do birefringence bands in rod outer segments record daily disk synthesis? Inv. Ophthalmol. Vis. Sci. **22a:** 204.

(36) Kaplan, M.W. 1984. Disk membrane position versus age as a necessary condition for shedding. Inv. Ophthalmol. Vis. Sci. **25a:** 241.

(37) Kaplan, M.W.; Defanbach, M.E.; and Liebman, P.A. 1978. Birefringence measurements of structural inhomogeneities in Rana pipiens rod outer segments. Biophys. J. **23**: 59-69.

(38) Kinney, M.S., and Fisher, S.K. 1978a. The photoreceptors and pigment epithelium of the adult Xenopus retina: morphology and outer segment renewal. Proc. Roy. Soc. Lond. B **201**: 131-147.

(39) Kinney, M.S., and Fisher, S.K. 1978b. The photoreceptors and pigment epithelium of the larval xenopus retina: morphogenesis and outer segment renewal. Proc. Roy. Soc. Lond. B **201**: 149-167.

(40) LaVail, M.M. 1973. Kinetics of rod outer segment renewal in the developing mouse retina. J. Cell Biol. **58**: 650-661.

(41) LaVail, M.M. 1976. Rod outer segment disc shedding in relation to cyclic lighting. Exp. Eye Res. **23**: 277-280.

(42) LaVail, M.M. 1976. Rod outer segment disc shedding in rat retina: relationship to cyclic light. Science **194**: 1071-1074.

(43) Lolley, R.N., and Farber, D.B. 1980. Cyclic GMP metabolic defects in inherited disorders of rd mice and RCS rats. In Neurochemistry of the Retina, eds. N.G. Bazan and R.N. Lolley, pp. 427-440. Oxford: Pergamon.

(44) Lolley, R.N.; Farber, D.B.; Rayborn, M.E.; and Hollyfield, J.G. 1977. Cyclic GMP accumulation causes degeneration of photoreceptor cells: simulation of an inherited disease. Science **196**: 664-666.

(45) Matsumoto, B., and Bok, D. 1984. Diurnal variations in amino acid incorporation into inner segment opsin. Inv. Ophthalmol. Vis. Sci. **25**: 1-9.

(46) Nir, I., and Papermaster, D.S. 1983. Differential distribution of opsin in the plasma membrane of frog photoreceptors: an immunochemical study. Inv. Ophthalmol. Vis. Sci. **24**: 868-878.

(47) Papermaster, D.S.; Converse, C.A.; and Zorn, M. 1976. Biosynthetic and immunochemical characterization of a large protein in frog and cattle rod outer segment membranes. Exp. Eye Res. **23**: 105-116.

(48) Papermaster, D.S.; Schneider, B.G.; and Besharse, J.C. 1979. Assembly of rod photoreceptor membranes: immunocytochemical and autoradiographic localization of opsin in smooth vesicles of the inner segment. J. Cell Biol. **83**: 275a.

(49) Papermaster, D.S.; Schneider, B.G.; Zorn, M.A.; and Kraehenbuhl, J.P. 1978. Immunocytochemical localization of opsin in outer segments and Golgi zones of frog photoreceptor cells. J. Cell Biol. **77**: 196-210.

(50) Peters, R.; Palade, G.E.; Schneider, B.G.; and Papermaster, D.S. 1983. Fine structure of a periciliary ridge complex of frog retinal rod cells revealed by ultrahigh resolution scanning electron microscopy. J. Cell Biol. **96**: 265-276.

(51) Roof, D.; Applebury, M.; and Kirsch, J. 1984. Localization of calmodulin and characterization of calmodulin binding proteins in the vertebrate rod outer segment. Biophys. J. **45**: 1a.

(52) Roof, D.J., and Heuser, J.E. 1982. Surfaces of rod photoreceptor disk membranes: integral membrane components. J. Cell Biol. **95**: 487-500.

(53) Schmidt, S.Y. 1985. Retinal degenerations. In Handbook of Neurochemistry, ed. A. Lajtha, vol. 10, pp. 461-507. New York: Plenum.

(54) Schnapf, J.L. 1983. Dependence of the single photon response on longitudinal position of absorption in toad rod outer segments. J. Physiol. **343:** 147-159.

(55) Steinberg, R.H.; Wood, I.; and Hogan, M.J. 1977. Pigment epithelium ensheathment and phagocytosis of extrafoveal cones in human retina. Phil. Trans. Roy. Soc. Lond. B **277:** 459-474.

(56) Tabor, G.A.; Anderson, D.H.; Fisher, S.K.; and Hollyfield, J.G. 1982. Circadian rod and cone disc shedding in mammalian retina. In The Structure of the Eye, ed. J.G. Hollyfield, pp. 67-73. Amsterdam: Elsevier.

(57) Tierstein, P.S.; Goldman, A.I.; and O'Brien, P.J. 1980. Evidence for both local and central regulation of rat rod outer segment disc shedding. Inv. Ophthalmol. Vis. Sci. **19:** 1268-1273.

(58) Usukura, J., and Yamada, E. 1981. Molecular organization of the rod outer segment. A deep-etching study with rapid freezing using unfixed frog retina. Biomed. Res. **2:** 177-193.

(59) Williams, D.S., and Fisher, S.K. 1984. Effects of hyperpolarising and depolarising agents on the shedding and phagocytosis of rod outer segment discs. In International Cell Biology, eds. S. Seno and Y. Okada, p. 372. Tokyo: Academic Press.

(60) Williams, D.S.; Wilson, O.; and Fisher, S.K. 1984. Activation of rod outer segment shedding by sodium substitution and ouabain. Inv. Ophthalmol. Vis. Sci. **25a:** 242-247.

(61) Williams, T.P. 1984. Some properties of old and new rhodopsin in single Bufo rods. J. Gen. Physiol. **83:** 841-852.

(62) Young, R.W., and Bok, D. 1969. Participation of the retinal pigment epithelium in the rod outer segment renewal process. J. Cell Biol. **42:** 392-403.

(63) Young, R.W., and Droz, B. 1968. The renewal of protein in retinal rods and cones. J. Cell Biol. **39:** 169-184.

The Molecular Mechanism of Photoreception, ed. H. Stieve, pp. 303-326. Dahlem Konferenzen 1985. Berlin, Heidelberg, New York, Tokyo: Springer-Verlag.

Turnover of Photoreceptor Membrane and Visual Pigment in Invertebrates

J. Schwemer
Institut für Tierphysiologie
Ruhr-Universität Bochum
4630 Bochum 1, F.R. Germany

Abstract. Numerous ultrastructural studies are dealing with the problem of membrane turnover in invertebrate photoreceptors, with special attention to changes in the structure of the rhabdom, rhabdomeres, and microvilli and to other cytological changes within the cell. Both synthesis and degradation of photoreceptor membrane involve the formation of sequences of as yet poorly defined organelles. The underlying mechanisms are not understood. Even the microvillar cytoskeleton may, amongst other functions, play an important role in both membrane assembly and degradation. The daily cycle of light and darkness sets off changes - in some species quite drastic - in the size of the rhabdom. Generally, the size of the rhabdom decreases in light and increases again in the dark, indicating an imbalance between the rates of membrane breakdown and synthesis and/or assembly. The possible mechanisms whereby membrane turnover is triggered by light or by darkness are still obscure. The diurnal effects may be modified by endogenous factors.

Relatively few studies, however, have dealt with the problem at a molecular level. Autoradiographic studies using labeled amino acids demonstrated a random distribution of the label in the rhabdomeric membrane. Although in most cases the labeled proteins were not identified, it is assumed that the bulk of proteins labeled is the visual pigment. The general observation that more label was associated with the rhabdom in the dark than in the light was taken as an indication that the breakdown of rhabdomeric proteins was more rapid in the light than in the dark. Additional information on the turnover of visual pigment has come from quantification of rhodopsin and metarhodopsin by spectrophotometry. Progress has been made by studies on blowfly photoreceptors, revealing a selective breakdown of metarhodopsin at a rate inversely proportional to the intensity of the ambient light. The other part of the visual pigment cycle, the biosynthesis of rhodopsin, has been shown to depend on the presence of the 11-cis chromophore, in other words, the all-trans isomer resulting from the degradation of metarhodopsin has first to be isomerized in order to induce the synthesis of rhodopsin. This isomerization, which occurs through a light reaction, plays a key role in the turnover of visual pigment in that it links its two aspects, the breakdown and the biosynthesis. The following is an attempt to show how ultrastructural findings can be related to the molecular events which underlie the turnover of rhabdomeric membrane and of the visual pigment.

INTRODUCTION

Photoreception in animals is generally subserved by highly specialized cells in which the area of the plasma membrane is so greatly enlarged that they have a high probability of absorbing an incident photon. Light is absorbed by molecules of visual pigment which consist of a protein moiety and a chromophore. These molecules are located within the specialized regions of the cell membrane. The absorption of light by visual pigment molecules triggers a sequence of reactions that lead finally to the photoreceptor cell giving an electrical response which is propagated to higher centers.

It is also true to say that light destroys the visual pigment which absorbs it: thus the isomeric configuration of the chromophore is changed upon light absorption, and photons eventually destroy the functional integrity of the protein moiety. Consequently, continuous vision depends fundamentally on two processes, the re-isomerization of the chromophore and the continuous renewal of the protein moiety - in other words, on a turnover. These two processes are essentially independent of each other in the case of vertebrates, by virtue of the instability of the visual pigment once it has absorbed a photon. In invertebrates, however, the photoproduct of the visual pigment is relatively more stable, with the consequence that the turnover of chromophore is intimately linked to the turnover of the protein moiety. In both vertebrates and invertebrates, however, the nature of the protein moiety as an integral membrane protein has the consequence that its renewal may necessarily involve the renewal of the cell membrane.

So, having introduced the cart, it remains to present the proverbial horse. Since membranes can readily be visualized by electron microscopy, whereas the chromophore and protein moiety cannot, the turnover of chromophore and of the protein moiety were originally seen (literally and conceptually) as a single process only: membrane turnover - which includes the breakdown of "old" membrane and the synthesis and assembly of "new" membrane. Investigations of membrane breakdown and assembly in vertebrate and invertebrate photoreceptors began in the 1960s and there now exists an extensive literature relating to each of these groups of animals (for vertebrates see (86); see also Roof, this volume). Despite this, our understanding of even the morphological aspects of membrane turnover in invertebrate photoreceptors remains dogged by the wide diversity of phenomena observed, which is perhaps only to be expected of such a breathtakingly diverse group of animals. By comparison, our knowledge of the physiological processes associated with the turnover of membrane and visual pigment is still patchy, and our knowledge of the mechanisms whereby turnover is controlled is scant indeed.

STRUCTURAL AND MOLECULAR ORGANIZATON

The photoreceptors that have been studied extensively belong to the rhabdomeric type in which an array of microvilli forms a light-capturing structure, the rhabdomere (review in (25)). The microvilli vary in length from about 0.5 µm to 5.0 µm and in diameter from 0.02 µm to 0.1 µm in different animals and are typically packed tightly together, being most highly ordered in arthropods. The grouping together of photoreceptor cells within the retina varies greatly.

The molecular composition of the photoreceptor membrane has been studied in detail in only a few animals (cephalopods (1, 2, 51); insects (87)). These studies show that the membrane is composed of roughly equal proportions of lipid and protein (55:45). The majority of lipids are phospholipids, principally phosphatidylethanolamine and phosphatidylcholine in which the predominant fatty acids are highly unsaturated. In contrast to vertebrates, the photoreceptor membrane of invertebrates contains a large amount of cholesterol (approximately 2-3 times that of vertebrate photoreceptor membrane). The major integral protein of the membrane is visual pigment (blowfly: 65% (50); cephalopods: 70-80% (42, 51); crayfish: 80% (33)) which has an apparent molecular weight of from 43,000 to 51,000 in cephalopods, 32,500 to 37,000 in insects (see review in (62)), and 35,000 in crayfish (33). As in vertebrate rhodopsin, the 11-cis retinal chromophore can be assumed to be bound to the protein moiety (opsin) by a Schiff's base linkage. Recent evidence indicates that the chromophore in some insect visual pigments is probably 3-hydroxyretinal (70, 71). The absorbance maxima of invertebrate visual pigments span a wide range of wavelengths from about 350 nm to 580 nm (62). Following the absorption of light, the pigments undergo a sequence of reactions which include isomerization of the 11-cis chromophore into the all-trans form and conformational changes in the opsin. The sequence usually results in the formation of metarhodopsin which is more stable than that of vertebrates and can be reconverted by light to the parent visual pigment, i.e., photoregeneration (for reviews see (34, 35, 64)). The absorbance maximum of metarhodopsin can be either close to that of the parent pigment or shifted towards shorter or longer wavelengths.

Freeze-fracture preparations of photoreceptors reveal the presence in the membrane of particles (7-9 nm in diameter) which are considered to be molecules of visual pigment. There is, however, a great variation in their density, ranging from about 3,000/μm^2 (e.g., (15)) to about 8,000/μm^2 (28). The visual pigment content of fly photoreceptors can be drastically reduced by experimental manipulation without affecting the dimensions of microvilli or rhabdomeres, but the density of membrane particles is significantly reduced (15, 37, 44, 58). Correspondingly, there is also a reduction in the amount of opsin that can be extracted from the membrane (49, 50).

The high level of polyunsaturated phospholipids suggests that the fluidity of the membrane is comparable to that of vertebrate rod outer segments. However, considerations of the size and shape of the microvilli (45) and spectroscopic measurements (30) imply that the mobility of rhodopsin molecules within the membrane is restricted. Moreover, ESR studies support the view that the microvillar membrane is less fluid than that of frog photoreceptors (51), and this is attributed in part to the high cholesterol content of the membrane (e.g., (1, 51)).

Recently, interest has been directed to the cytoskeleton associated with the microvilli of many photoreceptors. In the squid and in arthropods, each microvillus appears to possess a single axial filament which is linked to the microvillar membrane by side arms (13, 14, 55). The cytoskeleton of the squid is composed mainly of two polypeptides with molecular weights of 145,000 and 42,000, respectively, the latter having been identified as actin (55). The microvillar cytoskeleton presumably stabilizes the geometry of the microvillus. The bridges between neighboring microvilli (13, 55) may maintain the structural integrity of the microvillar array as a whole. The cytoskeleton may also restrict the diffusion of visual pigment within the microvillar membrane (30, 55). Finally, it could play a significant role in the degradation and assembly of microvilli ((12, 14); see also discussion in (64)) or, alternatively, in a molecular exchange of visual pigment during turnover or the selective removal of membrane components (12).

MORPHOLOGICAL CORRELATES OF TURNOVER

In a great many studies of invertebrate photoreceptor cells, special attention has been paid to structural changes of the rhabdom, rhabdomeres, and microvilli and to cytological changes within the cells. This has necessarily involved the interpretation of "static" pictures in terms of continuous "dynamic" processes such as membrane breakdown and synthesis. The main body of evidence relating to the turnover of photoreceptor membrane comes from ultrastructural studies of arthropod receptors.

Membrane Breakdown

From early studies on the mosquito Aedes (75) and the crab Libinia (27) a picture of membrane breakdown emerged involving the internalization of membrane by endocytosis (Fig. 1A). Small pinocytotic vesicles (PV) are formed at the base of the microvilli, or in some cases from invaginations of the cell membrane at the base of the microvilli (79). After pinching off from the cell membrane, the coated vesicles (CV) subsequently lose their clathrin coats (becoming smooth vesicles, SV) and form multivesicular bodies (MVB) by secondary endocytosis (e.g., (26, 73)). MVB have been classified as nascent or compact, implying a progression from one form to

FIG. 1 - Schematic representation of membrane breakdown. A: The intracellular, endocytotic route of membrane breakdown (shown on the right) is typical of many different photoreceptors. PV, nascent pinocytotic vesicle; CV, coated vesicle; SV, smooth vesicle; n-MVB and c-MVB, nascent and compact multivesicular body; PRIM. LYSOS., primary lysosomes; DB, dense body; RB, residual body (may undergo total lysis or may be exocytosed); GO, Golgi body; RER, rough endoplasmic reticulum; N, nucleus. B-D: Different types of extracellular degradation (drawn after (12)). B: At the edges of rhabdomeres microvilli are broken down. The arising vesicles (V) may be internalized by the same or neighboring cell (e.g., in Leptograpsus, sometimes observed in Diptera). PH.V, phagocytotic vesicle. C: The tips of microvilli are pinched off in endocytotic pits (EP) in the glial membrane (GL.M) and transformed into vesicles (V), the fate of which is unclear (jumping spiders). D: The tips of microvilli are shed into an extracellular space (SZ, shedding zone) after the dissolution of the axial cytoskeleton. Membrane vesicles (V) are phagocytosed at an area where pseudopodia (PS) are formed (tipulid flies).

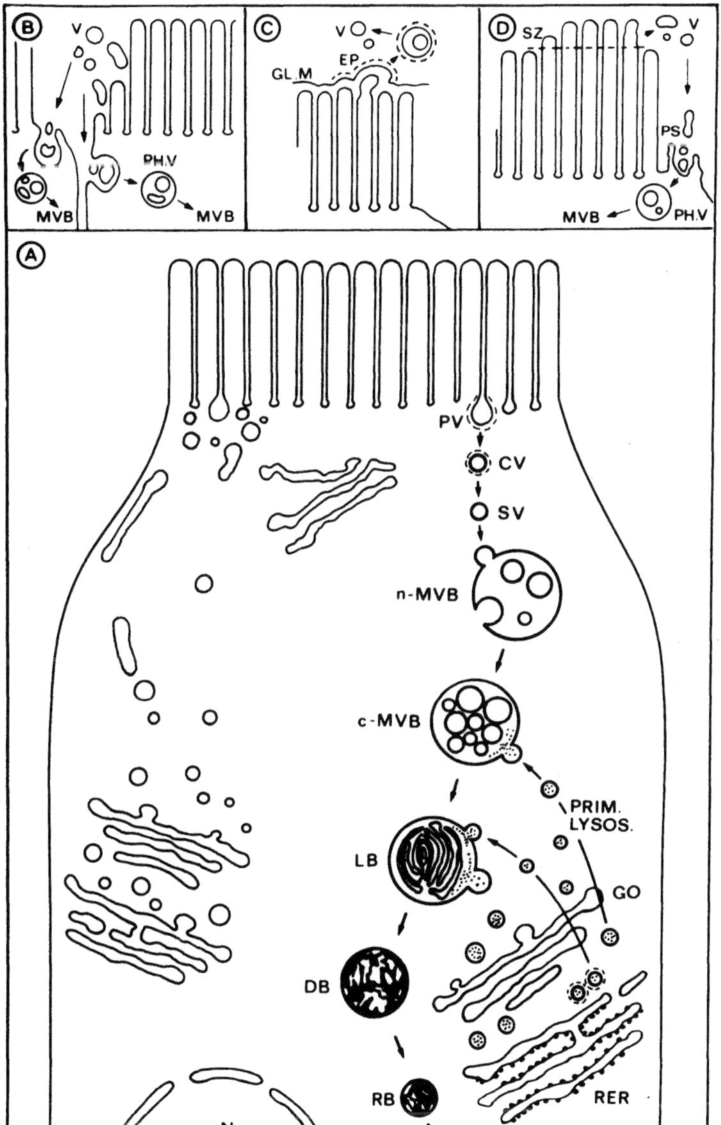
B
V
PH.V
MVB
MVB
C
V
EP
GL.M
D
SZ
V
PS
MVB
PH.V
A
PV
CV
SV
n-MVB
c-MVB
PRIM.
LYSOS.
LB
GO
DB
RB
RER
N
EXOCYTOSIS

the other. At some stage, other small vesicles representing primary lysosomes and presumably originating from the Golgi bodies fuse with the MVB which then, as secondary lysosomes, undergo a sequence of degradation changes corresponding to their transformation into bodies variously referred to as combination bodies, lamellar bodies (LB), dense and residual bodies (DB, RB) which may be completely lyzed or exocytosed: the process is seen as being comparable to that postulated for other cell types (38).

The verity of the (reconstructed) sequence of events just described has been strengthened by a number of studies. Thus ferritin applied extracellularly can subsequently be found in PV (75, 77). The difference in the density of particles in the MVB membranes (approximately 8,000/μm^2) and in those of lamellar-vesicular bodies (about 4,500/μm^2) has been seen as an indication of progressive breakdown of microvillar membrane (28). Quantitative analysis of the abundance of various cell organelles provides further support for the proposed sequence (77). Histochemical studies have demonstrated the presence of acid phosphatase, a marker enzyme for lysosomal activity, in compact MVB and bodies derived from them but not at earlier stages (9, 12, 16, 18, 28, 74). The sequence of events described above applies, with only slight modification, to most invertebrate photoreceptor cells (12, 74). Little mention has been made of the association between PV and the organelles which derive from them and elements of the smooth endoplasmic reticulum (SER) which is apparent in some cases (e.g., (6, 79)).

In addition to pinocytosis, evidence for extracellular degradation has been found in some animals (Fig. 1B-D); there is, however, a bewildering variety of different mechanisms. They involve the shedding of small groups of microvilli (39) or their tips (83) into the extracellular medium. Once in the extracellular milieu, the microvillar debris may take one of several routes. It may be phagocytosed by the photoreceptor cell (9), not necessarily the cell from which the microvillar membranes originated (12, 68). In one case resorption occurs at the level of the receptor intermediate segments (10). The debris is subsequently degraded within the cell in the conventional way. Alternatively, debris may be phagocytosed by glial cells (9, 12). In Procambarus, debris is phagocytosed not only by glial cells but also by granular hemocytes which reportedly invade the retina. The latter may also be a source of acid hydrolases (74).

Pinocytosis could suffice for a relatively low rate of membrane removal, whereas the different forms of shedding could potentially remove large amounts of membrane over a short period of time (9). A massive disruption of microvillar membrane appears to occur in some cases, but how the membrane is later internalized by the photoreceptor cells is not clear. Pictures of disrupted rhabdomeres have been presented by Winterhager and Stieve (85), but in this case they result from the effects of osmotic imbalance.

Membrane Synthesis and Assembly of Microvilli

Membrane synthesis and assembly of microvilli must occur and must in some cases compensate for the breakdown of massive amounts of membrane, but the underlying mechanisms remain for the most part obscure. In general, formation of microvilli involves the following sequence of cell organelles: rough ER (RER), transporting vesicles, Golgi bodies, and storage and transport organelles from which microvilli are finally assembled (Fig.2). Correspondingly, RER and Golgi bodies are a common feature of photoreceptor cells. The most likely candidate for the storage and transport of new membrane destined to become microvilli is some sort of vesicle (SER element), any number of which can usually be found in the cytoplasm of photoreceptor cells. Such inclusions should be considered as distinct from the specialized forms of SER which are a feature of many photoreceptors. Whittle (79) concludes that reticular specializations (agranular cisternae or vacuoles surrounding the rhabdom in arthropods, also referred to as perirhabdomal cisternae or the palisade; myeloid bodies in cephalopods, and photic vesicles in gastropods) probably do not fuse with the cell membrane, which can also be seen from the observations of Stowe (64) regarding the palisade. Thus, whereas reticular specializations are arguably involved in the metabolism of visual pigment (79), they are evidently not involved in the storage and transport of new membrane to the rhabdomere. Nevertheless, exactly this role has been attributed to the photic vesicles of gastropods (17,26), despite the rarity with which they appear to fuse with the rhabdomeric membrane (26); these inclusions are better interpreted as being reticular specializations (40,79). CV, normally taken to indicate membrane removal, have also been suggested to transport membrane to the rhabdomere. Another possibility is that membrane is delivered by more complex configurations of ER, such as concentric ellipsoids (39), condensed whorls of double membranes, and doublet ER (64), the latter being found adjacent to the base of the microvilli. In Limulus, however, similarly arranged membrane whorls are taken to indicate membrane breakdown (23). In the developing larval eye of an insect, a large number of cisternae have been observed at a stage prior to the formation of the microvilli (72, 76), and these endomembranes are assumed to contain a "prerhabdomeric" visual pigment (76).

However newly synthesized membrane is transported to the rhabdomere, the exact mechanism by which it is transformed into microvilli is an aspect of membrane turnover which remains the weakest and most elusive link in the chain of events leading to the renewal of microvilli. In particular, although it is often assumed that membrane elements (vesicles, doublet ER, etc.) fuse with the cell membrane, few reports provide evidence of this phenomenon which is commonly described in uncertain terms. Attention has been given almost exclusively to the cytoplasm underlying the rhabdomere, but in Calliphora, for example, there are indications that the cell membrane adjacent to the rhabdomere may be the site where new membrane is inserted (Whittle and Schwemer, unpublished observations). The possible

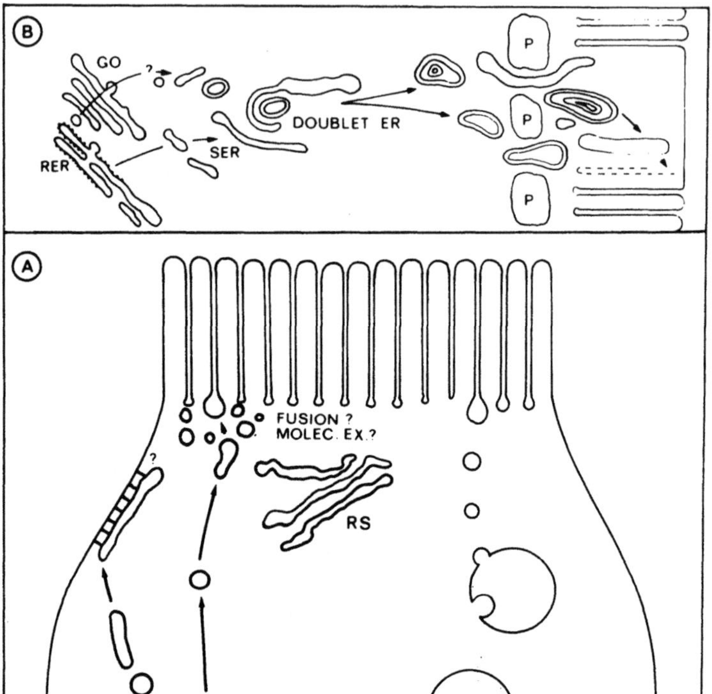
B
GO
?
RER
SER
DOUBLET ER
P
P
P
A
FUSION ?
MOLEC. EX.?
?
RS
MV
GO
RER
N

involvement of the cytoskeleton in the assembly of the microvilli is discussed by Stowe (64) amongst others.

Effects of Light and Darkness

In addition to those considered above, a further morphological correlate of membrane breakdown and synthesis is changes in the size of the rhabdomeres or rhabdom. In general, light has been found to lead to the breakdown of membrane and a decrease in rhabdom size, whereas membrane synthesis and an increase in rhabdom size occur in darkness.

Gross changes in the size of the rhabdom presumably occur when there is an imbalance between the rates of membrane breakdown and synthesis or assembly of microvilli; when these processes are in balance, no change in the size of the rhabdom is expected. In only a few of the large variety of animals studied to date does the size of the rhabdomeres remain constant under changing light conditions. In the fly, for example, the daily cycle of light and darkness produces no significant change in the cross-sectional area of the rhabdomeres and, correspondingly, the abundance of MVB, taken as a measure of membrane breakdown, does not change (82). In other cases the size of the rhabdom remains unchanged but the rate of membrane breakdown increases in the light, as indicated by the abundance of PV (27) or abscised microvilli (11).

The bulk of studies report, however, that moderate or drastic changes in rhabdom size occur upon the onset of light and darkness, respectively (e.g., (46)), indicating an imbalance between the rates of membrane breakdown and synthesis and/or assembly. A common finding of studies in echinoderms, cephalopod molluscs, crustaceans, chelicerates, and insects is that the size of the rhabdomeres decreases in the light, accompanied by signs of endocytosis or of extracellular breakdown (see above), and regrows in the dark by elongation of the existing microvilli and the formation of new microvilli at the periphery of the rhabdomere. Changes in microvillus diameter are reportedly small.

Investigations of the larval ocelli of a mosquito indicate that the volume of the rhabdom is inversely proportional to the intensity of the adapting light. The growth of rhabdom, which follows within 30 min of setting the animals

---

FIG. 2 - Schematic representation of membrane synthesis and assembly of microvilli. A: Hypothetical route taken by photoreceptor membrane synthesized on RER, transported through the Golgi bodies (GO), and thence delivered either to the base of the microvilli or to the cell membrane adjacent to the rhabdomere. The new membrane vesicles (MV) may fuse with the existing membrane, or its components may be inserted by means of some molecular exchange mechanism (MOLEC. EX.) N, nucleus. B: Membrane synthesis and transformations in Leptograpsus (drawn after (12)). RER forms smooth saccular membranes (SER) which are transformed into doublet endoplasmic reticulum (DOUBLET ER). SER and doublet ER are transported across the palisade (P) towards the region where new microvilli are assembled from these membrane sheets. A transport of the saccular membranes through the Golgi body is uncertain.

in darkness, appears to be due primarily to a reduction in the rate of membrane breakdown as reflected by the number of PV, whereas the rate of membrane synthesis (for which there is no morphological correlate) remains relatively constant and is estimated to be up to 100 times faster than the continuous growth of frog rod outer segments (78).

In contrast, the results obtained from other arthropods indicate that the rates of both membrane degradation and synthesis can vary independently. In Leptograpsus, for example, precursor membrane (e.g., doublet ER) is synthesized and microvilli assembled within a short time (30 to 60 min, but usually within a few hours in other animals) after the onset of darkness, thereby completely replacing a "day rhabdomere" with a new, somewhat larger "night rhabdomere" (64). After the onset of light at dawn the rhabdom becomes "disorganized" for a time, during which photoreceptor membrane is rapidly broken down. Eventually a smaller "day rhabdom" is formed. Interestingly, in locusts exposed to a gradually increasing intensity of light (natural sunrise) the microvilli do not become disorganized but remain orderly during the dawn period and the reduction of rhabdom size by endocytosis occurs more slowly than in the laboratory, where sudden switching on of the fluorescent lights apparently causes the rhabdoms to become disorganized and large segments of rhabdomeric membrane to be internalized within the cell (80). The pattern of breakdown and assembly just described has been shown to occur in many invertebrates (for references see (12, 74, 77, 80)).

Whereas the studies mentioned so far concern the effects of diurnal cycles of light and darkness on the photoreceptor membrane, a few studies have looked at the intramembranous particles which are thought to represent molecules of visual pigment. Thus, no difference between the density of particles in light- and dark-adapted animals was found in locust (80), although the size of the rhabdom changed. On the other hand, light was found to result in an increase in particle density in Astacus (84) but a decrease in Helix (19). Recent evidence indicates that whereas all visual pigment molecules are probably particles, not all particles are visual pigment, and that some other protein(s) are present in the membrane (58).

The effects of light and darkness and in particular of their normal daily cycle are in most cases sufficient to account for the observed changes in the breakdown and synthesis of photoreceptor membrane; i.e., the timing of membrane turnover is controlled by a single exogenous factor: light. Nevertheless, endogenous factors may also be involved in some cases (74). For example, in Leptograpsus the synthesis and assembly of photoreceptor membrane occurs earlier if the animals are transferred into the dark several hours before dusk, but the rhabdoms formed are smaller than the usual night rhabdom ((65); the same has been shown in locust (80)), which Stowe thinks likely to be due to a shortage of precursor membrane (doublet ER, large SER tubules). Similarly, if the light period is extended beyond the usual onset of darkness, membrane synthesis still occurs (65, 80), but again the rhabdoms are smaller than usual. Conversely, if darkness is extended beyond the usual onset of light, the breakdown of photoreceptor membrane does not occur in a sudden burst, rather the size of the rhabdom decreases slowly until it is, as usual, smaller than the night rhabdom (80).

Further studies on Locusta have demonstrated that membrane breakdown and assembly can be induced locally within the retina (81). Masking part of the eye during the day leads to the early synthesis of membrane and an increase in rhabdom size in the masked ommatidia, whereas the unmasked ommatidia retain their day rhabdoms. On the other hand, the "normal" breakdown of membrane following the onset of light is prevented only in the masked ommatidia. These results show not only that membrane turnover is unlikely to be primarily under endogenous control in Locusta but also that it is a local phenomenon confined to each ommatidium and possibly to each individual photoreceptor cell.

In general, little is known concerning the possible mechanisms whereby membrane turnover in a photoreceptor cell is triggered by light or by darkness. Membrane breakdown can be induced in preparations of Leptograpsus by experimental conditions that cause the photoreceptor to depolarize (67).

The possibility that membrane turnover is influenced by extraretinal mechanisms (neural and/or hormonal) remains contentious (12, 66, 74). In Limulus, the turnover of rhabdomeric membrane is blocked if the efferent input to the retina is cut, but it can be restored by stimulation of the optic nerve ((4, 22); see also (5) and discussion in (65)). As reported above, light causes a reduction in the size of the rhabdom in the lateral eye of Limulus (5, 6). According to a recent paper (23), photoreceptor membrane is shed at dawn in the form of membrane whorls which then undergo a sequential transformation into MVB, combination bodies, and LB. The formation of CV has not been observed during the process of shedding (6, 23). Within four hours of the shedding burst, the area of the rhabdom is again as great as that of the dark-adapted rhabdom. Aside from the early morning shedding, additional cyclical fluctuations occur in the length of the microvilli which represent a diurnal rhythm that is independent of efferent activity. Since keeping the animals in the dark delays the shedding burst until the eventual onset of light, it is concluded that light initiates the shedding of photoreceptor membrane. Furthermore, Chamberlain and Barlow (23) have shown that the onset of light must be preceded by at least three hours of efferent activity in the optic nerve, otherwise the shedding burst is abolished. It is therefore concluded that the efferent activity generated by a circadian clock in the brain (and probably mediated by octopamine) primes the shedding burst. The shedding burst is reportedly synchronized across the retina, but this finding contrasts with the results of Behrens (5) and Behrens and Krebs (6) which show that the rhabdom size varies from ommatidium to ommatidium and from one retinular cell to the next. These authors have suggested that photomechanical changes in Limulus' lateral eyes are triggered by local events. Other than this, no concrete evidence exists to show that membrane turnover follows a circadian rhythm, although other aspects of adaptation are clearly influenced by it (e.g., (3)).

VISUAL PIGMENT TURNOVER

Autoradiographic studies using radioactive amino acids to follow protein synthesis have demonstrated the presence of labeled protein in the cytoplasm and, sometime later, in the rhabdomeric membrane where it is randomly distributed (Limulus (21, 43); Apis (53); Oniscus (69); labeled vitamin A in Helix (17)). The general observation that more label is associated with the rhabdomeres in the dark than in the light has been interpreted in terms of there being an increased breakdown or, alternatively, a reduced synthesis of protein in the light. Biochemical studies have also indicated that the amount of protein labeled is greater in the darkness than in the light (21, 52). A renewal of proteins has also been revealed in crayfish (31). Both the cytoplasm and rhabdoms were rapidly labeled (5 min after the injection of (^{3}H)-leucine), suggesting that labeled proteins are synthesized by the RER directly underlying the rhabdomeres. The diffuse labeling of the rhabdomeres increased to a maximum after 12 hours and gradually declined thereafter. This labeling was uniform over the length of the rhabdomeres of light-adapted animals and was overall less intense than in dark-adapted rhabdomeres (as above) where there was a distinct gradient of labeling, decreasing proximally. The fact that dark-adapted rhabdoms showed a higher incorporation of label than light-adapted ones was taken as indication that the breakdown of rhabdomeric proteins was more rapid in the light than in the dark. These authors also found a very rapid labeling of MVB when the radioactivity associated with the rhabdomeres was still low: the labeling of MVB was at all times (except 2 min after injection) greater than that associated with any other structure. This could, in part, be due to the labeling of newly synthesized hydrolytic enzymes introduced into MVB from primary lysosomes (31). Since the peak of MVB labeling coincided with that of the cytoplasm and rhabdomeres, contrary to the expectation from ultrastructural studies that MVB form only after the breakdown of microvilli, it is unlikely that the MVB label was due to the incorporation of photoreceptor membrane. This emphasizes the need for complementary biochemical, physiological, and morphological studies.

In the studies mentioned so far, the labeled proteins were not identified but were thought to be visual pigment, and this view is supported by the demonstration that opsin is one of the labeled proteins which can be isolated in mosquito (63). Furthermore, more labeled opsin was found in dark-adapted animals than in those adapted to light, which led to the conclusion that light adaptation enhances the turnover of opsin, which is consistent with the morphological data.

The renewal of photoreceptor membrane in Octopus was investigated by light and electron microscopical autoradiography and biochemical techniques using both (^{3}H)-leucine and (^{3}H)-retinol. Robles et al. (54) found that 15 min after injecting leucine, label was associated with RER and, after 1 hour, with Golgi bodies: the latter probably add carbohydrate to the visual pigment protein (cephalopod rhodopsins have been shown to be glycoproteins, e.g., (41)). In this respect the path of label accords with the generally accepted pathway of protein synthesis (see section on Membrane Synthesis and Assembly of Microvilli), although the involvement of the Golgi has been doubted for certain arthropods (e.g., (39, 64, 74)). From the Golgi bodies, labeled proteins were transported from the inner segment of the cell to the central core of the cytoplasm of the outer segment where label was found 2 hours after the injection. After 17 hours a significant and diffuse labeling of the rhabdomeres was observed. Cephalopod photoreceptors contain a second retinal protein, retinochrome (e.g., (36, 47)), in addition to rhodopsin, and Robles et al. (54) have shown that both are labeled. The biochemical data suggest that the synthesis and turnover of retinochrome occur at a high rate.

Spectrophotometry has been widely used to characterize visual pigments and investigate their photochemistry and mechanism of photoregeneration (34, 35, 62). It has also been used to determine the visual pigment content of rhabdomeres in order to study regeneration of rhodopsin and turnover of rhodopsin and metarhodopsin. Comparing the results of different studies, however, it is evident that the time taken for the rhodopsin content to recover to its normal level following exposure of the animal to light varies a great deal. In the moth Galleria, 80% rhodopsin is recovered in about 5 days (29). A similarly slow recovery of rhodopsin has been found in Procambarus (24) and Astacus (32), whereas a faster recovery, with a half-time ranging from 5 min to more than 4 hours, has been reported for the lobster (20) and some insects (7, 8, 61). In general, a "slow" recovery has been thought to involve the de novo synthesis of rhodopsin (20, 29), whereas "fast" recovery involves the all-trans retinal in metarhodopsin being either isomerized enzymatically or exchanged for 11-cis retinal.

The molecular aspects of visual pigment turnover have been studied most intensively in blowflies. Vitamin A-deficient flies whose rhabdomeres contain very little visual pigment or opsin (49, 50) were shown to synthesize visual pigment in the dark after injection of 11-cis retinal into the eye: the half-time of this process was about 10 hours (56). Injecting all-trans retinal did not lead to the formation of visual pigment in darkness, indicating first that de novo synthesis of visual pigment requires the chromophore to be in the 11-cis configuration, and second, that all-trans retinal is not converted to the 11-cis isomer in the dark (56). The biosynthesis of visual pigment was shown to result in an increase in the number of particles in the microvillar membrane (58) and an increase in opsin content (50).

Evidence has been presented to show that the chromophore of the visual pigment in flies is not retinal but 3-hydroxyretinal, and it was proposed that the visual pigment should be called "xanthopsin" (70, 71). The hydroxylation of exogenously supplied retinal could be demonstrated by high-performance liquid chromatography (Schwemer, unpublished). To obviate the necessity of specifying in each case whether the visual pigment referred to is rhodopsin or xanthopsin, the more familiar terminology (i.e., rhodopsin, metarhodopsin, etc.) is retained for the purposes of this paper.

Spectrophotometric investigations of the breakdown of visual pigment in the dark have shown that rhodopsin is broken down only very slowly, whereas there is a much further and selective degradation of metarhodopsin (half-time about 2 hours). A breakdown was manifest in these cases because, in the dark, there is no compensatory synthesis of rhodopsin. Simply placing the flies in the light, however, results in rhodopsin synthesis and the restoration of the original, high level of visual pigment. This observation indicates that the all-trans chromophore resulting from the breakdown of metarhodopsin is not lost from the retina but stored and that light is responsible for converting it into the 11-cis isomer, which then supports rhodopsin synthesis (see Fig. 3A). These observations explain why it is that flies maintain their visual pigment content even if kept for a week in continuous light (57).

However, not all colors of light support rhodopsin synthesis. Thus, continuous yellow and green light for 120 hours resulted in a reduction of the total visual pigment content to about 40% and 2%, respectively, the rate constants being directly proportional to the fraction of metarhodopsin present (13% in yellow and 48% in green light). In contrast, blue light, which establishes the highest fraction of metarhodopsin (70%) and therefore would be expected to result in the fastest breakdown, caused no change in the total visual pigment content after 120 hours. These results led to the

FIG. 3 - Schematic representation of visual pigment turnover. A: Rhodopsin R (●) is converted by light into metarhodopsin M (○), which in turn converts into unstable metarhodopsin M_u(□). Both forms of metarhodopsin can be reconverted to R by light. All-trans retinal (AT) is bound to a protein to form a retinal-protein complex. The chromophore is isomerized by light into the 11-cis form, which then leads to opsin synthesis. Wavy lines = light reactions; straight lines = dark reactions; dashed lines = hypothetical reactions (after (57)). B: A hypothetical projection of the visual pigment cycle (shown in Fig. 3A) onto the morphology of the photoreceptor cell. R (●) is converted into M (○) and then into M_u (□). M_u is transported to the base of the microvilli, removed, and degraded by the endocytotic route (right-hand side) shown in Fig. 1A. The all-trans chromophore (AT) is released from M_u either while still in the microvillus (①) or during the breakdown of M_u in the various degradation bodies (②). The AT released is bound to a protein to form a pigment (possibly retinochrome) which may be localized within reticular specializations (RS). Light converts (③) retinochrome (▲) into metaretinochrome (Δ) which contains the chromophore in the 11-cis form. The 11-cis chromophore is then transported to the RER (④) where it serves as one of the prerequisites for opsin synthesis together with the appropriate mRNA. After its synthesis rhodopsin is delivered (probably via Golgi bodies) to the base of the microvilli or to the lateral cell membrane (left-hand side). If the 11-cis chromophore is not available for visual pigment synthesis, the membrane vesicles contain neither rhodopsin nor opsin. Another possible way of rhodopsin regeneration is that there is a mutual exchange of chromophore between metaretinochrome and and metarhodopsin (⑤; see (60)). The desmosome (D) represents a diffusion barrier for visual pigment molecules.

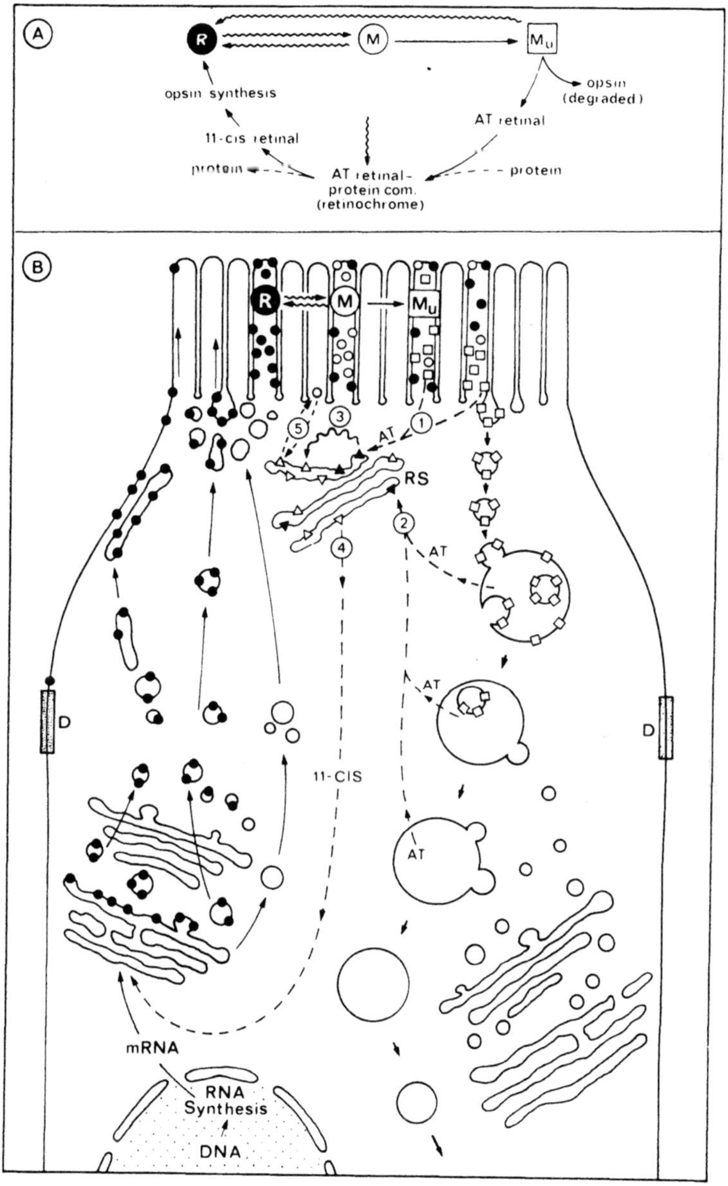
A
R
M
Mu
opsin synthesis
opsin (degraded)
11-cis retinal
AT retinal
protein
AT retinal-protein com. (retinochrome)
protein
B
R
M
Mu
AT
1
2
3
4
5
RS
AT
AT
D
D
11-CIS
AT
mRNA
RNA Synthesis
DNA

conclusion that blue light is necessary for the isomerization of the chromophore to the 11-cis form, which in turn is required for the biosynthesis of rhodopsin (see Fig. 3A). Furthermore, the rate of rhodopsin synthesis in blue light must at least compensate for the breakdown of visual pigment at the highest rate. That opsin synthesis is supported by blue light but not by red light has been demonstrated using labeled amino acids (50).

The rate of visual pigment breakdown is also influenced by the intensity of the light, the rate of breakdown being inversely proportional to the light intensity (57). Given that metarhodopsin is selectively degraded, this inverse proportionality was assumed to be due to two forms of metarhodopsin with spectrally identical properties, one of which is less stable than the other and more accessible to degradation. It is possible that the unstable state corresponds to metarhodopsin which is fully phosphorylated (57). The reversible phosphorylation of opsin has recently been demonstrated to occur in the blowfly (48). The visual pigment cycle is summarized in Fig. 3.

An exponential decrease of metarhodopsin in the dark has also been observed in butterflies (7, 8), although this occurs at a much higher rate (half-time about 10 min) than it does in the fly. The reported rate of rhodopsin recovery is similarly faster than in the fly. Whether this dark recovery is due to the replacement of the whole rhodopsin molecule or to an exchange of the chromophore is not clear. In Procambarus, the recovery of rhodopsin in the dark takes several days, but only about 2 days if preceded by exposure to blue light. This recovery differs from that in flies and butterflies in that metarhodopsin molecules are thought to be replaced stoichiometrically by rhodopsin molecules (24). The mechanism by which rhodopsin is restored in Astacus remains to be elucidated (32).

The isomerization of all-trans retinal plays a key role in the turnover of fly visual pigment in that it links the two aspects of turnover: breakdown and synthesis ((57); see Fig. 3A). For several reasons it is unlikely that isomerization is due simply to the absorption of light by free retinal (57). Rather, the available evidence points to its being a process mediated by a pigment similar to retinochrome of cephalopods, which has been intensively studied (e.g., (36, 47, 60)). Such a pigment exists in the retina of the honeybee. This complex serves primarily to ensure that the isomerization of all-trans chromophore by light is highly stereospecific and leads to the formation of only the 11-cis isomer (59). An essentially similar system is postulated to exist in the crayfish (24). Besides isomerization, retinochrome is also postulated to be involved in the reconstitution of rhodopsin ((60); see Fig. 3B). Moreover, such retinal-binding proteins may also serve as a store for chromophore. In contrast to vertebrates, little is known concerning the storage of chromophore in invertebrates, whether this be in the form of retinal, retinol, or esters; exceptionally, the lobster contains a large store of 11-cis retinol (73).

CONCLUDING REMARKS

A major source of difficulty in any attempt to present a coherent picture of turnover in photoreceptors is the diverse nature of the relevant evidence, which has been gained by the use of many different experimental techniques, each with its own limitations.

Results obtained from the fly indicate that membrane turnover is independent of the turnover of visual pigment, but not the other way round; in that case these processes are tightly coupled. Significantly, these two processes appear to share the same cellular machinery so that it is difficult to distinguish between them on the basis of morphology alone. The turnover of membrane appears to be directly proportional to the light intensity in a mosquito (78). In contrast, the rate of breakdown of visual pigment in a fly was found to be inversely proportional to the light intensity (57). Clearly, this dependency is of crucial importance with respect to the regulation of visual pigment turnover and the expected consequences of adapting an animal to different intensities of light. The studies on the fly also show that without a detailed knowledge of the photochemistry of the visual pigment system the results of simply adapting an animal to light or darkness can be misleading.

In the fly it has been conclusively demonstrated that the breakdown of visual pigment is a highly selective process whereby metarhodopsin is preferentially removed from the rhabdomeres (57). This selectivity represents a basis for the control of visual pigment breakdown. It must involve an as yet unknown mechanism by which visual pigment is removed from the microvillus. Regarding this mechanism, and the mechanism whereby synthesized rhodopsin is inserted into the rhabdomere again at the base of the microvilli, a comparable problem exists in that both appear to require the visual pigment to move over the length of the microvillus, to or from its base. However, there is strong circumstantial evidence that the microvillar membrane is relatively rigid and that the mobility of visual pigment molecules in the plane of the membrane is severely restricted (see section on STRUCTURAL AND MOLECULAR ORGANIZATION). This problem would be obviated if individual molecules could be inserted or removed, and not just at the base but over the entire length of the microvillus. Such a molecular exchange seems unlikely on energetic grounds since the visual pigment is an integral, transmembrane protein, although recent developments in cell biology indicate that this line of reasoning may be unfounded (for example, see Findlay, this volume).

The mechanism whereby metarhodopsin is selectively removed from the membrane could also regulate the rate of membrane breakdown, insofar as these processes are coupled together. That this may indeed be the case is suggested by a study on the fly (58) which, in essence, shows that light per se does not cause membrane breakdown; the latter occurs only when rhodopsin is present in the membrane and the animal is illuminated. A high rhodopsin content in the dark results in minimal breakdown of membrane, indicating that rhodopsin cannot regulate this process. In short, the most parsimonious explanation of the results is that unstable metarhodopsin determines not only the rate of visual pigment breakdown but also the rate of membrane breakdown. Further research is clearly needed to confirm this, yet it does provide a hypothetical framework which may explicate the disparate effects of light and darkness on different animals.

If membrane breakdown is regulated by unstable metarhodopsin, how then is membrane synthesis regulated? It is not known whether it varies or remains constant. Considered from a more general, cell biological point of view, it is likely that membrane synthesis remains constant but that newly synthesized membrane is constantly degraded within the cell, except that part required for the maintenance of the rhabdomere.

This article has attempted to emphasize the distinction between the various aspects of turnover in photoreceptors, namely, the turnover of photoreceptor membrane, of visual pigment protein, and of chromophore, as well as their relationships to light and dark adaptation. A contingent difficulty is not only the question of how all these processes are linked to each other, but also how the different rates of synthesis and degradation are regulated and their limits determined. In light of the emerging picture of the different aspects of turnover in fly photoreceptors, it is evident that the various processes involved and their regulation may be very different in other animals. Nevertheless, the principles learned from investigations on the fly may provide a stepping-stone to future research and a deeper understanding of the processes that together constitute the means by which vision is extended in time from a single event to a continuous process.

LIST OF ABBREVIATIONS

CV	coated vesicle
DB	dense body
ER	endoplasmic reticulum
ESR	electron spin resonance
LB	lamellar body
PV	pinocytotic vesicle
RB	residual body
RER	rough endoplasmic reticulum
SER	smooth endoplasmic reticulum
SV	smooth vesicle

Acknowledgements. The author is grateful to A.C. Whittle for his extensive discussions and help in the preparation of this manuscript and to K. Kirschfeld, H. Stieve, and A.D. Blest for their critical reading of the manuscript and their comments. This work was supported by the Deutsche Forschungsgemeinschaft (Heisenberg-Stipendium and SFB 114).

REFERENCES

(1) Akino, T., and Tsuda, M. 1979. Characteristics of phospholipids in microvillar membranes of octopus photoreceptor cells. Biochim. Biophys. Acta **556:** 61-71.

(2) Anderson, E.R.; Benolken, R.M.; Kelleher, P.A.; Maude, M.B.; and Wiegand, R.D. 1978. Chemistry of photoreceptor membrane preparations from squid retinas. Biochim. Biophys. Acta **510:** 316-326.

(3) Barlow, R.B., Jr. 1983. Circadian rhythms in the Limulus visual system. J. Neurosci. **3:** 856-870.

(4) Barlow, R.B., Jr.; Chamberlain, S.C.; and Levinson, J.Z. 1980. Limulus brain modulates the structure and function of lateral eyes. Science **210:** 1037-1039.

(5) Behrens, M.E. 1974. Photomechanical changes in the ommatidia of Limulus lateral eye during light and dark adaptation. J. Comp. Physiol. **89:** 45-57.

(6) Behrens, M.E., and Krebs, W. 1976. The effect of light-dark adaptation on the ultrastructure of Limulus lateral eye retinular cells. J. Comp. Physiol. **107:** 77-96.

(7) Bernard, G.D. 1983. Bleaching of rhabdoms in eyes of intact butterflies. Science **219:** 69-71.

(8) Bernard, G.D. 1983. Dark processes following photoconversion of butterfly rhodopsins. Biophys. Struct. Mech. **9:** 277-286.

(9) Blest, A.D. 1980. Photoreceptor membrane turnover in arthropods: Comparative studies of breakdown processes and their implications. In The Effect of Constant Light on Visual Processes, eds. T.P. Williams and B.N. Baker, pp. 217-245. New York: Plenum Press.

(10) Blest, A.D., and Day, W.A. 1978. The rhabdomere organization of some nocturnal pisaurid spiders in light and darkness. Phil. Trans. Roy. Soc. Lond. B **283:** 1-23.

(11) Blest, A.D., and Maples, J. 1979. Exocytotic shedding and glial uptake of photoreceptor membrane by a salticid spider. Proc. Roy. Soc. Lond. B **204:** 105-112.

(12) Blest, A.D.; Stowe, S.; and de Couet, H.G. 1984. Turnover of photoreceptor membranes in arthropods. Sci. Progr. **69:** 83-100.

(13) Blest, A.D.; Stowe, S.; and Eddey, W. 1982. A labile Ca^{2+}-dependent cytoskeleton in rhabdomeral microvilli of blowflies. Cell Tiss. Res. **223:** 553-573.

(14) Blest, A.D.; Stowe, S.; Eddey, W.; and Williams, D.S. 1982. The local deletion of a microvillar cytoskeleton from photoreceptors of tipulid flies during membrane turnover. Proc. Roy. Soc. Lond. **215:** 469-479.

(15) Boschek, C.B., and Hamdorf, K. 1976. Rhodopsin particles in the photoreceptor membrane of an insect. Z. Naturforsch. **39c:** 762.

(16) Brandenburger, J.L. 1977. Cytochemical localization of acid phosphatases in regenerated and dark-adapted eyes of a snail, Helix aspersa. Cell Tiss. Res. **184:** 301-313.

(17) Brandenburger, J.L., and Eakin, R.M. 1970. Pathway of incorporation of vitamin A $^3H_2$ into photoreceptors of a snail, Helix aspersa. Vision Res. **10:** 639-653.

(18) Brandenburger, J.L., and Eakin, R.M. 1980. Cytochemical localization of acid phosphatase in ocelli of the seastar Patiria miniata during recycling of photoreceptor membranes. J. Exp. Zool. **214:** 127-140.

(19) Brandenburger, J.L.; Reed, C.T.; and Eakin, R.M. 1975. Freeze-fracture studies of photoreceptors of dark and light-adapted snails. Am. Zool. **15:** 782.

(20) Bruno, M.S.; Barnes, S.N.; and Goldsmith, T.H. 1977. The visual pigment and visual cycle of the lobster, Homarus. J. Comp. Physiol. **120:** 123-142.

(21) Burnel, M.; Mahler, H.R.; and Moore, W.J. 1970. Protein synthesis in visual cells of Limulus. J. Neurochem. **17:** 1493-1499.

(22) Chamberlain, S.C., and Barlow, R.B., Jr. 1979. Light and efferent activity control rhabdom turnover in Limulus photoreceptors. Science **206:** 361-363.

(23) Chamberlain, S.C., and Barlow, R.B., Jr. 1984. Transient membrane shedding in Limulus photoreceptors: Control mechanisms under natural lighting. J. Neurosci. **4:** 2792-2810.

(24) Cronin, T.W., and Goldsmith, T.H. 1984. Dark regeneration of rhodopsin in crayfish photoreceptors. J. Gen. Physiol. **84:** 63-81.

(25) Eakin, R.M. 1972. Structure of invertebrate photoreceptors. In Photochemistry of Vision, Handbook of Sensory Physiology, ed. H.J.A. Dartnall, vol. VII/1, pp. 625-684. Berlin, Heidelberg, New York: Springer-Verlag.

(26) Eakin, R.M., and Brandenburger, J.L. 1975. Understanding a snail's eye at a snail's pace. Am. Zool. **15:** 851-863.

(27) Eguchi, E., and Waterman, T.H. 1967. Changes in the retinal fine structure induced in the crab Libinia by light and dark adaptation. Z. Zellforsch. **79:** 209-229.

(28) Eguchi, E., and Waterman, T.H. 1976. Freeze-etch and histochemical evidence for cycling in crayfish photoreceptor membranes. Cell Tiss. Res. **169:** 419-434.

(29) Goldman, L.J.; Barnes, S.N.; and Goldsmith, T.H. 1975. Microspectrophotometry of rhodopsin and metarhodopsin in the moth Galleria. J. Gen. Physiol. **66:** 383-404.

(30) Goldsmith, T.H., and Wehner, R. 1977. Restrictions of rotational and translational diffusion of pigment in the membranes of a rhabdomeric photoreceptor. J. Gen. Physiol. **70:** 453-490.

(31) Hafner, G.S., and Bok, D. 1977. The distribution of [^{3}H] leucine labeled protein in the retinula cells of the crayfish retina. J. Comp. Neurol. **174:** 397-416.

(32) Hamacher, K. 1981. Absorptionsspektropische Analyse des Astacus Rhodopsinsystems und Nachweis einer metabolischen Regeneration des Rhodopsins nach Helladaptation. Thesis, Berichte der Kfa Jülich Nr. 1718, Jülich, F.R. Germany.

(33) Hamacher, K., and Stieve, H. 1984. Spectral properties of the rhodopsin-system of the crayfish Astacus leptodactylus. Photochem. Photobiol. **39**: 379-390.

(34) Hamdorf, K. 1979. The physiology of invertebrate visual pigments. In Vision in Invertebrates, Handbook of Sensory Physiology, ed. H. Autrum, vol. VII/6A, pp. 145-224. Berlin, Heidelberg, New York: Springer-Verlag.

(35) Hamdorf, K., and Schwemer, J. 1975. Photoregeneration and the adaptation process in insect photoreceptors. In Photoreceptor Optics, eds. A.W. Snyder and R. Menzel, pp. 263-289. Berlin, Heidelberg, New York: Springer-Verlag.

(36) Hara, T., and Hara, R. 1982. Cephalopod retinochrome. In Methods in Enzymology. Part H: Visual Pigments and Purple Membranes I, ed. L. Packer, vol. 81, pp. 827-833. New York: Academic Press.

(37) Harris, A.W.; Ready, D.F.; Lipson, E.D.; Hudspeth, A.J.; and Stark, W.S. 1977. Vitamin A deprivation and Drosophila photopigments. Nature **266**: 648-650.

(38) Holtzman, E., and Mercurio, A.M. 1980. Membrane circulation in neurons and photoreceptors: some unresolved issues. Int. Rev. Cytol. **67**: 1-67.

(39) Itaya, S.K. 1976. Rhabdom changes in the shrimp, Palaemonetes. Cell Tiss. Res. **166**: 265-273.

(40) Kataoka, S. 1975. Fine structure of the retina of the slug, Limax flavus L. Vision Res. **15**: 681-686.

(41) Kito, Y.; Naito, T.; and Nashima, K. 1982. Purification of squid and octopus rhodopsin. In Methods in Enzymology. Part H: Visual Pigments and Purple Membranes I, ed. L. Packer, vol. 81, pp. 167-171. New York: Academic Press.

(42) Kito, Y.; Seki, T.; and Hagins, F.M. 1982. Isolation and purification of squid rhabdoms. In Methods in Enzymology. Part H: Visual Pigments and Purple Membranes I, ed. L. Packer, vol. 81, pp. 43-48. New York: Academic Press.

(43) Krauhs, J.H.; Mahler, H.R.; Minkler, G.; and Moore, W.J. 1976. Synthesis and degradation of protein of visual receptor membranes in lateral eyes of Limulus. J. Neurochem. **26**: 281-283.

(44) Larrivee, D.C.; Conrad, S.K.; Stephenson, R.S.; and Pak, W.L. 1981. Mutation that selectively affects rhodopsin concentration in the peripheral photoreceptors of Drosophila. J. Gen. Physiol. **78**: 521-545.

(45) Laughlin, S.B.; Menzel, R.; and Snyder, A.W. 1975. Membranes, dichroism and receptor sensitivity. In Photoreceptor Optics, eds. A.W. Snyder and R. Menzel, pp. 237-259. Berlin, Heidelberg, New York: Springer-Verlag.

(46) Nässel, D.R., and Waterman, T.H. 1979. Massive diurnally modulated photoreceptor membrane turnover in crab light- and dark adaptation. J. Comp. Physiol. **131**: 205-216.

(47) Ozaki, K.; Hara, R.; and Hara, T. 1983. Squid retinochrome. Biophys. J. **44**: 127-137.

(48) Paulsen, R., and Bentrop, J. 1984. Reversible phosphorylation of opsin induced by irradiation of blowfly retinae. J. Comp. Physiol. A **155:** 39-45.

(49) Paulsen, R., and Schwemer, J. 1979. Vitamin A-deficiency reduces the concentration of visual pigment protein within blowfly photoreceptor membranes. Biochim. Biophys. Acta **557:** 385-390.

(50) Paulsen, R., and Schwemer, J. 1983. Biogenesis of blowfly photoreceptor membranes is regulated by 11-cis retinal. Eur. J. Biochem. **137:** 609-614.

(51) Paulsen, R.; Zinkler, D.; and Delmelle, M. 1983. Architecture and dynamics of microvillar photoreceptor membranes of cephalopods. Exp. Eye Res. **36:** 47-56.

(52) Pepe, I.M., and Baumann, F. 1972. Incorporation of ^{3}H-labelled leucine into the protein fraction in the retina of the honeybee drone. J. Neurochem. **19:** 507-512.

(53) Perrelet, A. 1972. Protein synthesis in the visual cells of the honeybee drone as studied with electron microscope radioautography. J. Cell Biol. **55:** 595-605.

(54) Robles, L.J.; Cabebe, C.S.; Aguilo, J.A.; Anyakora, P.A.; and Bok, D. 1984. Autoradiographic and biochemical analysis of photoreceptor membrane renewal in octopus retina. J. Neurocytol. **13:** 145-164.

(55) Saibil, H.R. 1982. An ordered membrane-cytoskeleton network in squid photoreceptor microvilli. J. Molec. Biol. **158:** 435-456.

(56) Schwemer, J. 1983. Pathways of visual pigment regeneration in fly photoreceptors. Biophys. Struct. Mech. **9:** 287-298.

(57) Schwemer, J. 1984. Renewal of visual pigment in photoreceptors of the blowfly. J. Comp. Physiol. **154:** 535-547.

(58) Schwemer, J., and Henning, U. 1984. Morphological correlates of visual pigment turnover in photoreceptors of the fly. Cell Tiss. Res. **236:** 293-303.

(59) Schwemer, J.; Pepe, I.M.; Paulsen, R.; and Cugnoli, C. 1984. Light-activated trans-cis isomerization of retinal by a protein from honeybee retina. J. Comp. Physiol. **154:** 549-554.

(60) Seki, T.; Hara, R.; and Hara, T. 1980. Reconstitution of squid rhodopsin in rhabdomal membranes. Photochem. Photobiol. **32:** 469-479.

(61) Stavenga, D.G. 1975. Dark regeneration of invertebrate visual pigments. In Photoreceptor Optics, eds. A.W. Snyder and R. Menzel, pp. 290-295. Berlin, Heidelberg, New York: Springer-Verlag.

(62) Stavenga, D.G., and Schwemer, J. 1984. Visual pigments of invertebrates. In Photoreception and Vision in Invertebrates, ed. M.A. Ali, pp. 11-61. New York: Plenum Press.

(63) Stein, P.J.; Brammer, J.O.; and Ostroy, S.E. 1979. Renewal of opsin in the photoreceptor cells of the mosquito. J. Gen. Physiol. **74:** 565-582.

(64) Stowe, S. 1980. Rapid synthesis of photoreceptor membrane and assembly of new microvilli in a crab at dusk. Cell Tiss. Res. **211:** 419-440.

(65) Stowe, S. 1981. Effects of illumination changes on rhabdom synthesis in a crab. J. Comp. Physiol. **142:** 19-25.

(66) Stowe, S. 1982. Rhabdom synthesis in isolated eyestalks and retinae of the crab Leptograpsus variegatus. J. Comp. Physiol. **148:** 313-321.

(67) Stowe, S. 1983. Light-induced and spontaneous breakdown of the rhabdoms in a crab at dawn; depolarization versus calcium levels. J. Comp. Physiol. **153:** 365-375.

(68) Stowe, S. 1983. Phagocytosis of rhabdomeral membrane by crab photoreceptors (Leptograpsus variegatus). Cell Tiss. Res. **234:** 463-467.

(69) Tuurala, O., and Lehtinen, A. 1974. Inkorporierung des tritiummarkierten Leucins in den Sehzellen von Oniscus asellus L. (Isopoda, Oniscoidea). Ann. Zool. Fennici **11:** 135-140.

(70) Vogt, K. 1983. Is the fly visual pigment a rhodopsin? Z. Naturforsch. **38c:** 329-333.

(71) Vogt, K., and Kirschfeld, K. 1984. Chemical identity of the chromophores of fly visual pigment. Naturwiss. **71:** 211-212.

(72) Waddington, C.H., and Perry, M.M. 1960. The ultrastructure of the developing eye of Drosophila. Proc. Roy. Soc. Lond. B **153:** 155-178.

(73) Wald, G., and Burg, S. 1957. The vitamin A of the lobster. J. Gen. Physiol. **40:** 609-625.

(74) Waterman, T.H. 1982. Fine structure and turnover of photoreceptor membranes. In Visual Cells and Evolution, ed. J.A. Westfall, pp. 23-41. New York: Raven Press.

(75) White, R.H. 1968. The effect of light and light deprivation upon the structure of the larval mosquito eye. III. Multivesicular bodies and protein uptake. J. Exp. Zool. **169:** 261-278.

(76) White, R.H.; Brown, P.K.; Hurley, A.K.; and Bennett, R.R. 1983. Rhodopsins, retinal cell ultrastructure, and receptor potentials in the developing pupal eye of the moth Manduca sexta. J. Comp. Physiol. **150:** 153-163.

(77) White, R.H.; Gifford, D.; and Michaud, N.A. 1980. Turnover of photoreceptor membrane in the larval mosquito ocellus: rhabdomeric coated vesicles and organelles of the vacuolar system. In Effects of Constant Light on Visual Processes, eds. T.P. Williams and B.N. Baker, pp. 271-296. New York: Plenum Press.

(78) White, R.H., and Lord, E. 1975. Diminution and enlargement of the mosquito rhabdom in light and darkness. J. Gen. Physiol. **65:** 583-598.

(79) Whittle, A.C. 1976. Reticular specializations in photoreceptors: a review. Zool. Scripta **5:** 191-206.

(80) Williams, D.S. 1982. Ommatidial structure in relation to turnover of photoreceptor membrane in the locust. Cell Tiss. Res. **225:** 595-617.

(81) Williams, D.S. 1982. Photoreceptor membrane shedding and assembly can be initiated locally within an insect retina. Science **218:** 898-900.

(82) Williams, D.S. 1982. Rhabdom size and photoreceptor membrane turnover in a muscoid fly. Cell Tiss. Res. **226:** 629-639.

(83) Williams, D.S., and Blest, A.D. 1980. Extracellular shedding of photoreceptor membrane in the open rhabdom of a tipulid fly. Cell Tiss. Res. **205**: 423-438.

(84) Winterhager, E.; Dahl, G.; and Stieve, H. 1981. Ultrastructural changes of microvilli of the Astacus retina depending on the state of adaptation. Verh. Anat. Ges. **75**: 959-960.

(85) Winterhager, E., and Stieve, H. 1982. Effect of hyper- and hypoosmotic solutions on the structure of the Astacus retina. Cell Tiss. Res. **223**: 267-280.

(86) Young, R.W. 1976. Visual cells and the concept of renewal. Inv. Ophthalmol. **9**: 700-725.

(87) Zinkler, D. 1974. Lipid and fatty acid composition of a rhabdomeric retina. Verh. Dtsch. Zool. Ges. **67**: 28-32.

The Molecular Mechanism of Photoreception, ed. H. Stieve, pp. 327-351. Dahlem Konferenzen 1985. Berlin, Heidelberg, New York, Tokyo: Springer-Verlag.

Quantitative Models of Phototransduction

A. Borsellino*, L. Cervetto**, and V. Torre*
*Dipartimento di Fisica, Istituto di Scienze Fisiche
16146 Genova, Italy
**Istituto di Neurofisiologia, CNR
56100 Pisa, Italy

Abstract. Two main subjects of photoreceptor physiology are discussed: a) the properties of the light-sensitive channel, and b) the kinetics of the intervening events between photoisomerization and electrical response. The ionic nature of the photocurrent is discussed and evidence that ions other than Na^+ also contribute light-sensitive current in rods is reviewed. Both electrical measurements and determinations with radioactive tracers indicate that K^+ and divalent cations permeate through the light-sensitive channel, thus contributing dark current. It is suggested that the light-sensitive channel may exist in three different conductive states which are controlled by both divalent cations and cyclic nucleotides. The cascade model of phototransduction is briefly reviewed and compared with an alternative scheme. The role of cyclic nucleotide metabolism in the phototransductive process is evaluated on the basis of electrical measurements carried out in the presence of phosphodiesterase inhibitors. It is suggested that the hydrolytic activity of phosphodiesterase may be involved in controlling the kinetics of photoresponses by accelerating the flux of cGMP, which in turn may affect all the rate constants of the multistage model. Ca^{2+} extrusion from the visual cell is considered; we propose that this process may play a role in some of the phenomena associated with light adaptation and be responsible for the recovery of responsiveness during light saturation.

INTRODUCTION

An understanding of the process of phototransduction requires a detailed knowledge of the intervening events between photon absorption and the electrical response. The current idea is that calcium or cyclic nucleotides, alone or in combination, act as "second messengers" after a photon has been absorbed by a rhodopsin molecule. However, the role of these putative transmitters of phototransduction has not yet been defined. In fact, before attempting a comprehensive interpretation on how calcium and cyclic nucleotides interact in transforming absorption of photons into electrical signals, there are several questions which need to be formulated and answered. The turnover of the internal calcium, the mechanism of its extrusion, and the buffering capacity of the visual cell, for instance, are poorly understood. Similarly, amount, location, and kinetics of the light-induced hydrolysis of cGMP are controversial. In this article we shall attempt to discuss some quantitative remarks on the ionic nature of the photocurrent and on the properties of its control by light, including the kinetics of photoresponse and the control of the rod sensitivity during light adaptation.

PHOTOVOLTAGE AND PHOTOCURRENT

The electrical signal which conveys the visual information from photoreceptors to second-order neurons is a change in membrane potential. This photoresponse (photovoltage) is the result of at least three distinct ionic currents flowing in different regions of the visual cell: a) the photocurrent ($I_{h\nu}$) is an ionic current which is controlled by light. The current flows inwardly at the outer segment and outwardly at the inner segment. b) Passive ionic currents flow across specific membrane channels driven by their electrochemical gradient μ. These currents are located at the inner segment, cell body, and synaptic terminals. c) Ionic currents originate by active processes, such as electrogenic pumps or ion/ion exchanges.

In what follows we shall refer to rod photoreceptors from the retina of the toad (Bufo marinus or Bufo bufo) or of the tiger salamander (Ambystoma tigrinum). In these species rods have relatively large size and typically may have a total length of 80 µm and an average diameter of 8 µm, giving a volume of about 4×10^{-12}l. Given the presence of several intracellular structures (disks, nucleus, myoid, etc.), we estimate that the real free intracellular space v could be about 1.5×10^{-12}l. The free intracellular space v in tiger salamander rods is likely to be smaller, about 10^{-12}l.

Recently Baylor, Matthews, and Nunn (9) have shown that both the size and time course of the current recorded with a suction electrode are similar to the change in membrane current recorded under voltage-clamp conditions. The current recorded by these methods is usually called photocurrent and consists of an outwardly directed change in membrane current, which actually corresponds to a reduction in the inward current flowing through light-sensitive channels. Several experiments have shown that the peak level of current during the response to a saturating light corresponds to an absolute level of zero current, since under these conditions all light-sensitive channels are closed. The baseline level in darkness represents the steady inward photocurrent, commonly called "dark current" (34, 42).

Light-sensitive Current

The dark current flows in at the outer segment in the form of Na^+ ions and leaves the cell at the inner segment as K^+ ions. In rods of the toad this current amounts to about 20 to 30 pA (per rod) while larger values (40 to 80 pA) are usually measured in the rods of the salamander. The ionic nature of the photocurrent will be discussed in the section on NATURE OF THE PHOTOCURRENT, the relation between light intensity and suppression of photocurrent in the section on AMPLITUDE OF PHOTORESPONSE AND LIGHT INTENSITY, and its time course in the section on KINETICS OF PHOTORESPONSE.

The time course of current and voltage responses differ somewhat because of the presence of a rapid initial peak in the voltage response to bright flashes followed by decay to a smaller plateau. This initial oscillation is attributed to the activation of a voltage-dependent conductance which seems to be present in the cell's inner segment (29).

Passive Ionic Currents

There are at least three passive ionic currents which shape the voltage response. In toad rods the addition of 2 mmol/l Cs^+ to the extracellular medium completely blocks the sag of the voltage from the initial peak to the subsequent plateau (29). Such an effect is obtained with 1 mmol/l Cs^+, and with higher Cs^+ concentrations a second hyperpolarizing component appears which may reach unusually negative values of membrane potential (up to -135 mV) (50).

In rods of both toad and turtle, a high-pass filtering property has been described (25, 51). This high-pass filtering is thought to be caused by a voltage- and time-dependent K^+ conductance (1, 44). Moreover, under appropriate conditions regenerative responses have been observed in rods (30). These currents have been thoroughly studied by voltage-clamp methods (2) which showed the existence of three distinct voltage-activated currents blocked by Cs^+, TEA^+, and Co^{2+}, respectively.

Active Ionic Currents

The membrane resistance of a single rod in the toad or tiger salamander retina is about 1000 MΩ. Therefore, a current of about 1 pA gives a contribution of about 1 mV to the rod membrane resting potential V.

The Na^+-K^+ pump is, in general, electrogenic to some degree (32, 47). From Thomas (49) the contribution of the Na^+-K^+ pump to the membrane potential ΔVp in the steady state is given by

$$\Delta Vp = \frac{RT}{F}\, ln \frac{1}{r}\, \frac{r\left[K^+\right]_o + \frac{P_{Na}}{P_K}\left[Na^+\right]_o}{\left[K^+\right]_o + \frac{P_{Na}}{P_K}\left[Na^+\right]_o}, \tag{1}$$

where r is the number of Na^+ pumped out per K^+ pumped in and P_{Na} and P_K are total membrane permeability in darkness to Na^+ and K^+, respectively. Since in darkness $P_{Na} \sim P_K$ (14, 15) with the usual stoichiometry of $3Na^+/2K^+$, ΔVp is about -9 mV. One would indeed expect to observe a much larger contribution to membrane potential from electrogenic current during transient periods of increased membrane resistance. At the peak of the light response to a bright flash, when the membrane resistance has largely increased, the electrogenic current due to the Na^+-K^+ pump contributes about -15 mV to the photoresponse (50).

Another possible source of electrogenic current that may be present in the rod is the Na^+-Ca^{2+} exchange (28) if the stoichiometry of the exchange is different from $2Na^+/1Ca^{2+}$. However, neither the exact stoichiometry of the Na^+/Ca^{2+} exchange in rods nor the net size of this current is known.

INTRACELLULAR IONIC COMPOSITION

Rods are rather complex neurons with unusual electrical properties. Thus it is not immediately obvious whether their internal ionic composition is similar to that of other nerve cells. It is therefore important to have direct measurements or at least good estimates of the intracellular concentration of Na^+, K^+, and Ca^{2+}.

The Intracellular Level of K^+

At the peak of the response to a bright flash of light, the membrane potential is about - 80 mV, that is, close to the presumed value of the Nernst equilibrium potential for $K^+(E_K)$. This notion is supported by the observation that the peak of the light response to a bright flash behaves like the voltage of a K^+ electrode.

Figure 1 shows the relationship between the peak amplitude of the initial fast component and $[K^+]_o$ in the presence of (at least) 2 mmol/l of external Cs^+. The continuous line was computed from the equation (see (35, 36))

$$V = 58\,mV\,log_{10} \frac{\left[K^+\right]_o + \frac{P_{Na}}{P_K}\left[Na^+\right]_o}{\left[K^+\right]_i}, \qquad (2)$$

with the values of P_{Na}/P_K = 0.023 and $[K^+]_i$ = 100 mmol/l. The data of Fig. 1 show that the membrane potential at the peak of the light response to a bright flash behaves rather similarly to a K^+ electrode with an internal K^+ concentration of about 100 mmol/l.

The Intracellular Level of Na^+

In previous experiments the extracellular concentration of cations was about 135 mmol/l. Therefore, assuming osmotic equilibrium and electroneutrality, it turns out that the intracellular Na^+ concentration must be lower than 40 mmol/l. When the extracellular Na^+ concentration is about 110 mmol/l, E_{Na} has to be at least +30 mV. A better estimate of $[Na^+]_i$ gives values between 8 and 16 mmol/l (50) leading to a value for E_{Na} close to +50 mV.

The Intracellular Level of Ca^{2+}

At present there is no evidence as to whether the level of intracellular Ca^{2+} in dark-adapted rods is around 10^{-7} mol/l, as in neurons in the resting state. Since in darkness rods are thought continuously to release synaptic vesicles,

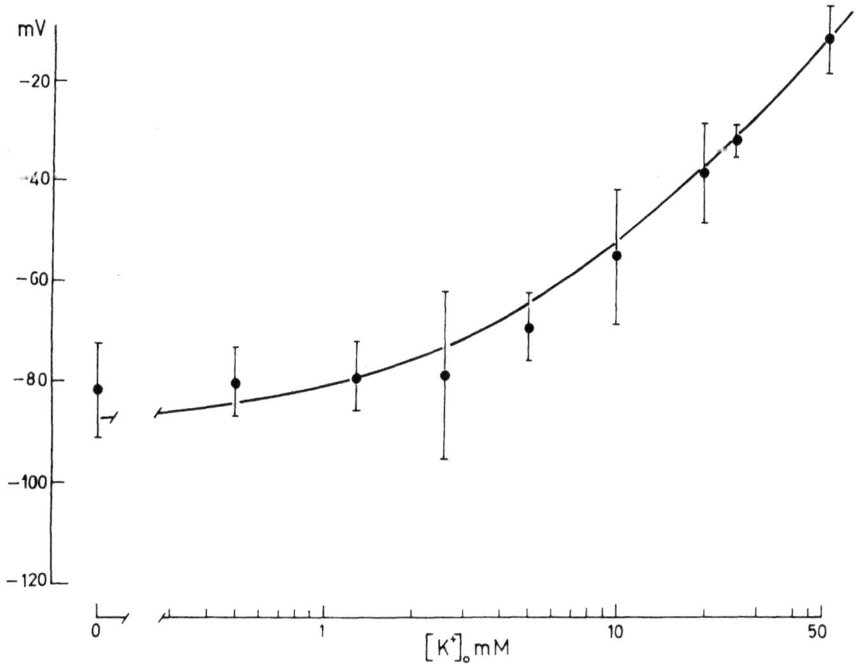

FIG. 1 - Dependence on $[K^+]_o$ of the absolute membrane potential of the initial fast component in the presence of external Cs^+. Data collected from twenty-eight cells. The stimulus was a 50 ms flash of monochromatic light (498 nm) equivalent to 1400 Rh* (photoisomerization). Continuous line drawn from Eq. 2.

it is conceivable that the intracellular Ca^{2+} is as high as 10^{-5} to 10^{-6} mol/l, as in active presynaptic terminals. It has been reported that extracellular Ca^{2+} increases during illumination in the proximity of outer segments (34, 35). However, it is not clear whether this transient Ca^{2+} accumulation in the extracellular space is caused by an increased Ca^{2+} efflux or by a decreased Ca^{2+} influx through the light-sensitive channel.

INTERNAL IONIC TURNOVER

Na^+ Turnover

The ionic current that flows inward at the rod outer segment in darkness amounts to about 30 pA (8) in the toad rods and under normal circumstances is carried mostly by Na^+. Considering a volume of the intracellular space v of 1.5×10^{-12} l, we have an influx of about 0.2 mmol/l Na^+ s. A bright light can suppress the current completely by closing ionic channels of the outer segment. It has been estimated that a single photoisomerization reduces the Na^+ current by about 1 pA (see (8, 15)).

The rate of change of internal sodium $[\dot{Na}^+]_i$ can be expressed as

$$\left[\dot{Na}^+\right]_i = \frac{I_{Na}}{vNq} - {}_pJ_{Na} \cdot \tag{3}$$

where I_{Na} is the passive Na^+ current, N is the Avogadro number, q the electron charge, and $_pJ_{Na}$ the active Na efflux driven by the Na^+-K^+ pump. In most tissues the Na pump is stimulated by $[Na^+]_i$ and by $[K^+]_o$ (32). Therefore, in first approximation when the extracellular K^+ is kept constant, we have

$${}_pJ_{Na} = \frac{{}_MV_{Na}\left[Na^+\right]_i}{K_{Na} + \left[Na^+\right]_i}, \tag{4}$$

that is, a usual enzymatic transport, where $_MV_{Na}$ is the maximal Na^+ extrusion and K_{Na} is the half-activation of the pump.

In toad rods, there is evidence (50) suggesting that

$$\frac{{}_MV_{Na}}{K_{Na}} \sim .03\ s^{-1} \tag{5}$$

and that K_{Na} is not lower than about 20 mmol/l.

As a result of these values we expect significant changes in Na^+ and K^+ internal concentration during prolonged exposure to bright light or when I_{Na} is altered by changes in $[Na^+]_o$ and/or $[Ca^{2+}]_o$.

Ca^{2+} Turnover

The turnover of internal Ca^{2+}, which may play a fundamental role in phototransduction, is at present poorly understood. It has been suggested that Ca^{2+} extrusion through the Na^+/Ca^{2+} exchange plays a major role (28, 33, 39) and may obscure the interpretation of experiments in which Na^+ is replaced with other monovalent cations (38, 54).

Recently (23, 55) the existence of a light-insensitive current has been shown which appears in the presence of extracellular Na^+ after loading the rod with Ca^{2+}. It has been proposed that this current is caused by an electrogenic Na^+/Ca^{2+} exchange. The stoichiometry of the exchange has been estimated to be $3Na^+/1Ca^{2+}$. In toad and salamander rods the maximal current thought to be driven by the Na^+/Ca^{2+} exchange is about 30 pA and decays with a time constant τ of about 0.5 s. Assuming that the Ca^{2+} extrusion $J_{C/N}$ via the Na^+/Ca^{2+} exchange occurs by the usual enzymatic transport

$$J_{C/N} = \frac{{}_MV_{C/N}\left[Ca^{2+}\right]_i}{K_{C/N} + \left[Ca^{2+}\right]_i} \tag{6}$$

given an exchange of $3Na^+$ for $1Ca^{2+}$, then ${}_MV_{C/N}$ can be estimated to be about 600 μmol/l s. If the decline of this current reflects the fall of intracellular Ca^{2+}, then we obtain

$$\frac{{}_MV_{C/N}}{K_{C/N}} \sim \frac{1}{\tau} \tag{7}$$

and, therefore,

$$K_{C/N} \sim \tau \, {}_MV_{C/N} \, .$$

A value for the half-activation of the Na^+/Ca^{2+} exchange by intracellular Ca^{2+} of around 300 μmol/l is higher than that measured in other tissues (26).

It has been proposed (54) that a Ca^{2+} current of about 2 pA flows in darkness through the light-sensitive channel. This Ca^{2+} entry is thought to be balanced by an equivalent Ca^{2+} extrusion via the Na^+/Ca^{2+} exchange. A Ca^{2+} current of about 2 pA is equivalent to a flux of about 20 μmol/l Ca^{2+} s. In order to extrude 20 μmol/l Ca^{2+} per second, the Na^+/Ca^{2+} exchange must be 1/30-activated. Given the estimate of $K_{C/N}$ (~300 μmol/l), the level of intracellular Ca^{2+} able to stimulate an extrusion of 20 μmol/l Ca^{2+} per second must be about 10 μmol/l.

In essence, the present knowledge of properties of the Na^+/Ca^{2+} exchange in rods suggests that a) the Na^+/Ca^{2+} exchange is a potent mechanism for Ca^{2+} extrusion, but its activation from intracellular Ca^{2+} is high (~10 to 30 μmol/l). b) The Ca^{2+} extrusion mediated by the Na^+/Ca^{2+} exchange is relevant in darkness only if the intracellular Ca^{2+} is in the micromolar range.

Very little is known about the existence of a Ca^{2+}-ATPase in rods, and its presence seems localized in the cell body and synaptic terminals. Similarly, the intracellular buffering mechanisms are poorly understood and only guesses can be made as to their functional role.

NATURE OF THE PHOTOCURRENT

Early experiments in which extracellular Na^+ was replaced by alkali ions, by choline, or by sucrose showed that the photocurrent rapidly disappeared (12, 15, 19). These results suggested that the photocurrent was essentially carried by Na^+ ions and that the light-sensitive channel was selective for Na^+.

However, when Na^+, Ca^{2+}, and Mg^{2+} are removed from the extracellular medium a photocurrent with inverted polarity can be observed (53). Figure 2 shows that, when using a buffer $3EDTA/1Ca^{2+}/1Mg^{2+}$ in the absence of extracellular Na^+, light flashes evoke responses with inverted polarity, and that the membrane potential depends on $[K^+]_o$ more in darkness than in light. These results and data with radioactive K^+ (and Rb^+ as well) show that in these conditions K^+ easily permeates through the light-sensitive channel (13).

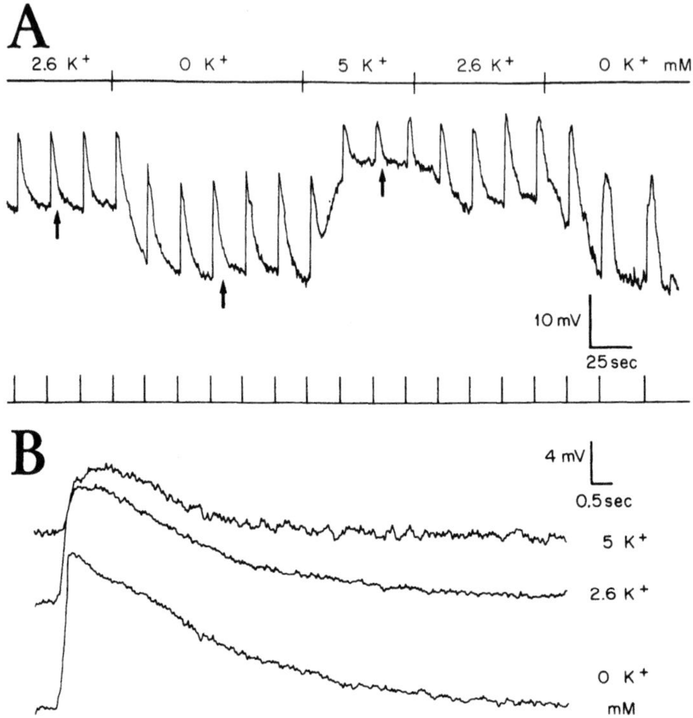

FIG. 2 - Effect of extracellular K^+ on the depolarizing photoresponse observed in an Na^+-, Ca^{2+}, Mg^{2+}-free 3 mmol/l-EDTA solution. A: chart recording of the experiment. Light stimulus as in Fig. 1. Dark membrane potential in the Na^+-free, 2.6 K^+ solution was -7.5 mV. B: responses indicated by arrows reproduced in greater detail.

When the external Ca^{2+} is reduced below 10^{-6} mol/l, photoresponses with normal polarity can be observed (4, 53) if millimolar amounts of $Mg^{2+}$are present. Figure 3 shows that when 0.5 mmol/l IBMX was added to the bathing medium, a photoresponse of about 8 mV could be observed when Na^+ was replaced by sucrose, even in the presence of millimolar amounts of Ca^{2+} in the extracellular medium.

Summarizing these results show that a) in physiological conditions, stable photocurrents are only observed in the presence of extracellular Na^+ or Li^+; and (b) when small amounts of PDE inhibitors are added to the bathing medium, a large variety of monovalent and divalent cations can carry photocurrent for several minutes.

The selectivity of the light-sensitive channel is best studied under those conditions in which stable currents carried by different ions can be observed

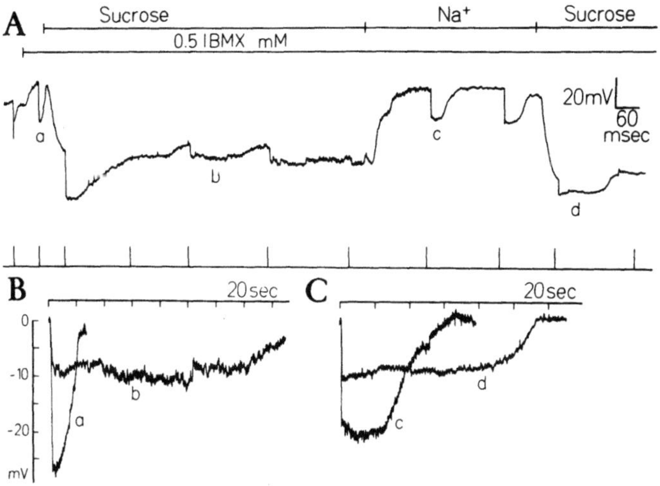

FIG. 3 - Effect of 0.5 mmol/l IBMX on the rod electrical activity when NaCl was replaced by sucrose. pH was buffered to 7.6 with choline bicarbonate and 5% CO_2. Dark membrane potential in Ringer's solution was -4.2 mV. Light stimulus as in Fig. 1.

during several minutes. In a nominal Na^+-, Ca^{2+}-, Mg^{2+}-free extracellular medium and in the presence of 0.5 mmol/l IBMX, voltage responses of about 30 to 40 mV are observed if 1 mmol/l Mn^{2+} is added to the bathing medium (13, 52).

Figure 4 shows the stoichiometry of this effect on voltage responses of a toad rod when Ba^{2+} is added to an Na^+, Ca^{2+}, Mg^{2+} nominally free medium, but in the presence of 0.5 mmol/l IBMX. These results show that in the described conditions a) a voltage response of about 7 mV can be observed when 10 mmol/l Na^+, 0.3 mmol/l Mn^{2+}, or 0.4 mmol/l Ba^{2+} is present in the extracellular medium. Therefore, the permeability ratio P_{Mn}/P_{Na} is about 16 and P_{Ba}/P_{Na} is about 12. b) The amplitude of the photocurrent carried by divalent ions (Mn^{2+} or Ba^{2+}) saturates at 1:3 mmol/l.

These are typical features of ionic channels selective for divalent cations rather than for monovalent cations. It is not clear, however, if this state of the channel is the only possible mode in which ionic permeation occurs.

When rods are bathed in a physiological extracellular medium it is very difficult to observe photocurrents carried by divalent cations lasting more than a few seconds. It has been proposed that removal of extracellular Na^+ with the subsequent blockage of the Na^+/Ca^{2+} exchange leads to a large increase of intracellular Ca^{2+}, which in turn causes the closure of the light-sensitive channel (28, 38, 54). However, this view is difficult to reconcile with properties of the Na^+/Ca^{2+} exchange discussed in the section on Ca^{2+} Turnover, mainly because a large increase of intracellular Ca^{2+} cannot be expected within 1-2 s following blockage of the Na^+/Ca^{2+} exchange.

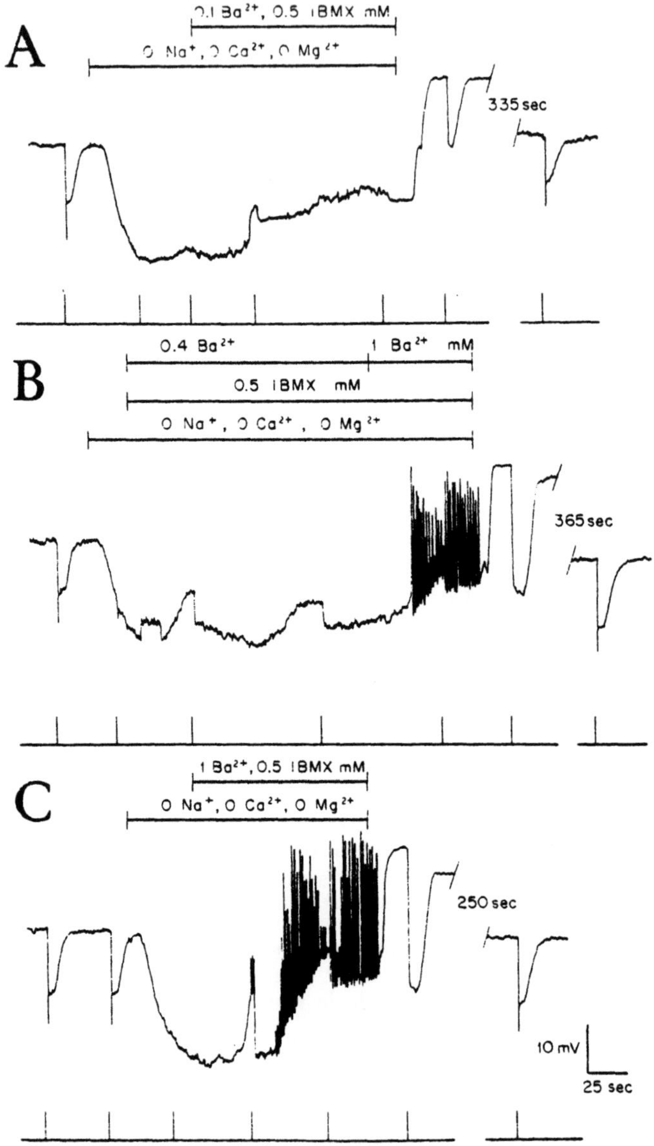

FIG. 4 - Effect of Ba^{2+} on the electrical activity of a rod in an Na^+-free solution in the presence of 0.5 mmol/l IBMX. Light stimulus as in Fig. 1. Dark level of the membrane potential in control Ringer was -53 mV. Observe spikes in the presence of 1 mmol/l Ba^{2+} (30).

We favor the idea of the existence of different conductive states of the light-sensitive channel and a direct role of external Na^+ in keeping the channel open. (For alternative explanations see Owen, this volume.)

Reversal Potential of the Photocurrent

In rods of tiger salamander the reversal potential of the photocurrent is between 0 and 10 mV ((3) and Baylor and Nunn, in preparation). Since E_{Na} is about +40 mV (see section on The Intracellular Level of Na^+), Na^+ ions that flow down the electrochemical gradient μ_{Na} do not account completely for the photocurrent. This result may be explained in different ways.

1 - Na^+ movements mediated by a neutral carrier:

We may suppose that extracellular Na^+ binds to a carrier A at the external surface and that intracellular Na^+ binds to a carrier B at the internal surface, according to the scheme

$$Na_o^+ + A \leftrightarrows Na_i^+ + B . \tag{8}$$

The reversal potential V_{Rev} for this reaction is

$$V_{rev} = E_{Na} + \frac{\mu_A^o - \mu_B^o}{F} + \frac{RT}{F} \ln \frac{[A]}{[B]} , \tag{9}$$

where

μ_A^o , μ_B^o

are the standard free energy of formation of A and B, and [A] and [B] are, respectively, the steady-state concentrations of A and B. If the carrier translocates passively between the two states A and B, Eq. 9 reduces to

$$V_{Rev} = E_{Na} . \tag{10}$$

2- Coupled Na^+ and K^+ movements:

If the permeation through the light-sensitive channel obeys the scheme

$$nNa_o^+ + mK_o^+ \leftrightarrows nNa_i^+ + mK_i^+ , \tag{11}$$

it is easy to see that

$$V_{Rev} = \frac{nE_{Na} + mE_K}{n + m} . \tag{12}$$

With E_{Na} of +50 mV and E_K of -90 mV, if we have n = 2 and m = 1, V_{Rev} is about +3 mV, in agreement with the experimental result.

3 - The light-sensitive channel as an Na^+/Ca^{2+} exchange:

If the permeation through the light-sensitive channel is a sort of Na^+/Ca^{2+} exchange, we have the scheme

$$nNa_o^+ + qCa_i^{2+} \leftrightarrows nNa_i^+ + qCa_o^{2+} . \tag{13}$$

In this case,

$$V_{Rev} = \frac{nE_{Na} - 2qE_{Ca}}{n - 2q}, \quad (14)$$

and with an exchange of $4Na^+/1\ Ca^{2+}$, we have a reversal potential close to the measured one.

The First Conductive State

At the usual concentrations of Ca^{2+} and Mg^{2+} (1 or 2 mmol/l), the light-sensitive current $I_{h\nu}$ is proportional to $[Na^+]^2_o$ (39) and depends on membrane potential approximately as exp (-2 FV/RT) (Baylor and Nunn, in preparation). These results can be explained assuming that monovalent cations permeate the light-sensitive channels in pairs. If the modes of permeation in what we call the first conductive state are

$$2\,Na^+_o \leftrightarrows 2\,Na^+_i, \quad (15)$$

and

$$Na^+_o + K^+_o \leftrightarrows Na^+_i + K^+_i, \quad (16)$$

then it is possible to explain the previous observation on the dependence of $I_{h\nu}$ on $[Na^+]^2_o$ and V, as well as the V_{Rev} close to +5 mV.

However, it has been suggested that the dependence of $I_{h\nu}$ on $[Na^+]^2_o$ is due to the effect of external Na+ on the level of intracellular Ca2+ (39). This effect would be mediated by an Na^+/Ca^{2+} exchange. It was also suggested that external changes of Na^+, K^+, and Ca^{2+} have predominant effects on the Na^+/Ca^{2+} exchange and therefore on the level of internal Ca^{2+}.

In the first conductive state the light-sensitive channel is clearly permeable to Na^+ and Li^+. A decrease of K^+ efflux during illumination has been measured with radioactive tracers (13, 18). This observation also suggests that in the first conductive state K^+ permeates through the light-sensitive channel; moreover, there is now evidence to suggest that K^+ and Rb^+ can also permeate through the light-sensitive channel in this state (38, 54).

The Second Conductive State

When extracellular Ca^{2+} is reduced below 10^{-6} mol/l or when a small amount of PDE inhibitors are added to the bathing medium (with physiological levels of Ca^{2+} and Mg^{2+}), photoresponses due to currents carried by divalent ions can be observed for a few minutes. In the presence of 50 µmol/l IBMX, a photocurrent of about 5 pA presumably carried by Ca^{2+} can be observed for a few minutes. This observation strongly suggests a genuine change of selectivity of the light-sensitive channel (13). In the presence of PDE inhibitors, the light-sensitive channel appears to be permeable to a variety of divalent ions with a selectivity sequence of

$$Mn^{2+} > Ba^{2+} > (Ca^{2+}, Co^{2+}, Mg^{2+}, Sr^{2+}). \quad (17)$$

This state of the channel will be called the second conductive state.

Hodgkin et al. (38) have shown that when Ca^{2+} was reduced to 10^{-6} mol/l the permeability ratio of Na^+ and Ca^{2+} P_{Ca}/P_{Na} through the light-sensitive channel was about 10:1.

As mentioned above (see the sections on Ca^{2+} Turnover and NATURE OF THE PHOTOCURRENT), there is a fair amount of evidence to suggest that calcium permeates through the light-sensitive channel, and if one assumes that calcium contributes a photocurrent of about 2 pA (38), it follows that about 15 μmol/l Ca^{2+} enter a rod outer segment every second. This raises questions as to how an ion which passively diffuses through a channel may become capable of blocking it by a simple electrostatic action. Moreover, under the conditions in which the light-sensitive channel is in the second conductive state and the cell produced photoresponses in the absence of external sodium, calcium cannot be extruded by the Na^+/Ca^{2+} exchange mechanism. Under these conditions the internal level of calcium must increase by 10-100 times without blocking the channel, unless mechanisms of extrusion other than the Na^+/Ca^{2+} exchange are already present or become activated. In the second conductive state, a light-modulated efflux of K^+ can be detected only in the presence of external Na^+, no matter how low the Ca^{2+} level is set. This may suggest that Na^+ and K^+ movements are coupled and that the Na^+/Ca^{2+} exchange mechanism plays no crucial role in this process.

The Third Conductive State

When the levels of external Ca^{2+} and Mg^{2+} are buffered at or below 10^{-6} mol/l, the light-sensitive channel appears to be in a third conductive state. Such a condition is characterized by the following properties:

1 - The efflux of K^+ and of Rb^+ through the light-sensitive channel is independent of the level of extracellular Na^+. This observation may suggest that in the third conductive state K^+ movements are no longer coupled to Na^+ (as they appear to be in the first and second conductive state).

2 - The efflux of K^+ and of Rb^+ is blocked by extracellular K^+ and, in this state, K^+ movements show the classical behavior of a single-file channel (37, 40).

The Structure of the Light-sensitive Channel

The basic properties of the light-sensitive channel are unknown. It is not clear, for instance, whether it is a pore or a carrier. Patch-clamp measurements of the photocurrent (24) suggest the single-channel conductance to be on the order of 50 fS, that is, much smaller than the conductance of usual membrane ionic pores. It is perhaps useful to note that:

1 - The light-sensitive channel appears to be permeable to Na^+, Li^+, K^+, and Rb^+ and is poorly permeable to Cs^+. It seems also permeable to a variety of divalent ions such as Mn^{2+}, Ba^{2+}, Ca^{2+}, Mg^{2+}, Co^{2+}, and Sr^{2+}. These observations suggest that if the light-sensitive channel is a

pore, its dimensions are such that an ion with a crystal radius substantially larger than a 1.67 Å cannot permeate through it.

2 - Na^+ movements through the light-sensitive channel in its first conductive state can be blocked by Ca^{2+} and by other divalent ions, which are perhaps acting on both sides of the membrane. Such a blocking action, however, may be different when the light-sensitive channel is in the second conductive state.

3 - The dual action of extracellular Ca^{2+} in carrying current and in reducing the current carried by Na^+ can be explained simply by a competition between Ca^{2+} and Na^+, specifically: a) the channel has a site at the external mouth with a high affinity for Ca^{2+} (A_{Ca} ~.1 mmol/l) and a low affinity for Na^+ (A_{Na} >100 mmol/l), so that in normal conditions 90% of the channels are occupied by Ca^{2+}. b) The mobility of Na^+ ions through the channel is about 1000 times higher than that of Ca^{2+} ions, so that it is possible to observe a fraction of photocurrent carried by Ca^{2+} while extracellular Ca^{2+} powerfully controls the current carried by Na.

4 - Ionic movements through the light-sensitive channels do not obey the independence principle, as shown by the fact that in the third conductive state K^+ movements conform to the single-file behavior.

5 - The transition from the first to the second conductive state, that is, the shift of selectivity towards divalent cations, seems to suggest a conformational change of the channel. If the channel is represented by a sequence of energy wells and barriers (35), then in the first conductive state Na^+ and Li^+ are likely to be selected because of their low-energy barrier (selectivity filter), while in the second conductive state divalent cations could be selected by deep-energy wells (strength of binding sites).

6 - The transition from the second to the third conductive state does not necessarily imply a conformational change of the channel. In the third conductive state internal K^+ can leak out through the light-sensitive channel because the binding sites are no longer occupied by divalent cations.

AMPLITUDE OF PHOTORESPONSE AND LIGHT INTENSITY

The size of the photoresponse (photovoltage or photocurrent) increases upon increasing the flux of absorbable light. In many cases it has been observed (see (3, 5)) that the relation between the measured response (ΔR) and light flux (I) is well fitted by the equation

$$\frac{\Delta R}{\Delta R_{max}} = \frac{I}{I + I_o}, \tag{18}$$

where ΔR_{max} is the maximal response to a bright flash of light and I_o a constant called the half-saturating intensity. This equation indicates that for low intensities of the light stimulus the amplitude of the photoresponse is directly proportional to the flash intensity. Equation 18 is a rectangular hyperbole and is commonly referred to as the "Michaelis-Menten" relation. This equation suggests (7, 22) that

$$Z + C_o \overset{\bar{K}}{\rightleftharpoons} C_c \tag{19}$$

where C_o is an open channel, C_c a closed channel, Z the internal transmitter capable of blocking ion channels, and $\bar{K}$ the equilibrium constant of the reaction. If Z is proportional to the light intensity I (assuming that unbound Z greatly exceeds bound Z and that Eq. 19 reaches equilibrium almost instantaneously), then the fraction of blocked channels C_c/C is

$$\frac{C_c}{C} = \frac{I}{I + I_o} \tag{20}$$

Equation 18 fails to describe experimental data in the presence of relatively high concentrations of IBMX (above 30 µmol/l) (17) or when external Ca^{2+} is kept below 100 µmol/l (44). Under these conditions doubling the intensity of a dim flash causes the response to increase more than twofold (supralinearity). The relation between amplitude of voltage photoresponse (ΔV) and light intensity (I) can be expressed as

$$\Delta V = C_1 I + C_2 I^2 + \ldots + C_n I^n \tag{21}$$

With 50 µmol/l IBMX, the value of n is between 3 and 4 (16). Supralinearity may simply be explained as being due to a large increase of the total light-sensitive conductance if the relation between the fraction of blocked channels and light intensity is steeper than the relation described by the Michaelis-Menten equation (see (43, 44)), namely,

$$\frac{\bar{G}e - Ge}{\bar{G}e} = 1 - e^{-I}, \tag{22}$$

where $\bar{G}_e$ is the total light-sensitive conductance and G_e is the light-sensitive conductance during illumination. Voltage responses scaled in amplitude depend on I as

$$\frac{\Delta V}{\Delta V_{max}} = \frac{1 - e^{-I}}{1 + K_o e^{-I}}, \tag{23}$$

where $K_o = \bar{G}_e/G$ is the ratio between the total light-sensitive conductance $\bar{G}_e$ and the shunting conductance G. Supralinearity occurs when the value of K_o exceeds 3. In the presence of 50 µmol/l IBMX, amplitude/intensity relations are fitted by Eq. 23 with values for K_o of over 50, an unusually high value which is not supported by input impedance measurements (51). This may suggest that in the presence of IBMX supralinearity is not completely accounted for by electrical events, but interactions between the effects of absorbed photons should be postulated. If we assume, for instance, that light produces a series of Poisson distributed events with an average λ proportional to I, so that the occurrence of at least one event leads to the closure of a light-sensitive channel, we obtain Eq. 22, but if the occurrence of at least n events is required in order to close a channel, we obtain

$$\frac{C_c}{C} = 1 - e^{-\gamma I}\left(1 + \gamma I + \frac{\gamma^2 I^2}{2} + \ldots + \frac{\gamma^2 I^2}{n!}\right). \quad \lambda = \gamma I \tag{24}$$

Equation 24 provides an alternative explanation of supralinearity and does not necessarily require the assumption of a high value for the light-sensitive conductance; for this reason one may call it a chemical model for supralinearity while Eq. 23 gives an electrical description of the phenomenon.

KINETICS OF PHOTORESPONSE

The analysis of the time course of the response to light may provide useful information on the kinetics of the phototransductive processes. In this respect it may be helpful to keep the analysis of dim flash responses distinct from that of bright flash responses.

Responses to Dim Flashes

In a variety of visual cells of both vertebrate and invertebrate animals, the kinetics of photoresponse are well described by assuming that following photon absorption a substance Z, able to block ionic channels, is produced through a series of delay stages (7, 8, 11, 20, 31, 46), that is,

$$h\nu \rightarrow Y_1 \xrightarrow{\alpha_1} Y_2 \xrightarrow{\alpha_2} \ldots \rightarrow Y_n \xrightarrow{\alpha_n}. \tag{25}$$

Recently Payne and Howard (45) have proposed a different model which performs remarkably well in invertebrate photoreceptors. Their scheme is based upon the assumption that after light absorption the concentration of the internal transmitter rises logarithmically with time, and that the thresholds of target channels are normally distributed. This is formally expressed by a log-normal function with a mean given by $\mu \exp(-\sigma^2)$ and a standard deviation σ. The parameter μ describes the rate of release of the internal transmitter and, since it is affected by temperature and light adaptation, it is supposed to represent the rate constant of a single transduction process. Such a model, however, is inadequate in describing photoresponses in vertebrate photoreceptors. The main difficulty is that the rising phase of the photoresponse in vertebrate photoreceptors is independent of the light intensity (as shown by log-log plots of the time course measured before allowing compression for saturation, see Fig. 11 in (8)), a feature which is not accounted for by the above-mentioned model. On the other hand, a sequence of transformation of the type indicated in Eq. 25 and characterized by delayed stages with values of about 2 s provides an excellent description of rod responses to flashes of dim light (8). This consideration, however, does not exclude the existence of other different explanations. A further step in understanding the process of phototransduction would be the identification of the molecular nature of the transformations involved in the cascade model.

Effect of PDE Inhibitors and Background Light

Let us first consider some experimental observations:

1 - When a small dose of IBMX (or any other PDE inhibitors) is applied to rods, the time course of the photoresponse is drastically prolonged (with 30 µmol/l IBMX, the time course of voltage responses to dim flashes increases from 0.7 s to 6 s (17)). The changes in response kinetics may be simply described by a linear expansion of the temporal scale of response. Adopting Eq. 25, this is equivalent to saying that IBMX acts by decreasing the value of all the rate constants by the same factor. Accordingly, 5 µmol/l IBMX would halve the values of α_i.

2 - As is well known, background illumination decreases the rod sensitivity to a test flash and accelerates the time course of the photoresponse (6, 8, 21, 27, 31). The time-to-peak (t_{Pk}) of dim flash responses decreases from about 0.8 s, the value measured in dark-adapted conditions, to 0.33 s when the test flash is superimposed on a steady illumination equivalent to about 100 photoisomerizations/s (Rh*/s) (see, however, (10)).

3 - IBMX antagonizes the effects of background illumination on the time course of dim flash responses (17). Considering the measured values of t_{Pk}, the experimental points can be fitted by the equation

$$t_{Pk} = t_{min} \frac{B + A_1 \left[1 + \frac{[IBMX]}{A_2} \right]}{B} \qquad B = 1 + A_3 I_{Bkg} , \tag{26}$$

where t_{min} = 0.33 s, A_1 = 0.2 s/Rh*, A_2 = 3 µmol/l, A_3 = 1.1, and I_{Bkg} is the intensity of the steady background illumination expressed in Rh*/s. A remarkable feature of Eq. 26 is that with a sufficiently bright background illumination, the time course of photoresponses does not depend on IBMX concentration. This suggests, perhaps, a competitive interaction between light and IBMX. In fact, Eq. 26 describes a process in which

a) the rate of reactions α_i is proportional to the velocity V of an enzymatic reaction with maximal velocity V_{max} and half-activation contant K_M and whose substrate is S, that is,

$$E + S \underset{}{\overset{K_m}{\rightleftharpoons}} ES \overset{K_3}{\rightleftharpoons} E + P \; ; \tag{27}$$

b) the enzyme E is competitively blocked by IBMX with a dissociation constant K_I, that is,

$$E + IBMX \underset{}{\overset{K_I}{\rightleftharpoons}} EIBMX \; ; \tag{28}$$

c) light produces a substance R*, which activates the enzyme by increasing V_{max} as follows:

$$h\nu \rightarrow R^* \tag{29}$$

$$E + R^* \underset{}{\overset{K_R}{\rightleftharpoons}} ER \; ; \tag{30}$$

$$ER^* + S \underset{}{\overset{K_M}{\rightleftharpoons}} ER^* \, S \underset{}{\overset{\widetilde{K}_3}{\rightleftharpoons}} ER^* + P. \tag{31}$$

If ${}_DV_{max} = E_T K_3$ and ${}_LV_{max} = E_T \widetilde{K}_3$ are the maximal velocities of the enzymatic reaction in darkness and during light, respectively, and E_T represents the total concentration of E, at the steady state we have from Eqs. 27-31

$$V = \frac{S\,\widetilde{V}_{max}}{S + \widetilde{K}_M}\,, \quad \widetilde{V}_{max} = \frac{{}_DV_{max} + R^*\,{}_LV_{max}}{1 + R^*}\,, \quad \widetilde{K}_M = K_M \left[1 + \frac{[IBMX]}{[1 + R^*]K_I}\right] \tag{32}$$

Now if we assume α_i to be proportional to V, we obtain a relation to the type described by Eq. 26.

It is tempting to suggest that the enzyme is the light-activated PDE of vertebrate rods (48). In this framework, the mechanism underlying changes in the kinetics during light adaptation could be the activation of PDE. It is also worth noting that the effects obtained by decreasing the temperature (41) could be explained by slowing down all the rate constants of Eq. 25 by the same amount.

Effect of Divalent Cations

Another clue to the understanding of the molecular nature of the transformations involved in phototransduction can be obtained by the analysis of the effect of a variety of divalent and trivalent cations on the photoresponse to dim flashes. In Fig. 5 we see the effect of 100 µmol/l Co^{2+} on the photoresponse to a dim flash of light. Here Co^{2+} prolongs the time course of the photoresponse but leaves the rising phase unaltered. This effect cannot be explained by a simple expansion of time scale as with PDE inhibitors. The crossed line in Fig. 5 has been obtained after appropriate scaling using Eq. 26 with $\alpha_1 = \alpha_2 = \alpha_3 = 2.5\ s^{-1}$ and $\alpha_4 = 0.8\ s^{-1}$. The change of time course of photoresponse caused by Co^{2+} can be explained by simply reducing the value of α_4 from $0.8\ s^{-1}$ to $0.4\ s^{-1}$, leaving unaltered the value of α_1, α_2, α_3 (see dotted line).

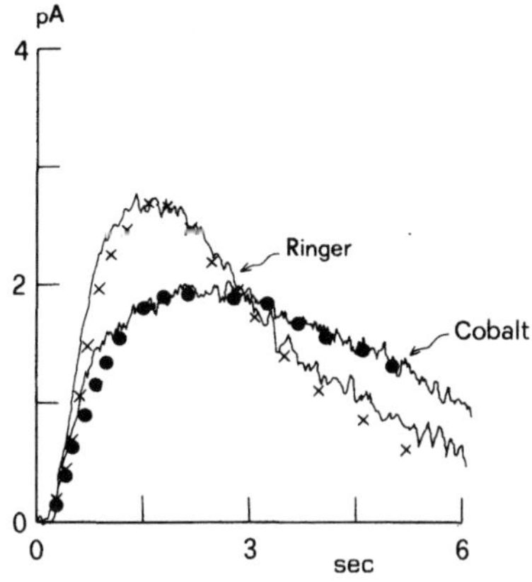

FIG. 5 - The Co^{2+} effect on the dim light response continuous traces: current responses from a toad rod with its inner segment drawn into a recording pipette and the outer segment projecting into a flowing solution containing normal Ringer's with 100 µmol/l Co^{2+}. Symbols represent the computed responses for both control (crosses) and test conditions (dots) obtained using Eq. 25 with the following parameter values:

$\alpha_1 = \alpha_2 = \alpha_3$ 2.5 s^{-1} α_4 = 0.8 s^{-1} (crosses);
$\alpha_1 = \alpha_2 = \alpha_3$ 2.5 s^{-1} α_4 = 0.4 s^{-1} (dots).

A similar effect on the time course of the photoresponse is observed under voltage-clamp conditions, when the membrane is depolarized (Baylor and Nunn, in preparation). This result suggests that at least (and possibly only) one of the slow stages involved in phototransduction is affected by divalent ions such as Co^{2+} and Mn^{2+}, by trivalent ions such as La^{3+}, and by membrane potential.

Responses to Bright Flashes

In this section we shall consider responses to bright flashes which produce a saturating hyperpolarization or a transient suppression of the photocurrent during a period of time of variable length, depending on the intensity of the stimulus. Such a component of the photoresponse is commonly referred to as plateau. Accordingly, the analysis of the kinetics of response to bright flashes focusses mainly on the duration of the plateau and on the decay of the voltage or current to the original dark level.

In previous studies (7, 20) it was noticed that the rate of decay of voltage responses to bright flashes decreases through increasing the intensity of light stimulation. In contrast, the analysis of current response reveals that, after the initial suppression, the rate by which current returns to the dark level bears little dependence on the intensity of the stimulus, and the records run almost parallel (see Fig. 3 in (39)). This feature may indicate that there is no need to suppose that removal of blocking particles from ionic channels occurs in several steps as originally supposed (see (7 , 20)).

In rods of lower vertebrates maintained in normal Ringer, the time course of responses to a wide range of light intensities is satisfactorily described by the four-delays scheme whose rate constants are accelerated after about 1 s from the onset of a background illumination. Such an action cannot be reproduced by a change of time scale as in the case of PDE inhibitors, but rather by slowing down only the rate constant of the last step which in Eq. 25 accounts for removal of the blocking agent.

It is perhaps worth noting that the same change in the value of the rate of removal successfully fits both dim and bright flash responses.

Hodgkin et al. (39) have shown that the plateau is prolonged when extracellular $[Na^+]_o$ is reduced below 50 mmol/l or when extracellular Na^+ is replaced by Li^+. Figure 3 shows that in the presence of 0.5 mmol/l IBMX in the extracellular medium the total duration of the light response doubles when NaCl is substituted for sucrose. The duration of the plateau and the time course of the recovery of the falling phase may be associated with the rate of extrusion of Ca^{2+} by the Na^+/Ca^{2+} exchange (39).

However, the effect of PDE inhibitors on the kinetics of response to dim and bright flashes cannot be explained in a simple way. Concentrations of IBMX sufficient to prolong the time course of photoresponse to dim flashes by 5-10 times do not prolong the plateau and falling phases by the same amount (see (13, 17)).

Light Desensitization and Recovery

The effects of background illumination on the kinetics of photoresponse to a test flash are completed within 1.5 s after the onset of the background light (21). With time, however, sensitivity may be partially restored, for instance, during a background light equivalent to about 1000 Rh*/s the flash sensitivity was initially reduced to 3 μV/Rh*, and after 25.6 s it rose to 15 μV/Rh*.

Size and time course of recovery from light desensitization can easily be visualized using repetitive flashes of different light intensity and time intervals. Figure 6 shows such an experiment in which the response to a flash of light equivalent to 1400 Rh* per rod delivered each 1.5 s was recorded.

It can be seen that the second and third flashes fail to produce any response. However, if the stimulation with the same flash delivered at the same rate is sufficiently prolonged, the response is gradually restored up to a certain level. If the rod is exposed to an Na^+-free Ringer before the exposure to the flash series, the size of the recovery is greatly enhanced. In other experiments, addition of micromolar amounts of La^{3+} greatly reduced or completely suppressed recovery from saturation. These phenomena may be explained supposing that a) during light adaptation the influx of Na^+ into the cell is reduced. Since the turnover of $[Na^+]_i$ occurs at a rate of about 0.2

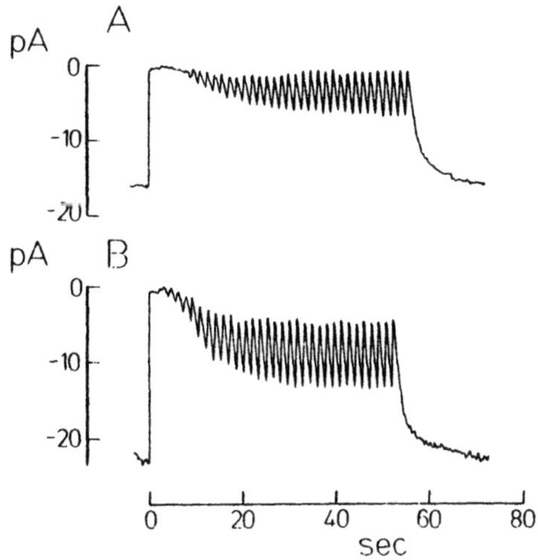

FIG. 6 - Effect of exposure to an Na+-free Ringer on the recovery from saturation during a repetitive flash stimulation. Light intensity as in Fig. 1. The time interval between successive flashes was 1.5 s. Upper trace A is the control response and the lower trace B is the response after 3 min of exposure to an Na+-free Ringer.

mmol/l/s (see the section on INTERNAL IONIC TURNOVER), we may expect an appreciable decrease if $[Na^+]_i$ within 10-20 seconds; and b) as a consequence of the decrease of $[Na^+]_i$, the Na^+ gradient increases and the Ca^{2+} extrusion mediated by the Na^+/Ca^{2+} exchange is stimulated.

Therefore, if we assume that one of the slow stages of Eq. 25 is controlled by Ca^{2+} extrusion, we have an obvious interpretation for the results shown in Fig. 6. In this framework, the Na^+/Ca^{2+} exchange would play an essential role in the recovery from desensitization and saturation during light adaptation.

CONCLUSION

We will assume that the basic physicochemical processes involved in phototransduction are a) a light-induced increase in the intracellular level of an internal transmitter (possibly a negative one) X; and b) a light-induced activation of the cGMP phosphodiesterase (PDE).

In this view a rise of X triggers a chain of chemical reactions of the type

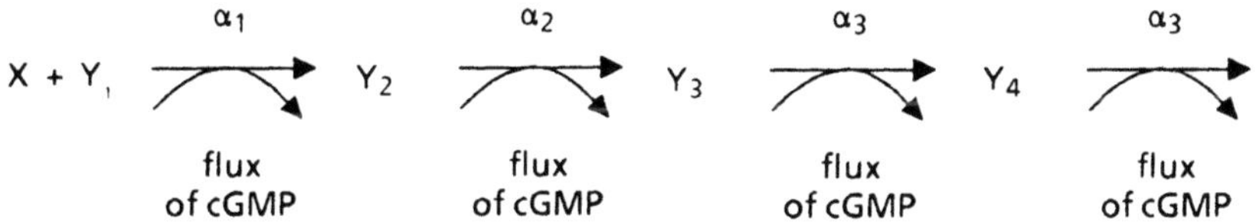

(33)

where Y_4 is the event leading to the closure of a light-sensitive channel and the rates of reaction are coupled to the flux of cGMP, that is, to the rate of hydrolysis of cGMP (see Miller et al., this volume). The mechanism of channel closure is represented by Eq. 24, where the value of n depends on the conductive state of the channel.

In the first conductive state n is probably equal to 1, but in the second conductive state n could be equal to 3 or even higher, so that the blocking effect of internal transmitter depends on the state of the channel. It is implicit in this model that for dim light the time course of the increase of X is fast, so that for responses to dim flashes the kinetics are essentially controlled by Eq. 33, which explains all the known experimental results on the kinetics of photoresponse.

We assume also that one reaction of Eq. 33 can be controlled by membrane voltage or by extracellular divalent or trivalent cations and may be associated with Ca^{2+} extrusion from the cell. Thus, we have, schematically,

1 - the increase of X and the activation of PDE initiates phototransduction;

2 - the activation of PDE controls the kinetics of photoresponse during light adaptation; and

3 - the Na^{+}/Ca^{2+} exchange controls the recovery from desensitization during light adaptation.

REFERENCES

(1) Attwell, J., and Wilson, M. 1980. Behaviour of the rod network in the tiger salamander retina mediated by membrane properties of individual rods. J. Physiol. **309:** 287-316.

(2) Bader, C.R.; Bertrand, D.; and Schwartz, E.A. 1982. Voltage-activated and calcium-activated currents studied in solitary rod inner segements from the salamander retina. J. Physiol. **331:** 253-284.

(3) Bader, C.R.; MacLeish, P.R.; and Schwartz, E.A. 1979. A voltage-clamp study of the light response in solitary rods of the tiger salamander. J. Physiol. **296:** 1-26.

(4) Bastian, B.L., and Fain, G.L. 1982. The effect of sodium replacement on the responses of the toad rods. J. Physiol. **330:** 575-591.

(5) Baylor, D.A., and Hodgkin, A.L. 1973. Detection and resolution of visual stimuli by turtle photoreceptors. J. Physiol. **234:** 163-198.

(6) Baylor, D.A., and Hodgkin, A.L. 1974. Changes in time scale and sensitivity in turtle photoreceptors. J. Physiol. **242:** 729-758.

(7) Baylor, D.A.; Hodgkin, A.L.; and Lamb, T.D. 1974. Reconstruction of the electrical responses of turtle cones to flashes and steps of light. J. Physiol. **242:** 759-791.

(8) Baylor, D.A.; Lamb, T.D.; and Yau, K.-W. 1979. The membrane current of single rod outer segments. J. Physiol. **288:** 589-611.

(9) Baylor, D.A.; Matthews, G.; and Nunn, B. 1984. Location and function of voltage-sensitive conductances in retinal rods of the salamander Ambystoma tigrinum. J. Physiol. **354:** 203-223.

(10) Baylor, D.A.; Nunn, B.J.; and Schnapf, J.L. 1984. The photocurrent, noise and sprectral sensitivity of rods of the monkey Macaca fasciculari. J. Physiol. **357:** 575-607.

(11) Borsellino, A., and Fuortes, M.G.F. 1968. Responses to single photons in visual cells of Limulus. J. Physiol. **196:** 507-539.

(12) Brown, J.E., and Pinto, L.H. 1974. Ionic mechanism for the photoreceptor potential of the retina of Bufo marinus. J. Physiol. **236:** 575-591.

(13) Capovilla, M.; Caretta, A.; Cervetto, L.; and Torre, V. 1983. Ionic movements through light-sensitive channels of toad rods. J. Physiol. **343:** 295-310.

(14) Capovilla, M.; Cervetto, L.; and Torre, V. 1980. Effects of changing the external potassium and chloride concentrations on the photoresponses of Bufo bufo rods. J. Physiol. **307:** 529-551.

(15) Capovilla, M.; Cervetto, L.; and Torre, V. 1980. The sodium current underlying responses to light of rods. J. Physiol. **317:** 223-242.

(16) Capovilla, M.; Cervetto, L.; and Torre, V. 1982. Antagonism between steady and phosphodiesterase inhibitors on the kinetics of rod photoresponses. Proc. Natl. Acad. Sci. USA **79:** 6698-6702.

(17) Capovilla, M.; Cervetto, L.; and Torre, V. 1983. The effect of phosphodiesterase inhibitors on the electrical activity of toad rods. J. Physiol. **343:** 277-294.

(18) Cavaggioni, A.; Sorbi, R.T.; and Turrini, S. 1973. Efflux of potassium from isolated rod outer segments: a photic effect. J. Physiol. **232:** 609-620.

(19) Cervetto, L. 1973. Influence of sodium, potassium and chloride ions on the intracellular responses of turtle photoreceptor. Nature **241:** 401-403.

(20) Cervetto, L.; Pasino, E.; and Torre, V. 1977. Electrical responses of rods in the retina of Bufo marinus. J. Physiol. **267:** 17-51.

(21) Cervetto, L.; Torre, V.; Pasino, E.; and Capovilla, M. 1984. Recovery from desensitization and saturation in toad rods. **In** Photoreceptors, eds. A. Borsellino and L. Cervetto. New York: Plenum Press.

(22) Cone, R.A. 1973. The internal transmitter model for visual excitation: some quantitative implications. **In** Biochemistry and Physiology of Visual Pigments, ed. H. Langer, pp. 275-282. Berlin, New York: Springer-Verlag.

(23) Cook, R.; Hodgkin, A.L.; McNaughton, P.A.; and Nunn, B.J. 1984. Rapid change of solutions bathing a rod outer segment. J. Physiol.**357:** 2p.

(24) Detwiler, P.B.; Conner, J.D.; and Bodoia, R.D. 1982. Gigaseal patch clamp recordings from outer segments of intact retinal rods. Nature **300:** 59-61.

(25) Detwiler, P.B.; Hodgkin, A.L.; and McNaughton, P.A. 1980. Temporal and spatial characteristics of the voltage response of rods in the retina of the snapping turtle. J. Physiol. **300:** 213-250.

(26) Di Polo, R., and Beaugé, L. 1983. The calcium pump and sodium-calcium exchange in squid axons. Ann. Rev. Physiol. **45:** 313-324.

(27) Fain, G.L. 1976. Sensitivity of toad rods: dependence on wavelength and background illumination. J. Physiol. **261:** 71-101.

(28) Fain, G.L., and Lisman, J.E. 1981. Membrane conductances of photoreceptors. Prog. Biophys. Molec. Biol. **37:** 91-147.

(29) Fain, G.L.; Quandt, F.H.; Bastian, B.L.; and Gershenfeld, H.M. 1978. Contribution of a caesium-sensitive conductance increase to the rod photoresponse. Nature **272:** 467-469.

(30) Fain, G.L.; Quandt, F.H.; and Gershenfeld, H.M. 1977. Calcium-dependent regenerative responses in rods. Nature **269:** 707-710.

(31) Fuortes, M.G.F., and Hodgkin, A.L. 1964. Changes in time scale and sensitivity in the ommatidia of Limulus. J. Physiol. **172:** 239-263.

(32) Glynn, I.M., and Karlish, J.D. 1975. The sodium pump. Ann. Rev. Physiol. **37:** 13-53.

(33) Gold, G.H. and Korenbrot, J.I. 1980. Light-induced calcium release by intact retinal rods. Proc. Natl. Acad. Sci. USA **278:** 5557-5561.

(34) Hagins, W.A.; Penn, R.D.; and Yoshikami, S. 1970. Dark current and photocurrent in retinal rods. Biophys. J. **10:** 380-412.

(35) Hille, B. 1973. Potassium channels in myelinated nerve: selective permeability to small cations. J. Gen. Physiol. **61:** 669-686.

(36) Hodgkin, A.L., and Horowicz, P. 1959. The influence of potassium and chloride ions on the membrane potential of single muscle fibres. J. Physiol. **148:** 127-160.

(37) Hodgkin, A.L., and Keynes, R.D. 1955. The potassium permeability of a giant nerve. J. Physiol. **128:** 61-88.

(38) Hodgkin, A.L.; McNaughton, P.A.; and Nunn, B.J. 1985. The ionic selectivity of light-sensitive channels in retinal rods from Bufo marinus. J. Physiol. **358:** 447-468.

(39) Hodgkin, A.L.; McNaughton, P.A.; Nunn, B.J.; and Yau, K.-W. 1984. Effect of ions on retinal rods from Bufo marinus. J. Physiol. **350:** 649-680.

(40) Horowicz, P.; Gage, P.W.; and Eisenberg, R.S. 1968. The role of the electrochemical gradient in determining potassium fluxes in frog striated muscle. J. Gen. Physiol. **72:** 405-442.

(41) Lamb, T.D. 1984. Effects of temperature changes on toad rod photocurrents. J. Physiol. **346:** 557-578.

(42) Lamb, T.D. 1984. Electrical response of photoreceptors. **In** Recent Advances in Physiology, vol. 10, pp. 29-66. New York: Churchill Livingstone.

(43) Lamb, T.D.; McNaughton, P.A.; and Yau, K.-W. 1981. Longitudinal spread of activation and background desensitization in toad rod outer segments. J. Physiol. 319: 463-496.

(44) Owen, W.G., and Torre, V. 1983. High-pass filtering of small signals by retinal rods: ionic studies. Biophys. J. **41:** 325-340.

(45) Payne, R., and Howard, J. 1981. Responses of an insect photoreceptor: a simple log-normal model. Nature **290:** 415-416.

(46) Penn, R.D., and Hagins, W.A. 1972. Kinetics of the photocurrent of retinal rods. Biophys. J. **12:** 1073-1094.

(47) Robinson, J.D., and Falshner, M.S. 1979. The (Na^{+}-K^{+}) activated ATPase. Enzymatic and transport properties. Biochim. Biophys. Acta. **549:** 145-176.

(48) Robinson, W.E.; Kawamura, K.S.; Abramson, B.; and Bownds, D. 1980. Control of the cyclic GMP phosphodiesterase of frog photoreceptor membranes. J. Gen. Physiol. **76:** 631-645.

(49) Thomas, R.C. 1972. Electrogenic sodium pump in nerve and muscle cells. Physiol. Rev. **52:** 563-594.

(50) Torre, V. 1982. The contribution of the electrogenic sodium-potassium pump to the electrical activity of toad rods. J. Physiol. **333:** 315-341.

(51) Torre, V., and Owen, W.G. 1983. High-pass filtering of small signals by the rod networks in the retina of the toad, Bufo marinus. Biophys. J. **41:** 305-324.

(52) Torre, V.; Pasino, E.; Capovilla, M.; and Cervetto, L. 1981. Rod photoresponses in the absence of external sodium in retinae treated with phosphodiesterase inhibitors. Exp. Brain Res. **44:** 427-430.

(53) Yau, K.-W.; McNaughton, P.A.; and Hodgkin, A.L. 1981. Effect of ions on the light-sensitive current in retinal rods. Nature **292:** 502-505.

(54) Yau, K.W., and Nakatani, K. 1984. Cation selectivity of light sensitive conductance in retinal rods. Nature **309:** 352-354.

(55) Yau, K.-W., and Nakatani, K. 1984. Electrogenic sodium-calcium exchange in retinal rod outer segment. Nature **311:** 661-663.

The Molecular Mechanism of Photoreception, ed. H. Stieve, pp. 353-367. Dahlem Konferenzen 1986. Berlin, Heidelberg, New York, Tokyo: Springer-Verlag.

Experimental Results and Physical Ideas towards a Model for Quantum Bumps in Photoreceptors

J. Schnakenberg and W. Keiper
Institut für Theoretische Physik
Rheinisch-Westfälische Technische Hochschule Aachen
5100 Aachen, F.R. Germany

Abstract. A statistical analysis of voltage-clamp quantum bump data in the ventral photoreceptor of Limulus is presented. The analysis is performed in terms of appropriately defined bump parameters. On the basis of this analysis, model elements for an overall scheme of the bump phenomenon are proposed and evaluated by various methods. The comparison with the experimental data particularly suggests that a) latency and amplification are separate processes, b) latency might be caused by a cooperative mechanism, c) individual properties of cells enter into the chain of events as a factor in the "current parameters" of bumps and are caused by individual distributions of density and conductance of the light-activated channels, and d) the possibility of a conclusive comparison of bump data with the results of different experiments crucially depends on the knowledge of or assumptions on the opening and closing mechanism of the light-activated channels.

QUANTUM BUMPS AND THEIR PARAMETERIZATION

The transduction process in invertebrate photoreceptors is yet an unsolved problem. The first quantitative model for the biochemical nature of the single-photon response of ventral photoreceptors of Limulus was proposed by Borsellino and Fuortes (2) - a chain of enzymatic amplifiers. Kramer (11) put forward an autocatalytic reaction as a model candidate. More recently, Goldring and Lisman (7) gave a review of the shortcomings of the Borsellino-Fuortes model and proposed a modified reaction scheme. The models to be presented in this paper are based upon experimental data of voltage-clamp "quantum bumps" in the ventral photoreceptor of Limulus. A quantum bump is the electrical response of the cell to the absorption of a single photon, namely, a transient inward compensation current across the cell membrane. Besides these light-evoked bumps, there are also spontaneous bumps in the dark which will be discussed in the section on THE PROBLEM OF THE SPONTANEOUS OR "DARK" BUMPS. The bump experiments to which this paper predominantly refers have been performed by Stieve and co-workers in the Institute of Neurobiology of the KFA Jülich. The experimental techniques of the measurements have been described in detail by Stieve and Bruns ((14, 15); cf. also Stieve, this volume).

The analysis of the experimental bump data is appropriately realized by introducing bump parameters as illustrated in Fig. 1:

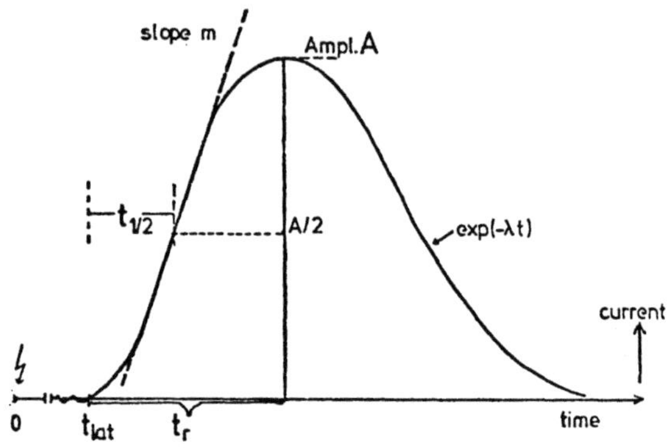

FIG. 1 - Definitions of bump parameters.

We distinguish two kinds of parameters, namely 1) "time-like" parameters expressed by units of time or inverse time, and 2) "current" parameters involving the unit of a current.

1) Time-like parameters:
   - t_{lat} latency time in flash experiments: time period between the stimulating flash and the onset of the bump current, (ms);
   - t_r rise time: time period between the onset and the maximum of the bump current, (ms);
   - λ decay rate of an exponential decay phase of the bump current, (1/ms).

2) Current parameters:
   - A maximum amplitude of the bump current, (nA);
   - F total area of the bump, i.e., the time integral of the bump current, (pC);
   - m slope of the linear portion of the rising phase of the bump current, (nA/ms).

The main difference between our choice of parameters and that of the "adapting bump model" as proposed by Wong, Knight, and Dodge (19) is our observation of two different time scales for the rise and the decay phase of a bump expressed by t_r and λ instead of only one in the "adapting bump model."

We have developed a FORTRAN-program which determines the numerical values of the parameters for each bump from the experimental data. First of all, this program identifies a current fluctuation as a bump if a) the maximum amplitude of a current fluctuation exceeds the value of 100 pA and b) its area exceeds the value of 2 pC. We have checked the results of this identification procedure for a large number of current fluctuations of different experiments by eye and found satisfactory agreement.

The determination of the numerical values of the above-listed parameters from the experimental data by our computer program turns out to be unambiguous except for m and λ. The definitions of both m and λ involve a differential quotient. After having tried several methods for determining m and λ, we eventually chose a linear regression method for m and an integration procedure for λ. The corresponding programs determine not only the numerical values of m and λ but also the optimal time intervals in which the rise phase of a bump can be approximated by a linear function of time and its decay phase by an exponential $\propto \exp(-\lambda \cdot t)$. We have compared the findings of the program with approximations by eye and found satisfactory agreement. It turns out that for bumps of reasonable size ($A \geq 0.3$nA) the approximate linear part of the rise phase constitutes no less than 50% of the total rise phase. The time interval of the approximate exponential decay extends from about 15-30 ms after the time of the current maximum up to the time region where the current signal becomes indistinguishable from the baseline noise.

A SHORT REVIEW OF THE RESULTS OF THE BUMP ANALYSIS

For a detailed presentation of the methods and results of our bump analysis, the reader is referred to Keiper, Schnakenberg, and Stieve (10). Figure 2a-c shows the distributions of the time-like parameters t_{lat}, t_r, and λ for two different dark-adapted cells.

Figure 2 shows that the time-like parameters except for λ are distributed very much in the same way for the two different cells, i.e., with the same mean values and with the same width. The deviations are within the range of the statistical spread due to the finite numbers of bumps evaluated (576 and 414 bumps for the left and right column, respectively). The λ-distributions of the two cells are similar in shape and width but show a slight difference of their means which may be just beyond the range of the statistical spread.

A particular result of the analysis which will imply decisive consequences for our model considerations is the fact that the mean bump duration t_D (time period during which the bump current significantly deviates from the baseline, roughly $t_r + 2/\lambda$) is less than half the mean latency time t_{lat}. This relationship holds not only for the means but also for the majority of the individual bumps. Note also that the relative width of the t_{lat}-distribution, i.e., its absolute width divided by its mean, is on the order of unity. This will turn out to be another important point for our model considerations.

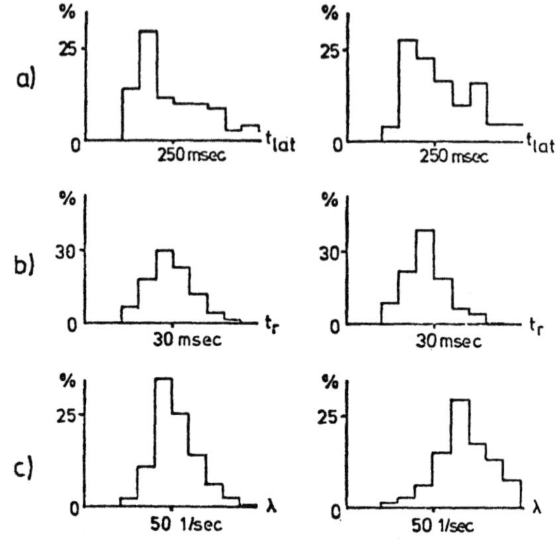

FIG. 2 - Distributions of the time-like parameters a) t_{lat}, b) t_r, and c) λ for two different cells (left and right column, respectively).

In contrast to the time-like parameters, the distributions of the current parameters A, F, and m of the same two cells as in Fig. 2 show distinct differences as is evident in Fig. 3a-c.

Not only the means and widths but also the shapes of the distributions are different for the two cells. These differences are thus markedly beyond the range of the statistical spread due to the finite number of bumps evaluated. In order to investigate the nature of this appearance of individual cell properties, we have also evaluated the distributions of the "scaled" ratios A/m and F/m of the current parameters as shown in Fig. 4a,b.

Obviously, the distributions of the scaled parameters of the two cells are again very similar, i.e., the deviations do not exceed the range of the statistical spread due to the finite number of bumps evaluated, as for the time-like parameters. This finding would imply that the individual properties of the cells enter into the current parameters as a factor.

We have also evaluated the correlation coefficients between the bump parameters on the basis of flash experiments on ten different cells. The results are shown in Table 1.

Note that there is apparently no correlation of t_{lat} with any of the other bump parameters, and the corresponding correlation coefficients are within the statistical spread. On the other hand, we observe a strong correlation of A with m and a weak correlation of A with t_r. The correlation coefficents

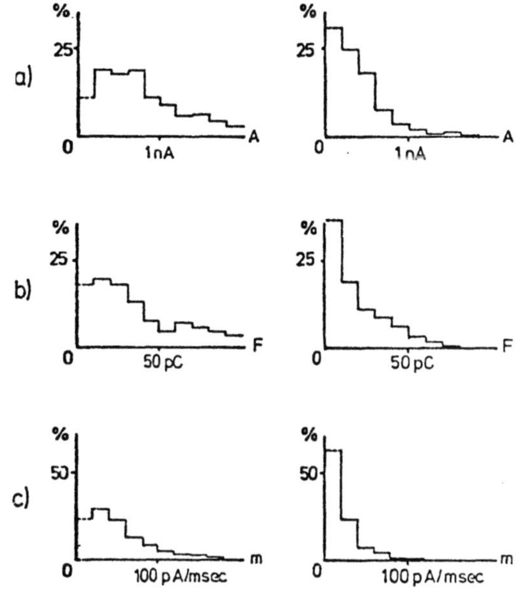

FIG. 3 - Distributions of the current parameters a) A, b) F, and c) m for the same two cells as in Fig. 2 (left and right column, respectively).

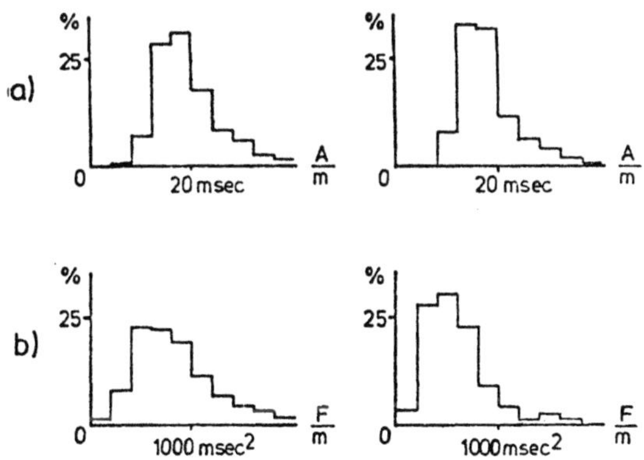

FIG. 4 - Distributions of the scaled current parameters a) A/m and b) F/m for the same two cells as in Figs. 2 and 3 (left and right column, respectively).

TABLE 1 - Correlation coefficients of the bump parameters t_{lat}, t_r, λ, m, and A.

	t_{lat}	t_r	λ	m	A
t_{lat}	1				
t_r	-0.08 +0.09	1			
λ	-0.26 +0.23	-0.11 -0.28	1		
m	+0.06 +0.12	-0.18 +0.22	-0.10 -0.36	1	
A	+0.13 +0.22	+0.15 +0.41	-0.03 -0.26	+0.83 +0.95	1

of λ with t_r and of λ with m may indicate some very weak anti-correlation. The "adapting bump model" by Wong, Knight, and Dodge (19) with only one time scale for the rise and the decay phase, would predict a strong anti-correlation between t_r and λ which is by no means justified by our results.

Under light adaptation, some of the bump parameters show a distinct variation compared to the dark-adapted state of the cell. Figure 5 shows the variation of the mean values of t_{lat}, t_r, λ, and m as a function of the energy of the light-adapting flash.

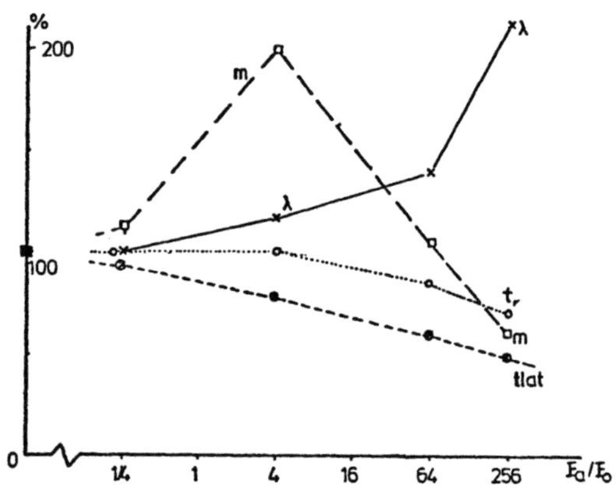

FIG. 5 - Bump parameters t_{lat}, t_r, λ, and m as functions of the energy E_a of a light-adapting flash; E_o = energy of the stimulating flash.

The slope m first starts to increase at moderately increasing light-adapting energies Ea, usually referred to as "facilitation." This increase is followed by a sharp decrease at larger E_a. Due to the strong correlation of the maximum amplitude A with m, the same variation is observed for A. Regarding the time-like parameters t_{lat}, t_r, and λ, the whole bump phenomenon becomes faster at increasing light adaptation.

Besides the light-stimulated bumps which we have described so far, the ventral photoreceptor of Limulus also shows spontaneous or dark bumps without stimulation by light. In agreement with earlier investigations (16, 20), our analysis of the dark bumps shows that they are "smaller" than light-evoked bumps. In terms of our parameters, this is expressed by smaller mean values of the current parameters A, F, and m, whereas their t_r- and λ-distributions are very much the same as those of the light-evoked bumps. The question as to whether the dark bumps are markedly different from light-evoked bumps will turn out to be crucial for constructing bump models.

AMPLIFICATION AND LATENCY: INTEGRATED OR SEPARATED PHENOMENA?

A conservative estimate of about 18 pS for the conductance and 18.7 ms for the mean open time of a light-activated single channel from Wong's noise analysis data (17) puts the number of channel openings in a medium-sized dark-adapted bump to more than 1000. If Bacigalupo and Lisman's (1) single-channel values of 35 pS for the conductance and 4 ms for the mean open time in sonificated and light-adapted membranes of the ventral photoreceptor of Limulus are translated to bumps, the number of channel openings is closer to 10,000, all triggered lastly by a single photon.

When constructing bump models, one first has to decide whether this amplification and the latency behavior should be described as integrated or separate phenomena. The model proposed by Fuortes and Hodgkins in 1964 (6) is an integrated scheme consisting of a cascade of a number of n = 9 enzymatic reactions of the type

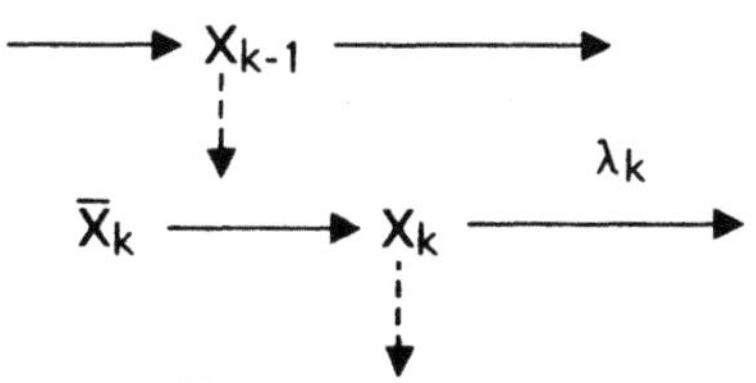

k = 1,2,...n, (1)

where $\overline{X}_k$ and X_k denote the inactive and active configurations of an enzyme in the kth stage of the cascade, respectively, and each X_k acts as a catalyst for its successor X_{k+1}. X_o represents a photoexcited rhodopsin molecule and X_n an open channel. The λ_k's are rate constants for a spontaneous decay of X_k. Upon integration of scheme 1 with $X_o = 0$ for $t < 0$ and X_o = constant for $t > 0$ as corresponding to a bright steady light stimulation, the channel openings start with a power law of $\propto t^n$

("macroscopic" response). For the situation of a "microscopic" bump event, scheme 1 has to be evaluated by a stochastic treatment as performed by Borsellino and Fuortes (2) and by Goldring and Lisman (7). The t^n-power law of the macroscopic event is thereby translated into a latency behavior of fluctuating time periods as intended.

At a first glance, model scheme 1 seems to predict a strong correlation between t_{lat} and the degree of amplification expressed, e.g., by the maximum amplitude A which would be in contrast to the experiments. We have therefore evaluated this correlation by a computer simulation of scheme 1, but we did not actually find any indication of the expected correlation. The reason is simply that the fluctuations of the X_k's along the cascade of reactions mask this correlation. Another consequence of model scheme 1, however, clearly contradicts the experimental results. As Goldring and Lisman (7) observed by computer simulations, scheme 1 predicts bump durations t_D which are markedly longer than t_{lat}, if amplification is assumed to be distributed more or less uniformly over the cascade stages. Remember that the experimental value of t_D/t_{lat} is always less than 0.5, as pointed out in the section on A SHORT REVIEW OF THE RESULTS OF THE BUMP ANALYSIS. Goldring and Lisman found that this value of t_D/t_{lat} could be realized by model scheme 1 if the first six of the nine cascade stages were made gainless, i.e., if amplification were concentrated in the last three stages. We have generalized this particular numerical result by an analytical evaluation of model scheme 1. We find that for n = 9, a ration $t_D/t_{lat} > 2...3$ is to be expected whenever amplification is assumed to take place in the first of the cascade stages, irrespective of how amplification and decay rates λ_k's are distributed over the stages of the cascade. The mathematical argument is so general that this result is not only a peculiarity of model scheme 1 but can be expected to apply also to other model types which start amplification in the very beginning of the process. We take this result as an argument for separating the phenomena of latency and amplification in our models, to be presented in the following sections. A very similar mathematical argument has been brought forward by M.A. Goldring (unpublished Ph.D. Thesis, Brandeis University, 1982). The observation of different temperature dependencies of latency and amplification led Wong, Knight, and Dodge (18) to suggest that the two phenomena should be considered as separate processes.

ALTERNATIVE MODELS FOR LATENCY

In our opinion, model scheme 1 with six initial gainless latency stages is not a satisfactory model description for latency since the very character of a cascade of enzymatic reactions is amplification. Instead of this scheme, we would like to propose two alternative latency models. The first one is a chain of configurational changes of some molecule X:

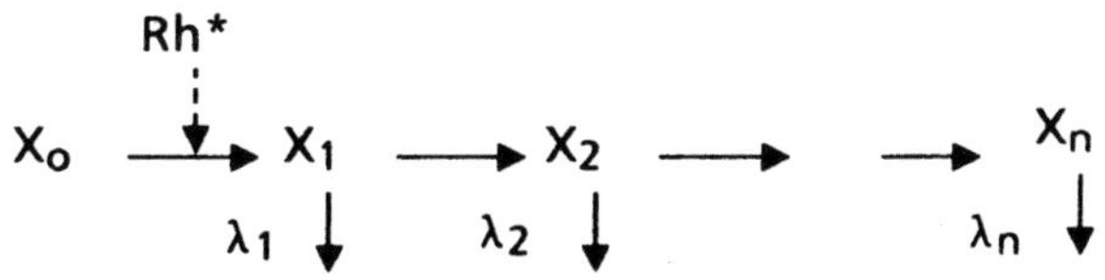

(2)

The symbol X_k denotes the kth configurational stage of X, k = 1,2,... The first step from X_0 to X_1 is assumed to be initiated by a photoexcited rhodopsin Rh*, and the final configuration X_n is assumed to initiate some amplification process not shown in scheme 2. For each configurational stage X_k, there is a finite probability for a spontaneous decay at a rate λ_k which turns out to yield a better fit to the experimental distributions than $\lambda_k = 0$. The crucial question is how many configurations n one needs to obtain t_{lat}-distributions from scheme 2 which resemble those experimentally obtained. For this comparison, the relative width of the distribution turns out to be the critical parameter. Its experimental value is nearly unity (cf. section on A SHORT REVIEW OF THE RESULTS OF THE BUMP ANALYSIS). This value can be realized by scheme 2 only for $n > 8$; lower values of n would make the t_{lat}-distributions too broad as compared with the experimental ones. A possible alternative would be to assume that Rh initiates two molecules of the X_1-configuration and two X_n- configurations are needed to start the amplifier. For this assumption, $n > 4$ would be sufficient. We have also evaluated other combinations. The general rule is: the higher the "cooperativity," i.e., the number of X_n- configurations needed to start the amplifier, the lower the number n of stages required for obtaining realistic t_{lat}-distributions.

This latter remark may serve for introducing our second alternative model for latency. It is well-known that cooperativity (or threshold mechanisms) cause latency behavior (e.g., critical slowing down in phase transitions) - cf. (8, 13). Our second model makes use of cooperativity in the following way:

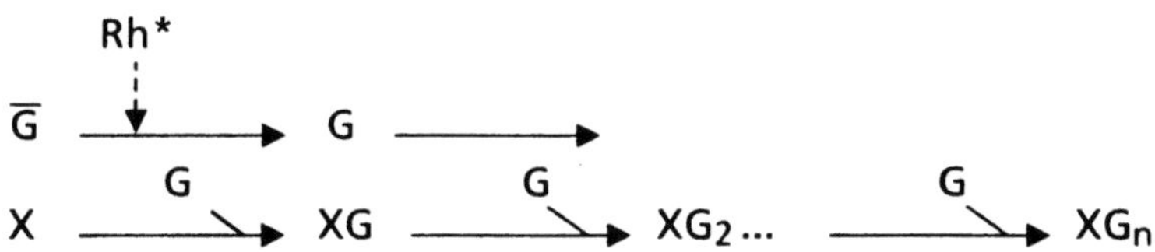

(3)

The photoexcited rhodopsin Rh* is assumed to activate enzymatically a molecule from an inactive configuration $\overline{G}$ to an active configuration G with some finite lifetime. A number n of active G's is bound successively to a molecule X such that the complex XG_n starts the amplifier or already constitutes its first stage. We have evaluated this model scheme by computer simulation and found that $n = 4$ yields realistic t_{lat}-distributions, provided that only one X-molecule is available per photoexcited rhodopsin. A further aspect of model scheme 3 is the possibility that the activity of G could be modified by light adaptation or by low external Ca^{2+} in such a way that the t_{lat}-time scale is thereby contracted or expanded, respectively, as observed in the experiments. We have chosen the symbol G in model scheme 3 in order to indicate that the G-protein which is known to be activated by a photoexcited rhodopsin could be involved in a mechanism of this kind which causes latency (12). Model scheme 3 should not be taken in a literal sense, but rather as a shorthand of a much more complex system of reactions, the essential point of which, however, is a cooperative process as in scheme 3.

A common property of our configurational and cooperative models is the obvious insensitivity to incident fluctuations somewhere along the chains in the dark, i.e., without the triggering action of a photon. For the configurational model presented in scheme 2, this property would be even more pronounced if all configurational transitions between the X_k's were enzymatically activated by the photoexcited rhodopsin. This consideration suggests that latency might be the price for the high performance of the system in detecting extremely dim light signals such as single photons without being misled by incident, e.g., thermal fluctuations. This speculation fits the observation that the latency times are markedly shortened in the light-adapted state of the cell.

THE PROBLEM OF THE SPONTANEOUS OR "DARK" BUMPS

If one agrees that latency and amplification should be thought of as separate processes and if one takes into account that the spontaneous or dark bumps are smaller in size than light-evoked bumps as pointed out in the section on A SHORT REVIEW OF THE RESULTS OF THE BUMP ANALYSIS, one inevitably has to make one of the following assumptions: a) dark bumps are processed by another amplifier than light-evoked bumps, or b) dark bumps are processed by the same amplifier as light-evoked bumps, but they are triggered via a different input.

Goldring and Lisman (private communication) have proposed a model with two differently active states of photoexcited rhodopsin (see also (8)). Dark bumps are assumed to be caused by spontaneous reversals of the inactivation reaction of the less active state of the rhodopsin. This model follows assumption b) above, but it does not include a consistent description of latency of the light-evoked bumps. On the other hand, Goldring and Lisman give an explanation for the individual distributions of the bump areas F of different cells (cf. section on A SHORT REVIEW OF THE RESULTS OF THE BUMP ANALYSIS) in their model by assuming that different chemical pathways along the active states of rhodopsin occur with different probabilities in different cells. This explanation, however, does not agree with our finding that the individual properties of cells very likely enter into the current parameters only as a factor.

THE TRANSMITTER HYPOTHESIS AND THE INDIVIDUALITY OF CELLS

A bump is a local event in the rhabdomeral part of the photoreceptor cell. From receptor current saturation experiments, Brown and Coles (3) estimated the size of the membrane region which changes its conductance during a bump event to be about $10(\mu m)^2$. With a value of 0.1 m for the diameter of a microvillus (4), this means that the primary excitation of some single rhodopsin molecule somehow propagates to the light-activated channels in an area encompassing about 1000 microvilli. We follow the hypothesis that this propagation is most likely realized by an internal transmitter (5, 9). We propose that the final stage of the latency chain triggers an amplifier which releases or activates the transmitter molecules. This amplifier may consist of only one or a few stages of an enzyme cascade. The transmitter molecules are assumed to diffuse parallel to the cell membrane. It may well be that the cisternae (or the palisade) as observed by Calman and Chamberlain (4) act as a barrier against the interior of the cell. It is well-known that the square of the mean diffusion radius (and thus its effective area) increases linearly with time, at least within the lifetime of the diffusing molecule. It may be that this linear increase is reflected by the linear portion of the rising phase of the bumps.

The total current across the light-activated channels opened by the transmitter molecules is not only proportional to the number of transmitter molecules released or activated by the amplifier or to their effective diffusion area, but also to the local densities c and conductances σ of the channels in the membrane region where the bump occurs. We conjecture that the product $c \cdot \sigma$ may show different distributions in different cells. This conjecture would explain why different cells show different distributions of the current parameters and why this individuality is cancelled for the scaled rations of current parameters A/m and F/m as reported above. Further support for this conjecture are Calman and Chamberlain's (4) observations that the rhabdomeral membrane of the ventral photoreceptor cells of Limulus appears to be rather inhomogeneous, including folds and locally fluctuating directions of the microvilli. It seems plausible to us that at this point individual structural properties of the cells enter into the bump event.

CHANNEL OPENING AND CLOSING MECHANISMS

We have investigated three models for the opening of light-activated channels by the transmitter. In each of the models, the transmitter T is assumed to be released or activated by the amplifier and to diffuse along the interior side of the membrane as described in the immediately preceding section. Moreover, we assume that the lifetime of the transmitter is limited by a spontaneous inactivation reaction

$$T \xrightarrow{\lambda_T} T', \tag{4}$$

where T and T' denote the active and inactive configuration of the transmitter, respectively, and λ_T is the time or rate constant of the inactivation reaction, i.e., $1/\lambda_T$ has the unit of time and the meaning of the mean lifetime of the active transmitter T as far as the spontaneous inactivation is concerned. Our three models differ by the way the channels are opened by T:

Model A: (5)

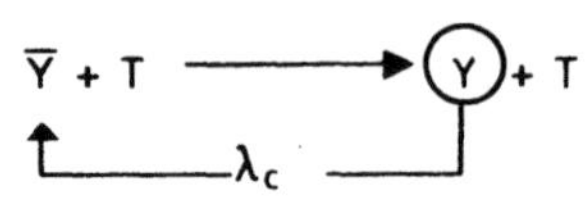

Model B1: (6)

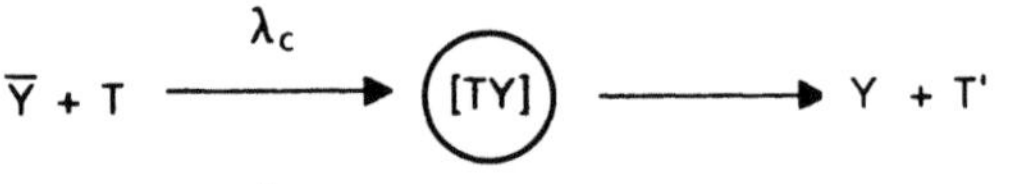

Model B1: (7)

In these model schemes, the closed configuration of the channel is denoted by Y, the open configurations by the encircled symbols Y in model A and [TY] in model B, respectively. In model A, T is assumed to open a channel "enzymatically" with a very short binding time, whereas in model B1 the bound complex [TY] is assumed to be the open channel configuration. Our third model, B2, is derived from B1 by writing T instead of T' after debinding of [TY] on the right-hand side of scheme 6, i.e., T may open more than one channel in the same way as A. In all three models, λ_c denotes the time or rate constant of the spontaneous closing or inactivation reaction of a single channel, i.e., $1/\lambda_c$ has the unit of time and the meaning of a mean open time of single channels.

Our three models represent only three particularly simple versions of a much larger variety of conceivable channel mechanisms. We have calculated the decay rates λ of bumps as being expected on the basis of our models. Due to the fact that our models involve two variables T and Y, we find two candidates for λ for each of the models. Evidently, the one with the lower value should be identical with the experimentally observed λ of bumps. At first glance, one might expect that the calculated λ is simply the smaller value of the two time constants λ_T and λ_c for all models. This simple relationship, however, only holds for model A, whereas for models B1 and B2 (and very likely for even more complex models) more complex algebraic combinations of λ_T, λ_C, and further model parameters enter into the expressions for the expected λ of bumps. This result has important consequences for a comparison of the λ-values of bumps with the mean open times of single channels as observed by Bacigalupo and Lisman (1) in patch-clamp experiments. As mentioned above, the mean open time of single channels is given by $1/\lambda_C$ for each of our three models. Thus, a coincidence of $1/\lambda$ of bumps with the mean open time of single channels cannot be expected except for the rather unrealistic model version A if $\lambda_C < \lambda_T$. Bacigalupo and Lisman found mean open times of single channels between 1.2 ms and 4.2 ms, whereas $1/\lambda$ for dark-adapted bumps is typically about 20 ms. To conclude from this disagreement that the decay phase of bumps is controlled by the time course of the transmitter (and not by that of the channels) means to jump at conclusions. Moreover, the above-mentioned mean open times of single channels were observed under steady light stimulations of light intensities which would cause marked light adaptation effects on bumps if properly translated to the situation of bump experiments. On the other hand, we have already observed that the $1/\lambda$ values of bumps markedly decrease for less intense light adaptations (cf. Fig. 5). The observed disagreement may thus even be due to different adaptation conditions in the two experiments.

We have also calculated the characteristic corner frequencies of the noise power spectrum as being expected on the basis of our three simple models. The transmitter release or activation by the amplifier is assumed here to occur at some finite and constant mean rate with Poisson-like fluctuations around it. For none of our models do any of the calculated corner frequencies coincide either with the channel closure rate constant λ_c or with the calculated bump decay rate constants λ. The corner frequencies markedly depend on the mean steady-state transmitter concentration level, which in turn will depend on the intensity of the steady light stimulus. This latter quantity, however, neither influences the spontaneous closing of channels nor does it play any role for bumps, since bumps are single-photon events. We therefore suspect that the λ-value of bumps and the characteristic corner frequency of a noise power spectrum, as measured by Wong (17), will coincide only by accident.

CONCLUSION

On the basis of an analysis of experimental bump data, we have proposed model elements for an overall scheme of the bump phenomenon. In view of our results, we suggest treating the bump parameters t_{lat}, m, t_r, and λ as "primary" parameters. First of all, these parameters are mutually uncorrelated (cf. Table 1). t_{lat} and λ reflect the properties of the latency chain and of the transmitter-channel interaction, respectively, i.e., of the head and tail ends of the chain of events. In contrast to these elements, the experimental data give less insight into the connecting link which we suggest is an amplifier. The rise time t_r would then mainly be determined by the time the amplifier is active, e.g., by the lifetime of the final stage of the latency chain. The slope m of the linear portion of the bump's rise phase is interpreted in our model scheme as a combined quantity. It includes the local values of the product $c \cdot \sigma$, i.e., of the density and conductance of the channels, the diffusion coefficient of the transmitter, and the degree of amplification (rate of transmitter production, activation, or release) as factors. We prefer to choose m as a primary parameter rather than the maximum amplitude A or the bump area F, since m has the physical nature of a rate whereas A and F are integrated quantities. Indeed, Table 1 shows that A is highly correlated with m, the correlation coefficient being about 0.9. The relatively moderate correlation of A with t_r (correlation coefficient between 0.15 and 0.41) is masked by the broad fluctuations of m. We have therefore evaluated the correlation of A/m with t_r and obtained correlation coefficients between 0.83 and 0.92. In other words, $A/(m \cdot t_r)$ can be treated as a dimensionless-form parameter. Its distribution (not shown here) turns out to be relatively narrow with a mean value of 0.70. A similar treatment can be applied to the bump area F. An appropriate dimensionless-form parameter is $F/(m \cdot t_r^2)$, which again has a relatively narrow distribution (not shown here) with a mean value of 0.85.

We have put together our model elements into mathematical models which are evaluated by various methods on a computer. We have calculated bump shapes as well as distributions of bump parameters and compared the results with the experimental data. Our next step will be the inclusion of bump data obtained under light adaptation and low external Ca^{2+}-concentration.

Acknowledgements. We are indebted to H. Stieve and co-workers for providing us with experimental data and many elucidating discussions and suggestions. This work was supported by the Deutsche Forschungsgemeinschaft (SFB 160).

REFERENCES

(1) Bacigalupo, J., and Lisman, J.E. 1983. Single-channel currents activated by light in Limulus ventral photoreceptors. Nature **304:** 268-270.

(2) Borsellino, A., and Fuortes, M.G.F. 1968. Responses to single photons in visual cells of Limulus. J. Physiol. **196:** 507-539.

(3) Brown, J.E., and Coles, J.A. 1979. Saturation of the response to light in Limulus ventral photoreceptors. J. Physiol. **296:** 373-392.

(4) Calman, B., and Chamberlain, S. 1982. Distinct lobes of Limulus ventral photoreceptors. II. Structure and ultra-structure. J. Gen. Physiol. **80:** 839-862.

(5) Cone, R.A. 1973. The internal transmitter model for visual excitation: Some quantitative implications. In Biochemistry and Physiology of Visual Pigments, ed. H. Langer, pp. 273-282. Berlin: Springer-Verlag.

(6) Fuortes, M.G.F., and Hodgkins, A.L. 1964. Changes in time scale and sensitivity in the ommatidia of Limulus. J. Physiol. **172:** 239-263.

(7) Goldring, M.A., and Lisman, J.E. 1983. Single photon transduction in Limulus photoreceptors and Borsellino-Fuortes model. IEEE Trans. SMC **13:** 727-731.

(8) Hamdorf, K. 1979. The physiology of invertebrate visual pigments. In Invertebrate Photoreceptors. Handbook of Sensory Physiology, ed. H. Autrum, vol. VII/6A, pp. 145-224. Berlin, Heidelberg, New York: Springer-Verlag.

(9) Hillman, P. 1982. The biophysics of intermediate processes in photoreceptor transduction: "Silent" stages, non-localities, single-photon responses and models. In Proceedings of the Symposium on The Biology of Photoreceptor Cells, pp. 443-475. Cambridge: Cambridge University Press.

(10) Keiper, W.; Schnakenberg, J.; and Stieve, H. 1984. Statistical analysis of quantum bump parameters in Limulus ventral photoreceptors. Z. Naturforsch. **39c:** 781-790.

(11) Kramer, L. 1975. Interpretation of invertebrate photoreceptor potentials in terms of a quantitative model. Biophys. Struct. Mech. **1:** 239-257.

(12) Liebman, P.A., and Pugh, E.N., Jr. 1982. Gain, speed and sensitivity of GTP-binding vs. PDE-activation in visual excitation. Vision Res. **22:** 1475-1480.

(13) Payne, J., and Howard, J. 1981. Response of an insect photoreceptor: a simple log-normal model. Nature **290:** 415-416.

(14) Stieve, H., and Bruns, M. 1980. Dependence of bump rate and bump size in Limulus ventral nerve photoreceptor on light adaptation and calcium concentration. Biophys. Struct. Mech. **6:** 271-285.

(15) Stieve, H., and Bruns, M. 1983. Bump latency distribution and bump adaptation of Limulus ventral nerve photoreceptor in varied extracellular calcium concentrations. Biophys. Struct. Mech. **9:** 329- 339.

(16) Stieve, H.; Bruns, M.; and Klomfass, J. 1982. Statistics of bump parameters (Limulus ventral nerve photoreceptor). Inv. Ophthalmol. Vis. Sci. **22 (Suppl.):** 275.

(17) Wong, F. 1978. Nature of light-induced conductance changes in ventral photoreceptors of Limulus. Nature **276:** 76-79.

(18) Wong, F.; Knight, B.W.; and Dodge, F.A. 1980. Dispersion of latencies in photoreceptors of Limulus and the adapting-bump model. J. Gen. Physiol. **76:** 517-537.

(19) Wong, F.; Knight, B.W.; and Dodge, F.A. 1982. Adapting bump model for ventral photoreceptors of Limulus. J. Gen. Physiol. **79:** 1089-1113.

(20) Yeandle, S., and Spiegler, J.B. 1973. Light-evoked and spontaneous discrete waves in the ventral nerve photoreceptor of Limulus. J. Gen. Physiol. **61:** 552-571.

The Molecular Mechanism of Photoreception, ed. H. Stieve, pp. 369-387. Dahlem Konferenzen 1986. Berlin, Heidelberg, New York, Tokyo: Springer-Verlag.

Potentials and Limitations of Noise Analysis of Light-induced Conductance Changes in Photoreceptors

J. Schnakenberg* and F. Wong**
* Institut für Theoretische Physik, RWTH
Aachen, F.R. Germany
** The Marine Biomedical Institute, University of Texas
Galveston, TX 77550, USA

Abstract. The potentials and limitations of noise analysis applied to light-induced currents in photoreceptors are discussed in this paper. Two limiting cases are presented for which noise analysis may yield some reliable information on the underlying mechanisms. The "bump approach" is based upon the assumption that the "elementary events" underlying the noisy current are the quantum bumps which obey Poisson statistics. From the experimentally observed power spectrum of the noisy current one then draws conclusions for the properties of the single bumps, including their variation under different conditions of light adaptation. The method is usually extended to values of the stimulating light intensity for which single bumps are no longer directly observable due to light adaptation.

On the other hand, the "steady-state approach" is based upon the assumption that the "elementary events" underlying the noisy current are the incoherent opening and closing events of individual ionic channels. This approach assumes a light-dependent, stationary but fluctuating intracellular level of some transmitter molecules which cause the opening (or closing) of the channels. From the experimentally observed power spectrum one then draws conclusions for the kinetics of the single channels and its control by the transmitter.

For each of the two approaches, a detailed model is presented and discussed. Qualitatively, one would expect the bump approach to be appropriate for "low" intensities of the stimulating light and the steady-state approach for "high" intensities. The crucial question, however, as to whether there is a sharp border line between the two approaches, an overlapping region of simultaneous applicability or a gap where both approaches fail, cannot be answered rigorously by the method of noise analysis itself since it is an indirect method which may continue to give answers even if the underlying assumptions become unrealistic on physical or physiological grounds. For any particular case of noise analysis applied to photoreceptors, we would like to evaluate the experimental data by means of both the bump and the steady-state approach and to discuss and compare the conclusions very carefully.

INTRODUCTION

Noise analysis of electric currents across cell membranes is a frequently applied method to analyze the elementary events by which the currents are generated. For a general survey of this method, the reader is referred to the review article by Chen (4) and the monograph by De Felice (8). Usually it is assumed that the currents are superpositions of independent stochastic opening and closing events of single-conductance channels. With this assumption, the opening and closing rates of the channels, i.e., their mean inverse opening and closing times, and their conductances can be derived or at least estimated from measurements of the current spectral density and of the mean and the variance of the current.

The application of noise analysis to photoreceptor cells cannot follow this simple scheme of evaluation since the conductance change of the membrane is controlled by some only partially known transduction process which is triggered by the stimulating light. Even if one assumes that the elementary events of the conductance change are openings and closures of single channels, those events are no longer stochastically independent but should be considered the final steps of the transduction process. It is now generally accepted that the opening (or closing) of a single light-sensitive channel is due to the interaction with some hypothetical transmitter molecule which is activated or released at the end of the transduction process (6, 12). Thus, noise analysis of photoreceptor currents would have to include the total transduction process and hence be at present subject to a large variety of perhaps conceivable but still rather arbitrary models. There are two limiting cases, however, for which we think noise analysis may yield some useful information about the underlying mechanism of the conductance changes, namely, a) the "bump approach," and b) the "steady-state approach."

The bump approach to noise analysis is based upon the observation that many photoreceptors exhibit large discrete responses (so-called "bumps") to the absorption of single photons. The experimental technique of bump measurements and the statistical analysis of individually observed bumps in the ventral photoreceptor of Limulus at very low light intensities are presented by Stieve and by Schnakenberg and Keiper (both this volume; cf. also (14)). Bumps are considered to be generated by local conductance changes of the rhabdomeral membrane due to the coordinated opening and closing of quite a number of channels in the vicinity of the rhodopsin molecule which has been photoexcited by a single photon. The transduction process for a single bump is the chain of events linking one photoexcited rhodopsin molecule to the opening of many channels, very likely by the interaction with transmitter molecules. For the dark-adapted state of the Limulus ventral photoreceptor, the number of discrete channel openings has been estimated between 1,000 and 10,000 within a membrane region of a few μm in diameter. With increasing intensities of the stimulating light, light adaptation, which makes the individual bumps smaller and speeds up their time course, comes into play (7). At about 10^{14} photon/m^2 of the stimulating flash, the bumps cease to be observable as discrete events. The noisy light-induced current, however, looks convincingly like a superimposition of bumps seen in dim light although smaller in size. Therefore, it seems reasonable to assume that the noisy current at high light intensities is due to the summed responses of small bumps. With this assumption, the measured noisy currents can be analyzed using methods of noise analysis to extract information about the properties of the individual bumps. With this approach ("adapting bump model" (21-23)), properties of a "bump" at high light intensities were obtained.

Although the bump approach of noise data continues to yield bump-like parameter values, the meanings of these values become harder to interpret, particularly at very bright light when the size of the bumps as reconstructed from the noise data approximates the size of a current due to the opening of only a very few ionic channels (or eventually of one single channel). In this situation, we expect that the bump approach ceases to be an adequate description since the spatial membrane regions of conductance changes, each triggered by single photons ("bump specks"), will start to overlap such that the openings (or closings) of a particular channel have to be considered the final steps of transduction processes initiated by different photoexcited rhodopsin molecules. An alternative, perhaps better, approach to analyze the noise data in this situation is to assume a stochastically fluctuating transmitter level with some finite mean value. This "steady-state approach" to noise analysis, however, requires knowledge or some assumptions on how the transmitter interacts with the channel to open (or close) it. A very simple assumption is that of an "enzymatic transmitter," i.e., a transmitter which interacts with the channel or is bound to it only for a very short time as compared to the time course of the whole transduction process. This means that the bound configuration of transmitter and channel can be eliminated in the kinetic scheme of the opening (or closing) process (cf. section on STEADY-STATE APPROACH: TWO-STATE CHANNELS INTERACTING WITH A TRANSMITTER, Eq. 31). Very recently, Gray and Attwell (11) published noise data obtained from whole patch-clamps of vertebrate photoreceptors (rods isolated from the axolotl retina) which can be interpreted on the basis of this assumption (cf. also (23)). Of course, this agreement is far from being something like an experimental proof since one cannot exclude that much more complex interaction models also agree with the experimental results.

MATHEMATICAL DESCRIPTION OF THE GENERAL METHOD

Let x(t) be a noisy signal measured in a steady state of a system. The time-dependence of x denotes the microscopic noise fluctuations of x. The macroscopic control variables of the system are assumed to be kept constant. For the case of noise analysis applied to photoreceptors, x(t) will be the light-induced noisy electric current J(t) crossing the membrane under a constant clamp voltage, constant level of illumination, constant temperature, etc. Let us assume that the signal x(t) is observed during a time interval $-t_0 \leq t \leq +t_0$ as illustrated in Fig. 1. The length $2t_0$ of this interval

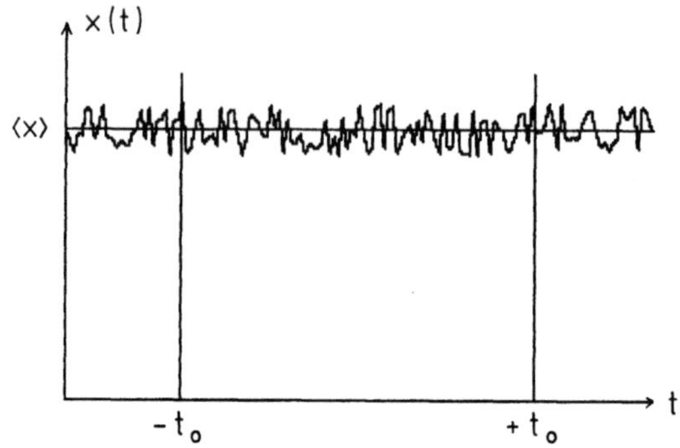

FIG. 1 - Scheme of the noisy signal x(t).

determines the minimum frequency $\omega_0 = 2\pi f_0 = \pi/t_0$ for which reliable conclusions can be drawn from the data.

The following definitions are useful for the description of the method:

a) the time average or the mean of the signal

$$<x> = \lim_{t_0 \to \infty} \frac{1}{2t_0} \int_{-t_0}^{+t_0} dt\, x(t), \tag{1}$$

b) the fluctuations

$$\xi(t) = x(t) - <x> \tag{2}$$

c) the variance

$$\sigma = <\xi^2> = \lim_{t_0 \to \infty} \frac{1}{2t_0} \int_{-t_0}^{+t_0} dt\, (\xi(t))^2, \tag{3}$$

d) the autocorrelation function

$$C(t) = \lim_{t_0 \to \infty} \frac{1}{2t_0} \int_{-t_0}^{+t_0} dt'\, \xi(t'+t)\,\xi(t'), \tag{4}$$

e) the time-limited process

$$\xi_{t_0}(t) = \begin{cases} \xi(t) \text{ for } -t_0 \leq +t_0, \\ 0 \quad \text{else,} \end{cases} \tag{5}$$

f) its Fourier transform

$$\tilde{\xi}_{t_0}(\omega) = \int_{-\infty}^{+\infty} dt\, e^{-i\omega t}\, \xi_{t_0}(t), \tag{6}$$

with $\omega = 2\pi f$, f being the frequency, ω being the angular frequency.

The spectral density or power spectrum of the process is defined and experimentally observed as

$$S(\omega) = \lim_{t_o \to \infty} \frac{1}{t_o} \left| \tilde{\xi}_{t_o}(\omega) \right|^2 . \tag{7}$$

The possibility to obtain information about the underlying microscopic process from S(ω) is based upon the Wiener-Chintschin theorem (4, 8) which relates S(ω) to the autocorrelation function C(t):

$$S(\omega) = 4 \int_0^{\infty} dt \cos \omega t \cdot C(t). \tag{8}$$

The aim of this general method is to assume a model for the x- or ξ-process, to calculate the autocorrelation function C(t) from the model, to compare it with the experimentally observed s(ω), and to determine the model parameters from this comparison if possible. We would like to emphasize that the hazard of this method lies in the fact that it may give answers in terms of model parameters even if the model is totally inadequate by physical or physiological reasons. Nevertheless, noise analysis has been used successfully in studying a number of membrane transport mechanisms, e.g., the acetylcholine-induced noise (1, 13) or ionic transport across epithelial membranes (16, 17).

THE BUMP APPROACH TO NOISE ANALYSIS

As discussed in the introduction, the essential assumption of the bump approach of noise analysis applied to photoreceptors is to interpret the light-sensitive current J(t) as a superposition of single bump events. In general, this assumption is formulated as

$$J(t) = \sum_i B_i(t - t_i - \vartheta_i), \tag{9}$$

where the summation extends over all bumps i = 1,2,3, ... within the time period of observation. By $B_i(t)$ we denote the time-dependent current of the single bump No. i starting at t = 0, i.e., $B_i(t) = 0$ for $t < 0$, and decreasing for $t > d_i$ where d_i is the duration of the bump No. i. The subscript i in $B_i(t)$ takes into account that in general the bump shapes vary from bump to bump. t_i is the time at which the stimulating photon is absorbed. Since the photons are independent (for incoherent light), the t_i obey a Poisson distribution. ϑ_i denotes the latency time of the bump No. i which elapses between the absorption of the photon and the onset of the bump current. For bumps in the ventral photoreceptor of Limulus, the distribution of latency times and of bump shapes have been studied and discussed in detail (e.g., (14, 22), and see Stieve and Schnakenberg and Keiper, both this volume). From these studies, there is convincing evidence that latency and the bump shape, i.e., bump amplification, should be considered as separate and uncorrelated processes.

Since noise analysis enables the determination of only a very few model parameters, it is impossible in general to derive information on all bump parameters from the spectral density of J(t) as defined in Eq. 9. In particular, one cannot derive the statistical distributions of bump parameters from noise data. For this reason, one has to make further approximations. A very incisive but inevitable approximation is the neglection of the variation of the bump shapes, i.e., replacing Eq. 9 by

$$J(t) = \sum_i B(t - t_i - \vartheta_i), \tag{10}$$

where B(t) without subscript i is now to be interpreted as a mean bump shape. The superposition (Eq. 10) of identical mean bumps can then be evaluated by means of the well-known shot noise theory, provided that the latency times ϑ_i of the bump events are mutually uncorrelated. With this assumption, the times $t_i + \vartheta_i$ at which the individual bump currents start are again independent, random events and thus obey the same Poisson distribution as the t_i, irrespective of the type of the latency time distribution. The assumption of uncorrelated latency times ϑ_i is far from being trivial since with increasing bump frequency, i.e., with increasing intensity of the stimulating stationary illumination, an interaction between the bumps will come into play due to light adaptation. One of the consequences of light adaptation is a marked decrease of the latency times. We shall come back to this point below. With the above assumption of uncorrelated latency times, we can make use of Campbell's theorems (8, 19) to derive expressions for the mean, the variance, and the spectral density of J(t):

$$<J> = \lambda \cdot \int_0^\infty dt\, B(t), \tag{11}$$

$$<(\delta J)^2> = \lambda \int_0^\infty dt\, (B(t))^2, \tag{12}$$

$$S_J(\omega) = 2\lambda \left| \widetilde{B}(\omega) \right|^2, \tag{13}$$

where λ is the frequency of bump events, i.e., of successfully absorbed photons, $\delta J(t) = J(t) - <J>$ and $\widetilde{B}(\omega)$ denotes the Fourier-transform of the bump shape B(t) defined by

$$\widetilde{B}(\omega) = \int_0^\infty dt\, e^{-i\omega t} B(t). \tag{14}$$

Equations 11-13 provide a first possibility for determining bump parameters from the experimental data for $<J>$, $<(\delta J)^2>$, and $S_J(\omega)$ without making any further assumption on the detailed type of the bump shape B(t). To this purpose, we define the "height" and the "duration" of a bump by

$$h = \frac{\int_0^\infty dt\,(B(t))^2}{\int_0^\infty dt\,B(t)}\,,\; d = \frac{\left[\int_0^\infty dt\,B(t)\right]^2}{\int_0^\infty dt\,(B(t))^2}\,. \tag{15}$$

First of all, we notice that h has the unit of current and d that of time if B(t) is a current. For any reasonable bump shape B(t) which starts at t = 0 and decays to zero after some time period, h and d as defined in Eq. 15 will yield some rough measure of its maximum amplitude and the time period between its start at t = 0 (i.e., not including the latency phase) and the time at which the bump signal disappears in the background noise, respectively. From Eqs. 13 and 14 we also immediately conclude that

$$S_J(o) = 2\,\lambda \left[\int_0^\infty dt\,B(t)\right]^2. \tag{16}$$

It is now a matter of simple algebraic calculation to express the bump parameters frequency λ, height h, and duration d in terms of the measurable quantities as

$$\lambda = \frac{2<J>^2}{S_J(o)}\,, \tag{17}$$

$$h = \frac{<(\delta J)^2>}{<J>}\,,\quad d = \frac{S_J(o)}{2<(\delta J)^2>}\,. \tag{18}$$

The consistency of this rough evaluation of bump parameters is demonstrated by the fact that

$$h\cdot\ d = \int_0^\infty dt\,B(t) = A\,, \tag{19}$$

i.e., the product of the bump's height and duration yields the bump area A which is the total net charge transfer during a bump.

For a more detailed determination of bump parameters from noise analysis, one has to make use of the information contained in the detailed structure of the spectral density $S_J(\omega)$ as a function of the angular frequency ω. Experimentally, the spectral density of the light-sensitive current J(t) of the Limulus ventral photoreceptor was found (cf. (22)) to be well fitted by an expression of the form

$$S_J\,(\omega) = \frac{S_J\,(o)}{[1 + (\omega t)^2]^{n+1}}\,, \tag{20}$$

i.e., $S_J(0)$, τ, and n can be obtained from the experimental data. By making use of Eqs. 13 and 14, the bump shape B(t) corresponding to $S_J(\omega)$ of Eq. 20 is derived as

$$B(t) = \frac{A}{n! \cdot \tau} \left(\frac{t}{\tau}\right)^n \cdot e^{-t/\tau} \tag{21}$$

(cf. also (13)), where τ is the typical time scale of the bump, n is an integer number determining the initial phase of the bump as $B(t) \sim (t/\tau)^n$, and A is the bump area or net charge transfer,

$$A = \int_o^\infty dt\, B(t). \tag{22}$$

The time course of the bump shape B(t) as given by Eq. 21 is determined by only one single time-scale parameter τ. This is an incisive simplification since the analysis of directly observed bumps shows at least two uncorrelated time scales for the rise (not including latency) and the decay phase of the bumps, respectively (cf. Schnakenberg and Keiper, this volume, and (14)). With this simplification, the bump parameters λ, A, d, and h as introduced before can be calculated from the fitting parameters $S_J(0)$, τ, and n and from the experimental value $<J>$ of the mean current in the following way: λ is obtained from Eq. 17,

$$A = \frac{S_J(o)}{2<J>}, \qquad d = c_n \cdot \tau, \tag{23}$$

where c_n is a number given by

$$c_n = \frac{2^{2n+1} \cdot n!}{(2n)!}, \tag{24}$$

and h = A/d (cf. Eq. 19). This analysis has been performed for the Limulus ventral photoreceptor by Wong, Knight, and Dodge (22). Figure 2 shows the fit of expression (Eq. 20) to the experimental values of $S_J(\omega)$ for four different light intensities.

To summarize this analysis, it has been shown that the frequency λ of bumps increases directly proportional to the light intensity, up to 10^5 bumps per second. Over the same range of light intensities, the mean bump amplitude h decreases as the - 0.7 power of the light intensity, and from the lowest to the highest light intensities the mean bump duration d was found to decrease by a factor of 0.5 (23). Therefore, as a first approximation, this analysis has yielded a self-consistent model of bump summation even in rather bright light. In addition, it provides a description, based on the properties of bumps, for the well-known observation that light adaptation causes the reduction in amplitude and the shortening of the time scale of the light response (cf. (10)). For instance, the decrease in amplitude is due mainly to the reduction of the bump size, and the shortening of the time scale of the response is due in part to the decreases in bump duration (9).

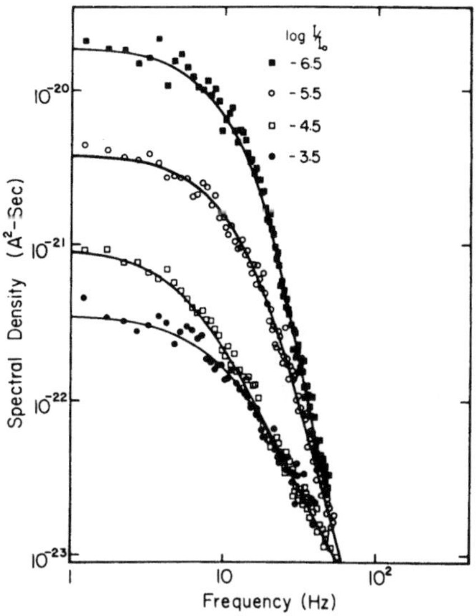

FIG. 2 - Power spectrum $S_J(\omega)$ of the type of Eq. 20 fitted to experimental data obtained at four different light intensities ($\log I/I_0$) with the following choice of values for the parameter n: n = 2 for $\log I/I_0 = -6.5$; n = 1 for $\log I/I_0 = -5.5$; n = 0 for $\log I/I_0 = -4.5$ and - 3.5 (from (20)).

Further interpretation of the observed shapes of the power spectra (Fig. 2) which are related to the bump shape at different light levels will depend on specific assumptions of the underlying mechanisms of bump generation. For example, it is noted that at high light intensities ($\log I/I_0 = -4.5$ and -3.5 in Fig. 2), the shape of the power spectra can be fitted by Eq. 20 with n of 0. This form of Eq. 20 is the well-known lorentzian which is expected for exponential relaxation phenomena. One particular interpretation of this observation is as follows. At these high light intensities, the individual bumps are made up of discrete two-state channels which open and close approximately independently of each other. The size and average opening time of an individual channel can then be estimated from the power spectrum (obtained at $\log I/I_0 = -3.5$) to be 18 pS and 18.7 ms, respectively (20). We emphasize again that this is one of many possible interpretations of the data. However, in the absence of any detailed knowledge of the underlying mechanisms, it provides a good working model for the organization of various observations (cf. (2, 20)).

In spite of the "self-consistency," one might argue that because the bumps adapt and therefore cannot be independent, the assumption of the Poisson shot noise model is in error. This is a valid argument and, specifically, Campbell's theorem for uncorrelated shot noise would be changed. The effect or errors due to the application of the theory of uncorrelated shot noise cannot be calculated without assuming specific models for light adaptation. Such theoretical analyses have been performed based on detailed models (5, 15). A common conclusion from these models is that the effect of correlation would reduce the variance of the signal by a multiplicative factor and that this factor may be estimated from the data. It is not appropriate to review here the detailed calculation of models but to point out that this factor has been estimated, in the extreme case, to be less than 3 (21). Although this argument could not provide any information on the underlying mechanisms of adaptation, it nevertheless suggests that the assumption of independent bumps in the steady state is not seriously in error.

STEADY-STATE APPROACH: THE TWO-STATE CHANNEL

Let us consider the case that transport across membranes is mediated by channels which can exist in an open and a closed state:

$$\text{closed} \underset{\alpha}{\overset{\beta}{\rightleftharpoons}} \text{open} \tag{25}$$

The kinetic equation for this simple model reads

$$\frac{dX}{dt} = \beta \cdot (N - X) - \alpha X , \tag{26}$$

where X is the number of open channels, N is total number of channels, and α and β are the rate constants for the closing and opening of the channels, respectively. From Eq. 26 we obtain the mean number of open channels in the steady state as

$$<X> = \frac{\beta}{\alpha + \beta} N = p \cdot N , \tag{27}$$

where $p = \beta/(\alpha + \beta)$ is the steady-state probability of a channel's being in the open state. Let i be the current of a single channel in the open state such that

$$<J> = p \cdot N \cdot i \tag{28}$$

is the mean steady-state current of all channels.

The noise properties of the two-state channel are readily obtained by interpreting the transitions between the open and closed state in Eq. 25 as Markow-processes of the type of one-step chemical reactions. The calculation of the mean variance and the spectral density of the current fluctuations $\delta J(t) = J(t) - <J>$ is straightforward (cf. (8)), yielding

$$<(\delta J)^2> = N \cdot p \cdot (1 - p) \cdot i^2 , \tag{29}$$

$$S_J(\omega) = \frac{4 \cdot N \cdot p(1-p) \cdot i^2}{\alpha + \beta} \frac{1}{1 + \left(\frac{\omega}{\alpha + \beta}\right)^2} \tag{30}$$

The spectral density is usually evaluated in a double logarithmic plot as shown in Fig. 3.

If the experimental data fit to this type of $S_J(\omega)$, one can evaluate from them $S_J(O)$ and the characteristic shoulder frequency $\omega_c = 2\pi f_c = \alpha + \beta$ as defined by $S_J(\omega_c) = S_J(O)/2$. The details of the further evaluation depend on the particular properties of the system under investigation.

This simple version of noise analysis based on the assumption of a two-state channel can in general not be applied to the light-sensitive current in photoreceptors since at least one of the transition rates α and β is controlled by the stimulating light, very likely by the action of a transmitter. This means that α or β or both are time-dependent including the fluctuations due to the transmitter kinetics. Sigworth (18) has extended the method to the case where α and β vary with time on a time scale which is slow compared to the fluctuations of the channel opening and closing. If, on the contrary, the transmitter fluctuations were fast compared to that of the channels, one would be inclined to treat α and β as functions of the mean transmitter level, which in turn will depend on the intensity of the stimulating light. In the following section we shall present a more detailed study of how the transmitter-channel interaction may influence the noise power spectrum. This treatment will include the two limiting cases mentioned above.

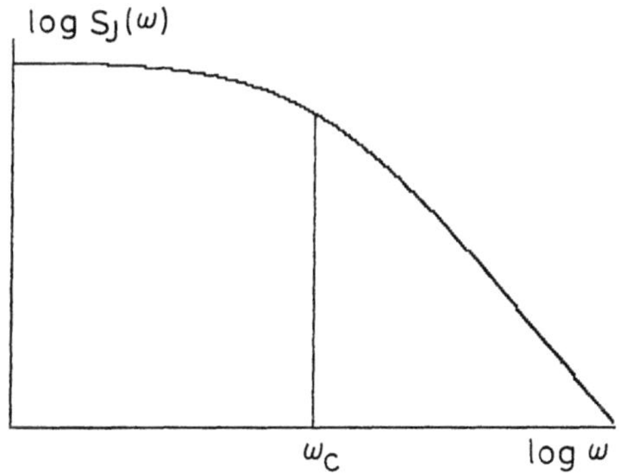

FIG. 3 - Double logarithmic plot of the spectral density $S_J(\omega)$ for the two- state channel model.

STEADY-STATE APPROACH: TWO-STATE CHANNELS INTERACTING WITH A TRANSMITTER

Let us consider a particularly simple model for the interaction of a two-state channel with a transmitter, namely,

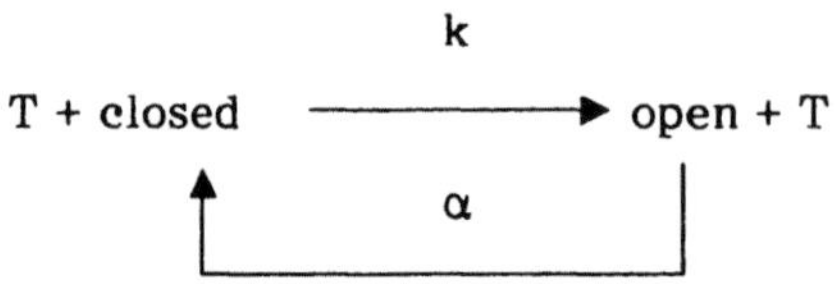

(31)

The kinetic equation for this model reads

$$\frac{dX}{dt} = k \cdot T \cdot (N - X) - \alpha X. \tag{32}$$

T is the number of the transmitter molecules present near the position of the channel, X is the number of open channels, and N is the total number of channels. As in the simple two-state model of the section on STEADY-STATE APPROACH: THE TWO-STATE CHANNEL, α denotes the rate constant for the random closing of channels. In contrast to the model in that section, the opening rate of channels is now assumed to be catalyzed by the action of the transmitter T, i.e., the product k·T replaces the rate constant β of the section on STEADY-STATE APPROACH: THE TWO-STATE CHANNEL. Since we want to include the fluctuations of the transmitter T as well, we have to formulate some kinetic equation for T of the type

$$\frac{dT}{dt} = \phi(T, \ldots), \tag{33}$$

where the dots in the argument of ϕ indicate that the activation and possibly also the deactivation of T depend on some prior step in the transduction process. Fortunately, for the calculation of the spectral density of the channels or of the current J we need not know the detailed structure of the function ϕ in Eq. 33. All we have to assume is that under stationary light stimulation there will evolve some stationary level $\langle T \rangle$ of the transmitter molecules and that the fluctuations $\delta T = T - \langle T \rangle$ obey some linearized equation of the type

$$\frac{d}{dt}\delta T = -s \cdot \delta T \tag{34}$$

with some finite variance $\langle (\delta T)^2 \rangle = \sigma_T \cdot s > 0$ denotes the time constant of the transmitter fluctuations which will depend on $\langle T \rangle$. Of course, Eq. 34 includes the assumption that the transmitter fluctuations can be treated as being independent of the fluctuations of prior steps in the transduction process. If this assumption is not satisfied, we would have to include the fluctuations of those prior steps into the linearized scheme as well. Equation 34 would then be replaced by a system of linear equations with a set of time constants s_1, s_2 ... instead of only one s-value. In any case, Eq.

34 may be interpreted as a shorthand expression of the fluctuations along the transduction process prior to the channel kinetics and s as a typical time constant of the transmitter at the end of the process. In view of the present knowledge of the transduction process, it seems inappropriate to discuss more detailed models.

The calculation of the mean variance and the spectral density of the current fluctuations in our model follows the same line as in the case of the previous section. The final result for the spectral density reads

$$S_J(\omega) = \frac{4 \cdot N \cdot p \cdot (1-p)\, i^2}{r} \cdot Z(\omega)\,, \tag{35}$$

$$Z(\omega) = \left[1 + \frac{p(1-p)rs}{c^2(s^2-r^2)} \frac{\sigma_T}{N}\right] \frac{1}{1+\left(\frac{\omega}{r}\right)^2} - \frac{p(1-p)r^3}{c^2 s(s^2-r^2)} \frac{\sigma_T}{N} \frac{1}{1+\left(\frac{\omega}{s}\right)^2}\,. \tag{36}$$

The result now shows two characteristic shoulder frequencies r and s for the channels and for the transmitter, respectively. The channel frequency r is defined as

$$r = \alpha + k \cdot \langle T\rangle\,, \tag{37}$$

in analogy to $\alpha + \beta$ of the preceding section. The transmitter frequency s has been defined already in context with Eq. 34. The terms "transmitter" and "channel frequency" should be taken as abbreviations for saying that 1/s and $(\alpha + k \langle T\rangle)^{-1}$ characterize the time scales on which the transmitter is active and channels open and close, respectively. As in the preceding section, i denotes the current of a single open channel and p is the steady-state probability of a single channel's being in the open state as defined by

$$p = \frac{k \cdot \langle T\rangle}{r} = \frac{k \cdot \langle T\rangle}{\alpha + k \cdot \langle T\rangle}\,. \tag{38}$$

The constant c denotes the steady-state number of transmitters T available per channel, i.e., $c = \langle T\rangle/N$. c may be larger or smaller than unity.

First of all, let us notice that the spectral density $S_J(\omega)$ of Eqs. 35 and 36 reduces to the result of the previous section as given in Eq. 30 with β replaced by $k \langle T\rangle$ if either $s \gg r$ (very fast transmitter) or $c \gg 1$ (large surplus of transmitters per channel) or $\sigma_T/N \ll 1$ (very small transmitter fluctuations). For the further discussion of our result, Figs. 4 and 5 show double logarithmic plots of $S_J(\omega)$ for two extreme cases, namely, s/r = 100 in Fig. 4, i.e., the transmitter kinetics being 100 times faster than that of the channels, and vice versa, s/r = 0.01 in Fig. 5.

The constants c, p, and σ_T/N have been chosen such that the structure of $S_J(\omega)$ becomes as marked as possible. The numerical values in the figures give the approximate values of the slopes in that particular region, i.e., 2 means $S_J(\omega) \propto \omega^{-2}$ and 4 means $S_J(\omega) \propto \omega^{-4}$. For the fast transmitter (s/r = 100) in Fig. 4, we find $S_J(\omega) \propto \omega^{-2}$ for $r < \omega < s$, and $S_J(\omega) \propto \omega^{-4}$ for $s < \omega$.

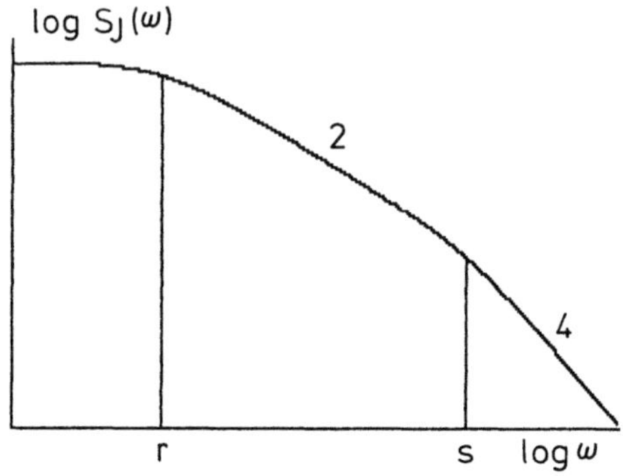

FIG. 4 - Double logarithmic plot of $S_J(\omega)$ for s/r = 100 (cf. text).

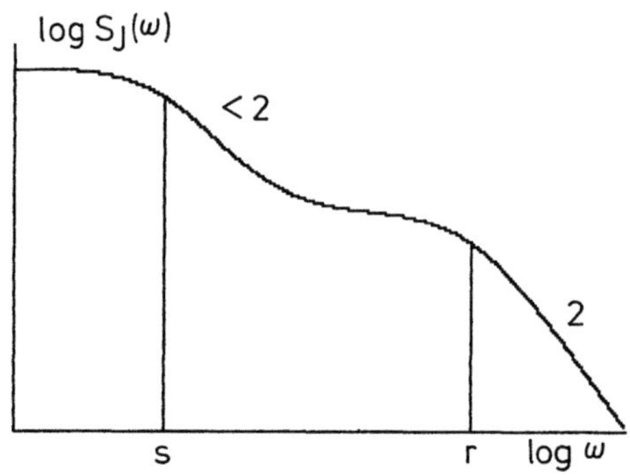

FIG 5 - Double logarithmic plot of $S_J(\omega)$ for s/r = 0.01 (cf. text).

At very high frequencies (not shown in Fig. 3), we find another bend and again a ω^{-2}-behavior beyond. The whole structure, however, is only little pronounced and seems hardly to be detectable if the experimental data include some uncertainty. For the slow transmitter (s/r = 0.01) in Fig. 5, $S_J(\omega)$ shows a much more pronounced structure, in particular a plateau in the frequency range $s < \omega < r$. The slope of the curve between the transmitter frequency s and the plateau is less than 2. We would have expected an $S_J(\omega) \propto \omega^{-2}$ behavior in that frequency range as well; however, the transition from $\omega = s$ to the plateau is too narrow to evolve a distinct ω^{-2} behavior. Regarding the plateau value of $S_J(\omega)$ between s and r, one can conclude from Eqs. 35 and 36 that it is given by

$$S_J(\omega_p) = \frac{4Np(1-p) \cdot i^2}{r} , \tag{39}$$

where ω_p denotes the plateau frequency. This means that the plateau and the high-frequency component ($\omega > \omega_p$) of $S_J(\omega)$ can be evaluated as if the transmitter were present at a constant level without any fluctuations.

A spectral density of the type shown in Fig. 5 has been reported recently by Gray and Attwell (11) for noise data from whole cell patch-clamps of vertebrate photoreceptors (rods isolated from axolotl retina). One may argue that for vertebrate photoreceptors the transmitter should be assumed to cause the closing of channels rather than their opening as in our model. It is clearly evident that an interchange of the words "closed" and "open" in our model scheme (Eq. 31) and of the probabilities p and 1-p does not affect any of our results. One may thus state that the analysis as given by Gray and Attwell and our model at least do not contradict each other. We would like to emphasize, however, that such an agreement should not be taken as experimental proof. In fact, the low-frequency shoulder in those experimental results is not very pronounced. Also, our model implicitly involves a number of assumptions which need not be satisfied, and we cannot exclude that other model versions produce results for $S_J(\omega)$ which are very similar to that of Fig. 5. To begin with, our model assumes a two-state channel. Second, and even more seriously, we assumed that the transmitter opens (or closes) a channel without being bound for a finite time, i.e., for a time which is comparable with the time scale of all other events. It seems more realistic that a bound transmitter-channel complex represents an open (or closed) channel and that the transmitter is deactivated after having opened (or closed) a channel. It could even be possible that a channel has to bind more than one transmitter molecule to be opened (or closed). We have performed some preliminary calculations for such modifications of the model (7). We found that if the transmitter interacts with the channel for a finite time period or if it is deactivated by the interaction, the characteristic shoulder frequencies of $S_J(\omega)$ become algebraic combinations of the time constants of the transmitter and channel kinetics. This would mean that even if the transmitter were fast or slow compared to the channel kinetics, one would expect two shoulder frequencies which are not well separated. Thus Gray and Attwell's results could be interpreted as some indication that the transmitter-channel interaction is "weak" in the sense mentioned before.

Unfortunately, there is no general rule regarding how to draw conclusions from an experimental $S_J(\omega)$-curve which deviates from the simple $S_J(\omega)$-curve of uncorrelated and independent channels as shown in Fig. 3. Such deviations may be a) appearance of more than one shoulder; b) an $S_J(\omega) \propto \omega^{-n}$-behavior with $n > 2$ at high frequencies; c) a "low-frequency attenuation," i.e., $S_J(\omega)$ has a maximum not at $\omega = 0$ but at some finite frequency ω_m, $S_J(\omega)$ increases moderately for $\omega < \omega_m$, and $S_J(\omega) \propto \omega^{-n}$ at high frequencies for $\omega > \omega_m$ (cf. Minke, this volume).

In all such cases, there are obviously several processes involved in the noise generation of the light-induced current, very probably the channel itself and at least one further control process. As far as shoulder frequencies are distinguishable, it is clear that their values characterize the time scales of the involved processes. All further speculations as to how the involved processes are coupled and what their biochemical nature is are beyond the border line of rigorous implications from noise analysis and rely on perhaps obvious model assumptions such as that of the bump hypothesis in the section on THE BUMP APPROACH TO NOISE ANALYSIS or the transmitter hypothesis in this section.

CONCLUSION

We have tried to show that a careful interpretation of noise analysis data for photoreceptors may give some helpful but limited information about the underlying mechanism. To repeat the critical point of noise analysis once more: noise analysis is a rather indirect method. The interpretation of noise data requires knowledge of our assumptions on the detailed structure of the underlying mechanism. It may happen that the noise data fit to the calculated spectral density based on a specific model although the model is a rather bad description of the system. This uncertainty particularly applies to the question as to under which conditions the bump approach or the steady-state approach should be made use of when analyzing or interpreting the experimental noise data obtained in photoreceptors. One might argue that as long as the "bump frequency" calculated from Eq. 17 is proportional to the light intensity, the bump approach is appropriate. This is not a conclusive argument since the right-hand side of Eq. 17 coincides with the "bump frequency" only if it is assumed that the noisy current consists of bump events at the corresponding intensity of the stimulating light. The same right-hand side of Eq. 17, however, could also be interpreted, e.g., in terms of uncorrelated and independent channels as presented in the section on STEADY-STATE APPROACH: THE TWO-STATE CHANNEL. Making use of Eqs. 28 and 30, it is a simple matter of algebraic transformation to obtain under this assumption

$$\frac{\beta \cdot (a+\beta)}{2a} N = \frac{2<J>^2}{S_J(o)}, \tag{40}$$

with the same definitions for α, β, and N as in the above-mentioned section. If the right-hand side of Eq. 40 is found to be proportional to the light intensity, this could now be interpreted by saying that the opening (or closing) rate β is proportional to the intensity of the stimulating light and $\alpha \gg \beta$, i.e., the system is far from saturation. The same right-hand side of Eqs. 17 or 40 may be submitted to even further interpretations if use is made of still other model assumptions of noise generation such as, e.g., the mechanism presented in the section on STEADY-STATE APPROACH: TWO-STATE CHANNELS INTERACTING WITH A TRANSMITTER. To conclude this discussion, we would like to repeat our suggestion to try different ways of interpreting the experimental noise data in terms of different models which seem reasonable in the corresponding situation, to discuss the conclusions very carefully, and to find arguments independent of the noise analysis results which favor one of the approaches. As far as the analysis of bumps or transmitter-controlled channels is concerned, a direct observation of those events (cf. (14) for bumps and (2) for single channels in the ventral photoreceptor of Limulus) provides a more reliable picture of the underlying mechanisms in any case.

Acknowledgements. The authors would like to express their gratitude to H. Stieve and to B. Minke and S. Barash (Jerusalem) for careful reading of the manuscript, many valuable allusions, and helpful critical remarks.

LIST OF ABBREVIATIONS

A	bump area (net charge transfer)
α	rate constant for closing of a single channel
B(t)	bump shape (time course of a bump)
$\tilde{B}(\omega)$	Fourier transform of B(t)
β	rate constant for opening of a single channel
c	average number of transmitters available per channel
C(t)	autocorrelation function of a noisy signal
d	bump duration
f	frequency
ϕ	transmitter kinetics
h	bump height (amplitude)
i	current through a single open channel
I/I_0	ratio of reduction of the stimulating light
J(t)	noisy electric current
$\langle J \rangle$	time average of J(t)
$\delta J = J(t) - \langle J \rangle$	fluctuations of J(t)
k	rate constant for channel opening by a transmitter
λ	bump frequency
n	order of the initial rise phase of a bump B(t)
N	total number of channels
$\omega = 2\pi f$	angular frequency
ω_c	shoulder frequency
ω_p	plateau frequency
p	probability of the open state of a channel
r	time constant of channel kinetics involving interaction with transmitters
s	time constant of transmitter fluctuations
$S(\omega)$	spectral density
$S_J(\omega)$	spectral density of the electric current
σ	variance
σ_T	variance of transmitter fluctuations
t_0	half observation time for a noisy signal
T	fluctuating number of transmitters
$\langle T \rangle$	time average of T
ϑ	latency time
τ	time-scale parameter of bumps
x(t)	arbitrary noise signal

$<x>$	time average of x(t)
X	number of open channels
$\xi(t)=x(t)-<x>$	fluctuations of x(t)
$\xi_{t_0}(t)$	fluctuations ξ (t) within the time interval of observation $-t_0 < t < t_0$
$\xi_{t_0}(\omega)$	Fourier transform of $\xi_{t_0}(t)$

REFERENCES

(1) Anderson, C.R., and Stevens, C.F. 1973. Voltage clamp analysis of acetylcholine produced end-plate current fluctuations at frog neuromuscular junction. J. Physiol. **235:** 655-691.

(2) Bacigalupo, J., and Lisman, J.E. 1983. Single-channel currents activated by light in Limulus ventral photoreceptors. Nature **304:** 268-270.

(3) Baylor, D.A.; Matthews, G.; and Yau, K.-W. 1980. Two components of electrical dark noise in toad retinal rod outer segments. J. Physiol. **309:** 591-621.

(4) Chen, Y.-D. 1978. Noise analysis of kinetic systems and its applications to membrane channels. Adv. Chem. Phys. **37:** 67-97.

(5) Celasco, J., and Stepanescu, A. 1977. Power spectrum of pulse sequence with correlation between pulse shape and pulse separation time. J. Appl. Phys. **48:** 3635-3638.

(6) Cone, R.A. 1973. The internal transmitter model for visual excitation: some quantitative implications. In Biochemistry and Physiology of Visual Pigments, ed. H. Langer, pp. 273-282. Berlin: Springer-Verlag.

(7) De Felice, L.J. 1981. Introduction to Membrane Noise. New York and London: Plenum Press.

(8) Dirnberger, G.; Keiper, W.; Schnakenberg, J.; and Stieve, H. 1985. Comparison of time constants of single channel patches, quantum bumps and noise analysis in Limulus ventral photoreceptors. J. Membr. Biol. **83:** 29-43.

(9) Dodge, F.A.; Knight, B.W.; and Toyoda, J. 1968. Voltage noise in Limulus visual cells. Science **160:** 88-90.

(10) Fuortes, M.G.F., and Hodgkin, A.L. 1964. Changes in time scale and sensitivity in the ommatidia of Limulus. J. Physiol. **172:** 239-263.

(11) Gray, P., and Attwell, D. 1985. Kinetics of light-sensitive channels in vertebrate photoreceptors. Proc. Roy. Soc. Lond. B **223:** 279-388.

(12) Hillman, P. 1982. The biophysics of intermediate processes in photoreceptor transduction: "silent" stages, non-localities, single photon responses and models. In Proceedings of the Symposium on The Biology of Photoreceptor Cells, pp. 443-475. Cambridge: Cambridge University Press.

(13) Katz, B., and Miledi, R. 1972. The statistical nature of the acetylcholine produced potential and its molecular components. J. Physiol. **224:** 665-699.

(14) Keiper, W.; Schnakenberg, J.; and Stieve, H. 1984. Statistical analysis of quantum bump parameters in Limulus ventral photoreceptors. Z. Naturforsch. **39c**: 781-790.

(15) Knight, B.W. 1973. A stochastic problem in visual neurophysiology. In American Mathematical Society Symposium on Stochastic Differential Equations, eds. J. Keller and H.P. McKean. Providence, RI: American Mathematical Society.

(16) Lindemann, B. 1980. The beginning of fluctuation analysis of epithelial ion transport. J. Membr. Biol. **54**: 1-11.

(17) Lindemann, B., and Van Driesche, W. 1977. Sodium-specific membrane channels of frog skin are pores: current fluctuations reveal high turnover. Science **195**: 292-294.

(18) Sigworth, F.J. 1981. Interpreting power spectra from nonstationary membrane current fluctuations. Biophys. J. **35**: 289-300.

(19) van Kampen, N.G. 1981. Stochastic Processes in Physics and Chemistry. Amsterdam, New York, Oxford: North-Holland Publishing Company.

(20) Wong, F. 1978. Nature of light-induced conductance changes in ventral photoreceptors of Limulus. Nature **175**: 76-79.

(21) Wong, F., and Knight, B.W. 1980. Adapting bump model for eccentric cells of Limulus. J. Gen. Physiol. **76**: 539-557.

(22) Wong, F.; Knight, B.W.; and Dodge, F.A. 1980. Dispersion of latencies in photoreceptors of Limulus and the adapting bump model. J. Gen. Physiol. **76**: 517-537.

(23) Wong, F.; Knight, B.W.; and Dodge, F.A. 1982. Adapting bump model for ventral photoreceptors of Limulus. J. Gen. Physiol. **79**: 1089-1113.

The Molecular Mechanism of Photoreception, ed. H. Stieve, pp. 389-399. Dahlem Konferenzen 1986. Berlin, Heidelberg, New York, Tokyo: Springer-Verlag.

Molecular Genetic Approach to the Study of Phototransduction in Drosophila

F. Wong
Marine Biomedical Institute and
Dept. of Physiology and Biophysics
University of Texas Medical Branch
Galveston, TX 77550, USA

Abstract. It has become apparent that phototransduction is a complex process involving multiple molecular components. Therefore, a major task in unravelling phototransduction is to define precisely the functional roles of these molecular components and their interrelationships. The newly developed techniques of molecular genetics may be useful for achieving these goals because they make available a systematic approach to study these problems. In this paper, a specific example of studies utilizing these techniques to analyze a Drosophila mutation (trp) which directly affects phototransduction is discussed. A general plan to study the trp gene is also outlined in order to provide a framework for examining the potential contributions of this approach.

INTRODUCTION

In the past few years, recombinant DNA technology has proved useful in providing the powerful techniques needed for the study of many scientific problems. In the field of neuroscience, some fundamental principles are beginning to emerge as a result of application of recombinant DNA technology. For example, novel brain-specific genes have been identified, and in the rat brain, most of these genes have been found to contain an identifier nucleotide sequence which shares homology with a brain-specific cytoplasmic RNA (12, 22). These sequences may control the expression of brain-specific genes and, therefore, are fundamentally important for understanding the operation of differentiation as well as neuronal function. Other novel mechanisms involved in the determination of neuronal phenotype have also been identified. Selective RNA splicing was shown to be the mechanism that gives rise to the tissue-specific production of alternate polypeptide products from the calcitonin gene; namely, the precursor of a novel neuropeptide which predominates in the brain and the precursor to the hormone calcitonin which predominates in thyroid C cells (1, 19). These observations suggest general principles which may be applicable to the study of phototransduction. I shall discuss this matter in greater detail later in this paper.

In the area of phototransduction, the rod opsin gene from bovine and from human have been cloned (15, 16). There are indications that the human cone opsin genes have also been identified. This provides the necessary first step to test the hypothesis that human rod and cone pigments are encoded by different opsin genes and that hereditary alterations in color vision are due to defects in different cone opsin genes (15, 16). The outcome of these studies could have immediate practical applications. In addition, comparison of several opsins, including a Drosophila opsin, would enable us to speculate about sequence homologies which may have implications for function ((17, 28); see also Findley and Applebury et al., both this volume.) There is little doubt that the study of phototransduction could benefit from the direct use of these new techniques and concepts.

PHOTOTRANSDUCTION AND THE MOLECULAR GENETIC APPROACH

As evidenced from the several articles on messenger substances presented in this volume, a major task in unravelling the mechanisms of phototransduction is to define concisely the functional roles of the underlying molecular components (proteins) and their interactions. The list of proteins includes rhodopsin, G-protein (transducin), cGMP phosphodiesterase, rhodopsin kinase, the "48K protein," the molecules with which calcium ions interact, and the "light-dependent" conductance channels. The list will grow to include the still unidentified components which may be necessary for the complete phototransduction system. The exact functional roles in excitation and adaptation of each of these proteins have to be determined and their interactions have to be defined.

A common approach to study a multi-component system is to isolate the individual parts and study the dissociated protein components. A major drawback of this approach is that it does not allow the study of interactions of the various components. If several of these components are isolated, their interactions could be studied in a reconstituted system. In the case of phototransduction, a major problem is that the relative importance of the various components is still unclear. As pointed out by Bownds and Brewer (this volume), experiments measuring light-initiated physiology and biochemistry in chemically defined living photoreceptor cells are very much needed.

The newly developed techniques of molecular genetics may be appropriate for studying phototransduction because they offer novel and powerful ways to manipulate proteins. This is achieved through efficient methods to isolate and manipulate genes. For instance, procedures are available for the rapid determination of nucleotide sequences. Once the sequence of a gene is known, the protein sequence and often the structure may be deduced easily. This information could be used to "design" antibodies targeted at particular regions of the protein and thus could serve as specific markers for studying the developmental expression and cellular localization of the protein (24) as well as specific probes for studying physiological function. Most importantly, genes can be modified in vitro and the foreign DNA introduced into an organism so that its functional expression can be studied (20). These techniques would yield a physiologically and biochemically defined system within which the structure of a protein important to transduction can be modified systematically, through precise alterations, while the consequences of such manipulations are measured. This approach would not only provide the ideal system to establish the function of a protein that is important for phototransduction, but also would circumvent the requirement of having to identify all the components before studying their interactions. The power of these new techniques and their wide range of applications have been amply demonstrated and argued for in various reviews (see, for example, (21)); they should be directly useful for a systematic approach to study phototransduction.

A first step in this molecular genetic approach is to isolate the genes that are known to be important for phototransduction. The fruitfly Drosophila offers an ideal system for these studies because techniques for manipulating Drosophila genes are available that are virtually unparalleled in other model systems. Furthermore, there are several available mutations in Drosophila that are thought to affect some components of phototransduction (18). The identified mutations fall into two general categories. The first category includes the mutations known to affect rhodopsin; these are the PDA mutants ((18), and Minke, this volume). The second category includes the mutations that are thought to affect intermediate steps of transduction; these are no receptor potential A (norp A), transient receptor potential (trp), and retinal degeneration B (rdg B). In fact, norp A and trp are thought to affect the mechanisms underlying bump occurrence. (The significance of bumps and the bump parameters which are affected by the norp A and trp mutations, respectively, are discussed by Stieve, this volume.) The major defect caused by the rdg B mutation appears to be retinal degeneration. However, rdg B is thought to be involved in phototransduction because it interacts (genetically) with the norp A mutation in an allele-specific manner (7). To my knowledge, out of the mutants in the second category, only the trp gene has been isolated. In this paper, I shall describe some recent studies of this gene as a framework for discussing this general approach and its potential contributions.

STUDIES OF THE trp LOCUS

The trp Phenotype

The mutant trp was first isolated on the basis of its behavioral phenotype. The mutants behave normally in low ambient light but behave as though blind in bright light (4). When studied with the intracellular recording technique, it was found that the receptor potential decayed to near baseline during an intense, prolonged stimulus. This decay is attributed to a decline in the rate of bump occurrence in bright light (14). However, the mutation appears to affect steps of transduction prior to the occurrence of bumps because in the mutant, the individual bumps seen in dim light resemble those of the wild type in amplitude and duration (14). The defect does not appear to be related directly to internal calcium ion concentration which is known to affect adaptation of invertebrate photoreceptors (13).

In order to define the site of the mutational effect, properties of the photopigment in the mutant have been examined. The content and absorption characteristics of rhodopsin in this mutant were found to be normal (14), suggesting that the mutation is likely to affect steps subsequent to the pigment phototransition. Therefore, the mutation is likely affecting a step in the mechanisms linking the light-induced changes in rhodopsin and the subsequent membrane conductance changes. More specifically, it is affecting the mechanisms underlying bump generation.

The defective visual response is observable at eclosion when no other defects caused by this mutation can be observed. However, beginning shortly after eclosion, a progressive degeneration of the retina can be observed (5, 13). Since a defective receptor potential can be observed in the mutant prior to retinal degeneration, it can be concluded that the defective mechanism of bump generation is a primary rather than a secondary result of the mutation. Because the mutational effect is highly specific, namely, in altering the rate of bump occurrence, it is likely that the trp gene product would play a direct role in transduction. Therefore, it is hopeful that identification of the products encoded by the trp gene and knowledge of the molecular basis of the defect caused by the trp mutation would lead to an understanding of the normal bump-generating mechanisms.

Genetic Studies

The trp mutation first appeared spontaneously in a highly inbred line of a wild-type strain. Since then, several EMS-induced alleles of trp have been isolated. These mutations are recessive and all mapped to a single locus near the tip of the right arm of the third chromosome, at map position 100 (18). The cytological location of this locus is defined by two duplication-deletions with virtually identical breakpoints in the 99C5-6 chromosomal bands (26, 27). This precise localization limits the trp locus to a region of less than 60 kilo-basepairs (kb) of DNA. This stretch of approximately 60 kb of DNA shall be referred to as the trp region and should encompass the trp locus.

Molecular Cloning

A 16-kb fragment of DNA (559) which had previously been isolated (9) was found to overlap with the trp region. Therefore, a viable strategy for cloning the trp gene is to isolate the entire trp region using the available fragment as a probe to identify, by homology, partially overlapping genomic clones that extend into the adjacent DNA. By reiterations of this procedure, each time using as probes DNA fragments that are farthest from the fragment that was used in the previous step as the probe, a long stretch of DNA extending from both sides of the original fragment used to initiate the process can be obtained. This procedure is known as chromosomal walking (2). A long stretch (about 100 kb) of DNA which includes the trp region has been isolated by this method (26). The following discussion will focus on the analysis of this stretch of DNA.

Size of the trp Gene

In order to identify the coding sequence (exons) in the trp region, the set of genomic clones spanning the trp region was used to hybridize to RNAs isolated from normal flies in Northern analyses (10). Four RNA species of 6.0, 3.5, 1.5, and 0.8 kb, respectively, were thus identified (27). They are referred to as RA, RB, RC, and RD, respectively. These four RNAs are polyadenylated and, except for RA (which is currently under study), have been shown to direct the synthesis of proteins in a cell-free in vitro translation system (Wong et al., unpublished results).

Since the total length of the RNAs identified is about 12 kb, approximately 1/8 of the length of the corresponding DNA region, it is likely that the exons are separated by relatively long non-coding regions (introns). In other words, the gene(s) encoding for these RNAs is (are) split such that the RNAs identified are each made up of multiple exons. In fact, some of the exons from the different RNAs were found to intermingle (Wong et al., unpublished results). That is, two adjacent but nonoverlapping DNA fragments, A and C, would share homology with the same RNA species, RB and RC. Such observations would suggest that they must identify (share homology with) different portions of the RNA molecules, which in turn implies that exons located far apart in the genomic DNA have been joined together in forming the RNAs. More interestingly, a third DNA fragment B, which is located between fragments A and C but does not overlap with either A or C, was shown to share homology with the RNA species RA. Taken together, these results would imply that the exons of RNAs RA, RB, and RC are intermingled. Similarly, the 0.8 kb RNA species RD appears to have an exon located between those of RA and RB. Given the present knowledge of gene structure and mechanisms of RNA processing, the simplest interpretation of these results is that the four RNAs are all derived from the same gene, probably by some form of alternate RNA processing. A corollary of this conclusion is that the trp locus may encompass more than 60 kb of DNA and, therefore, is relatively large.

Molecular Basis of Phenotype

With the isolated DNA, which should contain the trp gene, questions concerning the molecular basis of the mutant phenotype can be asked. Previously, two-dimensional electrophoresis gel analysis had revealed that multiple proteins are altered in the trp mutant (25). Such results were difficult to interpret because it was difficult to distinguish the primary defect on the protein level from the pleiotropic effects of a single mutation. On the other hand, defects identified in the RNA derived from the trp gene must be the primary effect of the mutation and thus would establish the causal relationship in the multiple changes observed on the protein level. A common problem of these experiments is that the defects may not be easily identified. Conventionally, single point mutations are thought to involve basepair substitutions or small deletions leading to frame-shift mutations. Such mutations could cause dramatic effects on the production of the protein products but may not alter the size of the RNA sufficiently so that the changes are difficult to detect. On the other hand, if the mutant defect in DNA occurs in a control region, the resulting RNA may have a large detectable change. In the case of the trp mutation, some obvious changes were in fact observed (in the original Cosens-Manning allele). The 3.5 and 1.5 kb RNA species identified from the trp region in the wild type were missing in the mutant (Fig. 1). This is a specific defect of the mutant and therefore could be the result of the trp mutation. The simplest conclusion from these studies is that the 3.5 and 1.5 kb sequences are part of the trp gene and that the lack of one or both of these two RNAs may be responsible for the molecular basis of the trp phenotype.

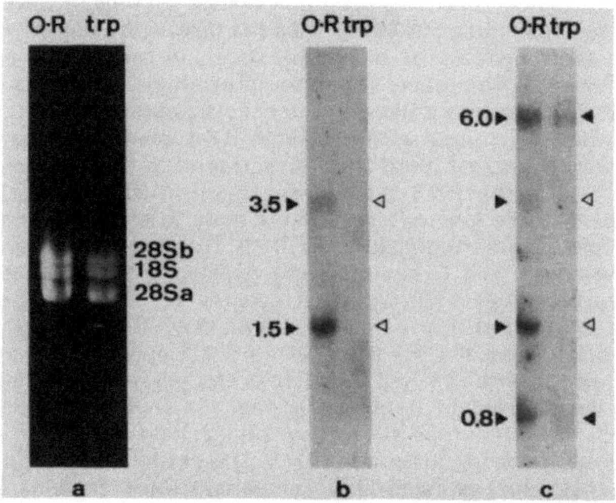

FIG. 1 - a) Under ultraviolet illumination, the most prominent bands appearing on the agarose gel were the components of the 28S and 18S ribosomal RNAs. The sizes of these bands are: 1.93, 1.85, and 1.64 kb. O-R: wild-type Oregon-R strain. b) Northern blot analysis with a DNA segment from the trp region as a probe revealed two RNA bands (3.5 kb and 1.5 kb) in the wild type. These two bands were missing entirely in the mutant. c) Northern blot analysis with an adjacent, non-overlapping DNA segment as a probe revealed four RNA bands in the wild type: the two RNA bands identified previously, the 6.0 kb and the 0.8 kb bands. The last two bands were also present in the mutant (from (27)).

Molecular Basis of Mutation

Determination of the molecular basis of the mutation may, in this case, have direct relevance to phototransduction because the mutation appears to occur in a region of DNA which controls the appearance of two RNAs. By comparing the restriction map of the mutant DNA and that of the wild type, a 2.3 kb deletion was found in the mutant (Wong et al., unpublished results). The relatively large deletion detected is unusual for a spontaneous mutation, according to the conventional theories of mutation. However, more modern examples and theories are available to account for large deletions based on the interactions of the mechanisms of replication and repair and DNA structure. For example, large deletions could arise through misalignments due to distant repeated sequences or to DNA palindromes - sequences in which one strand read in the usual 5' to 3' direction is identical to that of the complementary strand, also read in its 5' to 3' direction. Misalignment occurs when DNA sequences pair correctly in an incorrect context during replication, leaving many intervening bases unpaired. The unpaired sequence is removed enzymatically and, as a result, a large deletion occurs in the mutant. Combinations of repeated sequence misalignment and palindrome misalignment have been proposed to explain a spontaneous deletion in the E. coli lac I gene (6).

Plans for Future Studies

Studies of the trp gene will be pursued along three main lines: to establish the identity of the trp products, to establish the functions of the trp products, hence their roles in transduction, and to determine the control of the expression of the trp gene.

1) One way to determine the identities of the various trp products is to compare their amino acid sequences with all known protein sequences. For example, it has been found recently that a portion of the amino-terminal peptide sequence of the transducin alpha subunit is highly homologous with the corresponding region in the ras protein which is a protooncogene product (8). Such comparisons may determine if the trp products are familiar or novel proteins. If the trp products turn out to have homologous peptide sequences with some familiar proteins, they may also have analogous functions. The sequence information would also allow the design of antibodies targeted at unique sites of the proteins for studying the localization of the various products.

2) The functional roles in transduction of the trp products can also be studied by a combination of transformation and site-directed mutagenesis experiments. First, the wild-type trp gene will be introduced into the mutant to see if it would rescue the defect - to produce bumps in bright light. If successful, this will provide the necessary system to carry out the studies of function. For example, different regions of the trp gene may be modified and the resulting effects on the various bump parameters can be studied.

3) The 2.3 kb deletion identified in the trp region is most likely the site of the DNA defect. This fragment of DNA will be sequenced. This region of DNA may contain a promoter that is required for the proper expression of the trp gene. Alternatively, it may contain an important splice junction needed for proper mRNA processing.

POTENTIAL CONTRIBUTIONS TO UNDERSTANDING PHOTOTRANSDUCTION

The framework and techniques covered so far in this paper are quite general. Therefore, it is reasonable to discuss the potentials of the molecular genetic approach in the context of studying the trp gene. Furthermore, isolation of the genes encoding for other important components of transduction will not be long in coming (8). In this regard, the Drosophila system may serve as a general vehicle to study phototransduction. Recently, it has been observed that many monoclonal antibodies specific for molecules in the fly brain cross-reacted with specific molecules in the human brain, suggesting that close homology, hitherto unsuspected, may exist between the two systems (11) or, for that matter, in many other systems. The proteins which are important for transduction and have been studied extensively in other systems (8) may share homology in structure and function with their counterparts in the fly. Therefore, the homologous genes isolated from these systems could be introduced into the fly where their functions can be studied, perhaps more efficiently.

The two proteins derived from the 3.5 and 1.5 kb RNAs must be important to transduction since the lack of which would prevent bump occurrence. Since both proteins are missing in the mutant, it is not clear if one or both are important for transduction. In fact, if the trp gene indeed yields four distinct proteins, the other two proteins (which are present in the mutant) may also be related to transduction. Information concerning the exact roles which they play and their differential regulations would also be relevant to the understanding of phototransduction.

Finally, one could speculate along the line of coordinated expression of retinal genes. In the differentiation of photoreceptors, a battery of photoreceptor-specific genes must be coordinately expressed. These genes which exhibit common regulation must contain common control sequences. This simple idea is part of a generally accepted model of eukaryotic gene expression (3). More recently, a lineage-based gene control model has been proposed (23). The model is based primarily on the discovery in the rat of a common sequence called an identifier sequence, found in most adult brain-specific mRNA precursors. It is argued that these sequences must be involved in the coordinated control of neuronal gene expressions. It seems logical that some such common regulators would exist for photoreceptor- (or eye-)specific genes. If such common sequence can be found it may lead to the identification of many, and ultimately all, of the necessary components of transduction. So far, the opsin gene sequence is the only one available among the components known to be important for phototransduction (15-17, 28). Further research of this kind on photoreceptor- (or eye-) specific genes might help to put these ideas to experimental tests. In this regard, it should be mentioned that several homologous sequences have been identified between the DNA flanking the 5' ends of the human and bovine opsin genes. These sequences may partake in the regulation of the opsin genes (16).

CONCLUSION

I have discussed in this paper a specific example of how the molecular genetic approach is used to study phototransduction. Techniques used to analyze the trp gene are typical of the kinds used in this general approach. Based on the current knowledge of the trp gene, I have also speculated on several theoretical possibilities. These speculations serve to illustrate the broad context, afforded by this approach, for a systematic multidisciplinary study of phototransduction.

Acknowledgements. The author thanks his collaborators with whom the research conducted in his laboratory was performed. This research was suppported by grant EY03308 from the U.S. National Institutes of Health and by the Marine Biomedical Institute General Budget.

REFERENCES

(1) Amara, S.G.; Jonas, V.; Rosenfeld, M.G.; Ong, E.S.; and Evans, R.M. 1982. Alternative RNA processing in calcitonin gene expression generates mRNAs encoding different polypeptide products. Nature **298**: 240-244.

(2) Bender, W.; Akam, M.; Karch, F.; Beachy, P.A.; Peifer, M.; Spierer, P.; Lewis, E.B.; and Hogness, D.S. 1983. Molecular genetics of the bithorax complex in Drosophila melanogaster. Science **221**: 23-29.

(3) Britten, R.J., and Davidson, E.H. 1969. Gene regulation for higher cells: a theory. Science **165**: 349-357.

(4) Cosens, D., and Manning, A. 1969. Abnormal electroretinogram from a Drosophila mutant. Nature **224**: 285-287.

(5) Cosens, D.J., and Perry, M.M. 1972. The fine structure of the eye of a visual mutant, A-type, of Drosophila melanogaster. J. Insect Physiol. **18**: 1773-1786.

(6) Farabough, P.J.; Schmeissner, U.; Hofer, M.; and Miller, J.H. 1978. Genetic studies of the lac repressor. VII. On the molecular nature of spontaneous hot spots in the lac I gene of E. coli. J. Molec. Biol. **126**: 847-857.

(7) Harris, W.A., and Stack, W.S. 1977. Hereditary retinal degeneration in Drosophila melanogaster: a mutant defect associated with the phototransduction process. J. Gen. Physiol. **69**: 261-291.

(8) Hurley, J.B.; Simon, M.I.; Teplow, D.B.; Robishaw, J.D.; and Gilman, A.G. 1984. Homologies between signal transducing G proteins and ras gene products. Science **226**: 860-862.

(9) Levy, L.S.; Ganguly, R.; Ganguly, N.; and Manning, J.E. 1982. The selection, expression, and organization of a set of head-specific genes in Drosophila. Devel. Biol. **94**: 451-464.

(10) Maniatis, T.; Fritsch, E.F.; and Sambrook, J. 1982. Molecular Cloning, a Laboratory Manual. Cold Spring Harbor: Cold Spring Harbor Laboratory.

(11) Miller, C.A., and Benzer, S. 1983. Monoclonal antibody cross-reactions between Drosophila and human brain. Proc. Natl. Acad. Sci. USA **80:** 7641-7645.

(12) Milner, R.J.; Bloom, F.E.; Lai, C.; Lerner, R.A.; and Sutcliffe, J.G. 1984. Brain-specific genes have identifier sequence in their introns. Proc. Natl. Acad. Sci. USA **81:** 713.

(13) Minke, B. 1982. Light-induced reduction in excitation efficiency in the trp mutant of Drosophila. J. Gen. Physiol. **79:** 361-385.

(14) Minke, B.; Wu, C.-F.; and Pak, W.L. 1975. Induction of photoreceptor voltage noise in the dark in Drosophila mutant. Nature **258:** 84-87.

(15) Nathans, J., and Hogness, D.S. 1983. Isolation, sequence analysis, and intron-exon arrangement of the gene encoding bovine rhodopsin. Cell **34:** 807-814.

(16) Nathans, J., and Hogness, D.S. 1984. Isolation and nucleotide sequence of the gene encoding human rhodopsin. Proc. Natl. Acad. Sci. USA **81:** 4851-4855.

(17) O'Tousa, J.E.; Baehr, W.; Martin, R.L.; Hirsh, J.; Pak, W.L.; and Applebury, M.L. 1985. The Drosophila ninaE gene encodes an opsin. Cell **40:** 839-850.

(18) Pak, W.L.; Conrad, S.K.; Kremer, N.E.; Larrivee, D.C.; Schinz, R.H.; and Wong, F. 1980. Photoreceptor function. **In** Development and Neurobiology of Drosophila, eds. O. Siddiqu, P. Babu, L.M. Hall, and J.C. Hall, pp. 331-346. New York: Plenum Publishing Corp.

(19) Rosenfeld, M.L.; Lin, C.R.; Amara, S.G.; Stolarsky, L.; Roos, B.A.; Ong, E.S.; and Evans, R.M. 1982. Calcitonin mRNA polymorphism: Peptide switching associated with alternative RNA splicing events. Proc. Natl. Acad. Sci. USA **79:** 1717-1721.

(20) Rubin, G.M., and Spradling, A.C. 1982. Genetic transformation of Drosophila with transposable element vectors. Science **218:** 348-353.

(21) Shortle, D.; DiMaio, D.; and Nathans, D. 1981. Directed mutagenesis. Ann. Rev. Genet. **15:** 265-294.

(22) Sutcliffe, J.G.; Milner, R.J.; Bloom, F.E.; and Lerner, R.A. 1982. Common 82-nucleotide sequence unique to brain RNA. Proc. Natl. Acad. Sci. USA **79:** 4942-4946.

(23) Sutcliffe, J.G.; Milner, R.J.; Gottesfeld, J.M.; and Reynolds, W. 1984. Control of neuronal gene expression. Science **225:** 1308-1315.

(24) Sutcliffe, J.G.; Milner, R.J.; Shinnick, T.M.; and Bloom, F.E. 1983. Identifying the protein products of brain-specific genes with antibodies to chemically synthesized peptides. Cell **33:** 671-682.

(25) Wong, F. 1980. Visual mutations in Drosophila melanogaster: a genetic and biochemical analysis of the trp mutants. Association for Research in Vision and Opthalmology. Southern Section Meeting.New Orleans.

(26) Wong, F.; Hokanson, K.M.; and Chang, L.-T. 1984. Molecular analysis of visual mutation in Drosophila. Neurosci. Abstr. **10:** 619.

(27) Wong, F.; Hokanson, K.M.; and Chang, L.-T. 1985. Molecular basis of an inherited retinal defect in Drosophila. Inv. Ophthalmol. Vis. Sci. **26:** 243-246.

(28) Zuker, C.S.; Cowman, A.F.; and Rubin, G.M. 1985. Isolation and structure of a rhodopsin gene from D. melanogaster. Cell **40:** 851-858.

Standing, left to right:
Heino Prinz, Benjamin Kaupp, John Findlay, Andrea Cavaggioni

Seated (middle), left to right:
Marc Chabre, Helen Saibil, Paul Liebman, Meredithe Applebury, Andreas Schleicher

Seated (front), left to right:
Lubert Stryer, Nelson Goldberg, Hermann Kühn

The Molecular Mechanism of Photoreception, ed. H. Stieve, pp. 401-429. Dahlem Konferenzen 1986. Berlin, Heidelberg, New York, Tokyo: Springer-Verlag.

Triggering and Amplification

Group Report

M.L. Applebury, Rapporteur
A. Cavaggioni
M. Chabre
J.B.C. Findlay
N.D. Goldberg
U.B. Kaupp
H. Kühn
P.A. Liebman
H. Prinz
H.R. Saibil
A. Schleicher
L. Stryer

INTRODUCTION

The molecular events which link photon absorption to membrane conductance changes in photoreceptor cells begin with light activation of rhodopsin. In vertebrates, activated rhodopsin (Rh*) triggers an enzymatic cascade resulting in hydrolysis of cyclic GMP (cGMP). Some 80-90% of the protein in the photoreceptor rod outer segment is devoted to the execution or regulation of this enzymatic cascade. We may, therefore, specify light-controlled cGMP metabolism as a major function of the vertebrate rod outer segment. Indeed, cGMP has been proposed to serve as the important messenger linking photon absorption to membrane conductance changes. It may well play this role and be involved in other important cellular functions. It has not been our goal to settle this issue. Rather, we set out to examine thoroughly the rhodopsin-triggered mechanisms controlling metabolism at the molecular level. We extended our examination of these mechanisms to invertebrates, drawing on recent evidence and underlying expectations which suggest that initial events in visual transduction will be conserved among diverse species.

The identified components in the cGMP enzyme cascade are Rh* (light-activated rhodopsin), G-protein (transducin), cGMP phosphodiesterase (PDE), rhodopsin kinase, and 48K-protein. To complete the metabolic pathway, a phosphoprotein phosphatase and a guanylate cyclase must participate; however, the latter two are poorly characterized. Amplification is inherent in the pathway since one Rh* activates many G-proteins by catalyzing GDP-GTP nucleotide exchange. In turn, G*-GTP can activate the phosphodiesterase enzyme achieving elevated cGMP hydrolysis. We emphasized that this amplification may be only a part of the amplification described as linking photon absorption to membrane conductance changes in the photoreceptor cell (see Fig. 1).

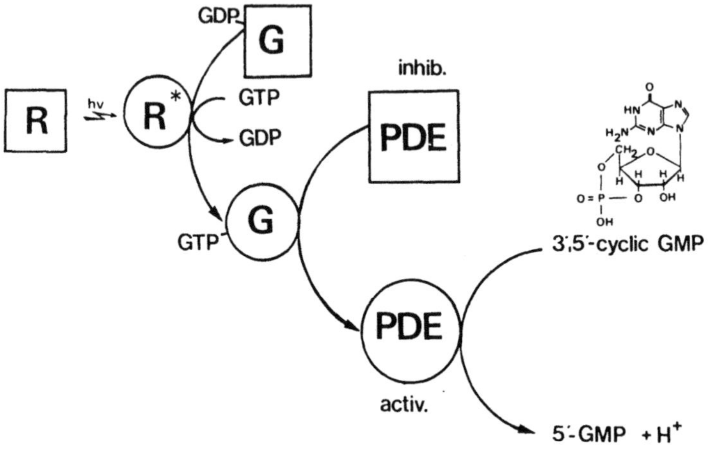

FIG. 1 - The cGMP cascade. One Rh* may activate as many as 500 G- proteins. Interaction of G*-GTP with PDE increases the PDE enzymatic activity by 200-1000-fold.

In the following sections, we record the highlights of our discussions about the interactions between the protein components of the cGMP cascade. We have tried to identify the important unsolved problems in this complex pathway. Finally, we explore the functional significance of cGMP in the photoreceptor and touch on new molecular investigations that are needed to further our understanding of visual transduction.

RHODOPSIN: EXAMINATION OF THE PHOTORECEPTOR IN MOLECULAR DETAIL

We have reached a new era in the study of visual pigments. The elucidation of the amino acid sequences of several rhodopsins has permitted the construction of tentative structural models for rhodopsins that are consistent with biophysical and chemical data. Such models are important because they stimulate the design of new experiments to help us address the significance of conserved and variant domains, the structural basis for rhodopsin activation, the basis of color perception, and the functional relationships between visual pigments in vertebrate and invertebrate systems.

Sequence Comparisons

Recently, the complete primary sequences of bovine, ovine, and human rhodopsins and the partial sequence of porcine and equine rhodopsins have appeared (Findlay, this volume). The mammalian proteins exhibit a remarkably high conservation (>90%) in linear sequence. Only two small regions of the polypeptide chain appear to have marked variability; one is in the region of the retinal binding site, the second is between residues 194 and 198. The significance of these limited variabilities is unknown. Partial sequence information on the chicken opsin (Findlay, this volume) suggests that the mammalian and avian proteins may also have a homology greater than 90%.

The first invertebrate visual pigment, Drosophila rhodopsin, has now been characterized by recombinant DNA techniques and gene sequencing (47). We compared the structure of this pigment with the mammalian opsins. About 36% of the Drosophila opsin amino acids are identical to those of the vertebrate group. Analysis of both primary and predicted secondary structure confirm that the protein is in principle like the vertebrate opsins, even though there is a significant degree of amino acid substitution. Interestingly, there is high conservation in the retinal binding domain and two other regions to which no functional roles have yet been ascribed (see below). Paradoxically, the C-terminal end of opsin, which is phosphorylated and is a region of defined functional significance in vertebrate opsins, is not conserved in Drosophila, although the latter does have potential sites for phosphorylation.

In addition, the first information about mammalian cone pigments has been generated by use of recombinant DNA techniques. Preliminary observations (J. Nathans, unpublished data) suggest that although similarities between rod opsins and cone opsins from the same species can be recognized, the variability is substantial. Red and green cone pigments are quite similar but vary from the blue cone pigment.

An Examination of Secondary and Tertiary Structure

The determination of a high-resolution, three-dimensional structure of opsin awaits further development of methods applicable to crystallizing membrane proteins. Until the exact structure is forthcoming, it is possible to arrive at folding patterns and secondary structure predictions which are consistent with all the available biochemical, chemical, and biophysical information (13, 49). Although care should be taken not to over-interpret the models, they can serve as points of reference.

The general usefulness of such models is illustrated by the relative ease with which the Drosophila sequence can be accommodated (Fig. 2) - even to the point of preservation of the putative regions of helix distortions which may have functional relevance. Comparative analyses of the Drosophila and mammalian opsins show that two conserved regions map at the first cytoplasmic and second intradiscal loop of the protein, respectively. The latter shows a potential EF-hand structure which may predispose it to Ca^{2+} or Mg^{2+} binding (33). A 15 residue insert is found in the third cytoplasmic loop previously predicted to form a large helical region extending into the aqueous phase (49).

The Transmembrane Nature of Rhodopsin

Fifty to sixty percent of the opsin mass is associated with the hydrophobic phase of the membrane bilayer and is probably organized as seven transmembrane helices. Functionally, these helices provide an intramembrane domain for the retinal chromophore. The retinal environment must facilitate cis-trans isomerization of 11-cis retinal, provide the necessary

A

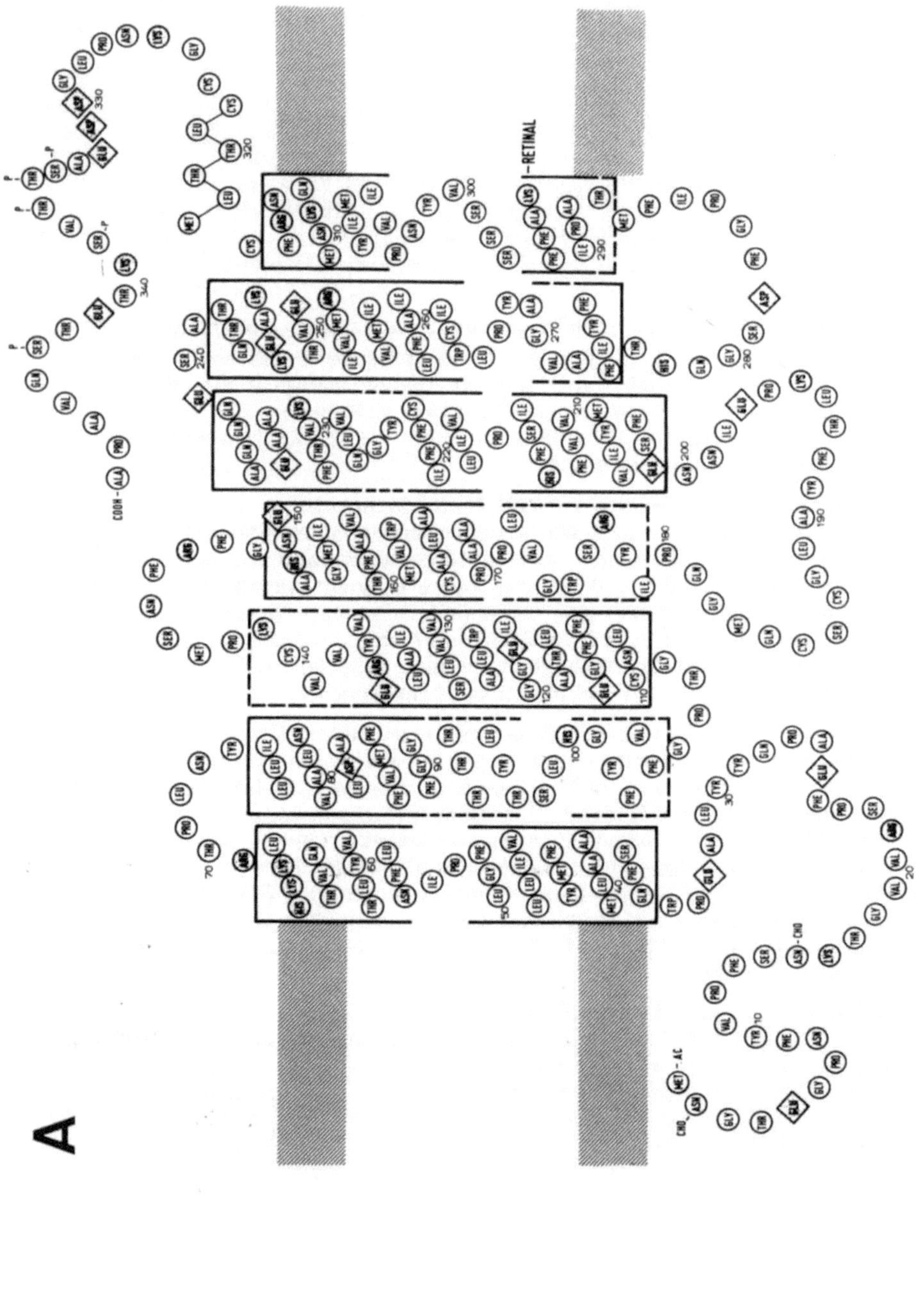

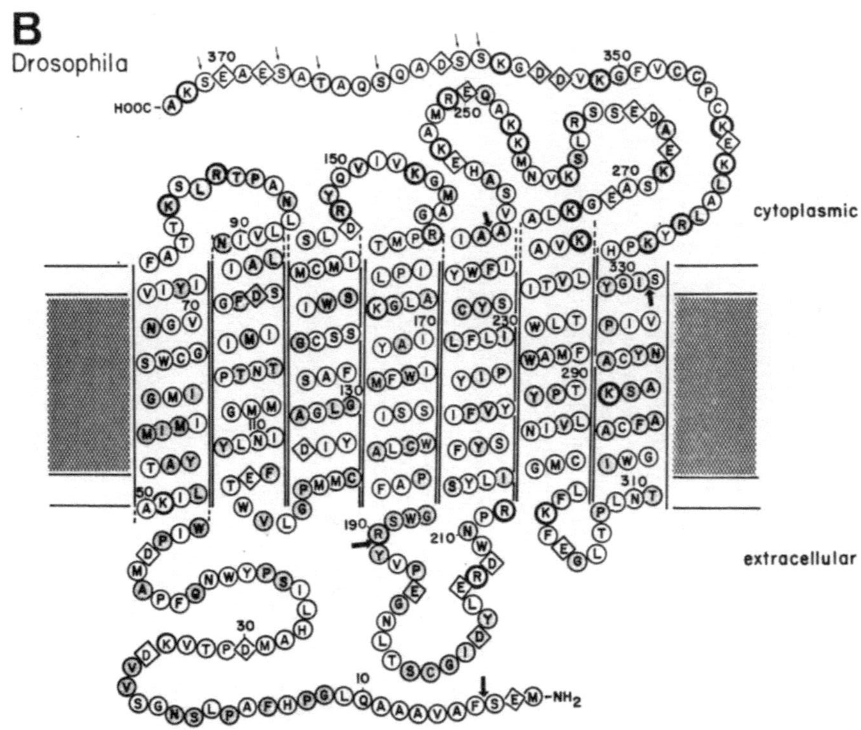

FIG. 2 - Models predicting the polypeptide chain folding of (A) Ovine opsin (from (49)) and (B) Drosophila opsin (from (47)). Both the mammalian opsins (71) and invertebrate opsins (52) undergo light-dependent phosphorylation. Potential sites (marked by P or thin arrows) are identified in the C- terminals of both proteins.

means of regulating the wavelength of absorption, and transmit conformational change following the absorption of light.

We questioned whether there are additional functional aspects of this organization. We drew on analogies of opsin with other relevant biological systems and noted the following features: a) an integral membrane protein (and by extrapolation, its rigidly bound chromophore) is functionally oriented. This state, however, does not require more than a single transmembrane helix. Moreover, the few proteins whose essential activities are extramembranous and whose sequences are known have only one or two intramembranous segments. b) The intramembranous domain in the bilayer clearly provides the binding site for the chromophore, but this should not be regarded as an obligatory requirement. Many proteins whose function involves the binding of a hydrophobic cofactor are not associated with a membrane. Thus, a hydrophobic site could equally be achieved in a peripheral membrane domain. c) Although the intramembranous protein structure provides the binding site for retinal, it is not impossible to envisage that it also provides other functional capacities like those found for bacteriorhodopsin and halorhodopsin (64). d) The intramembranous domain appears to possess charged and hydrophilic amino acid side chains, which in general do not appear to be exposed to the phospholipid milieu.

We concluded that since opsin conforms to the general organization of those few proteins classified as transport mediators rather than those which have a purely receptive, enyzmatic, and/or interactive function, it is not inconceivable that it may also have a transport role in addition to those already defined. Only further experimentation will test this hypothesis.

Peripheral Functional Domains

The transmembrane segments of opsin are interconnected by peripheral domains. At the cytoplasmic face, the predicted domains may interact with soluble components and peripheral membrane components. Likewise, predicted domains at the intradiscal surface in rods, or extracellular surface in cones and rhabdomeres, may provide interactive sites for other extracellular components. In our mechanistic examination of rhodopsin action, we will need to identify the functional domains on the cytoplasmic face which interact with the G-protein, rhodopsin kinase, and the 48K-protein. In invertebrates, for example, squid and crayfish, an additional cytoplasmic domain may be needed for immobilizing the rhodopsin.

For the intradiscal or extracellular face there is no currently known functional demand, despite the fact that about 25% of rhodopsin resides on this side of the membrane. Comparison of mammalian and Drosophila rhodopsins reveal that the 4-5 interhelical loop is the most strongly conserved domain in the protein structure, which is suggestive that it might have some potential function. The loop has the potential to serve as an imperfect Ca^{2+} or Mg^{2+} binding site, but such a role is yet to be experimentally confirmed. Such a function would be welcomed for the intradiscal rod space or the intermicrovillar space to provide a storage site for Ca^{2+}.

Summary
A major goal in the next few years for the examination of the structure and function of rhodopsin, and other visual pigments, will be to define the functional domains rigorously. In this respect, site-directed mutagenesis may provide a useful approach.

RHODOPSIN: STRUCTURAL ASPECTS OF THE R → Rh* TRANSITION AND Rh* DECAY

What Is Rh*?

In the visual system, the initial event in phototransduction may be specified as the photochemical conversion of rhodopsin to a new "activated" state. Cis-trans isomerization of the retinal chromophore is the driving force coupled to opsin conformational changes, but these changes are subtle and are not definitively established. For our considerations of the cGMP cascade, the light-activated rhodopsin (Rh*) must acquire a new capacity to interact with G-protein and this capacity may be regulated by the interaction with other proteins such as opsin kinase and 48K-protein.

We reviewed the evidence that suggests that there are distinct conformational differences between rhodopsin and light-activated rhodopsin: a) Rh* can be phosphorylated ~100 times faster than rhodopsin by an intrinsic rod outer segment rhodopsin-kinase (71). b) Exogenously added thermolysin cleaves a peptide bond near the carboxyl terminus (between residues 341-342) about 1.5 times faster in Rh* than in rhodopsin (41). These two observations probably reflect the same conformational change, although the intrinsic kinase appears more adept at recognizing this change than the protease. c) Several chemically detectable conformation changes occur; there is increased accessibility of one SH group (10); the retinal chromophore becomes accessible to $NaBH_4$; protonation changes are observed. d) Light-induced binding of G-protein, so prominent at low ionic strength in the absence of GTP, also indicates conformational change (36). The binding site for G-protein is probably separate from the bulk of phosphorylation sites since removal of most of the phosphorylation sites by treatment with thermolysin does not influence light-induced binding of G-protein (40).

There are some suggestions for specific cytoplasmic domains that bind G-protein. A first clue implicated the loop between helices 5 and 6 (40). Proteolytic removal of several amino acids from the loop between helices 5 and 6 abolishes light-induced binding of G-protein and activation of nucleotide exchange; a single clip in the loop without removal of amino acids has little effect (Wehmer, Hargrave, and Kühn, unpublished data). However, this loop is strikingly different in mammals and Drosophila, even though both opsins probably activate G-protein. Since loops 1-2 and 3-4 have regions of strong conservation, these domains may be good binding sites. There are several caveats in these interpretations. One must consider the possibility that a structural change induced locally may affect a distal structure of interest. Although conservation of a specific domain may imply that the region is an important functional structure, it need not be for G-protein activation. Moreover, the binding domain may involve more than one cytoplasmic loop. These considerations must be kept in mind for future experimental identification of domains.

Which Photoproduct Is Rh*?

Strong evidence from spectroscopic data (reviewed in (9, 37)) indicates that Meta II is the species that binds G-protein. In the absence of GTP, the Rh*-G complex is artificially stabilized. Under these conditions, G-protein binding shifts the MI-MII equilibrium in favor of MII (14, 15). In the presence of low GTP concentrations (<10 μmol/l), a transient equilibrium shift is observed with a decay proportional to [GTP] (K.P. Hofmann, unpublished data). Assays of light-activated cGMP phosphodiesterase activity carried out in the presence of GTP also suggest that MII is the appropriate Rh* candidate in accord with the following observations (reviewed in (46)): a) PDE activation via G-protein is tightly coupled to the thermal dependence of the MII formation rate. b) The pH-dependence of MII formation matches that of phosphodiesterase activation (via G-protein) and does not implicate MI. c) When measured using weak bleaches, activation of PDE persists only as long as the lifetime of MII. d) NH_2OH destroys MII and PDE activation at the same rate. Most evidence thus suggests a strict correlation between the MII photoproduct and the binding of G-protein. It is important to point out, however, that the functional Rh* is not necessarily related to the traditional spectroscopically defined intermediates. Other rate-limiting changes, not identified spectrally, are possible.

How Does Rh* Decay?

Spontaneous decay of Rh* (defined by its capacity to bind and activate G-protein) is identical with MII decay (reviewed in (29, 37)). The Rh* decay is certainly not related to the slow retinal dissociation from opsin. Although ATP seems necessary for the deactivation process, it is unclear whether phosphorylation does or does not accelerate Rh* decay. The spectral decay of MII measurable by absorption is not accelerated by phosphorylation. Phosphorylation, however, has been postulated to interfere with G-protein binding (43). The susceptibility of light-activated rhodopsin to phosphorylation by the kinase is not well correlated with a defined photoproduct. Both MII and MIII can be phosphorylated, and Paulsen and Bentrop (51) suggest that even MI may be phosphorylated. Opsin is not a good substrate for the kinase, nor is rhodopsin (62). Thus, it seems more appropriate to describe Rh* decay as the blocking or quenching of Rh*. For instance, the interaction of 48K-protein with phosphorylated Rh* could quench its capacity to interact with G-protein (see below).

Summary

Although we can generally correlate the light-activated state of rhodopsin with the spectrally identified MII, we know little about the structural features that "define" this state. The mechanism of Rh* decay is a significant, unanswered question. While it is clear that light-activated rhodopsin can slowly decay of its own accord to an inactive MIII and later photoproducts, Rh* is inactivated much faster by an ATP-dependent process in which rhodopsin kinase and 48K-protein seem to be involved.

EXAMINATION OF PHOTORECEPTOR MEMBRANE TOPOGRAPHY

The interactions between Rh* and peripheral proteins and their kinetics are governed by the membrane topography of the photoreceptor cell. We thus considered the photoreceptor membrane's structural organization and its implications for molecular encounters within a confined space.

Summary

In both vertebrates and invertebrates there is compartmentalization of photoreceptor units. In the rod cell, the unit is ~30 nm in total thickness, consisting of the interdiscal space confined by the two rhodopsin-containing membranes. The space is partitioned off by structural intermembrane linkages located every 7 nm around the disc rim (56). Confined within this restricted space are G-proteins, cGMP phosphodiesterase, rhodopsin kinase, the appropriate phospho-opsin phosphatase, presumably the 48K-protein, and possibly other important minor components not yet identified (Fig. 3A).

In invertebrates, the microvillar construction results in a different compartmentalization, where the unit is a 60 nm tube (Fig. 3B). G-proteins (59, 67),

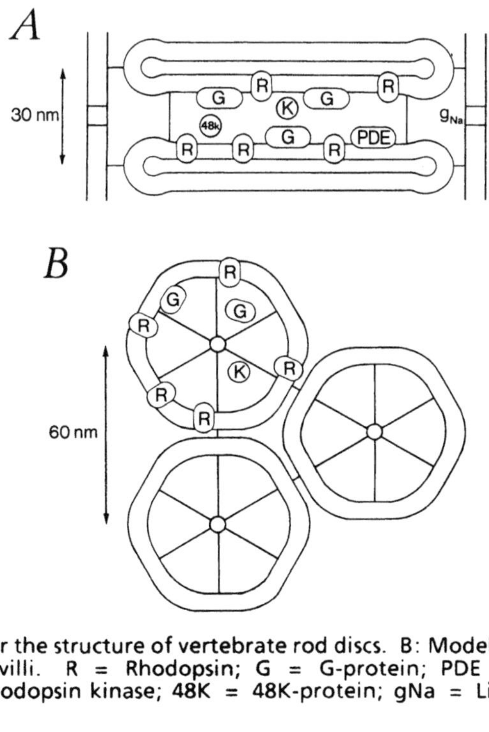

FIG. 3 - A: Model for the structure of vertebrate rod discs. B: Model for the structure of invertebrate microvilli. R = Rhodopsin; G = G-protein; PDE = cGMP phosphodiesterase; K = Rhodopsin kinase; 48K = 48K-protein; gNa = Light-sensitive cation conductance.

rhodopsin kinase (52, 67), and the phosphatase (52) must be confined within the microvillar space, which is also traversed by an internal cytoskeleton (57). Dibutyryl cGMP elevates the threshold of the octopus light response (75), and a light-activated increase in cGMP content has been found in squid (58). However, the site of action of G and the role of the cascade in invertebrate phototransduction have not yet been established.

EXAMINATION OF Rh*-PROTEIN INTERACTIONS

The interaction of Rh* with G-protein triggers the cGMP cascade and the interaction of Rh* with other proteins must regulate and turn off Rh*. We discussed the temporal sequence of events and the nature of the Rh* state that interacts with these proteins.

Rh*-R Interactions

Models have often been proposed in which multiple rhodopsins might form a pore, channel, or larger macroscopic surface with triggering activity (16). In these models, one Rh* must interact with neighboring unbleached molecules since a single light-activated rhodopsin suffices to trigger a visual response. There is no experimental evidence for such oligomer formation; moreover, recent data of Downer and Cone (12) show that the rotational diffusion constant of rhodopsin does not change upon bleaching. The latter provides the best evidence to rule out such models. We conclude that this receptor operates as a monomer.

Rh*-G-protein Interactions

The interaction of G-protein with Rh* is the best documented event in the cGMP cascade. Specific binding can be stabilized in the absence of GTP and measured by sedimentation of membranes or by retention of G-protein on affinity columns consisting of ConA-bound, bleached rhodopsin (reviewed in (37)). The binding is reversed and the nucleotide exchanged upon addition of GTP. The fact that G-protein influences the MI-MII equilibrium (15) and influences the MII-MIII conversion (54) suggests that the binding is direct and interactive. The binding constant of G-protein for Rh* is calculated to be $\sim 1 \times 10^7$ mol/l^{-1} if a 3-dimensional interactive system is assumed (30, 50); or if a 2-dimensional interactive system is considered in which action is confined to the membrane surface, the value becomes $\sim 1 \times 10^5$mol/l^{-1} (50). The lifetime of the complex in the presence of GTP/GDP would be a valuable addition to our understanding of mechanism. Kinetic determination of GTP turnover at room temperature suggests that an upper limit for the Rh*-G lifetime ranges between 500 and 2000 s^{-1} for a bleach of 10^{-5} (44). Fluorescence relaxation measurements might provide an estimate of such values, although these are fraught with experimental difficulties.

Rh*-Rhodopsin Kinase Interaction

The C-terminal end of rhodopsin is specified as one site of kinase interaction; seven serine and threonine residues may be phosphorylated in this region (27). Two additional sites of phosphorylation exist on one or more of the other cytoplasmic loops (71). Steric competition must exist between the kinase and the 48K-protein and G-protein. In the absence of GTP, G-protein indeed inhibits kinase action on Rh* (reviewed in (37)). Knowledge is lacking about the normal level of phosphorylation in vivo, about the action of the phosphatase, and whether different phosphorylation levels serve different functions.

Rh*-48K-protein Interaction

The 48K-protein was one of the first components shown to have marked light-induced interaction with the ROS membranes (35), but its mechanism of interaction has only recently been studied (39). Light-induced binding becomes much stronger in the presence of ATP (or GTP). In bovine preparations, 48K-protein binds to Rh* in the absence of ATP but only at high bleaches; this might be interpreted to be due to the presence of residual phosphorylated rhodopsin since binding is not observed in frog membranes. Recently acquired data suggest that 48K-protein binds phosphorylated rhodopsin and imply that the role of ATP (or GTP) in this process is to serve as a substrate for the kinase (39). Both phosphorylation by ATP and a light-induced conformational change are needed to achieve binding. 48K-protein will not bind phosphorylated rhodopsin in membranes which have been regenerated with 11-cis retinal until they are again bleached. Since non-hydrolyzable analogs of ATP do not promote binding, it seems clear that phosphorylation is necessary to provide the binding site.

Kinetic studies by Zuckerman et al. (76) have shown that the ATP quench is enhanced by 48K-protein and that the quench is severely impaired by antibodies to 48K-protein (or the equivalent S-antigen antibodies). Although other experiments lead Zuckerman to hypothesize a more elaborate role for 48K-protein, e.g., as an ATP-binding protein that interferes with G-protein-PDE interaction, the antibody results are certainly compatible with the notion that the phosphorylated MII state binds 48K-protein to block further G-protein interaction.

The stoichiometry of binding is less well understood. 48K-protein is a major protein present at ~3% of rhodopsin (54). It is a very soluble protein and easily lost in rod preparations. The binding affinity to phosphorylated rhodopsin and its dependence on the extent of phosphorylation need to be established.

Summary

Interactions of Rh* with itself or unbleached rhodopsin are ruled out; interactions of Rh* with G-protein, rhodopsin kinase, and 48K-protein have now been identified. The dynamic competition between these proteins for interaction with the rhodopsin cytoplasmic face and the specific roles they play need further investigation.

IDENTIFICATION OF PROBLEMS IN STUDYING RECONSTITUTED MEMBRANES

Reconstitution of functional photoreceptor membranes has played a key role in deciphering the mechanisms involved in the cGMP cascade. The systems, in which purified components were reassembled, identified the role of G-protein in activating cGMP phosphodiesterase (22, 66), in identifying light-scattering signals associated with binding of G-protein to rhodopsin (38), and most recently in identifying potential roles for 48K-protein ((39); Kühn et al., unpublished results; Liebman et al., unpublished results). In spite of the valuable information we have gathered, there are pitfalls in both the reassembly of the system and its study. The kinetics of G-protein binding and PDE activation observed in freshly isolated ROS membranes have never been reproduced in reconstituted systems. The following notes some of the problems in studying reconstituted systems and our reservations about the system.

In freshly isolated ("normal") ROS membranes, bright light-induced PDE activation occurs within milliseconds, whereas that in reconstituted membranes at similar protein concentrations takes 1-2 s. The delay between bleaching and PDE activation in "normal" bovine ROS membranes is 0.8 s for a 10^{-5} bleach, but that in reconstituted membranes is 5 to 7 s (reviewed in (45)). Some of this problem may reside in differences in the structure of membranes being studied where "normal" ROS membranes more closely resemble the native stacked configuration of the photoreceptor cell. The state of the reconstituted proteins, however, is also in question. Addition of purified G-protein to "normal" ROS membranes in which PDE is kinetically potent indicates that the G-protein is kinetically and thermodynamically abnormal. The binding affinity of purified G-protein for ROS membranes is ~10-fold lower than that calculated for G-protein before purification. G-protein activity decays more quickly on storage than PDE activity (45).

In reconstituted systems, addition of G-protein gives variable maximal levels of PDE activation upon membrane bleaching, and the activation at a saturating bleach often falls below the maximum PDE activity obtained by limited trypsin proteolysis (43). To achieve maximal levels, addition of a 20-fold excess of G-protein over PDE is often necessary (63). Since the natural ratio of G-protein to PDE is 6 to 10, this may suggest that there is a weak interaction between these two proteins. Such weak interaction may be designed to provide rapid kinetics of deactivation upon GTP-hydrolysis (45), but this conclusion awaits proof that all G-proteins are functional.

Summary

Even though reconstituted systems have provided us with insight about the sequence of interactions in the cascade, we have yet to find conditions under which they will yield insightful dynamic information for the native photoreceptor state. Proper reassembly demands reproducing the correct membrane topography, component stoichiometry, and component functional state. We have not yet achieved experimental control over these properties.

AMPLIFICATION AND CONTROL IN THE cGMP CASCADE

Amplification and regulation in the cGMP cascade are well documented (Chabre and Applebury, this volume), but the particular molecular mechanisms by which they are achieved are not well understood. The following discussions revolved about potential molecular mechanisms, controversies, and the need for more insightful data.

Mechanisms of G-protein Activation

The initial stage of amplification in the cGMP cascade is governed by the number of G-proteins that one Rh* can activate per unit time. This number depends upon the encounter rate of G GDP for Rh*, the lifetime of the Rh*-G complex, and the release rate of G*-GTP from Rh*. Considerable controversy surrounds the evaluation of these rates as well as the currently proposed molecular mechanisms associated with these processes. We identified and addressed selected aspects of G-protein activation:

"Hopping" versus "skating." The encounter of G-protein with its activator Rh* and, subsequently, its target phosphodiesterase may be accomplished by "hopping" from one membrane to the other in the interdiscal space or by "hopping" along the membrane surface. Alternatively, these two encounter processes may be confined to the membrane surface where activation was initiated and may take place by "skating." Sorting out these possibilities is problematic, since the interpretation of Rh*-G mechanisms demands that we consider kinetic and thermodynamic processes occurring in ordered photoreceptor discs (Fig. 1 and Fig. 3 in paper by Miller (this volume)). We acknowledged that local organization of the proteins may well facilitate reactions which might normally seem improbable.

A note of caution concerning experimental interpretation was suggested. Observed rate constants in fractured membranes may depend on the nature of the membrane preparations. If the proteins are arranged on the membrane as a dense array, the experimentally observed rates may appear too slow, since the rate-limiting step may be retarded diffusion of substrate along a matrix of binding sites, e.g., a "chromatography effect." The rates may appear too fast where there are only a few binding sites on a given patch of membrane if two-dimensional diffusion along the surfaces facilitates the rates, e.g., an "antenna effect." Additionally, the rate measured may not be directly proportional to the concentration of reactive species, since the number of species effectively employed (encountered) on a two-dimensional surface does not exceed the ratio of the membrane diameter to the collective diameter of the molecules capable of reaction. Any of these effects may lead to rate constants which are off by orders of magnitude (see also (2)).

Liebman and colleagues argue strongly for "skating" mechanisms (44, 45). These investigators suggest that if G-protein hopped to encounter its Rh* activator or its PDE target, dilution should slow activation because the bimolecular encounter rate between G-protein and PDE would be reduced. In experiments assessing the rate of light activation of PDE, dilution of the preparations has little effect on the level of activity and causes no change in the rate of PDE activation. There is, however, a rapid and continuous loss of initial activation velocity upon dilution which is interpreted to be loss of G-protein from the membrane once it dissociates from PDE.

Liebman and Sitaramayya (45) have tried to evaluate the rate constants for G-protein binding to Rh* membranes (k_1) and dissociation from unactivated membranes (k_{-1}) by mixing normal or reconstituted unactivated membranes with "freshly" light-flashed rhodopsin membranes containing only PDE in the presence of GTP and cGMP. When G-protein must transfer from reconstituted unactivated membranes, $k_1 = 2.5 \times 10^5$ mol/l^{-1}s^{-1} and $k_{-1} = 0.025$ s^{-1}; when G-protein must transfer from normal unactivated membranes, $k_1 = 4.0 \times 10^5$ mol/l^{-1}s^{-1} and $k_{-1} = 0.4$ s^{-1}. At first glance, the dissociation rates suggest that G-protein may transfer rapidly enough from reconstituted membranes to be consistent with PDE activation kinetics, but not so for transfer from normal unactivated membranes. These results argue that hopping of G-protein through the aqueous phase to activated rhodopsin is unlikely under conditions where G-protein has not been removed and attempts at reconstitution have been made. The caveat in this interpretation is that normal membranes may remain stacked and impede the dissociation of G-protein from finding the added membranes with Rh* and PDE, thus giving a rate irrelevant to the native photoreceptor membrane organization. These experiments do not address the question of whether already activated G-protein hops or skates to find PDE.

The effect of nucleotides on Rh*-G lifetime. The problem of determining the lifetime of G-protein interaction with Rh* is complex and may depend on the GTP, GDP, or analog nucleotide(s) present and the nature of the interactions of Gα and Gβγ with Rh*. Recent work of Yamanaka and Stryer (to be published) indicates that GTP and GTP analogs differ in their susceptibility to hydrolysis and their affinities for G-protein. In a step towards clarifying the dynamics of Rh*-G interactions, binding affinities and rates of exchange of bound nucleotides have been measured:

	Relative binding affinity	Relative exchange rate of the bound nucleotide
GDP	1	880
GppCh2p	1.7	640
GppNHp	5.9	200
GTP-γS	100	1

The exchange rates are inversely related to binding affinity. The free energy gap between G-GDP and Rh*-G-GDP is smallest, and that between G-GTPγS and Rh*-G-GTPγS is largest. This free energy difference (ΔG) determines the exchange rate. These relative rates help us understand differences in G-protein behavior during toxin modification (see below) and will be fundamental in determining the k_{on} and k_{off} of nucleotide interaction with G-protein and the k_{on} and k_{off} of G-protein with Rh*.

The release of G*-GTP from Rh*. Following nucleotide exchange catalyzed by Rh*, G-protein is released to seek its target. Discussions of mechanism revolved about two questions: a) Is Gα/Gβγ released from the membrane following rhodopsin-catalyzed GTP/GDP exchange? and b) Upon GTP/GDP exchange, does Gα separate from Gβγ?

Biochemical measurements, which have no temporal resolution, do show release of Gα-GTP from the membrane and separation from Gβγ at physiological ionic strength (reviewed in (37)). As detected by sedimentation of isolated membranes or by gel filtration on a rhodopsin affinity matrix, Rh* always releases Gα when GTP or GTPγS is added. The majority of Gβγ is retained under conditions of moderate ionic strength, but it is released at low ionic strength (37). The extent of release does depend upon the concentration of the membranes used with respect to the volumes of eluting wash; the time resolution of these sedimentation studies is no better than 5 min (36, 37).

Light-scattering measurements of discs and intact rods can address the binding and dissociation of G-protein on the time scale in which photoreceptor transduction occurs (reviewed in (9, 15)). In the presence of GTP, ROS fragments give a rapid "dissociation" signal which is interpreted to be due to the release of solubilized G-protein from membranes. The kinetics are dependent on the structural state of ROS fragments. In structurally intact frog ROS the kinetics are more complex. In the presence of GTP, two signals are apparent, a rapid "release" signal and a slower "loss" signal (68). The "dissociation" signal in ROS fragments or the "release" signal in intact ROS is kinetically rapid enough to account for activation of a significant number of G-proteins within a few milliseconds. The "loss" signal observed in intact ROS is directly analogous to the "dissociation" signal in fragments but is slowed significantly, $t_{1/2} = \sim 4$ s. Presumably, G-protein must diffuse along the interdiscal space before being released, thereby accounting for the slow time constant in intact ROS. We discussed the difficulties in quantitation and in molecular interpretation of these light-scattering signals. An absolute calibration for the amplitude of scattering signal (i.e., the number of G-protein undergoing reaction) is lacking, and it was questioned whether the signals could be due to the presence of broken ROS fragments from which G-protein is free to dissociate. Although the experimental system has a great deal of power, the interpretation is difficult (see (9)) and does not resolve the question of whether the G-protein "hops" or "skates" to find its target.

Effects of bacterial toxins. Investigations of bacterial toxin labeling of Gα have given additional insight into the nature of the Rh*-G complex. Bovine Gα can be ADP-ribosylated by both cholera toxin (CT) and pertussis toxin (PT). Different sites on Gα are labeled by these toxins. The CT site is far from either terminus, whereas the PT site is adjacent to the C-terminus. Moreover, different states of G-protein are targets for CT and PT. a) PT labels G-GDP irrespective of whether it is in solution or membrane-bound. b) CT labeling requires Rh* and GppNHp (or $GppCH_2p$). The form labeled is likely to be Rh*-G-GppNHp. Cholera toxin labels Gα only if it is in the G-GppNHp form and membrane (Rh*)-bound. GTP and GTPγS prevent labeling by CT because they effectively release $G\alpha$ from the membrane. In contrast, there is an appreciable concentration of Rh*-G-GppNHp and Rh*-G-$GppCH_2p$ because the free energy levels of the complexes of G-protein with GppNHp and $GppCH_2p$ are higher than those of the GTP and GTPγS complexes. c) In squid membranes, CT labels a 44 kD membrane-bound polypeptide in the absence of guanine nucleotides and in the presence of Rh* (67).

There are several consequences of ADP-ribosylation. PT labeling prevents G-GDP from interacting with Rh*; G-protein is blocked in the GDP form. Formation of $G\alpha$-GTP and activation of the PDE are completely blocked. CT labeling slows GTP-GDP exchange threefold and inhibits the GTPase hydrolytic step twentyfold. Consequently, CT leads to a persistently activated $G\alpha$-GTP. Since G-protein is a target for both toxins, a single G-protein can first be ADP-ribosylated by CT and then ribosylated at a different site by PT. Hormonal $G_s\alpha$ is the target for CT only, and $G_i\alpha$ is the target for PT only. The actions of these toxins on photoreceptor $G\alpha$ closely resemble their actions on G_s and G_i, further strengthening the homologies between these signal-coupling proteins.

Summary. The first stage of amplification in the cGMP cascade is controlled by Rh* interaction with G-protein. We considered whether G-protein diffuses to Rh* by a "hopping" mechanism or is confined to two-dimensional diffusion along one surface by a "skating" mechanism. We noted the importance of considering the structure of the sample being studied. The true interdiscal space is about 15 nm and provides a confined area of activity. Moreover, diffusion of proteins from this layer should be slowed by the presence of rim connecting proteins in the intact rod (Fig. 3). We did not resolve the question of "hopping" or "skating," but we did question whether it is necessary to be concerned about this problem. In the interdiscal confined space, the kinetics may not be affected one way or another. In contrast, the potential for $G\alpha$ to dissociate from $G\beta\gamma$ leaving $G\beta\gamma$ behind on the disc surface has far more striking implications about mechanism. Our "wish list" of important research projects includes experiments that will define the lifetime of interaction of Rh*-G, the rates of Rh*-G dissociation, and the state of quaternary interactions of $G\alpha$ and $G\beta\gamma$ in the first stage of amplification.

Mechanisms of PDE Activation

The second stage of the Rh*-triggered cascade is the activation of cGMP phosphodiesterase. We discussed the mechanism of activation and what constituted the "activated" state of PDE by considering the following questions.

Is $G\alpha$ alone sufficient to activate PDE? The original experiments of Fung et al. (22) and Uchida et al. (66) indicated that $G\alpha$-GppNHp alone is sufficient to turn on PDE. In dilute ROS membrane suspensions, $G\alpha$-GTP dissociates from $G\beta\gamma$. The hydrolysis of GTP is slow enough that persisting $G\alpha$-GTP might diffuse to interact with a given PDE. Fung et al. (22) and

Sitaramayya et al. (63) have measured a binding constant of activated G-protein for PDE of ~1 μmol/l. The study of G-GTP activation of PDE in reconstituted membranes indicates that the process is inefficient and that Gα-GTP binds PDE weakly (43, 65). It is difficult to reconstitute an activatable PDE with rhodopsin-containing membranes that shows the properties of the native system. The consensus of opinion is that Gα is sufficient to activate PDE, but no work has yet ruled out that PDE may be activated by the holocomplex Gαβγ in native membranes. Little is known about how the complex is formed even in fractured membrane systems.

How does the γ subunit of PDE regulate the activation of PDE? Hurley and Stryer's experiments (31) show that removal of γ by proteolysis leads to PDE activation; readdition of isolated γ inhibits proteolytically activated PDE. Thus, we accept the importance of this small subunit in regulating PDE activity. How does Gα-GTP activate the phosphodiesterase? One possibility is that γ is released when Gα-GTP binds to PDE. A second mechanism involves the binding of Gα-GTP to PDE without release of γ. A third mode of activation would involve binding of Gα-GTP to PDE followed by Gα-GTP dissociation carrying away γ. The dissociation constant of γ from trypsin-activated PDE is low, ~0.1 nmol/l, indicating that dissociation of γ may be too slow to account for PDE activation in the absence of an accelerating effect by Gα-GTP. Liebman reported work in press (63) showing that the Kd for inhibition of light-activated PDE by purified γ is about two orders of magnitude higher than that for inhibition of trypsin-activated PDE. These investigators favor the interpretation that the active complex is G-protein-PDEαβγ.

Some experiments suggest that PDE-γ may be dissociated under some conditions. Yamazaki et al. (73) have shown that washing frog ROS membranes with GTP leads to activation of membrane-bound PDE. The eluents may be heat-denatured and added back to invoke inhibition. Sorbi has confirmed these observations with bovine membranes (unpublished results). Unfortunately, no biochemical identification of the "inhibitor" in these preparations has been made, hence it is not possible to decide whether this factor is the γ-subunit or not. It remains to be determined whether γ leaves the PDE during physiological activation by Gα-GTP.

Are there other modes of control of PDE activity? We noted that little attention has been paid to other potential factors that might control PDE activity. In other systems, acidic phospholipids, especially phosphatidylinositol and fatty acids, affect the activity of PDE. Examination of such factors might prove interesting.

Can the Km for cGMP be a regulating factor in PDE activity? PDE displays an unusual substrate Km in the light-activated state. Trypsin-activated PDE has a Km = 70 μmol/l and is in the same range as that measured for basal unactivated PDE. The light-activated Km for PDE is measured to be 1-3 mmol/l (55, 63), considerably above the probable cGMP cytoplasmic concentrations. It is not unusual for phosphodiesterases in other systems to show multiple Km's, but usually the Km does not change. Liebman pointed out that such a high Km has marked implications for physiological studies in which cGMP is injected into photoreceptors. The rate of hydrolysis will depend upon the substrate level. If V = VmS/(S + Km) and S > Km, the rate of hydrolysis will not change with increasing substrate concentration, since V = Vm. But if Km > S, the rate of hydrolysis will increase as more substrate is injected. Under this condition, the steady-state level of cGMP will not change even though more and more substrate is added.

The measured Km may be compromised by considerations of kinetics confined to a two-dimensional surface where Rh*, G-protein, and PDE interact. The calculation of Km from kinetic data may not be straightforward if the concentration of cGMP is not uniform. A locally activated PDE may act as a cGMP sink changing the local concentration. A similar phenomenon was observed by Kühn in ROS with leaky membranes but with relatively well preserved structure. The GTP substrate for nucleotide exchange seems to be locally consumed and replenishment of GTP for G-protein release from Rh* becomes rate-limiting (37). Koch and Kaupp (32) report that cGMP is not accessible to disc sites in well preserved outer segments. Thus, the high Km values of 1-3 mmol/l may be artifacts due to the rate-limiting slow entry of cGMP into the organelles. The kinetics need to be reevaluated.

What is the nature of PDE active sites and regulatory sites? Neither the number of active sites nor the regulatory sites of PDE are well-defined. Photoaffinity labeling with azido-inosine monophosphate (72) has indicated that PDE has two high-affinity noncatalytic sites for cGMP. Similar sites can be labeled in a PDE enzyme found in platelets; methylxanthines and other inhibitors promote binding of cGMP to these noncatalytic sites (70).

Summary. The molecular mechanisms of PDE activation are unresolved. In addition, we posed several questions that have yet to be addressed. What is the role of the α and β PDE subunits? Do both subunits have active catalytic sites (and noncatalytic sites)? Is there negative cooperativity in this enzyme? Is the high-affinity site a regulatory site? Is the affinity for cGMP modified by light activation? Clearly, PDE is an enzyme that needs mechanistic examination.

Mechanisms of PDE Inactivation

To restore the resting state of the cell, PDE must be turned off. We discussed mechanisms of inactivation by addressing the following questions.

What is the role of Gβγ in deactivation of PDE? Two possible roles for Gβγ were considered; Gβγ may directly interact with Gα-PDE to produce an inactivated ternary complex or Gβγ may simply act as a sink for removal of Gα from PDE. Fung (21) has shown that Gβγ is not limiting for Gα activation, although it may limit the rate of turnover. A stoichiometry of 1 Gβγ: 10 Gα is sufficient to achieve maximal Gα GTPase activity. Gβγ is not necessary for GTPase activity, which indicates that the hydrolytic activity for GTP is inherent in Gα. Earlier work by Shinozawa and Bitensky (61), which suggested that helper (Gβγ) is necessary for activity, may be explained by noting that Gβγ is necessary to measure turnover of Gα-GTP and the sustained steady state of hydrolysis. Moreover, Gα-GTPγS can be isolated from Gβγ because the hydrolysis time of the nucleotide is so slow that recombination of the two subunits does not occur. Preliminary work indicates that the apparent rate of hydrolysis is the same in the presence or absence of Gβγ (Yamanaka and Stryer, to be published). Bearing these observations in mind, it was concluded that Gβγ is only necessary for Gα binding to Rh*. The GTP hydrolytic activity resides in Gα.

These observations are consistent with the model that Gβγ acts as a competitive sink for Gα-GDP binding. It is reasonable to consider that Gβγ recombines with Gα-GDP once the Gα-GTP has hydrolyzed its GTP. Analogies can be drawn from hormone systems in which Gβγ addition appears to deactivate the adenylate cyclase catalytic unit by competing for Gs, although Gβγ may directly but weakly inhibit the catalytic unit as well (24). No tests of the effects of Gβγ on PDE have been conducted. Clearly, the role for Gβγ in controlling the deactivation of PDE needs to be explored.

The obvious mechanism for turning off PDE activity is hydrolysis of GTP by Gα to form the ineffective Gα-GDP. Qualitative observations of PDE kinetics, as measured by H+ kinetic titration method, suggest that PDE turns off in 2.5 to 5 s (43). The generally accepted rate of GTP hydrolysis, as measured by release of ^{32}P, is 0.5-1.0 mol GTP/mol G-protein/min (1, 25) and is far too slow to account for deactivation of Gα-GTP by hydrolysis. Goldberg's metabolic flux measurements made in intact retinas by determination of ^{18}O labeling of photoreceptor quanine nucleotide α-phosphoryls indicate that the temporal resolution (on/off) of the light-activated phosphodiesterase is even faster, ~0.5 to 1 s (28). The rate of PDE turnoff does not indicate that GTP hydrolysis is the sole Gα turnoff mechanism. The GTP hydrolysis rate has never been measured in an intact system following a weak bleach. A provocative inquiry was posed as to whether Gα-GDP might serve to activate PDE as well as Gα-GTP. In this case, hydrolysis of GTP would not serve as a mechanism of inactivation of PDE. This latter hypothesis is worth testing.

Other problems exist considering control of PDE. ATP may have some effect on PDE activity other than deactivating the enzyme by phosphorylating Rh*. The presence of this nucleotide shifts the apparent velocity achieved by a given bleach to lower values, which is unaccounted for by Rh* blockage.

Summary. At the present time, no single molecular mechanism adequately explains the turnoff of PDE. Some key step is missing.

REGULATION OF Rh* ACTIVITY

Light-activated rhodopsin must be "deactivated" to restore the cell to the resting state. The lifetime of Rh*, as indicated by the spectrally observable MII state, is many seconds. Either Rh* is inactivated to some form that is spectrally similar to MII, or another mechanism for deactivation must be present. We explored the suggestion that phosphorylation of rhodopsin provides this deactivating mechanism.

48K-protein Interaction with Phosphorylated Membranes Provides a New Mode of Regulation of Rh* Activity

Kühn reported recent work indicating that phosphorylated bleached rhodopsin binds 48K-protein (39) and thus inhibits PDE activity (Kühn et al., to be published). In these experiments, ATP added during the bleaching of ROS membranes leads to an average of 5-7 Pi incorporated per opsin, and 1-2% of opsin remains unphosphorylated. The opsin is regenerated with 11-cis retinal, washed, and reconstituted with PDE and G-protein to measure light-activated PDE activity. Under these conditions, 48K-protein significantly inhibits PDE activity, even in the absence of ATP. Excess 48K-protein must be added to effect significant inhibition. This is consistent with the fact that to achieve concentrations proportional to the high local interdiscal concentrations in the intact rod, considerable amounts must be added since 48K-protein is not membrane-bound. The binding constant of 48K-protein to rhodopsin has not yet been measured.

It is interesting to note that phosphorylated rhodopsin still triggers PDE activation, although it is somewhat reduced, but it takes the addition of 48K-protein to turn off activation. Some of us questioned whether this is a specific effect, i.e., could other proteins inhibit PDE under these conditions as well? No influence of 48K-protein on PDE activity is observed, however, when unphosphorylated control membranes are used for activation.

Similar quenching effects of added 48K-protein on PDE activity are also observed in experiments with more intact ROS, provided ATP is present (37, 76). In this more complex system, the mechanism of action of 48K-protein is less clear and it is difficult to distinguish its role from the addition of ATP alone. Added ATP generally decreases initial velocity and accelerates turnoff (42). 48K-protein seems simply to facilitate the action of ATP. This might be explained by considering that some residual 48K-protein is always present in most preparations and that additional 48K-protein restores the preparation to full action extending the effect of ATP-induced PDE quench to higher bleaching intensities. There may be other mechanistic effects, however, which are yet to be understood. Additional experiments will bring much clarification to this new area of exploration.

Phosphorylation of Rh* Is the Suggested Mechanism Providing Rapid Blockage of Rh* Activity

Data published by Sitaramayya and Liebman (62) indicate that phosphorylation of rhodopsin is fastest at low bleaches and that it is fast enough to account for Rh* turnoff. They speculate that at higher bleaches the opsin kinase available would become limiting and the rate would be slowed. With low bleaches Bownds and colleagues, however, do not see phosphorylation increases over the already high background of phosphorylation in RIS/ROS preparations. Moreover, they find the time course to be slow at low-light levels (see Bownds and Brewer, this volume). The kinetics and extent of phosphorylation await further exploration.

Summary
Surprisingly, phosphorylated, regenerated rhodopsin is able to activate G-protein upon bleaching. This indicates that phosphorylation per se does not prevent G-protein nucleotide exchange. Thus, something else must mediate the turnoff. 48K-protein which interacts with phosphorylated rhodopsin would be a good candidate to block Rh* action. It is clear that components are still missing which must provide regulation in the cascade. The kinetic constants that are used to calculate the Rh*-triggered turnover of cGMP in the cascade need to be reevaluated in the presence of 48K-protein. The general trend suggests that turnover will be slow compared to the original values given by Yee and Liebman (74). These latter studies examined an uncontrolled PDE activation. Current studies are contending with a partially controlled state. When all the components are identified, the real turnover may accelerate or decelerate.

THE FUNCTIONAL SIGNIFICANCE OF PDE ACTIVITY
The ultimate project on our "wish list" is the identification of the functional role of light-triggered cGMP flux. We surmised that the following topics of investigation are bringing us closer to an answer.

The cGMP Response Mechanism May Reflect the Rate of cGMP Turnover and Not the Absolute cGMP Concentration
In the intact retina, light induces a rapid acceleration of cGMP turnover, but the levels of cGMP remain nearly constant during a 20 s period of continuous (28) or intermittent (10 ms) (26) illumination. To achieve this, cGMP synthesis by guanylate cyclase (GC) must be tightly coupled to cGMP hydrolysis. Goldberg's data indicate that the resting flux is 33 pmol cGMP/s/mg retinal protein in the dark (estimated to be 240 pmol/s/mg ROS protein). This rate can increase incrementally with increasing intensities of continuous or flashing light to a maximum of 1130 pmol/s/mg ROS protein. From these measurements it has been calculated that a single photon promotes the hydrolysis of 9300 molecules of cGMP/s (28). At low light, or slow flashing light frequencies, rates of cGMP flux increase linearly in relation to increasing intensities or frequencies of photic stimulation and are independent of any cGMP steady-state level changes. To achieve these 5-fold increments in light-induced flux without altering the cGMP concentration, the guanylate cyclase activity must increase as PDE activity increases. Moreover, such coupling exists in the dark as well as with photic stimulation. For instance, addition of increasing concentrations of IBMX to isolated rabbit retinas causes progressive increases in photoreceptor dark

current along with progressive suppression of cGMP metabolic flux. cGMP concentration is affected little, if at all, with IBMX addition. The "mode" of cyclase/phosphodiesterase coupling is simply not understood.

The above comments illustrate that guanylate cyclase must play an important metabolic role in setting cGMP levels and/or maintaining metabolic flux. No useful characterization of enzyme activity, enzyme regulation, or molecular properties of guanylate cyclase in ROS has been achieved. Its activity appears to be low, but this is most probably because optimal assay conditions have not been defined. The enzyme has so far only been identified as an activity associated with axonemes (cilial stalk) (19). A putative role for Ca^{2+} regulation has been suggested (19, 34, 48), but it has not been verified nor has Ca^{2+} been shown to be effective at reasonable physiological concentrations.

The physiological role of stimulus-induced acceleration of cyclic nucleotide metabolic fluxes is not understood, but because of the correlations between cGMP flux induced by various photic stimuli and the changes measurable in photoreceptor dark current, Goldberg (26) has suggested that the metabolic flux is itself of functional importance in phototransduction.

There Is No Evidence for a cGMP-mediated Protein Kinase in ROS in Contrast to Other Cells where cGMP Controls Kinase Activities

A light-induced dephosphorylation of two small peptides has been observed to be affected by cGMP (see Bownds and Brewer, this volume). In theory, light-induced reduction of cGMP concentration would lower kinase activity or stimulate phosphatase activity. The reaction is observed only in somewhat disrupted cells, and the validity of this one known cGMP function is unsubstantiated at present.

A Direct Allosteric Role of cGMP Is a Possible Mechanism of Action for cGMP in Cell Function

Long-standing evidence of Cavaggioni and co-workers (6, 7) indicates that ROS membranes contain an ion conductance which is directly regulated by cGMP. Kaupp and Koch have recently confirmed these observations and have demonstrated that the ROS discs themselves contain cGMP-regulated cation conductances. Conductance opening has a cooperative dependence on cGMP, requiring at least 2 molecules of cGMP ((32), and Kaupp and Koch, this volume). The specificity for cGMP is explicit; phosphorylation by a cGMP-dependent kinase is not a mechanism regulating this conductance. cGMP enhances the permeability for the monovalent cations Na^{+}, K^{+}, and Rb^{+}, but not Cl^{-} ions, and enhances the permeability for divalent ions Ca^{2+}, Ba^{2+}, Sr^{2+}, and some transition metal ions. Mg^{2+} is poorly transported (5-8). It is possible that this conductance also exists in the plasma membrane and represents the light-sensitive conductance.

An exciting and similar role for cGMP in controlling photoreceptor conductances was reported by Miller et al. and Owen et al. (both this volume) in their coverage of the data of Fesenko et al. (18). Fesenko and colleagues have shown that cGMP directly affects ion conductances in patches pulled from rod outer segments ((18), and Miller et al. and Owen et al., both this volume). This documented role for cGMP is the first known instance of direct allosteric action of a cyclic nucleotide on ion channels.

ARE THERE PARALLEL OR BRANCHED PATHWAYS TRIGGERED BY Rh*?

Recent physiological investigations of photoreceptor function have suggested that rhodopsin may play more than one functional role. The implications of these new data were considered as follows.

Two Observations Suggest that the Pathways of Metabolism in Photoreceptor Segments May Be Branched

1) As discussed above, light-stimulated fluxes of cGMP in the absence of concentration changes in cGMP require that PDE activity is tightly coupled to guanylate cyclase activity. Can the same activation mechanism which stimulates PDE also stimulate the cyclase or are there side branches for cross talk between these enzymes?

2) Recent intracellular injections of IP_3 (inositol triphosphate) suggest that this metabolic product can affect ion conductances and that parallel pathways may be present in the photoreceptor. In vertebrates, Waloga and Anderson's work (69) shows that IP_3 induces transient hyperpolarization in salamander rods which "adapts" the response to light. Reciprocally, light adapts the response to IP_3. The results are consistent with a mechanism in which IP_3 mediates ion conductances.

In the Limulus ventral eye, both Fein (17) and Brown (4) have shown that IP_3 injections elicit transient membrane depolarizations, and Fein and colleagues (11) have shown that IP_3 injections correlate with the rise of Ca^{2+} concentration in the cytoplasmic space. These results are consistent with a mechanism in which IP_3 causes cytoplasmic Ca^{2+} release that subsequently controls channel conductance. An important observation, however, shows that injection of EGTA with IP_3 blocks the IP_3 response but <u>not</u> the light response (53). It must be concluded that there is more than one pathway for controlling ion conductances.

Biochemical Evidence Supports the Presence of a Phosphatidylinositol Turnover Pathway in Vertebrate and Invertebrate Photoreceptors

Schmidt investigated phosphatidylinositol metabolism in whole rat retina and demonstrated light-stimulated turnover of phosphatidylinositol. The magnitude of the effect is largest in the inner segment (60). Ghalayini and Anderson (23) have reported light-stimulated phosphatidylinositol turnover in frog, and IP_3 production was shown to occur within 5 s following the light stimulus. Yoshioka et al. (75) and Vandenberg and Montal (67) have demonstrated that light affects polyphosphoinositide metabolism, supporting a light-dependent generation of IP_3 in squid and octopus receptors. Similar data have been published for turnover in the Limulus ventral receptor (4, 17).

By Analogy to Better-known Cell Mechanisms, Rh* Is Implicated to Trigger Phosphatidylinositol Breakdown

Receptor-stimulated release of IP_3 from phosphatidylinositol is usually accomplished by the activation of a phospholipase C (reviewed in (3)). In ROS, there is as yet no molecular identification of this lipase. Currently, there are no known blockers or inhibitors for this activity which would biochemically or physiologically help to identify it. Moreover, the mechanism of lipase stimulation (hence IP_3 release) is unclear and needs investigation.

The α-adrenergic receptor which triggers Ca^{2+} release requires GTP for its action, and this action has been shown to be sensitive to pertussis toxin (20). Such experiments suggest that it would be worth testing G-protein for some additional role in activating phospholipase.

Several other regulatory metabolites are produced along with IP_3. Hydrolysis of phosphatidylinositol also yields diacylglycerol, which effects activation of protein kinase C in the presence of Ca^{2+}. Diacylglycerol serves as a source of production of the arachidonates and prostaglandins, which have transient regulatory functions in many cells. Moreover, the presence of phosphatidylinositol in membranes influences Ca^{2+} ATPase activities (3). The possibility that these metabolites might play a functional role in ROS needs to be explored.

Summary

There is metabolic precedence for IP_3 production in ROS as well as functional evidence that it regulates ion conductances in the plasma membranes of vertebrates and invertebrates. Exploration of this pathway at the molecular level would be welcome.

CONCLUSION

The unusual morphology of sensory visual tissue provides the biochemist with one of the most accessible neural systems existing for study. In the vertebrate retina, the rod photoreceptor cell type predominates. The cells house the molecular receptors and triggering proteins in separate and specialized rod outer segments that can be isolated by simple procedures. Thus, it has been easier for the visual biochemist to obtain abundant, purified protein components for study than for biochemists studying other receptor systems. In the visual system the mechanisms of receptor triggering, amplification, and second messenger "production" can be formulated in specific molecular detail, and kinetic and thermodynamic arguments can be made. Our discussions have drawn information provided by other systems, such as the adrenergic receptor-cAMP pathway, to suggest principles that need to be further examined. In return, the rhodopsin-triggered cGMP enzyme cascade serves as an excellent model for membrane receptor intracellular signalling processes that function in most cells.

Acknowledgements. I thank J.E. Brown, D.J. Roof, and P.J. Stein for their contributions to this report.

REFERENCES

(1) Baehr, W.; Morita, E.; Swanson, R.; and Applebury, M.L. 1982. Characterization of bovine rod outer segement G-protein. J. Biol. Chem. **252:** 6452-6460.

(2) Berg, H.C., and Purcell, E.M. 1977. Physics of chemoreception. Biophys. J. **20:** 193-219.

(3) Berridge, M.J., and Irvine, R.F. 1984. Inositol triphosphate, a novel second messenger in cellular signal transduction. Nature **312:** 315-321.

(4) Brown, J.E.; Rubin, L.J.; Ghalayini, A.J.; Tarver, A.P.; Irvine, R.F.; Berridge, M.J.; and Anderson, R.E. 1984. Myo-inositol polyphosphate may be a messenger for visual excitation in Limulus photoreceptors. Nature **311:** 160-163.

(5) Capovilla, M.; Caretta, A.; Cavaggioni, A.; Cervetto, L.; and Sorbi, R.T. 1983. Metabolism and permeability in retinal rods. In Progress in Retinal Research, eds. N. Osborne and G. Chader, vol. 2, pp. 233-247. Oxford: Pergamon Press.

(6) Caretta, A., and Cavaggioni, A. 1983. Fast ionic flux activated by cyclic GMP in the membrane of cattle rod outer segments. Eur. J. Biochem. **132:** 1-8.

(7) Caretta, A.; Cavaggioni, A.; and Sorbi, R.T. 1979. Cyclic GMP and the permeability of the disks of the frog photoreceptors. J. Physiol. **295:** 171-178.

(8) Cavaggioni, A., and Sorbi, R.T. 1981. Cyclic GMP releases calcium from disc membranes of vertebrate photoreceptors. Proc. Natl. Acad. Sci. USA **78:** 3964-3968.

(9) Chabre, M. 1985. Molecular mechanism of visual phototransduction in retinal rod cells. Ann. Rev. Biophys. Chem. **14:** 331-360.

(10) Chen, Y.S., and Hubbel, W.L. 1978. Reactions of the sulfhydryl groups of membrane-bound bovine rhodopsin. Membr. Biochem. **1:** 107-130.

(11) Corson, D.W.; Fein, A.; and Payne, R. 1984. Detection of an inositol 1,4,5-triphosphate-induced rise in intracellular free calcium with aequorin in Limulus ventral photoreceptors. Biol. Bull. **167:** 524-525.

(12) Downer, N.W., and Cone, R.A. 1985. Transient dichroism in photoreceptor membranes indicates that stable oligomers of rhodopsin do not form during excitation. Biophys. J. **47:** 277-284.

(13) Dratz, E.A., and Hargrave, P.A. 1983. The structure of rhodopsin and the rod outer segment disk membrane. Trends Biochem. Sci. **8:** 128-131.

(14) Emeis, D., and Hofmann, K.P. 1981. Shift in the relation between flash-induced metarhodopsin I and metarhodopsin II within the first 10% rhodopsin bleaching in bovine disc membranes. FEBS Lett. **143:** 29-34.

(15) Emeis, D.; Kühn, H.; Reichert, J.; and Hofmann, K.P. 1982. Complex formation between metarhodopsin II and GTP-binding protein in bovine photoreceptor membranes leads to a shift of the photoproduct equilibrium. FEBS Lett. **143:** 29-34.

(16) Fatt, P. 1982. An extended Ca^{2+}-hypothesis of visual transduction with a role for cGMP. FEBS Lett. **149:** 159-166.

(17) Fein, A.; Payne, R.G.; Corson, D.W.; Berridge, M.J.; and Irvine, R.F. 1984. Photoreceptor excitation and adaptation by inositol 1,4,5,-triphosphate. Nature **311:** 157-160.

(18) Fesenko, E.E.; Kolesnikov, S.S.; and Lyubarsky, A.L. 1985. Induction by cyclic GMP of cationic conductance in plasma membrane of retinal rod outer segment. Nature **313:** 310-313.

(19) Fleischman, D., and Denisevich, M. 1979. Guanylate cyclase in isolated bovine retinal rod axonemes. Biochemistry **18:** 5060-5066.

(20) Francisco, J.A.; Mills, I.; García-Sáinz, J.A.; and Fain, J.N. 1983. Effect of pertussis toxin treatment on the metabolism of rat adipocytes. J. Biol. Chem. **258:** 10938-10943.

(21) Fung, B.K.K. 1983. Characterization of transducin from bovine retinal rod outer segments. J. Biol. Chem. **258:** 10495-10502.

(22) Fung, B.K.K.; Hurley, J.G.; and Stryer, L. 1981. Flow of information in the light-triggered cyclic nucleotide cascade of vision. Proc. Natl. Acad. Sci. USA **78:** 152-156.

(23) Ghalayini, A., and Anderson, R.E. 1984. Phosphatidylinositol 4, 5-biphosphate: Light-mediated breakdown in the vertebrate retina. Biochem. Biophys. Res. Commun. **124:** 503-506.

(24) Gilman, A.G. 1984. G proteins and dual control of adenylate cyclase. Cell **36:** 577-579.

(25) Godchaux, W., and Zimmerman, W.F. 1979. Membrane-dependent guanine nucleotide binding and GTPase activities of soluble protein from bovine rod cell outer segments. J. Biol. Chem. **254:** 7874-7884.

(26) Goldberg, N.D.; Ames, A.; Gander, J.E.; and Walseth, T.F. 1983. Magnitude of increase in retinal cGMP metabolic flux determined by ^{18}O incorporation into nucleotide α-phosphoryls corresponds with intensity of photic stimulation. J. Biol. Chem. **258:** 9213-9219.

(27) Hargrave, P.A.; Fong, S.L.; McDowell, J.H.; Mas, M.T.; Curtis, D.R.; Wang, J.K.; Juszcazak, E.; and Smith, D.P. 1980. The partial primary structure of bovine rhodopsin and its topography in the retinal rod cell disc membrane. Neurochem. Int. **1:** 231-244.

(28) Heyman, R.; Ames, A.; Walseth, T.; Barad, M.; Graeff, R.; and Goldberg, N. 1985. Evidence that cGMP hydrolysis is causal in phototransduction. Biophys. J. **47:** 101a.

(29) Hofmann, K.P.; Emeis, D.; and Schnetkamp, P.M. 1983. Interplay between hydroxylamine, metarhodopsin II and GTP-binding protein in bovine photoreceptor membranes. Biochim. Biophys. Acta **725:** 60-70.

(30) Hofmann, K.P.; Reichert, J.; and Emeis, D. 1984. Light-scattering signal of G-binding in ROS depends on osmolarity. Inv. Ophthal. Vis. Sci. **25(6):** 156.

(31) Hurley, J.B., and Stryer, L. 1982. Purification and characterization of the gamma regulatory subunit of the cyclic GMP phosphodiesterase from retinal rod outer segments. J. Biol. Chem. **257:** 11094-11099.

(32) Koch, K.W., and Kaupp, U.B. 1985. Cyclic GMP directly regulates a cation conductance in membranes of bovine rods by a cooperative mechanism. J. Biol. Chem. **260**, in press.

(33) Kretsinger, R.H. 1980. Structure and evolution of calcium-modulated proteins. CRC Crit. Rev. Biochem. **8:** 119-174.

(34) Krishnan, N.; Fletcher, R.T.; Chader, G.J.; and Krishna, G. 1978. Characterization of guanylate cyclase of rod outer segments of the bovine retina. Biochim. Biophys. Acta **523:** 506-515.

(35) Kühn, H. 1978. Light-regulated binding of rhodopsin kinase and other proteins to cattle photoreceptor membranes. Biochemistry **17:** 4389-4395.

(36) Kühn, H. 1980. Light- and GTP-regulated interaction of GTPase and other proteins with bovine photoreceptor membranes. Nature **283:** 587-589.

(37) Kühn, H. 1984. Interactions between photoexcited rhodopsin and light-activated enzymes in rods. In Progress in Retinal Research, eds. N. Osborne and G. Chader, vol. 3, pp. 123-156. Oxford: Pergamon Press.

(38) Kühn, H.; Bennett, N.; Michel-Villaz, M.; and Chabre, M. 1981. Interactions between photoexcited rhodopsin and GTP-binding protein: kinetic and stoichiometric analysis from light-scattering changes. Proc. Natl. Acad. Sci. USA **78:** 6873-6877.

(39) Kühn, H.; Hall, S.W.; and Wilden, U. 1984. Light-induced binding of 48-kDa protein to photoreceptor membranes is highly enhanced by phosphorylation of rhodopsin. FEBS Lett. **176:** 473-478.

(40) Kühn, H., and Hargrave, P.A. 1981. Light-induced binding of GTPase to bovine photoreceptor membranes: Effects of limited proteolysis of the membranes. Biochem. **20:** 2410-2417.

(41) Kühn, H.; Mommertz, O.; and Hargrave, P.A. 1982. Light-dependent conformational change at rhodopsin's cytoplasmic surface detected by increased susceptibility to proteolysis. Biochim. Biophys. Acta **679:** 95-100.

(42) Liebman, P.A., and Pugh, E.N., Jr. 1979. The control of phosphodiesterase in rod disc membranes: kinetics, possible mechanisms and significance for vision. Vision Res. **19:** 375-380.

(43) Liebman, P.A., and Pugh, E.N., Jr. 1980. ATP mediates rapid reversal of cyclic GMP phosphodiesterase activation in visual receptor membranes. Nature **287:** 734-736.

(44) Liebman, P.A., and Pugh, E.N., Jr. 1982. Gain, speed and sensitivity of GTP binding versus PDE activation in visual excitation. Vision Res. **23:** 1475-1480.

(45) Liebman, P.A., and Sitaramayya, A. 1984. Role of G-protein/receptor interaction in amplified phosphodiesterase activation of retinal rods. Adv. Cyclic Nucl. Res. **17:** 215-225.

(46) Liebman, P.A.; Sitaramayya, A.; Parkes, J.H.; and Buzdygon, B. 1984. Mechanism of cGMP control in retinal rod outer segments. Trends Pharm. Sci. **5:** 293-296.

(47) O'Tousa, J.; Baehr, W.; Martin, R.; Hirsh, J.; Pak, W.L.; and Applebury, M.L. 1985. The Drosophila ninaE gene encodes an opsin. Cell **40:** 839-850.

(48) Pannbacker, R.G. 1973. Control of guanylate cyclase activity in the rod outer segment. Science **182:** 1138-1140.

(49) Pappin, D.J.C.; Eliopoulos, E.; Brett, M.; and Findlay, J.B.C. 1984. A structural model for ovine rhodopsin. Intl. J. Biol. Macromol. **6:** 73-76.

(50) Parkes, J.H., and Liebman, P.A. 1984. Metarhodopsin II is weakly bound to G-protein. Inv. Ophthal. Vis. Res. **25(9):** 156.

(51) Paulsen, R., and Bentrop, J. 1984. Activation of rhodopsin phosphorylation is triggered by the lumirhodopsin-metarhodopsin I transition. Nature **302:** 417-419.

(52) Paulsen, R., and Bentrop, J. 1984. Reversible phosphorylation of opsin induced by irradiation of blowfly retinae. J. Comp. Physiol. A **155:** 39-45.

(53) Payne, R.; Fein, A.; and Corson, D.W. 1984. A rise in intracellular calcium is necessary and perhaps sufficient for photoreceptor excitation and adaptation by inositol 1,4,5-triphosphate. Biol. Bull. **167:** 531.

(54) Pfister, C.; Kühn, H.; and Chabre, M. 1983. Complex formation between photoexcited rhodopsin and GTP-binding protein influences the post metarhodopsin II decay and the phosphorylation rate of rhodopsin in frog rods. Eur. J. Biochem. **136:** 489-499.

(55) Robinson, P.R.; Kawamura, S.; Abramson, B.; and Bownds, M.D. 1980. Control of the cyclic GMP phosphodiesterase of frog photoreceptor membranes. J. Gen. Physiol. **76:** 631-645.

(56) Roof, D.J., and Heuser, J.E. 1982. Surfaces of rod photoreceptor disk membranes; integral membrane components. J. Cell Biol. **95:** 487-500.

(57) Saibil, H.R. 1982. A membrane-cytoskeleton network in squid photoreceptor microvilli. J. Molec. Biol. **158:** 435-456.

(58) Saibil, H.R. 1984. A light stimulated increase of cyclic GMP in squid photoreceptors. FEBS Lett. **168:** 213-216.

(59) Saibil, H.R., and Michel-Villaz, M. 1984. Squid rhodopsin and GTP-binding protein cross-react with bovine photoreceptor enzymes. Proc. Natl. Acad. Sci. USA **81:** 5111-5115.

(60) Schmidt, S.Y. 1983. Phosphatidylinositol synthesis and phosphorylation are enhanced by light in rat retinas. J. Biol. Chem. **258:** 6863-6868.

(61) Shinozawa, T., and Bitensky, M.W. 1981. Purification and characterization of photoreceptor light-activated guanosine triphosphatase. Biochemistry **20:** 7068-7074.

(62) Sitaramayya, A., and Liebman, P.A. 1983. Phosphorylation of rhodopsin and quenching of cyclic GMP phosphodiesterase activation by ATP at weak bleaches. J. Biol. Chem. **258:** 12106-12109.

(63) Sitaramayya, A.; Parkes, J.H.; Harkness, J.; and Liebman, P.A. 1985. Kinetic studies suggest light activated cyclic GMP phosphodiesterase is a complex with G-protein subunit. J. Biol. Chem. **260,** in press.

(64) Stoeckenius, W., and Bogomolni, R.A.A. 1982. Bacteriorhodopsin and related pigments of halobacteria. Ann. Rev. Biochem. **51:** 587-616.

(65) Tyminski, P.N., and O'Brien, D.F. 1984. Rod outer segment phosphodiesterase binding and activation in reconstituted membranes. Biochemistry **23**: 3986-3993.

(66) Uchida, S.; Wheeler, G.L.; Yamazaki, A.; and Bitensky, M.W. 1981. A GTP-protein activator of phosphodiesterase which forms in response to bleached rhodopsin. J. Cyclic Nucl. Res. **7**: 95-104.

(67) Vandenberg, C.A., and Montal, M. 1984. Light-regulated biochemical events in invertebrate photoreceptors. Biochemistry **23**: 2339-2352.

(68) Vuong, T.M.; Chabre, M.; and Stryer, L. 1984. Millisecond activation of transducin in the cyclic nucleotide cascade. Nature **311**: 659-661.

(69) Waloga, G., and Anderson, R.E. 1985. Effects of inositol-1,4,5-triphosphate injections into salamander rods. Biochem. Biophys. Res. Commun. **126**: 59-62.

(70) Walseth, T.F.; Yuen, P.S.T.; Panter, S.S.; and Goldberg, N.D. 1985. Identification of a cGMP binding/cGMP phosphodiesterase (cGMP BP/PDE) in human platelets by direct photoaffinity labeling with ^{32}P-cGMP. Fed. Proc. **44**: 728.

(71) Wilden, U., and Kühn, H. 1982. Light-dependent phosphorylation of rhodopsin: Number of phosphorylation sites. Biochemistry **21**: 3014-3022.

(72) Yamazaki, A.; Bartucca, F.; Ting, A.; and Bitensky, M.W. 1982. Reciprocal effects of an inhibitory factor on catalytic activity and noncatalytic cGMP binding sites of rod phosphodiesterase. Proc. Natl. Acad. Sci. USA **79**: 3702-3706.

(73) Yamazaki, A.; Stein, P.J.; Chernoff, N.; and Bitensky, M. 1983. Activation mechanism of rod outer segment cyclic GMP phosphodiesterase. Release of inhibitor by the GTP/GDP-binding protein. J. Biol. Chem. **258**: 8188-8194.

(74) Yee, R., and Liebman, P.A. 1978. Light-activated phosphodiesterase of the rod outer segments. Kinetics and parameters of activation and deactivation. J. Biol. Chem. **253**: 8902-8909.

(75) Yoshioka, T.; Inoue, H.; Takagi, M.; Hayashi, F.; and Amakawa, T. 1983. The effect of isobutylmethylxanthine on the photoresponse and phosphorylation of phosphatidylinositol in octopus retina. Biochim. Biophys. Acta **744**: 50-55.

(76) Zuckerman, R.; Buzdygon, B.; Philp, N.; Liebman, P.; and Sitaramayya, A. 1985. Arrestin: An ATP/ADP exchange protein that regulates cGMP phosphodiesterase activity in retinal rod disk membranes (RDM). Biophys. J. **47**: 37a.

Standing, left to right:
Peter Stein, David Blest, Reinhard Paulsen, Vincent Torre

Seated (middle), left to right:
Deric Bownds, Kristin Baer, Dorothy Roof, Dorothea Neugebauer, Bill Miller

Seated (front), left to right:
Juan Korenbrot, King Yau, Gordon Fain

The Molecular Mechanism of Photoreception, ed. H. Stieve, pp. 431-449. Dahlem Konferenzen 1986. Berlin, Heidelberg, New York, Tokyo: Springer-Verlag.

Internal Messengers

Group Report

W. H. Miller, Rapporteur
K.M. Baer
A.D. Blest
M.D. Bownds
G.L. Fain
J.I. Korenbrot
D.-C. Neugebauer
R. Paulsen
D.J. Roof
P.J. Stein
V. Torre
K.-W. Yau

INTRODUCTION

The molecules that are links in the chain which connects the absorption of light in the photoreceptor organelle's rhodopsin with the channel that controls the receptor potential or current are termed internal messengers (reviewed by Pugh and Cobbs, McNaughton et al., and Miller, all this volume). The rival calcium and cyclic GMP internal messenger hypotheses have dominated speculation of mechanisms of vertebrate rod outer segment (ROS) phototransduction for the last decade. The discussion brought out new evidence favoring cyclic GMP as an internal messenger. Yau discussed Fesenko et al.'s (15) experiments which provide evidence that cyclic GMP controls a conductance in isolated patches of ROS plasma membrane which probably corresponds with the light-sensitive conductance in vivo. Nakatani and Yau (38) confirmed these findings. Not only is cyclic GMP a ROS internal messenger of transduction according to this, but evidence against the Ca hypothesis was also developed in further discussion: a) calcium has no effect on the patched conductance (15, 38) and ROS intracellular free Ca may decrease with illumination (64), b) Ca buffer experiments by several groups suggest that Ca may not directly control the light-sensitive conductance ((32), and Paulsen, personal communication), and c) Ca cannot directly control the light-sensitive conductance because it acts more slowly than light (McNaughton et al., this volume). This evidence, which was hotly debated, is appearing in print as the final draft of this report is being prepared. Readers can judge for themselves the significance of this work.

The discussion of vertebrate and invertebrate phototransduction was moderated by Bownds with the following agenda:

1. What physical constraints determine the properties a putative messenger must have?

2. What is the evidence for and against Ca as an internal messenger in rod transduction?
3. What is the evidence for and against cyclic GMP as an internal messenger in rod transduction?
4. What is the evidence for and against other processes or substances serving the messenger function?
5. Summary of the facts, agreements, and disagreements, and discussion of future experiments. What follows is a summary of the discussion based on the rapporteur's notes and written summaries of the participants (including Cervetto, Goldberg, Kaupp, Lamb, Pugh, and Torre who were members of other discussion groups).

MESSENGER PROPERTIES

The group commenced with a discussion of how the structural configuration of photoreceptor organelles constrains the nature of internal messengers. To summarize: transduction appears to consist of local events in anatomically discrete locations that must be connected by diffusible molecules in both vertebrate and invertebrate photoreceptor organelles. The absorption of light by rhodopsin in rod disks and cone sacs is transmitted by diffusible molecules to cause a surface membrane permeability change. One absorbed photon can shut off several percent of the rod membrane conductance. The shape of the rod response is stereotyped; therefore, a large number of particles must be involved and some mechanism must exist for erasing memory of the location of the absorption (such as diffusion of proteins in the disk or release of the messenger from the same spot regardless of the absorption site). Translational diffusion of rhodopsin in the plane of the disk of rods and cones gives a diffusion coefficient of about 4×10^{-9} cm^2 s^{-1}.

On the other hand, there is no evidence for a translational diffusion of rhodopsin in invertebrate rhabdomeres (22), perhaps as a result of the high concentration of membrane cholesterol (44, 49) and/or interactions with a sub-membrane cytoskeleton. While translational diffusion has been postulated to play an important role in activation of the vertebrate cyclic nucleotide enzymatic cascade (Chabre and Applebury, and Minke, both this volume), a similar mechanism has not been shown for invertebrates. The longitudinal spread of excitation and desensitization in the amphibian rod has a diffusion coefficient of at most $1\text{-}2 \times 10^{-7}$ cm^2 s^{-1}. There is disagreement as to whether the disk incisures counteract the obstructive effect of disk baffling, but it seems that one has to invoke some means (e.g., adsorption or larger molecules) to explain a diffusion that is about 100 times slower than would be expected by diffusion of small molecules in an aqueous medium. The high Q_{10} for time-to-peak of the rod response to dim flashes suggests that biochemical reactions rather than diffusion of the messenger are rate-limiting. A 0.1 s time-to-peak for the cone response can probably be explained by lateral diffusion of proteins in the membrane and a diffusible small molecule.

In invertebrate rhabdomeres up to 1000 channels are activated per photon, which suggests a diffusible messenger. A diffusion coefficient has not been measured, but different substances seem to be involved for excitation and adaptation. The general point is made that there are more cytoskeletal and extracellular protein links to the membrane surface of invertebrate microvilli than to the vertebrate disk membranes. However, whether there are sufficient side arms linking the microvillar axial filament to the plasmalemma to tether all the rhodopsins is still an open question (1, 13, 49). The structural constraints of cytoskeleton elements are discussed by Schwemer and by Roof (both this volume).

Neugebauer discussed the relation between the structural role of the photoreceptor organelle cytoskeleton of vertebrates and its physiological function:

> The cytoskeleton needs to be considered for physiological reasons since most of the reactions in vision take place on interfaces. The interface between the membrane and the cytosol is by no means smooth, since considerable parts of membrane-integrated proteins are sticking out into the liquid face and a small molecule approaching a disk membrane will see a lot of variations in electric field and steric obstacles. From recently gained knowledge of the acetylcholine-receptor membrane in postsynaptic membranes, microvilli in the intestine (18), and hair cells from the inner ear (39), the concept of membrane architecture may need to be revised. It appears that a clear distinction between the membrane proper and the cell coat is not always possible so that it might be useful to revive the concept of the "greater membrane" first suggested by Revel and Ito (47), which conceives of the membrane together with the intracellularly and extracellularly attached compounds (membrane cytoskeleton and glycocalix (33)) as a functional unit.

ROLE OF CALCIUM IN ROD TRANSDUCTION

There was lively discussion of recently published and unpublished data suggesting that the Ca hypothesis as originally proposed, by which ROS intracellular free Ca concentration is increased by illumination and directly decreases the light-sensitive conductance, is incorrect. Discussion of evaluations of data proposing that intracellular Ca decreases on illumination and that Ca does not directly affect the light-sensitive conductance are summarized below.

Ca Content and Localization in Rods

This topic was addressed by Fain:

> Large quantities of Ca have been observed in rod outer segments amounting to 4-5 millimoles per liter tissue volume or approximately 1-2 Ca per Rh. In darkness the Ca is largely inexchangeable with the Ca in the extracellular medium in spite of a rather high rate of transport of Ca across the plasma membrane. The Ca exchanges so slowly that, were the Ca to be stored within the rod cytosol bound to some buffer

substance, the lifetime of the Ca-buffer complex would have to exceed 10^3 s. This is an order of magnitude higher than the lifetimes of known Ca-binding proteins such as calmodulin, troponin, and parvalbumin. A more likely explanation is that nearly all the Ca in the outer segment is contained within the disks. The Ca in the disks does not exchange with extracellular (or cytosolic) Ca, probably because the disk membrane in darkness is nearly impermeable to Ca. In light, there is a specific increase in the transport or permeability of Ca across the disk membrane, and large amounts of Ca are released into the cytosol (52).

It was felt that the finding of 1-3 Ca/Rh is general in isolated outer segments (e.g., (51)). Yet recently Somlyo and Walz (54) found an order of magnitude less Ca/ROS in intact rods that were fast-frozen and cryosectioned. The physiological relevance of this observation is supported by their finding appropriate changes in intracellular concentrations of Na and K. They failed to detect changes in intracellular Ca concentration. However, their method does not have sufficient resolution to detect changes as small as those inferred from measurement of Ca fluxes in intact rods (19, 20, 64). This observation (54) raises an interesting issue that has yet to be resolved (also see section on Ca Efflux Experiments, below).

Ca Buffer Experiments

"In press" experiments by Matthews, Torre, and Lamb (32) were presented by Lamb:

The photoresponse of an isolated salamander rod was recorded simultaneously with a suction pipette and with a 'whole-cell' patch pipette under voltage-clamp. When the patch pipette was filled with artificial intracellular medium, normal responses could be recorded from both pipettes for many minutes. When, however, the whole-cell patch pipette filling solution included 10 mmol/l of the calcium chelator BAPTA (or EFTA), very significant changes were observed. Very soon after the patch was broken (and the BAPTA presumably began diffusing into the cell), the dark current began increasing somewhat, the responses to bright flashes lasted progressively longer, and a very pronounced overshoot developed in the recovery phase of the flash response. Despite these changes, the rising phase of responses to bright flashes was virtually unaltered.

The pronounced changes induced by BAPTA (particularly the overshoot) were interpreted to indicate that the chelator had diffused into the outer segment. The continued existence of responses to bright flashes, together with the unchanged rising phase, therefore showed that the response to light did not require elevation of cytoplasmic calcium levels.

However, the experiments give no direct measure of cytoplasmic calcium levels. Indeed, it cannot even be said that BAPTA led to a reduction in steady levels of free calcium, but simply that changes in calcium concentration were buffered.

This interpretation of the experiment was challenged in discussion: whether 10 mmol/l BAPTA is a high enough concentration of buffer was questioned by Lisman who said that 100 mmol/l buffer is required in other systems. Thus the 10 mmol/l BAPTA may not penetrate to all the disks. Furthermore, there is no correspondence with extracellular measurements of Ca, according to Korenbrot, who cited his experiments (35) showing that BAPTA and QUIN-2 increase both the photocurrent and extracellular Ca as measured with a Ca-sensitive electrode which sees three to four more Ca ions per photon than in the absence of Ca buffering. Ca may not control the channel but it has a big effect on the response, and there is still a question as to why QUIN slows the response when Ca diffuses faster because of the buffer (though buffer slows diffusion in vitro) (Korenbrot, personal communication).

Related experiments by Nicol et al. (40) were described by Kaupp:

> When isolated rods are perfused with a solution that contains a Ca ionophore (A23187), the dark current is suppressed and the responses to light disappear rapidly. The electrophysiological response reappears if the extracellular Ca concentration is lowered to less than 10^{-6} mol/l. In the presence of high concentrations of the ionophore, responses to light in a low Ca medium can be maintained for up to an hour. Measurement of Ca fluxes in this preparation after addition of A23187 indicates that about 1 Ca/Rh can be released at low external Ca concentrations. These results suggest that a change in intracellular Ca is not required to close the light-sensitive channels in the plasma membrane.

Two points that were raised in discussion were the questions as to whether Ca could still be the messenger under normal conditions and whether a trace of Ca remaining could still be partitioned in the presence of the ionophore.

Na/Ca Exchange

The ROS electrogenic Na/Ca exchange was evaluated by Torre:

> Let us suppose that a Ca extrusion through the Na/Ca exchange can be described by the usual Michaelis-Menten relation
>
> $$V_{Ca} = V_{max}\,[Ca_i]/K_M + [Ca_i] \tag{1}$$
>
> Equation 1 can be used when the Na/Ca exchange does not run backward (i.e., $[Na_e]$ higher than $[Na_i]$). Yau and Nakatani (64) have shown that the Na/Ca exchange is electrogenic with a ratio of 3 Na/Ca. The maximal current that can be driven by the Na/Ca exchange is about 10-30 pA. Given a ratio of 3 Na/Ca for the exchange and a volume, v, of 10^{-12} l for the rod, V_{max} is about 200-600 μmol/s:
>
> $$V_{max} = I_{exchange}/z\;q\;N\;v. \tag{2}$$
>
> The time constant, t, of the decline of the exchange current is about 0.5 s. If this decline fully reflects changes of $[Ca_i]$, then

$K_M = V_{max}$. (3)

Therefore, K_M = 100-300 μmol/l, which is 10-100 times larger than that reported in other preparations. A value for K_M of the same order of magnitude can be obtained from the Ca loading experiments of Yau and Nakatani (63). Given the high value of K_M, it is necessary to have an unusually high value of $[Ca_i]$ to activate the Na/Ca exchange.

It was pointed out by Kaupp that K_M values for Na/Ca exchange have been determined by Schnetkamp (51) to be about 1 μmol/l, which conflicts with the very high values calculated above.

Ca Efflux Experiments

Experiments by Cervetto and McNaughton (9) on inhibition of Ca extrusion by La^{3+} were described by Cervetto:

La slowly suppresses the light-sensitive current and slows down the falling phase of the current response to light. It is suggested that La directly interacts with the light-sensitive channel but also inhibits Ca extrusion. The La effects on the photocurrent observed in these experiments are consistent with the idea that the rate of calcium extrusion from the cell does not contribute directly to one of the time constants of the light response. The slowed decay of the current response in La might be explained by an indirect effect of increased Ca levels on the cyclase activity.

In other experiments, net Ca efflux from the ROS increased following illumination (20, 65). That the time course of increased Ca appeared to correspond with that of the photovoltage response has been one of the main pieces of evidence for the Ca hypothesis. These measurements were made in low Na Ringer's. Gold (in experiments that were not discussed at this workshop (19)) repeated the experiment in normal Na Ringer's and found that using bright light, extracellular Ca flux decreased while the photovoltage was maintained. He suggested that the dissimilarity in waveforms is inconsistent with the Ca hypothesis.

That ROS intracellular free Ca decreases on illumination because of Ca extrusion by the Na/Ca exchange mechanism as Ca influx through the Na conductance is shut off by light was hypothesized in experiments then in press reported by Yau (64):

1) Upon turning on the light, Ca influx stops immediately based on measurement of membrane current. The Ca efflux declines exponentially with a time constant of 0.5 s. This transient efflux does not increase with light intensity or duration as soon as the electrical response reaches saturation.

2) Incremental flashes (no matter how bright) superimposed on a light step produced no further transients of Ca efflux.

3) Estimate of dark Ca influx via the light-sensitive conductance in Ringer's is consistent with the transient Ca efflux seen in the light. Moreover, removal of dark Ca influx also removes transient Ca efflux in light, while an increase in dark Ca influx likewise increases transient Ca efflux in light.

4) The idea that light merely shuts off the dark Ca influx (and hence leads to an efflux) leads to the prediction that a) about 10^5 Ca ions efflux per Rh in agreement with Gold and Korenbrot's data (20), and b) a transient net Ca influx occurs during recovery of the light response in agreement with Gold's data (19).

Korenbrot raised two points: first, whether or not ROS intracellular Ca decreases with light would have to be confirmed by direct measurement, and second, that by Yau's data the Ca efflux stops shortly after the photocurrent saturates. Yet according to Gold and Korenbrot (20) it does not; rather, it continues. There is thus a conflict in the data that will have to be resolved.

Fain questioned the assumptions upon which Yau's conclusions were based:

> Yau apparently assumed that all of the Ca moving across the rod plasma membrane moves by way of an electrogenic component of the Na/Ca exchange. Furthermore, he assumes that the rods have essentially no ability to buffer intracellular Ca levels. Finally, he assumes that light has no effect on Na/Ca exchange in rods. None of these assumptions have been demonstrated by experiment, and some of them seem highly unlikely.

Evidence of Kaupp and Koch (25, 26) that cyclic GMP increases a disk's cationic conductance was described by Kaupp. About 11,000 Ca ions are released per disk at saturating concentrations of cyclic GMP. A minimum of 2 cyclic GMP's act cooperatively and the cyclic GMP-dependent Ca efflux is independent of light (26). The Ca efflux requires a Ca gradient, and the maximal efflux rate is circa 7 x 10^4 Ca/s per rod. These experiments (26) which describe a cyclic GMP, disk-cationic permeability controlled by cyclic GMP, suggest that the plasma-membrane, light-sensitive channel regulated by cyclic GMP (15) is also in the disk membranes. Kaupp also suggested that as one consequence, internal Ca may be decreased upon illumination because cyclic GMP, which increases disk permeability, is decreased, and hence less Ca may enter the cytosol via the cyclic GMP-sensitive conductance in the disk membrane.

Other data involving the interaction of cyclic GMP and Ca should be mentioned. Cavaggioni and Sorbi (8) suggested that cyclic GMP releases Ca from binding sites on disks; Caretta and Cavaggioni (6) found that cyclic GMP activates a Ca flux with no appreciable delay, while George and Hagins (16) reported a cyclic GMP-mediated uptake of Ca.

Ca Suppression of Dark Current

Experiments by McNaughton, Nunn, and Hodgkin (this volume) indicate that the suppression of the dark current caused by Ca is slower than the suppression of the dark current by light, and they conclude that Ca cannot directly block the dark current as proposed by the Ca hypothesis.

Ca as an Adaptation Messenger

Fain discussed the following points:

> An interpretation of the role of Ca during adaptation very much depends upon whether Ca concentration is thought to go up or down in illumination (see above). If Ca levels are thought to go up, then it is unlikely that Ca is a messenger substance in this process, since increases in Ca in the rod appear to have no influence on photoreceptor sensitivity. If Ca levels are thought to go down, then it is conceivable that decreases in Ca could alter cyclic GMP levels by modulating guanylate cyclase activity. In this way, changes in intracellular Ca could produce a kind of feedback loop, elevating cyclic GMP levels which have been depressed by light. Such a mechanism might account for part of the process of light adaptation, for example, the decline in the photocurrent during steady light which is observed at moderate illuminations. It is presently difficult to evaluate the contribution of this idea since insufficient data are available on the range of Ca sensitivity of the cyclase. Furthermore, it seems unlikely that such a mechanism could account for the large (3-4 orders of magnitude) changes in sensitivity which occur during exposure to continuous backgrounds.

MacLeish pointed out that even though Fain has shown that increasing Ca in a dark-adapted rod does not mimic light adaptation, a role of Ca in light adaptation cannot be ruled out since light initiates several processes which might be influenced by calcium. Fain agreed.

Polyphosphoinositides

The discussion of polyphosphoinositides, which are thought to be involved in the control of calcium in other cells, was summarized in relation to photoreceptors by Roof:

> Waloga (61) reports that pressure injection of IP_3 into salamander rods causes reversible hyperpolarizations of the rod membrane that are diminished by bright light and recover with dark adaptation. This is entirely analagous to experiments in invertebrates which show that injection of IP_3 into Limulus ventral photoreceptors both excites and adapts the cells (4, 14). Light has also been shown to mediate large, rapid changes in polyphosphoinositide metabolism in both Limulus and squid photoreceptor membrane preparations (24, 57, 59).
>
> The idea that IP_3 brings about these physiological effects through an alteration in internal Ca is attractive. Payne et al. (45) and Brown (3) have shown that IP_3 injections into Limulus photoreceptors are

accompanied by a transient rise in intracellular Ca as measured with aequorin. However, IP_3 and its effects on calcium cannot be part of a simple linear pathway for transduction because injection of calcium chelators completely blocks the effects of subsequent IP_3 injections but does not block subsequent light responses (45, 48). Conversely, calcium does not regulate IP_3 metabolism because Limulus photoreceptors loaded with Ca chelators continue to show light-induced changes in PIP_2/IP_3 (Brown, personal communication). However, since light responses are not normal in calcium-buffered cells, it is possible that an IP_3-Ca pathway is synergistic with another transmitter and that calcium is necessary for normal excitation.

ROLE OF CYCLIC GMP IN ROD TRANSDUCTION

As mentioned in the introduction, evidence that cyclic GMP directly controls the light-sensitive conductance was presented. The discussion of the role of cyclic GMP in both vertebrate and invertebrate phototransduction, including the controversy as to whether cyclic GMP exerts its effects by changes in flux or mass (concentration), is summarized:

Cyclic GMP Is a Rod Internal Messenger

Yau reported that two groups had studied the properties of ROS inside-out patches of plasma membrane (15, 38). Cyclic GMP, acting on the inside of the plasma membrane, reversibly increases the cationic conductance of such patches. Ca does not. According to Fesenko et al. (15):

> The cyclic-GMP induced conductance increase occurs in the absence of nucleoside triphosphates and, hence, is not mediated by protein phosphorylation, but seems rather to result from a direct action of cyclic GMP on the membrane. The effect of cyclic GMP is highly specific; cyclic AMP, and 2',3'-cyclic GMP are completely ineffective when applied in millimolar concentrations. We were unable to recognize discrete current steps that might represent single-channel closings modulated by cyclic GMP. Analysis of membrane current noise showed the elementary event to be 3 fA with 110 mmol/l Na^+ on both sides of the membrane at a membrane potential of -30 mV. If the initial event is assured to be the closure of a single cyclic GMP—sensitive channel, this value corresponds to a single-channel conductance of 100 fS. It seems probable that the cyclic GMP-sensitive conductance is responsible for the generation of the rod photoresponse in vivo.

The cyclic GMP-sensitive conductance showed all the properties of the light-sensitive conductance (0 reversal potential and "prominent" (38) outward rectification) except, of course, light sensitivity. The slope of a Hill plot n = ca. 1.8 suggested that several cyclic GMP molecules cooperate to open a channel. This evidence of Fesenko et al. (15) which was confirmed by Nakatani and Yau (38) suggests that cyclic GMP directly controls the light-sensitive ROS cationic conductance.

Pugh discussed the recent data of Cobbs and Pugh (10) which showed that no more than 10% of the maximal possible dark current is on in the dark. Cyclic GMP introduced via a whole-cell patch reversibly increased the dark current ca. ten times. That virtually all the ROS current caused by cyclic GMP is found to be light-sensitive is consistent with the role of cyclic GMP as an internal messenger. Cyclic GMP caused no large change in latency under the conditions of this experiment. Matthews, Torre, and Lamb (32) have similar data.

Evidence that cyclic GMP injected into ROS in the presence of background light of 100 Rh*/s does not increase photocurrent was presented by Torre (personal communication). He commented that it is puzzling to think that cyclic GMP is the internal transmitter when at the same time injection of exogenous cyclic GMP in light does not open channels.

MacLeish (personal communication) showed evidence that cyclic GMP injected into ROS increases dark current and showed that with a depolarizing step, influx of Ca decreases the light-sensitive current. But this is not true after cyclic GMP, which suggests that cyclic GMP may affect binding of Ca.

Data that cyclic GMP and light do not control the same conductances because their reversal potentials differ in his experiments was discussed by Brown (46). Cyclic GMP-induced currents reversed sign 20-45 mV more positive than the resting potential; light-induced currents did not reverse sign for voltages depolarized more than 55 mV from the dark, resting potential.

In the discussion there was a question as to whether in the experiments of Fesenko et al. (15) and of Nakatani and Yau (38) the patch might be from disk rather than plasma membrane. The similarity of the cation channels whose permeability is controlled by cyclic GMP in disks was noted (7, 25, 26). The probability that the plasma membrane and disk channels are identical was discussed by Kaupp:

> The light-sensitive conductance in the plasma membrane and the cyclic GMP-regulated channels in the disk membrane may be similar if not identical. The similarities are indeed striking. Both conductances are directly regulated by cyclic GMP in a cooperative manner (15, 26). Both conductances are permeable to most monovalent cations and even to Ca and Mg. The dissociation constant for the activation by cyclic GMP is about 50 μmol/l for both conductances.

Cyclic GMP in Invertebrate Photoreceptors

The light response in the ventral photoreceptor of Limulus is only slightly affected by very high concentrations of cyclic AMP and cyclic GMP (reviewed by Brown, this volume), and there is no evidence linking cyclic nucleotides to transduction in that photoreceptor. Paulsen (43) indicated that the concentration of cyclic AMP and GMP in the fly is not affected by light as in Limulus, but there are cyclic AMP PDE and cyclase activities associated with the rhabdomeric membranes.

Saibil (50) investigated homogenized squid photoreceptor and found that GTP causes a linear increase in cyclic GMP. GTP plus light causes a twofold faster increase in cyclic GMP content. While these data do not speak directly to whether cyclic GMP is involved in transduction, the parallel with vertebrates is consistent. Light increases cyclic GMP in this depolarizing receptor and decreases it in the hyperpolarizing vertebrate ROS.

Cyclic GMP Mass vs. Flux

Goldberg (21) presented evidence that cyclic GMP flux rather than mass (concentration) controls transduction:

> The evidence was obtained by monitoring cyclic GMP metabolic flux as a function of PDE-promoted hydrolysis of ^{18}O from ^{18}O water into endogenous guanine nucleotide α-phosphoryls. The initial experiments conducted with this technology showed that incremental increase in cyclic GMP metabolic flux, independent of changes in steady-state mass, correspond to increasing intensities of intermittent photic stimulation. Increasing the frequency of flashing light accelerates the flux in correspondence with the voltage-time integral of the electrical response without detectable changes in retinal cyclic GMP concentration. The electrical response saturates at a lower flashing frequency than the biochemical response. With increasing intensities of continuous light, the inhibition of dark current closely parallels increases in cyclic GMP flux independent of cyclic GMP mass. IBMX increases dark current while diminishing flux. These results demonstrate that the dark current and cyclic GMP flux are closely correlated whether transduction is activated or inhibited and that the biochemical event is temporally resolved as rapidly and takes place over the same range of stimulus intensities as the electrical response. The results support the hypothesis that light-evoked acceleration of cyclic GMP hydrolysis independent of change in cyclic GMP steady-state mass plays a causal role in phototransduction.

The notion that flux can play a role in phototransduction received support from Torre:

> Similar conclusions were drawn from electrophysiological experiments (5) recording membrane potential or current (9). These experiments showed that the kinetics of the rod photoresponse were slowed down by IBMX and speeded up by backgrounds of light. These observations and the biochemical observation that in the same conditions no clear change in the total mass of the cyclic GMP could be observed suggest that the flux of cyclic GMP may control the kinetics of the photoresponse; that is, the rate of hydrolysis of cyclic GMP controls the time scale of the photoresponse.

Goldberg added:

> An energy-linked process could 'drive' conductance changes directly or indirectly, and this idea is attractive because it is easier to control changes in flux than in mass.

Pugh commented, "Is the pool of cyclic GMP mass that is measured the light-sensitive pool? PDE may bind most of the cyclic GMP. This is a question that is not yet answered."

Stein discussed mass vs. flux as follows:

> The question of the relationship between cyclic GMP mass and cyclic GMP flux in the rod photoreceptor arises in large measure because of our inability to determine the activity of the nucleotide in the cell. In fact, studies to date of the light-dependence of cyclic GMP have measured only total cyclic GMP mass in the rod outer segment. It has become clear that cyclic GMP may have three binding locations with different binding constants, some of which may be influenced by light. These include the high-affinity binding sites on PDE (.009 μmol/l) (62), the catalytic site on PDE (50-70 μmol/l dark, 700 μmol/l light), and a cyclic GMP-dependent ionic permeability (75-100 μmol/l) (7, 25). The free cyclic GMP in the rod outer segment will reflect the net interaction of these binding sites, the synthetic activity of guanylate cyclase, and the hydrolytic activity of PDE. Differences in the time course of the enzymatic activations as a function of light could easily produce very rapid, transient changes in local concentration which would be difficult to detect with measures of total cyclic GMP mass. To date it has not been possible to correlate changes in cyclic GMP mass with dim bleaches of either intact photoreceptors or isolated inner-outer segment preparations. The critical question of whether rapid, transient changes in cyclic GMP mass occur cannot yet be resolved.
>
> Goldberg et al.'s (21) ^{18}O data do demonstrate increases in light-dependent cyclic GMP turnover and argue strongly for light-dependent guanylate cyclase activity, possibly coupled to PDE activity.

Physiological evidence does not support the conclusion that light-evoked acceleration of cyclic GMP hydrolysis independent of cyclic GMP mass controls transduction. If flux controlled transduction, an increase in cyclic GMP mass followed immediately by a flash of light to activate PDE and increase flux should result in a vigorous light response of decreased latency. Yet the opposite is true. Instead, the light response is transiently blocked as if the reduction in mass were obligatory for transduction (41). Furthermore, as described above, the application of cyclic GMP to the inside of a patch of the plasma membrane (most of ROS PDE is on the disks) controls the light-sensitive conductance (15). It seems likely that the light-sensitive channel is controlled by local changes in cyclic GMP mass, but a role for flux and for cofactors other than Ca has not been ruled out.

MESSENGER CANDIDATES OTHER THAN CYCLIC GMP AND Ca

Attempts to define a transduction messenger in the ventral photoreceptors of Limulus by internal dialysis with a Ca-free ethanol extract of squid retina were described by Lisman (56). The active extract did not mimic light but reduced the light response.

Methylation inhibitors do nothing when injected into vertebrate ROS and the Limulus ventral photoreceptor, according to Brown (personal communication).

Minke described experiments in fly photoreceptors (36) in which fluoride (without illumination) and hydrolysis-resistant GTP analogs (following illumination) induced noisy depolarization due to summation or unitary voltage responses (bumps). These experiments indicate that bumps can be produced for a long time in the dark by chemicals that are known to activate guanine nucleotide-binding protein. Shot noise analysis indicates that the shapes of the chemically and light-induced bumps are remarkably similar; however, the site of the chemically induced bumps is 3-5 times smaller than that of the light-induced bumps. This suggests an absence of an amplification step in the chemically induced bumps (see Stieve, this volume).

In fly photoreceptors it is possible to modulate membrane potential by photopigment conversions between rhodopsin (R) and metarhodopsin (M) (see Minke, this volume). Minke described additional experiments (Blumenfeld et al., submitted) in the fly eye in which GTPase activity could be modulated in strict quantitative correlation with the receptor potential - i.e., R-to-M pigment conversion induced up to a 20-fold increase in GTPase activity depending on the amount of pigment conversion, while M-to-R pigment conversion suppressed the GTPase activity back to the dark level. These results indicate that a transducin-like protein may be activated by invertebrate rhodopsin (see also results of Corson and Fein (11) and Bolsover and Brown (2) in Limulus ventral photoreceptors).

The possible role of opsin phosphorylation in controlling adaptation was discussed by Lisman. Liebman and Pugh (29) had suggested that phosphorylation of photolyzed rhodopsin would act as an off-switch by inhibiting its ability to activate PDE. Effects of opsin phosphorylation on PDE activation have been reported (34, 53). Lamb (28) had suggested that some photoproduct reverting back to the active state could be responsible for the "photon-like" event activity associated with "dark light" in the recovery following a strong bleach of vertebrate rods. Lisman (30) analyzed whether the reverse reaction from a photoproduct could be achieved by phosphorylation and suggested that multiple phosphorylations increase the reliability of the off-switch since one phosphorylation would not achieve the necessary reliability to account for the dark noise in the ultraviolet receptor of the median eye of Limulus. Paulsen and Bentrop (42) have analyzed the rhodopsin phosphorylation cycle in the fly showing that metarhodopsin is phosphorylated and becomes dephosphorylated when metarhodopsin is converted to rhodopsin. Two types of recent results may be viewed as physiological correlates of these biochemical reactions: a) lowering nucleotide levels increased the rate of spontaneous quantum bumps consistent with the idea that an ATP-utilizing phosphorylation reaction is part of the shut-off mechanism (55). b) Photo-reconverting phosphorylated metarhodopsin created phosphorylated rhodopsin which then became rapidly ($T_{1/2}$ less than 20 s) dephosphorylated in the dark (42). When inactive metarhodopsin is reconverted to rhodopsin by light, the newly created rhodopsin does not support transduction (31). Only with time does the rhodopsin become modified so that absorption of a photon yields a response. Thus it seems plausible that phosphorylated rhodopsin observed by Paulsen corresponds to the non-transducing rhodopsin described by Lisman et al. (31). The possible role of the 48K protein for inactivation of photolyzed rhodopsin is discussed by Applebury et al. (this volume).

Since the time that Tα remains active is long with respect to the single-photon response, an independent role (for Tα) as messenger was thought unlikely but has not been ruled out. Protons are thought unlikely to serve as messengers because of the cytosol buffering power. The arguments with regard to candidates other than Ca and cyclic GMP are covered by Pugh and Cobbs (this volume).

SUMMARY

The molecules that link the absorption of light in photoreceptor organelles with the physiological effects of illumination at another location are termed internal messengers. Internal messengers of invertebrate and vertebrate photoreceptors were discussed under an agenda that included consideration of the properties of putative messengers, the evidence for and against Ca as a messenger, the evidence for and against cyclic nucleotides as messengers, and the evidence for and against other molecular processes serving as internal messengers of transduction.

The group discussed findings that calcium is the internal messenger for adaptation of invertebrate photoreceptors and is possibly a synergistic agent for excitation in some invertebrate photoreceptors. A transducin-like molecule independent of the cyclic nucleotide cascade and phosphorylation reactions may be involved in invertebrate transduction, but the full chain of events is unknown.

New evidence that a) the Ca hypothesis of ROS phototransduction may be incorrect and b) cyclic GMP may be an internal messenger molecule in ROS was discussed. Lively discussion probed unanswered questions regarding these findings as reported above, and the discussants agreed that it would be valuable to obtain additional information regarding time-resolved measurements of intracellular Ca and cyclic GMP activities in ROS, to measure bound vs. free pools of these molecules, and to explore the biochemistry of the Na/Ca exchange mechanism.

Although the discussants were not polled, it seemed to this rapporteur that the introduction of the evidence suggesting that cyclic GMP, rather than Ca, directly controls the ROS light-sensitive conductance (15) marked a turning point in the development of our knowledge of phototransduction. This news seemed to foreshadow the end of an era that commenced with the discovery of rhodopsin by Boll and Kuhne (27), continued with the identification of the chromophore's molecular structure by Morton and Goodwin (37), the discovery by Hubbard and Kropf (23) of its light-induced stereoisomerization, the discovery of the vertebrate hyperpolarizing photoresponse by Tomita (58), and began to mature with the assembly of cyclic GMP as the probable molecular link connecting Rh* and the light-sensitive channel by Fesenko et al. (15) announced at this workshop.

Acknowledgements. Thanks to the participants who contributed written summaries and to M.D. Bownds for reviewing the manuscript.

REFERENCES

(1) Blest, A.D.; Stowe, S.; and de Couet, H.G. 1984. Membrane turnover in the photoreceptors of arthropods. Sci. Progr. **69:** 83-100.

(2) Bolsover, S.R., and Brown, J.E. 1982. Injection of guanosine and adenosine nucleotides into Limulus ventral photoreceptor cells. J. Physiol. **332:** 325-342.

(3) Brown, J.E., and Rubin, L.J. 1985. Inositol-triphosphate induces an increase in intracellular ionized calcium in intact and functioning Limulus photoreceptors. Biophys. J. **47:** 37a.

(4) Brown, J.E.; Rubin, L.J.; Ghalayini, A.J.; Tarver, A.P.; Irvine, R.F.; Berridge, M.J.; and Anderson, R.E. 1984. Myo-inositol polyphosphate may be a messenger for visual excitation in Limulus photoreceptors. Nature **311:** 160-163.

(5) Capovilla, A.; Cervetto, L.; and Torre, V. 1983. The effects of phosphodiesterase inhibitors on the electrical activity of toad rods. J. Physiol. **343:** 277-294.

(6) Caretta, A., and Cavaggioni, A. 1983. Fast ionic flux activated by cyclic GMP in the membrane of cattle rod outer segments. Eur. J. Biochem. **132:** 1-8.

(7) Caretta, A.; Cavaggioni, A.; and Sorbi, R.T. 1979. Cyclic GMP and the permeability of the disks of the frog photoreceptors. J. Physiol. **295:** 171-178.

(8) Cavaggioni, A., and Sorbi, R.T. 1981. Cyclic GMP releases Ca from disk membranes of vertebrate photoreceptors. Proc. Natl. Acad. Sci. USA **78:** 3964-3968.

(9) Cervetto, L., and McNaughton, P.A. 1985. The effects of PDE inhibitors and La on the light-sensitive current of retinal rods. J. Physiol., in press.

(10) Cobbs, W.H., and Pugh, E.N., Jr. 1985. Cyclic GMP can increase rod outer segment light sensitive current ten fold without delay of excitation. Nature **313:** 585-587.

(11) Corson, D.W., and Fein, A. 1983. Chemical excitation of Limulus photoreceptors I. J. Gen. Physiol. **82:** 639-657.

(12) Corson, D.W.; Fein, A.; and Payne, R. 1984. Detection of an inositol 1,4,5-triphosphate-induced rise in intracellular free calcium with aequorin in Limulus photoreceptors. Biol. Bull. **167:** 524-525.

(13) de Couet, H.G.; Stowe, S.; and Blest, A.D. 1984. Membrane-associated actin in the rhabdomeral microvilli of crawfish photoreceptors. J. Cell Biol. **98:** 834-856.

(14) Fein, A.; Payne, R.; Corson, D.W.; Berridge, M.J.; and Irvine, R.F. 1984. Photoreceptor excitation and adaptation by inositol 1,4,5-triphosphate. Nature **311:** 157-160.

(15) Fesenko, E.E.; Kolesnikov, S.S.; and Lyubarsky, A.L. 1985. Induction by cyclic GMP of cationic conductance in plasma membrane of retinal rod outer segment. Nature **313:** 310-313.

(16) George, J., and Hagins, W. 1983. Control of Ca in rod outer segment disks by light and cyclic GMP. Nature **303:** 344-348.

(17) Ghalayini, A.J., and Anderson, R.E. 1984. Brief light exposure stimulates IP2 breakdown in frog retina. Inv. Ophthalmol. Vis. Sci. **25:** 61.

(18) Glenney, J.R., Jr., and Glenney, P. 1984. The microvillus 11OK cytoskeleton protein is an integrated membrane protein. Cell **37:** 743-751.

(19) Gold, G.H. 1985. Plasma membrane calcium fluxes in intact rods are inconsistent with the "Ca hypothesis". Biophys. J. **47:** 356a.

(20) Gold, G.H., and Korenbrot, J.I. 1980. Light-induced calcium release by intact rods. Proc. Natl. Acad. Sci. USA **77:** 5557-5561.

(21) Goldberg, N.D.; Ames, A. III; Gander, J.E.; and Walseth, T.F. 1983. Magnitude of increase in retinal cGMP metabolic flux determined by ^{18}O incorporation into nucleotide-phosphoryls corresponds with intensity of photic stimulation. J. Biol. Chem. **258:** 9213-9219.

(22) Goldsmith, T.H., and Wehner, R. 1977. Restrictions on rotational and translational diffusion of pigment in the membranes of a rhabdomeric photoreceptor. J. Gen. Physiol. **70:** 453-490.

(23) Hubbard, R., and Kropf, A. 1958. The action of light on rhodopsin. Proc. Natl. Acad. Sci. USA **44:** 130-139.

(24) Irvine, R.F.; Anderson, R.E.; Rubin, L.J.; and Brown, J.E. 1985. Inositol 1,3,4-triphosphate concentration is changed by illumination of Limulus ventral photoreceptors. Biophys. J. **47:** 38a.

(25) Kaupp, U.B., and Koch, K.-W. 1984. Cyclic GMP releases calcium from leaky rod outer segments. Vision Res. **24:** 1477-1479.

(26) Koch, K.-W., and Kaupp, U.B. 1985. Cyclic GMP directly regulates a cation conductance in membranes of bovine rods by a cooperative mechanism. J. Biol. Chem. **260:** 6788-6800.

(27) Kuhne, W. 1878. On the Photochemistry of the Retina and on Visual Purple, ed. M. Foster. London: MacMillan.

(28) Lamb, T.D. 1981. The involvement of rod photoreceptors in dark adaptation. Vision Res. **21:** 1773-1782.

(29) Liebman, P.A., and Pugh, E.N., Jr. 1980. ATP mediates rapid reversal of cyclic GMP phosphodiesterase activation in visual receptor membranes. Nature **287:** 734-736.

(30) Lisman, J. 1985. The role of metarhodopsin in the generation of spontaneous quantum bumps in ultraviolet receptors of Limulus median eye. J. Gen. Physiol. **85:** 171-187.

(31) Lisman, J.; Levine, E.; Crain, E.; and Robinson, P. 1985. Nontransducing rhodopsin. Inv. Ophthalmol. **26:** 43.

(32) Matthews, H.R.; Torre, V.; and Lamb, T. 1985. Effects on the photoresponse of calcium buffers and cyclic GMP incorporated into the cytoplasm of retinal rods. Nature **313:** 582-585.

(33) McClosey, M., and Poo, M.-M. 1984. Protein diffusion in cell membranes: some biological implications. Intl. Rev. Cytol. **87:** 19-81.

(34) Miller, D.L., and Dratz, E.A. 1984. Phosphorylation at sites near rhodopsin's carboxyl-terminus regulates light-initiated cyclic GMP hydrolysis. Vision Res. **24:** 1509-1521.

(35) Miller, D.L., and Korenbrot, J.I. 1984. The effects of intracellular calcium buffer, Quin 2, on photocurrents and light-induced calcium release from individual rods. Inv. Ophthalmol. Vis. Sci. **26:** 168.

(36) Minke, B., and Stephenson, R.S. 1985. The characteristics of chemically induced noise in Musca photoreceptors. J. Comp. Physiol., in press.

(37) Morton, R.A., and Goodwin, T.W. 1944. Preparation of retinene in vitro. Nature **153:** 405-406.

(38) Nakatani, K., and Yau, K.-W. 1985. cGMP opens the light-sensitive conductance in retinal rods. Biophys. J. **47:** 356a.

(39) Neugebauer, D.-C., and Thurm, U. 1985. Interconnections between the stereovilli of the fish inner ear. Cell Tiss. Res. **240:** 449-453.

(40) Nicol, G.D.; Kaupp, U.B.; and Bownds, M.D. 1985. Phototransduction occurs in the absence of transmembrane calcium gradients in isolated frog rod photoreceptors. Biophys. J. **47:** 100a.

(41) Nicol, G.D., and Miller, W.H. 1978. Cyclic GMP injected into retinal rod outer segments increases latency and amplitude of response to illumination. Proc. Natl. Acad. Sci. USA **75:** 5217-5220.

(42) Paulsen, R., and Bentrop, J. 1984. Reversible phosphorylation of opsin induced by irradiation of blowfly retinae. J. Comp. Physiol. **A 155:** 39-45.

(43) Paulsen, R.; Bentrop, J.; and Peters, K. 1984. Photochemistry and biochemistry of blowfly photoreceptor membranes. Vision Res. **24:** 1700.

(44) Paulsen, R.; Zinkler, D.; and Delwelle, M. 1983. Architecture and dynamics of microvillar membranes of a cephalopod photoreceptor. Exp. Eye Res. **36:** 47-56.

(45) Payne, R.; Fein, A.; and Corson, D.W. 1984. A rise of intracellular Ca^{2+} is necessary and perhaps sufficient for photoreceptor excitation in Limulus photoreceptors. Biol. Bull. **167:** 531.

(46) Pinto, L.H., and Brown, J.E. 1985. Pressure injection of 3',5'-cyclic GMP into solitary rod photoreceptors of the tiger salamander. Brain Res. **304:** 197-200.

(47) Revel, J.P., and Ito, S. 1967. The Specificity of Cell Surfaces, eds. B.D. Davis and L. Warren, p. 211. Newark, NJ: Prentice Hall. Cited after Revel, J.P., and Goodenough, D.A. 1970. Cell coats and intercellular matrix. In Chemistry and Molecular Biology of the Intercellular Matrix, ed. E.A. Balazs, vol. 3, pp. 1361-1380.

(48) Rubin, L.J., and Brown, J.E. 1985. Intracellular injection of calcium buffers blocks IP3-induced but not light-induced electrical responses of Limulus ventral photoreceptors. Biophys. J. **47:** 38a.

(49) Saibil, H. 1982. An ordered membrane-cytoskeleton network in squid photoreceptor microvilli. J. Molec. Biol. **158:** 435-456.

(50) Saibil, H. 1984. A light-stimulated increase in cyclic GMP in squid photoreceptors. FEBS Lett. **168:** 213-216.

(51) Schnetkamp, P.P.M. 1979. Calcium translocation and storage of isolated intact cattle rod outer segments in darkness. Biochim. Biophys. Acta **554:** 441-459.

(52) Schroeder, W., and Fain, G.L. 1984. Light-dependent calcium release from photoreceptors measured by laser micro-mass analysis. Nature **309:** 268-270.

(53) Sitaramayya, A., and Liebman, P.A. 1983. Phosphorylation of rhodopsin and quenching of PDE activation by ATP at weak bleaches. J. Biol. Chem. **258:** 12106-12109.

(54) Somlyo, A.P., and Walz, B. 1985. Elemental distribution in Rana pipiens retinal rods: quantitative electron probe analysis. J. Physiol. **358:** 183-195.

(55) Stern, J.; Chin, K.; Robinson, P.; and Lisman, J. 1985. The effect of nucleotides on the rate of spontaneous quantum bumps in Limulus photoreceptors. J. Gen. Physiol. **85:** 157-169.

(56) Stern, J., and Lisman, J. 1982. Internal dialysis of Limulus ventral photoreceptors. Proc. Natl. Acad. Sci. USA **79:** 7580-7584.

(57) Szuts, E.; Reid, M.; Payne, R.; Corson, D.W.; and Fein, A. 1985. Biochemical and physiological evidence for the involvement of inositol 1,4,5-triphosphate in visual transduction. Biophys. J. **47:** 202a.

(58) Tomita, T. 1965. Electrophysiological study of the mechanisms subserving color coding in the fish retina. Cold S. H. Symp. Quant. Biol. **30:** 559-566.

(59) Vandenberg, C.A., and Montal, M. 1984. Light-regulated biochemical events in invertebrate photoreceptors. 2. Light-regulated phosphorylation of rhodopsin and phosphoinositides in squid photoreceptor membranes. Biochemistry **23:** 2347-2352.

(60) Waloga, G., and Anderson, R.E. 1985. Effects of IP3 injections into salamander rods. Biochem. Biophys. Res. Comm. **126:** 59-62.

(61) Waloga, G.; Anderson, R.E.; and Irvine, R.F. 1985. Modulation of vertebrate photoreceptor potentials by injection of inositol triphosphate. Biophys. J. **47:** 37a.

(62) Yamazaki, A.; Sen, I.; Bitensky, M.W.; Casnellie, J.; and Greengard, P. 1980. Cyclic GMP-specific, high affinity, non-catalytic binding sites on light-activated phosphodiesterase. Proc. Natl. Acad. Sci. USA **255:** 11619-11624.

(63) Yau, K.-W., and Nakatani, K. 1984. Electrogenic Na-Ca exchange in retinal rod outer segment. Nature **311:** 661-663.

(64) Yau, K.-W., and Nakatani, K. 1985. Light-induced reduction of cytoplasmic free calcium in retinal rod outer segments. Nature **313:** 579-582.

(65) Yoshikami, S.; George, J.S.; and Hagins, W.A. 1980. Light-induced calcium fluxes from outer segment layer of vertebrate retinas. Nature **286:** 395-398.

Standing, left to right:
Alfred Fahr, Hennig Stieve, Peter MacLeish, Luigi Cervetto

Seated (middle), left to right:
Geoff Owen, Denis Baylor, Jürgen Schnakenberg, John Lisman

Seated (front), left to right:
Wolfgang Hanke, Fulton Wong, Jonathan Coles

The Molecular Mechanism of Photoreception, ed. H. Stieve, pp. 451-464. Dahlem Konferenzen 1986. Berlin, Heidelberg, New York, Tokyo: Springer-Verlag.

Light-sensitive Channels, Pumps, and Carriers

Group Report

W.G. Owen, Rapporteur
D. Baylor
L. Cervetto
J.A. Coles
A. Fahr
W. Hanke
J.E. Lisman
P.R. MacLeish
J. Schnakenberg
H. Stieve
F. Wong

INTRODUCTION

In planning our discussion, a series of broad questions relating to the structural and functional properties of light-sensitive channels were formulated as follows:

1) What do the characteristics of the macroscopic light-dependent current tell us about the molecular architecture of the channel?
2) What are the properties of the Na^{2+}/Ca^{2+} exchange carrier and how do these influence both the operation of the transduction mechanism and out interpretation of results from experimental studies of the light-sensitive channel?
3) What do microscopic measurements (e.g., noise analysis, single-channel recordings) tell us about the light-sensitive channel and its gating?
4) How does the light-sensitive channel respond to possible transmitter substances?
5) What does the electrical fine-structure of the single-photon event (quantum bump) tell us about the mechanism of channel opening and closure?

Data from invertebrate and vertebrate photoreceptors were compared and contrasted: the highlights of our discussion are presented in this report. There were inevitably areas of our discussion that overlapped with the focus of other groups. Indeed, a joint session with the group discussing internal messengers (Miller et al., this volume) was arranged to establish what is known about internal transmitter substances. Some duplication of material presented in other reports is therefore unavoidable. No attempt has been made to cite all the literature upon which discussion was based as this would be cumbersome. Where a paper is cited, it is usually the one that was invoked by a discussant to support or counter an argument. Discussants frequently expressed opinions and offered speculation about a subject and this led to occasional disagreements. The consensus view is emphasized in this report, though opposing views are also presented in outline. It was agreed that "light-sensitive channel" would serve as an appropriate shorthand term during discussion for any plasma membrane conductance that is modulated as a direct consequence of photon absorption (rather than as a secondary consequence of a voltage change or a change in ionic concentration). It will be used throughout this report.

DISCUSSION

An important difference that may exist between Limulus photoreceptors and vertebrate rods is that in the former, there is more than one type of light-sensitive channel: a rapidly activating channel permeable primarily to Na^+ ("Na^+ channel") and a slowly inactivating K^+ channel which has been identified as the delayed rectifier. Though the delayed rectifier is strongly voltage-dependent, the effect of light is a direct one since it is seen under voltage-clamp. A consequence of the existence of these two light-sensitive channels is that the photoresponse does not exhibit a unique reversal potential. Also, barnacle and Hermissenda photoreceptors contain more than one class of light-sensitive channel.

In vertebrate rods the outer segments appear to contain only light-sensitive channels and there is as yet no evidence of more than one type, though it is possible that it can exist in more than one conductance state.

In Limulus ventral photoreceptors, the light-activated "Na^+ channel" has been shown to have a reversal potential between 0 and 10 mV. At negative potentials the steady-state conductance of this channel remains essentially constant but increases by a factor of ten at positive potentials. The instantaneous I-V relation measured in the barnacle is ohmic but relaxes with a time constant of about 14 ms to the steady-state form. Thus the observed outward rectification is not an intrinsic property of the channel but is time-dependent. This is confirmed by studies of single-channel currents in Limulus ventral photoreceptors using the patch-clamp (2). The single-channel current is a linear function of voltage. Depolarization of the patch causes no change in this relation, but both the mean lifetime of the channel and the probability of opening are increased. Thus the macroscopically observed outward rectification reflects a voltage-dependence of channel gating rather than a nonlinearity of the single-channel I-V relation.

This picture derived from the macroscopic measurement of the steady-state I-V relation of the light-sensitive channel in vertebrate rods differs in significant ways. First, the photoresponse exhibits a unique reversal potential near 0 mV. At potentials in the physiological range, the I-V relation is nearly flat, indicating that the channel behaves as a constant-current generator. At positive potentials a strong outward rectification develops (3). It was pointed out that the steepness of the I-V relation at these positive potentials - an e-fold increase in current for a potential change of +12.5 mV (Zimmerman and Baylor, personal communication) or 25 mV (3) - can be explained in terms of a rate-limiting energy barrier located towards the outer surface of the membrane, though the steepness of the curve measured by Zimmerman and Baylor also requires that charges cross the membrane in pairs; i.e., a divalent cation or two monovalent cations moving together (Baylor, personal communication). Alternatively, one could postulate that the channel is gated and that gating is voltage-dependent, an explanation that would be consistent with the known properties of ionic pores in other nerve membranes and similar to that obtaining in invertebrate photoreceptors. Any attempt to detect a time-dependence of rectification is necessarily limited by the time resolution of the measuring technique. Within the ~3 ms resolution of the best published measurements, no time-dependent behavior had been seen suggesting that any voltage-dependent gating must be very fast with a time constant of less than 3 ms. It was mentioned that recent studies by Bodoia and Detwiler (5), Attwell and Gray (1), and Zimmerman and Baylor (personal communication), in which the light-dependent noise measured under whole-cell patch-clamp was analyzed, all showed a relaxation time constant on the order of 1-2 ms which, if ascribed to channel gating, suggests that it is indeed very fast, an order of magnitude faster than channel gating in Limulus (see below, however).

The finding of a reversal potential near 0 mV, for both the "Na^+ channel" of Limulus and the light-sensitive channel in the vertebrate rod, implies a weak selectivity for Na^+ over K^+, given what we know about the internal and external ion concentrations. In Limulus, Brown and Mote (7) showed that Li^+ permeates as easily as Na^+ through the light-sensitive "Na^+ channel" and that K^+, too, is perhaps half as permeable. In addition, Stieve reported that the light-induced transient increase in intracellular Ca^{2+} in Limulus photoreceptors depends upon membrane voltage in a way that is consistent with a significant permeability to Ca^{2+} (21). An extensive study of ion selectivity of the Limulus channel remains to be done, however.

In discussions regarding the various studies of ion selectivity of vertebrate rods, a recurring question was whether or not there might exist more than one conductance state, the ion selectivity being different in the different states. A complicating factor is the existence of an active Na^+/Ca^{2+} exchange mechanism in the plasma membranes of both Limulus photoreceptors and vertebrate rods. Fain and Lisman (13) pointed out that in any attempt to study ion selectivity involving replacement of external Na^+ by other ions, Na^+/Ca^{2+} exchange is likely to be reduced or blocked. Since the light-sensitive channel may be influenced directly or indirectly by internal Ca^{2+}, this could profoundly affect the apparent permeability of the substituted ions. Yau and Nakatani (35) and Hodgkin, McNaughton, and Nunn (20) argued that very rapid changes in the external ion concentration allow the direct effects of the ion substitution to be separated from the slower effects mediated by the Na^+/Ca^{2+} exchanger. On this basis, Yau and Nakatani found that even with millimolar concentrations of Ca^{2+} in the bathing medium, the channel is permeable to Li^+, Na^+, K^+, Rb^+, Cs^+, and the divalent cations Ca^{2+} and Sr^{2+}. With nanomolar concentrations of Ca^{2+} in the external medium, these ions were seen to pass through the channel in the steady state. Thus the channel appears to be relatively nonselective. Since neither internal ionic concentrations nor transmembrane voltages were controlled in these experiments, however, one should perhaps remain cautious about drawing such a conclusion (Lisman, personal communication).

Other experiments were cited in which not only electrical measurements were made, but also determinations of unidirectional fluxes using radioactive tracers, and where controls were carried out to distinguish between effects due to permeability changes and those due to voltage changes (9, 33). In those experiments internal cGMP levels were measured whenever appropriate. They showed that even with micromolar concentrations of Ca^{2+} in the external medium, the permeability of Ca^{2+} through the channel is five times that of Na^{+}, and that with the addition of PDE inhibitors, Mn^{2+} passes through the channel fifteen times as easily as Na^{+}. Indeed, in the presence of PDE inhibitors and with only one permeable ion in the external medium, divalent cations in general appeared more permeable than monovalent ions at the same low concentration, suggesting that the channel might be selective for divalent cations.

It was generally agreed that ionic selectivity will best be worked out using an excised patch in which the ionic and cyclic nucleotide concentrations on each side of the membrane are controlled, though the question was raised whether after such drastic treatment one could be confident that the observed properties of the channel would necessarily be the same as those it possesses under normal conditions.

Experiments performed in Cervetto's laboratory (9) indicate that ion movements through the light-sensitive channel do not conform to the independence principle. Rather they suggest that if the channel is a pore, ions move through in single file, while if it is a carrier, two or more ions must bind in order to be transported. This is not inconsistent with Baylor's interpretation of the steepness of the channel I-V curve at positive potentials.

The proposition was questioned that the Na^{+}/Ca^{2+} exchanger is entirely responsible for the decrease in conductance seen upon reducing external Na^{+}. A quantitative evaluation of Na^{+}/Ca^{2+} exchange was presented (Torre, personal communication) which suggested that within one or two seconds of blocking the exchanger, internal $[Ca^{2+}]$ would be expected to rise no more than fourfold. Torre also presented arguments suggesting that $[Ca^{2+}]_{in}$ might normally be near 10 µmol/l in darkness. Hefurther argued that an increase to 40 µmmol/l would not be sufficient to cause the blockage of the light-sensitive channel since experiments in which Ca^{2+} was injected into the inner segment through a patch pipette suggest that more than 100 µmol/l $[Ca^{2+}]$ is required in the outer segment for full blockage to occur. In his view, therefore, an alternative explanation must be sought.

Torre also challenged the view of Yau and Nakatani (35) that the channel exists only in a single conductance state, arguing that, for example, in the presence of millimolar $[Ca^{2+}]_o$, activation of the channel could require the binding of Na^+ to some external site. He further suggested that the high selectivity of the channel for divalent cations in the presence of IBMX or cGMP represented a different conductance state of the channel (9). This was supported by the finding of MacLeish et al. (26) that in low (micromolar) $[Ca^{2+}]_o$ the injection of cGMP changed the shape of the I-V relation. MacLeish suggested that this might be because the addition of cGMP altered the degree of phosphorylation of the channel which, in turn, might change its conductance.

During this discussion it became clear that there are several lines of evidence consistent with the idea that the light-sensitive channel can exist in more than one conductive state. It was pointed out that this would not be unexpected in view of the properties of other known channels. In experiments on fragmented ROS membranes, reconstituted in lipid bilayers (19), measured single-channel conductances fell into two distinct populations with mean values of 20 pS and 100 pS, respectively. These could represent either two different species of channel or a single species in two different conductance states (Hanke, personal communication).

Recent studies of noise across the plasma membranes of both invertebrate and vertebrate receptors have yielded significant new information concerning the nature of the light-sensitive channel. In Limulus, analysis of single-channel recordings obtained under patch-clamp in the "cell-attached" configuration reveals two populations of channel events: frequent events due to a channel with a unitary conductance of 40 pS and less frequent events due to a channel with a conductance of 20 pS. Again, whether or not these represent two different channel species or a single species that can exist in two different conductance states is unknown (Lisman, personal communication).

In Limulus the power spectrum of the channel noise depends upon the level of illumination (34). At low intensities it can be fitted with the product of three Lorentzians, suggesting that channel opening involves several correlated events. As light intensity is increased, however, the required number of Lorentzians decreases until at high intensities the spectrum can be well fitted by a single Lorentzian, suggesting that gating then requires but a single event. In measurements on excised patches the results are qualitatively similar though they differ quantitatively. Specifically, whereas the relaxation time constant of the current noise measured in whole-cell voltage-clamped Limulus photoreceptors is ~15 ms, the mean open time of single channels seen in excised patches of the plasma membrane is only about 3 ms (Lisman, personal communication).

In vertebrate rods a rather different picture is beginning to emerge. The elementary conductance change estimated from recordings of noise obtained under patch-clamp with the cell attached is in the range 50 fS (12) to 100 fS (1). The power spectrum has a low-frequency component with a characteristic frequency near 1 Hz, closely similar to that of the "continuous noise" recorded by Baylor et al. (4) using a suction pipette and which is thought to reflect fluctuations in the biochemical process leading to channel closure. (Significantly, this component is not seen in the spectrum of noise induced in excised patches of plasma membrane, as will be discussed later). In addition, there is a high-frequency component which can best be fitted by a single Lorentzian with a characteristic frequency in the range 100-200 Hz. If attributed to channel gating, this implies a mean channel lifetime on the order of 1 ms. On the basis of an estimated single-channel flux of 1.25×10^4 monovalent ions per second (12), we would expect, at the resting potential of -40 mV, a net flux of only about twelve monovalent ions per single-channel event!

These properties suggest that the channel in the vertebrate ROS may be significantly different, at least in its normal, physiological form, from that in the invertebrate photoreceptor. Their implications for channel structure are discussed in some detail by Owen (this volume). It was agreed that if the channel is a pore, it is one of atypical structure. In order to account for its unusually low conductance, one might suppose that in its normal form it is partially blocked, though that possibility is not easily reconciled with the observed low selectivity of the channel. The nonindependence of ion movement through the channel would perhaps require that there be two or more binding sites within the pore. As mentioned earlier, the strong outward rectification could be explained if, in addition, there were a large energy barrier near the external face of the membrane. Alternatively, one could suppose that the channel is some kind of ion carrier though, if so, it is likely to be a hybrid type possessing some of the properties of a pore. By contrast, there seems little doubt that the channel in Limulus (in its functional state) has all of the characteristics of a pore.

We next turned our attention to the question of which substances regulate the conductance of the light-sensitive channel. In Limulus ventral photoreceptors, two substances have been identified as components of the transduction process. These are Ca^{2+} and inositol 1,4,5-triphosphate, ($InsP_3$) (see Lisman, this volume). The work of Lisman and Brown (25) established a primary role for Ca^{2+} in the mechanism of adaptation. Recent work by Payne, Fein, and Corson (30) suggests that it may also be capable of exciting the cell; a bolus of 1 mmol/l Ca^{2+} injected into the R-lobe of the receptor caused the transient activation of a conductance having the same reversal potential as the light-sensitive conductance. Ca^{2+} cannot be the only messenger of excitation, however, since injection of EGTA does not block the light response.

Repetitive illumination of the cell in the presence of low, external $[Ca^{2+}]_o$ seems to cause a change in its state (Brown, personal communication). Under normal conditions, a repetitive stimulus elicits an inward current that rapidly peaks and then decays to a lower maintained level. Aequorin fluorescence reveals a concurrent increase in internal free Ca^{2+}. Injecting Ca^{2+} reduces the cell's sensitivity to light. By contrast, after repetitive illumination the peak inward current occurs much later in the response than does peak aequorin fluorescence. Injection of Ca^{2+} now increases sensitivity. Brown suggested that there might be a "light-labile" compartment in which Ca^{2+} is normally sequestered and that repetitive illumination in the presence of $<10^{-7}$ mol/l Ca^{2+} depletes this internal Ca store. If Ca^{2+} increases the gain of some intermediate stage of the excitation pathway, these observations can be understood. This notion would be consistent with Lisman and Brown's observation (25) that EGTA blocks adaptation but not excitation and would imply, in addition, an indirect role for Ca^{2+} in the excitation process.

$InsP_3$ is known to be present in Limulus ventral photoreceptors. 20 μmol/l $InsP_3$ injected into the R-lobe induces the production of discrete waves of depolarization that closely resemble quantum bumps. Under voltage-clamp inward currents are seen, the reversal potential of which is the same as that of the light-induced current (8, 14). It thus appears that $InsP_3$ can activate the light-dependent current. Recent work by Payne et al. (30), however, suggests that it does so indirectly since it has no effect in the presence of EGTA. Various possibilities were discussed that might explain its role in excitation, the most plausible being that it causes an increase in internal free Ca^{2+} which in turn regulates the gain of a late stage in the direct excitation pathway. Other recent evidence from the laboratories of Brown and Fein (Brown, personal communication) indicates that $InsP_3$ does indeed mediate a rise in intracellular free Ca^{2+} in Limulus ventral photoreceptors. Lisman, however, playing devil's advocate, pointed out (personal communication) that $InsP_3$ need not act in this way at all but might simply destabilize the inactive rhodopsin so that the frequency of spontaneous isomerizations is increased.

It was generally agreed that on the basis of present evidence neither Ca^{2+} nor $InsP_3$ is a likely component in the direct pathway of excitation in Limulus photoreceptors.

There is evidence, however, to suggest that channel gating in Limulus may involve the competitive binding of Na^+ and Ca^{2+} (Stieve, personal communication). Experimental results are consistent with the idea that the binding of Na^+ opens the channel while the binding of Ca^{2+} closes it. Light appears to cause a transient reduction in the relative affinity of the binding site for Ca^{2+} over Na^+ which increases the probability of the channel being open. The normal affinity of the binding site is greater for Ca^{2+} than for Mg^{2+} and greater for Na^+ than for Li^+ (31, 32).

In the discussions concerning vertebrate rods several lines of evidence emerged, all pointing to a single major conclusion, namely, that the light-induced decrease in ROS conductance is not the direct consequence of a rise in the internal concentration of calcium. Kaupp described a series of experiments making use of the ionophore A23187 (29). While recording the ROS current using a suction pipette with the inner segment drawn in, A23187 was applied in the presence of 1 mmol/l Ca^{2+}. The dark current disappeared and with it, the light response. Subsequently reducing $[Ca^{2+}]_o$ to between 10^{-7}

and 10^{-8} mol/l caused light responses to reappear. Since ionophore was almost certainly partitioned into both plasma membrane and disc membranes, the clear implication is that significant light responses can be generated in the virtual absence of internal free Ca^{2+}. In discussion, Fain commented (personal communication) that this does not necessarily mean that excitation is calcium-independent, pointing out that the exocytosis of synaptic vesicles, which is known to be a calcium-dependent process, was shown by Tsien and co-workers to continue even in the virtual absence of all calcium. This reference was to a study in which exocytosis was stimulated by α-latrotoxin, without apparent change in $[Ca^{2+}]_{in}$ as measured by the cytosolic indicator quin2, and with EGTA in the external medium (28).

MacLeish described experiments in which Ca^{2+} was injected into a solitary rod through an intracellular electrode (26). Injected into the inner segment, it had no effect. Injected into the outer segment, a reduction in the inward current was seen, though this developed slowly. The largest reduction that could be achieved was 50%. (In analogous experiments by Lamb, Matthews, and Torre, injection of Ca^{2+} into the inner segment resulted in a decay of the inward current with a time constant of about 200 ms - Lamb, personal communication.) EGTA injected into the outer segment caused a rapid and reversible increase in the inward current (26). When 100 μmol/l cGMP was injected into the ROS in the presence of EGTA, an enormous, rapid increase in the inward current was seen. The hydrolysis-resistant analogue, 8-bromo-cGMP, also produced a large increase in inward current. The I-V curves measured in the presence of EGTA before and after addition of cGMP could not be superimposed when scaled, perhaps suggesting that cGMP causes a change in the degree of phosphorylation of the channel protein. These results are consistent with the idea that the light-sensitive channel is not blocked by a direct action of Ca^{2+}. The results obtained with cGMP and 8-bromo-cGMP further suggest that cGMP might act directly on the channel in a manner that does not involve hydrolysis.

Lamb described recent experiments with Matthews and Torre (24, 27) in which the Ca^{2+} chelator, BAPTA, was introduced into the cytosol through a patch pipette sealed to the inner segment of an isolated rod. The current flowing across the ROS membrane was simultaneously monitored using a suction pipette. The concentration of BAPTA in the patch pipette was 10 mmol/l. Soon after injection of BAPTA began, the light responses altered drastically in form, indicating that the buffer was entering the outer segment of the cell. The sensitivity to dim flashes increased and responses were slowed. The rising phase of responses to bright flashes remained unchanged. Given that the presence of the buffer would minimize changes in the internal free calcium concentration, Lamb argued that a significant change in internal $[Ca^{2+}]$ is unnecessary for the generation of the light response (27). A pronounced overshoot that developed in the recovery phase of the response, he suggested, could be due to a reduction in cytoplasmic calcium levels during the response. This would be possible if the internal Ca^{2+} level were normally set by a balance between the inward leakage of

Ca^{2+} through the light-sensitive channel and its extrusion by the Na^+/Ca^{2+} exchanger. In his analysis of Na^+/Ca^{2+} exchange, Torre had pointed out that in the absence of any native buffering this balance would likely result in a resting level of Ca^{2+} in darkness of about 10 mol/l. The argument in favor of this assertion is outlined in the report by Miller et al. (this volume).

The finding of a light-induced rise in external [Ca^{2+}] (18), rather than implying a release of Ca^{2+} from internal stores, could equally well result from a light-induced reduction in Ca^{2+} influx set against a constant rate of extrusion by the Na^+/Ca^{2+} exchanger (Lamb, personal communication). The comment was made that speculation along these lines will only be resolved by a direct measure of the dark level of internal free Ca^{2+} and that the achievement of such a measure must be considered a prime goal of future studies (Korenbrot, personal communication).

Experiments were described by Kaupp (personal communication) in which Ca^{2+} movements across the membranes of isolated rod discs in suspension were monitored. Similar experiments have been described by Cavaggioni and Sorbi (11) and Caretta and Cavaggioni (10). Kaupp presented clear evidence for the existence in the disc membrane of a channel, permeable to Ca^{2+}, which is activated by the direct action of cGMP (23). cGMP stimulated the release of Ca^{2+} from the discs, provided there were cations in the external medium with which Ca^{2+} could exchange. K^+, Li^+, Na^+, Rb^+, and Cs^+ were all effective in this regard and, therefore, passed through the channel. The activation curve of the channel was very steep, activation occurring over the range 30-70 µmol/l. A Hill plot had a slope of 2-3, implying that channel activation required the cooperative action of more than one molecule of cGMP. 8-bromo-cGMP was equally effective at 1/10 of the necessary cGMP concentration, implying that cGMP hydrolysis is not required for activation. Moreover, the removal of all cytosolic proteins had no effect upon cGMP action, suggesting that a protein kinase is not involved. Of especial interest was Kaupp's finding that this channel could be blocked by l-cis-Diltiazem. (The stereoisomer, d-cis-Diltiazem, is a known blocker of Ca^{2+} channels in other systems but had no effect on this channel.)

Kaupp speculated that, in view of the common origin of disc membrane and plasma membrane, this channel may also exist in the plasma membrane of the rod and could be identical with the light-sensitive channel. He pointed out that in darkness, where free cGMP levels should be highest, this conductance should be open. Cytosolic free Ca^{2+} should then be maximal. A light-induced fall in free [cGMP] should cause the channels to close and internal free [Ca^{2+}] to fall as Ca^{2+} extrusion mechanisms continue to act. Lamb questioned whether a free cGMP level of 30-70 µmol/l was physiologically feasible, given that the total cGMP content of the ROS is near this value and that the free cGMP level could be one or two orders of magnitude lower. Kaupp argued that only about half of the total cGMP was likely to be bound, which would mean a free concentration of about 20-35 µmol/l.

Yau described the then unpublished results of two studies, one by Fesenko, Kolesnikov, and Lyubarsky (16), the other by Yau and Nakatani (personal communication), in which current flow across excised, inside-out patches of ROS plasma membrane was analyzed. Transmembrane current was extremely small until application of cGMP to the cytoplasmic face of the patch. This induced a noisy current, though not the discrete current steps one might expect upon activation of gated channels. The induced current reversed at a potential near +10 mV, suggesting that the channel is relatively unselective. The conductance increase occurred in the absence of nucleoside triphosphates, implying that it does not depend upon phosphorylation by a protein kinase (16).

The most marked increase in conductance occurred with addition of cGMP in the range 20-60 µmol/l with saturation occurring at 300 µmol/l. (This is similar to the activation range of the disc membrane conductance described by Kaupp - see above.) Calcium added in the absence of cGMP in concentrations as high as 1 mmol/l had absolutely no effect; in the presence of cGMP, Ca^{2+} had a modest effect upon the magnitude of the cGMP-dependent conductance.

Fesenko et al. (16) also analyzed the noise in the cGMP-induced current. They estimated a unitary conductance change of about 100 fS, in good agreement with estimates of the conductance of a single light-sensitive channel made on the basis of whole-cell patch-clamp recordings by Detwiler and Bodoia (12) and Attwell and Gray (1). The power spectrum could be fitted by a single Lorentzian with a characteristic frequency of 316 Hz implying a process with a relaxation time constant of ~0.5 ms.

It was agreed that these findings strongly suggest that the ROS channel can be activated directly by cGMP and that Ca^{2+} is not necessary for channel inactivation. Indeed, it seems likely that Ca^{2+} may have no direct effect upon the channel though Yau agreed that the channel might exist in more than one conductance state, leaving open the possibility that Ca^{2+} might act directly under other experimental conditions. The similarity in both magnitude and kinetics of the elementary cGMP-dependent conductance and light-sensitive conductance, while not definitive, supports the conclusion that they are one and the same and argues for cGMP as the internal transmitter.

The discussion of these various findings was long and detailed. The consensus of the group was that the ROS light-sensitive channel can no longer be thought of simply as being blocked by Ca^{2+}. The dramatic effects of changing external $[Ca^{2+}]$ in all probability are mediated indirectly. There was some debate about what that might mean, much of it centering upon the Na^+/Ca^{2+} exchange. It is still possible that $[Ca^{2+}]_{in}$ regulates the effect of cGMP upon the channel, though how this might occur remains mysterious.

In discussing these various results we seemed to have been forced to discard our only candidate for a blocker of the light-sensitive channel. Perhaps a new one was acquired, however. If Kaupp was correct in his suggestion that the cGMP-activated channel in the ROS discs is identical with the light-sensitive channel of the plasma membrane (and the similarities are certainly striking), then 1-cis-Diltiazem is a channel blocker. This possibility offers hope for a new pharmacological assay of the light-sensitive channel.

Our final round of dicussions focused upon the mechanisms of generation of quantum bumps in invertebrate photoreceptors. The eyes of the honeybee drone never dark-adapt to a level where bumps can be discerned (15), nor are bumps seen in squid, crayfish, or barnacle eyes (Coles, personal communication). Though they are seen in the eyes of flies and locusts, most of the discussion concerned findings in Limulus.

In Limulus, quantum bumps appear with various latencies following a light flash. The latency of the bump-generating mechanism is set by the first seven stages of the Fuortes-Hodgkin (17) cascade model (34). Schnakenberg noted that the bump latency was uncorrelated with any other response parameter and exceeded the bump duration by a factor of at least two (22). This could come about only if the first seven stages are without gain (34). He identified several parameters that are related to the generator current: bump area, bump amplitude, and slope of the rising phase. These are determined by the last three stages of the cascade, which possess gain (34). Individual values of the "current" parameters were found to vary from cell to cell and could not be matched by linear scaling. If the ratio of any two of the current parameters is compared from cell to cell, however, the cell-to-cell differences disappear (22). Schnakenberg argued that the cell-to-cell variation might reflect differences in the product of channel density and channel conductance between cells. Lisman preferred the view that cell-to-cell variations in bump amplitude, area, rise time, etc., are more likely due to differences in cell viability reflecting, perhaps, different degrees of damage upon penetration of the cell. Schnakenberg reiterated the proposal (22) that since the bump parameters of latency, slope of the rising phase, rise time, and decay time constant are mutually uncorrelated, they could serve as "primary" parameters characterizing a bump with respect to its size and time course.

Hillman pointed out that the near-exponential frequency distribution of bump areas is consistent with a model in which one stage of the bump generator mechanism is active. He contrasted this with the quantum bumps of the vertebrate rod described by Baylor and his co-workers (4). The frequency distribution of bump areas was very narrow in that case, suggesting that one stage of the mechanism was saturated in some anatomical unit. The alternative, several active states of some agent with possible cooperativeness among them, is much less likely since the number of such states that would be required is very large.

Stieve noted that there is on the order of one channel per microvillus in Limulus photoreceptors (6) and questioned whether a channel might perhaps be located at the base of each microvillus. It was pointed out by Blest, however, that the base of the microvillus is the site of rapid membrane turnover involving pinocytosis (personal communication). In his view, this would make it an unlikely site for a functionally important channel protein.

CONCLUSION

Our discussions ended with some consideration of new directions that might be explored and new approaches to existing problems. There was general agreement that the study of excised patches of photoreceptor membrane should yield answers to many of the questions still remaining, e.g., whether the channel exists in multiple conductance states and, if so, what their ion selectivities might be, as well as the question of how the channel conductance is regulated. The great advantage of the excised patch is that it allows the channels to be uncoupled from the internal chemistry that controls them. The ionic and metabolic milieu can then be controlled on both sides of the patch and transmembrane potentials are easily regulated.

The development of new ways of incorporating highly purified membrane fragments into planar lipid bilayers offers a further approach to the study of channels under tightly controlled biochemical conditions, allowing precise pharmacological experiments to be performed.

Perhaps the most exciting possibilities were heralded by the experiments of Wong in which the techniques of molecular genetics and electrophysiology are being applied in a study of the bump-generating mechanism in Drosophila (see Wong, this volume). This combined approach is potentially of enormous power. If developed, it could open up new avenues to the study of membrane proteins, possibly affording a means of discovering which molecular components subserve a given functional property. The implications for our future understanding of membrane channels, pumps, and carriers are exciting indeed.

REFERENCES

(1) Attwell, D., and Gray, P. 1984. Patch-clamp recording from isolated rods of the salamander retina. J. Physiol. **351:** 9P.

(2) Bacigalupo, J., and Lisman, J.E. 1983. Light-activated channels in Limulus ventral photoreceptors. Biophys. J. **45:** 3-5.

(3) Bader, C.R.; MacLeish, P.R.; and Schwartz, E.A. 1979. A voltage-clamp study of the light response in solitary rods of the tiger salamander. J. Physiol. **298:** 1-6.

(4) Baylor, D.A.; Matthews, G.; and Yau, K.-W. 1980. Two components of electrical dark noise in retinal rod outer segments. J. Physiol. **309:** 591-621.

(5) Bodoia, R.D., and Detwiler, P.B. 1984. Patch-clamp study of the light response of isolated frog retinal rods. Biophys. J. **45:** 337a.

(6) Brown, J.E., and Coles, J.A. 1979. Saturation of the response to light in Limulus ventral photoreceptors. J. Physiol. **296:** 373-392.

(7) Brown, J.E., and Mote, M.I. 1974. Ionic dependence of the reversal potential of the light response in Limulus ventral photoreceptors. J. Gen. Physiol. **63:** 337-350.

(8) Brown, J.E.; Rubin, L.J.; Ghalyini, A.J.; Tarver, A.P.; Irvine, R.F.; Berridge, M.J.; and Anderson, R.E. 1984. Myo-inositol phosphate may be a messenger for visual excitation in Limulus photoreceptors. Nature **311:** 160-163.

(9) Capovilla, M.; Caretta, A.; Cervetto, L.; and Torre, V. 1983. Ionic movements through light-sensitive channels of toad rods. J. Physiol. **343:** 295-310.

(10) Caretta, A., and Cavaggioni, A. 1983. Fast ionic flux activated by cyclic GMP in the membrane of cattle rod outer segment. Eur. J. Biochem. **132:** 1-8.

(11) Cavaggioni, A., and Sorbi, R.T. 1981. Cyclic GMP releases calcium from disk membranes of vertebrate photoreceptors. Proc. Natl. Acad. Sci. USA **78:** 3964-3968.

(12) Detwiler, P.B.; Connor, J.D.; and Bodoia, R.D. 1982. Gigaseal patch clamp recordings from outer segments of intact retinal rods. Nature **300:** 59-61.

(13) Fain, G.L., and Lisman, J.E. 1981. Membrane conductances of photoreceptors. Prog. Biophys. Molec. Biol. **37:** 91-147.

(14) Fein, A.; Payne, R.; Corson, D.W.; Berridge, M.J.; and Irvine, R.F. 1984. Photoreceptor excitation and adaptation by inositol 1,4,5-trisphosphate. Nature **311:** 157-160.

(15) Ferraro, M.; Levi, R.; Lovisolo, D.; and Vadacchino, M. 1983. Voltage noise in honeybee drone photoreceptors. Biophys. Struct. Mech. **10:** 129-142.

(16) Fesenko, E.E.; Kolesnikov, S.S.; and Lyubarsky, A.L. 1985. Induction by cyclic GMP of cationic conductance in plasma membrane of retinal rod outer segment. Nature **313:** 310-313.

(17) Fuortes, M.G.F., and Hodgkin, A.L. 1964. Changes in time scale and sensitivity in the ommatidia of Limulus. J. Physiol. **172:** 239-263.

(18) Gold, G.H., and Korenbrot, J.I. 1980. Light-induced Ca release by intact retinal rods. Proc. Natl. Acad. Sci. USA **77:** 5557-5561.

(19) Hanke, W., and Kaupp, U.B. 1984. Incorporation of ion channels from bovine rod outer segments into planar lipid bilayers. Biophys. J. **46:** 587-595.

(20) Hodgkin, A.L.; McNaughton, P.A.; and Nunn, B. 1985. The ionic selectivity and calcium-dependence of the light-sensitive pathway in toad rods. J. Physiol. **358:** 447-468.

(21) Ivens, I., and Stieve, H. 1984. Influence of the membrane potential on the intracellular light-induced Ca^{2+} concentration change of the Limulus ventral photoreceptor monitored by Arsenazo III under voltage-clamp conditions. Z. Naturforsch. **39c:** 986-992.

(22) Keiper, W.; Schnakenberg, J.; and Stieve, H. 1984. Statistical analysis of quantum bump parameters in Limulus ventral photoreceptor. Z. Naturforsch. **39c:** 781-790.

(23) Koch, K.W., and Kaupp, U.B. 1985. Cyclic GMP directly regulates a cation conductance in membranes of bovine rods by a cooperative mechanism. J. Biol. Chem. **260:** 6788-6800.

(24) Lamb, T.D.; Matthews, H.R.; and Torre, V. 1985. Introduction of calcium buffers into rod photoreceptors of the salamander. J. Physiol. **369:** 20.

(25) Lisman, J.E., and Brown, J.E. 1972. The effects of intracellular iontophoretic injection of calcium and sodium ions on the light response of Limulus ventral photoreceptors. J. Gen. Physiol. **59:** 701-709.

(26) MacLeish, P.R.; Schwartz, E.A.; and Tachibana, M. 1984. Control of the generator current in solitary rods of the Ambystoma tigrinum retina. J. Physiol. **348:** 645-664.

(27) Matthews, H.R.; Torre, V.; and Lamb, T.D. 1985. Effects on the photoresponse of calcium buffers and cyclic GMP incorporated into the cytoplasm of retinal rods. Nature **313:** 582-585.

(28) Meldolesi, J.; Huttner, W.B.; Tsien, R.Y.; and Pozzan, T. 1984. Free cytoplasmic Ca^{2+} and neurotransmitter release: Studies on PC12 cells and synaptosomes exposed to α-latrotoxin. Proc. Natl. Acad. Sci. USA **81:** 620-624.

(29) Nicol, G.D.; Kaupp, U.B.; and Bownds, D. 1985. Phototransduction occurs in the absence of transmembrane calcium gradients in isolated frog rod photoreceptors. Biophys. J. **47:** 100a.

(30) Payne, R.; Fein, A.; and Corson, D.W. 1985. A rise in intracellular Ca^{2+} is necessary and perhaps sufficient for photoreceptor excitation and adaptation by inositol 1,4,5-trisphosphate. Biol. Bull., in press.

(31) Stieve, H., and Bruns, M. 1978. Extracellular calcium, magnesium, and sodium ion competition in the conductance control of the photosensory membrane of Limulus ventral nerve photoreceptor. Z. Naturforsch. **33c:** 574-579.

(32) Stieve, H.; Pflaum, M.; Klomfaß, J.; and Gaube, H. 1985. Calcium/sodium binding competition in the gating of light-activated membrane conductance studied by voltage clamp technique in Limulus ventral nerve photoreceptor. Z. Naturforsch. **40c:** 278-291.

(33) Torre, V.; Pasino, E.; Capovilla, M.; and Cervetto, L. 1981. Rod responses in the absence of external sodium in retinae treated with phosphodiesterase inhibitors. Exp. Brain Res. **44:** 427-430.

(34) Wong, F.; Knight, B.W.; and Dodge, F.A. 1980. Dispersion of latencies in photoreceptors of Limulus and the adapting bump model. J. Gen. Physiol. **79:** 517-537.

(35) Yau, K.-W., and Nakatani, K. 1984. Cation selectivity of light-sensitive conductance in retinal rods. Nature **309:** 352-354.

Standing, left to right:
Joachim Schwemer, Joel Brown, Bob Shapley, Kuno Kirschfeld

Seated (middle), left to right:
Ed Pugh, Trevor Lamb, Baruch Minke, Winfried Keiper, Peter Hillman

Seated (front), left to right:
Peter Hochstrate, Kurt Hamdorf, David Pepperberg

The Molecular Mechanism of Photoreception, ed. H. Stieve, pp. 467-488. Dahlem Konferenzen 1986. Berlin, Heidelberg, New York, Tokyo: Springer-Verlag.

Adaptation

Group Report

E.N. Pugh, Jr., Rapporteur
J.E. Brown
K. Hamdorf
P. Hillman
P. Hochstrate
W.J.M. Keiper
K. Kirschfeld
T.D. Lamb
B. Minke
D.R. Pepperberg
J. Schwemer
R. Shapley

INTRODUCTION

Adaptation is the modulation of the visual transduction process by prior illumination. Accordingly, its understanding depends on and contributes to understanding of the visual transduction process itself. Unfortunately, this mutual influence has not yet been very fruitful, perhaps because it calls for at least the beginnings of a complete model of the transduction process. Most of the work on adaptation, some highlights of which are reported here, have therefore been studies of the phenomenology of adaptation and of the chemicals which directly intermediate it. These studies are still largely groping in the dark, especially in vertebrates, and are therefore necessarily more tentative and less focussed than those of the transduction process itself, about which more is known. The formulation of open questions with which this report ends suggests, however, that the time has come when the state of our biochemical understanding of the transduction process (Chabre and Applebury and Applebury et al., both this volume) will make possible direct exploitation of the constraints imposed by adaptational data.

The Function of Adaptation

The essential task of adaptation is to enable the eye to function optimally in all normal illumination conditions - in particular, to have maximum sensitivity when there is little light and not to be dazzled when there is more light while exploiting the bright conditions to improve spatial and temporal resolution. The visual system fulfills this task admirably through use of all available mechanisms - mechanical, biochemical, and neural. Most of the adaptive modulations of the system make teleological sense, a possible exception being facilitation (the enhancement under certain circumstances of the system's sensitivity by prior illumination) whose behavioral role, if any, is as yet unclear.

Mechanisms of Adaptation

As a matter of good engineering practice, modulations of gain should be applied at the earliest possible stages. In the vertebrate visual system, first the pupil size is exploited and then the biochemical gain of the photoreceptor; though further modulations take place in neural network processing the biggest factor remains the photoreceptor gain, and this is the subject of this report.

Within the photoreceptor, good engineering practice would call for the gain control being applied to early stages of the transduction chain and not, for instance, at the level of channel opening or closing. Opposing views are presented here.

Much research has gone into the biophysical and biochemical characterization of photoreceptor adaptation. The most striking observations on this adaptation are: that light adaptation sets in fairly rapidly (but with a definite latency) and declines in the dark much more slowly with at least two separable phases; that light adaptation not only decreases the gain of the transduction process but also speeds it up (a confrontation presented here between kinetic models and electrophysiological data is shown to lead to strong constraints on possible biochemical mechanisms); that its effect is relatively localized; and that in invertebrates Ca^{2+} of both extra- and intracellular origin is an essential intermediate - in the vertebrate no essential intermediate is known (Ca^{2+} may still be a candidate). However, determination of some important characteristics of this intermediate is described below. Evidence that the Ca^{2+} acts on the outside of the invertebrate membrane is also given, but this result is very controversial. Finally, data are presented suggesting that a light-induced rise in cytoplasmic inositol triphosphate may be at least one source of the rise in intracellular free Ca^{2+} concentration.

It has become clear that the adaptive modulation of the photoreceptor itself has several facets. The primary process of gain reduction in light and recovery in darkness has now been shown to have at least two components, separable by their time courses and by their dependence on Ca^{2+}. A model presented here suggests that there is a component, which may be one of these two, which corresponds to localized depletion and then exhaustion of some transduction intermediate. Then there is a process of facilitation, which acts on the transduction process but apparently by an independent mechanism. Another mechanism of gain control is linked to the bleaching of visual pigment in vertebrates and to its conversion into an inactive state in invertebrates. A demonstration of the similarities between vertebrate and invertebrate "bleaching" adaptation is presented here. Support is given for the idea that bleaching adaptation in vertebrates corresponds to a "dark light," but other data are presented arguing against this idea and instead suggest depletion of some "agent." However, it is not clear that such depletion, if present, may not in fact be a product of the dark light. Finally, certain combinations of bright stimuli of different wavelengths induce prolonged afterpotentials in whose presence the cells are strongly desensitized. Suggestions are presented here that phosphorylation of the visual pigment may be responsible. A process which is not normally considered adaptive, and is therefore not discussed here, is that by which in a cell the presence of an electrical response potential modulates the cell's ionic conductances.

TERMINOLOGY

We agreed upon this working definition of photoreceptor adaptation: "any change in the functional state of a photoreceptor consequent to its history of illumination." In this definition "history" means time up to and including the present; "functional state" is operationally defined in terms of a photoresponse to an appropriately selected test probe. The probe, if possible, should be a stimulus which does not of itself alter the functional state.

"History of illumination" in the above definition often refers to one of two frequently used experimental paradigms: a) pulse/probe, and b) step/probe. In a pulse/probe experiment, Q quanta are delivered to a photoreceptor in a brief flash and sensitivity is probed thereafter, typically as a function of time after the pulse. In a step/probe experiment, light is delivered at a fixed rate of I quanta/s and sensitivity probed either over time beginning at step onset, or at some fixed moment after onset. A generic name for the pulse or step is conditioning light. Supposing the probe to be a pulse of q quanta or a step of intensity i quanta/s, absolute sensitivity is defined as the ratio r/q or r/i, where r = r(i) or r = r(q) is the photoresponse (photovoltage, photocurrent) peak amplitude, or perhaps some other response measure or distribution statistic. It should be noted that absolute sensitivity is not well (uniquely) defined unless r is a linear function of q or i in the range of response over which r is measured. Relative sensitivity usually means sensitivity relative to the absolute sensitivity measured in the fully dark-adapted state. In many experiments the intensity of the probe is varied to obtain a criterion response, r_{cr}, and sensitivity is simply given in units of the intensity (q or i) required to produce the criterion response. In studying adaptation it is often important to know whether or not a photoreceptor is in steady state with respect to a constant illumination I (which may be zero, in which case we mean dark-adapted steady state). One operational definition is this: a photoreceptor or photoreceptor subunit (e.g., a microvillus) whose responses to probes adequately spaced in time (non-interacting) are identical is said to be in steady state.

Figure 1 from Claßen-Linke and Stieve (11) shows data from a pulse/probe experiment, illustrating many of the definitions. In these experiments a conditioning pulse causing a 53 mV depolarization is given a Limulus ventral photoreceptor, and at varying time t_{DA} after the pulse onset, a probe is given and adjusted over trials to find the intensity of the probe that yields a 24 mV criterion response.

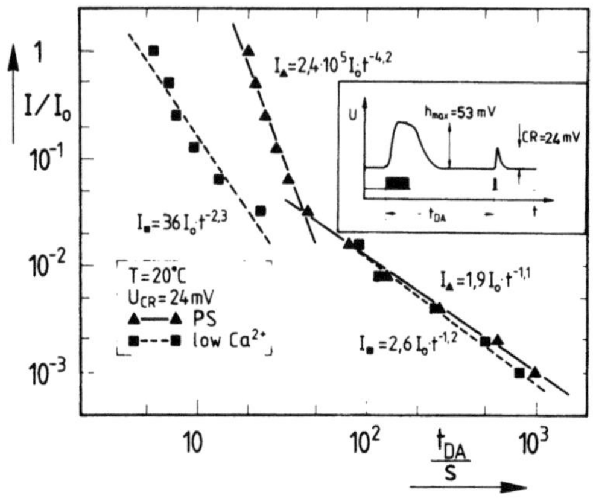

FIG. 1 - Sensitivity of a Limulus ventral photoreceptor in a pulse/probe (dark adaptation) experiment. The stimulus sequence is given in the inset. The abscissa is time after onset of the conditioning pulse (logarithmic scale); the ordinate is log-relative intensitiy of the test flash, with the highest intensity attainable arbitrarily set at 1. The equations labelling the curve segments represent linear regressions fit in the logarithmic coordinates of the graph to various segments of the curves. The calcium activity of the superfusate was changed from 10 mmol/l (physiological saline) to 250 µmol/l (low Ca saline). Note that only one region of the curves depends significantly on the calcium activity of the superfusate (from (11)).

Empirically, the type of adaptational phenomena observed in a photoreceptor depends in a pulse/probe experiment upon Q and the time after application of the pulse, and in a step/probe experiment upon I and the time after the onset of the light step - as well as upon other experimentally varied parameters (such as calcium - see Fig. 1). "Bleaching adaptation" in vertebrate photoreceptors is a particular example of this dependence of adaptational events on intensity level. We agreed to define bleaching adaptation to mean all adaptational phenomena consequent upon a pulse of Q quanta (or perhaps a step of intensity I and duration T such that IT = Q) which causes a fraction >1% of the photopigment in a vertebrate photoreceptor to be bleached. Although arbitrary, this definition attempts to capture the empirical fact that at such bleach levels adaptational phenomena occur (discussed below) that are not evident at lower levels. Bleaching adaptation per se does not occur in invertebrate photoreceptors (which have photointerconvertible and more or less thermally stable visual pigments); invertebrate photoreceptors, however, do show a prolonged depolarizing afterpotential (PDA) when a significant amount of rhodopsin is isomerized, and this and attendant phenomena bear some resemblance to events that occur in the vertebrate photoreceptor during bleaching adaptation (see below).

LIGHT ADAPTATION AT LOW LEVELS OF I

Stieve and others ((38), and see Stieve and Schnackenberg and Keiper, both this volume) have studied the effects of a conditioning light on the properties of the single-photon responses of Limulus ventral photoreceptors. With increasing conditioning light intensity I, the entire photocurrent time scale (including latency distribution) is shortened. Bump quantum efficiency (conditional probability that a bump is produced, given an isomerization), amplitude, and area are all enhanced at low I by a small factor before beginning to decline with increasing I.

A different, indirect approach using a dynamic signal noise analysis technique was applied by Grzywacz, Hillman, and Knight:

"The study of adaptation is the study of the nonlinearities in the transduction process. The biophysical approach to this study in the past has been based on two main methods: the construction of molecular chain models with nonlinear stages and the noise-analysis decomposition of responses into their elementary single-photon components ("bumps") in order to determine the effects of interactions among them. The above-mentioned authors have combined these approaches and applied them to observations on the ventral eye of Limulus to achieve several new insights into the mechanisms of adaptation (17).

"The dependence of the amplitude of the response on the intensity of a steady stimulus follows a Stevens power law (37) with an exponent of about 1/5 over more than four decades on intensity (43). The chemical chain model of Borsellino, Fuortes, and Smith (6) is widely supported today; Grzywacz, Hillman, and Knight have examined a variety of modifications of this model and have shown that the only one which predicts the Stevens Law is a negative feedback comprising an inhibitory binding of n molecules (of Ca?) on an enzyme of the chain, with the molecules arising from a later part of the chain. A power law of 1/(n+1) is predicted, which in Limulus gives n=4.

"In order to determine further details of the structure of this molecular chain, these authors have made observations on the temporal changes in mean, variance, and power spectrum of the "bump noise" which appear in consequence of time-variable light inputs. In order to apply this approach to rapidly varying signals, we have developed a new technique of dynamic noise analysis.

"The chemical chain may be conceptually divided into a non-amplifying "early" portion with dynamics characterized by bump latency, followed by an enzymatically amplifying "late" portion characterized by bump height and time course (see Stieve and Schnakenberg and Keiper, both this volume). Analysis of the noise over repeated trials effectively allows us to observe separately the discrete output of the "early" part (by measurement of the bump rate) which is the input to the "late" part, as well as the output of the entire chain.

"The amplitude of the responses of dark-adapted cells to flashes exhibits a supralinear dependence on intensity in a certain range (see Stieve, this volume). The dynamic noise analysis method has been applied to these observations to show that the supralinearity is a consequence of an increase in the amplitude of the component bumps (18). The authors also showed that this increase follows a time course roughly similar to that of the response itself, which is not inconsistent with the possibility that the supralinearity arises from a direct cooperativity in the "late" part of the transduction chain.

"Experiments with sinusoidally modulated steady light have led to further conclusions about the structure of the "late" part of the chain. At low modulation frequencies (<0.1 Hz) the common light adaptation effect is dominant; however, this adaptation is sluggish, and at higher frequencies it is dominated by a fast enhancement in the sinusoidal amplitude of the response signal.

"In order to determine the mechanism of this enhancement, Grzywacz, Hillman, and Knight applied their dynamic noise analysis to these observations and found that part of the enhancement arises from an increase in bump amplitude, the mechanism for which may be the same as that reponsible for the flash supralinearity. In addition to the bump amplitude enhancement, however, they were surprised to find a modulation of the rate of appearance of the bumps which was larger than that of the stimulus over a wide range of frequencies (however, see evidence consistent with this suggestion presented by Stieve, this volume). This hypermodulation corresponds to an effect on the "early" part of the chain, comprising either a modulation of the quantum efficiency about a mean value which remains constant over several decades of steady light intensity (43) or a very sensitive modulation of the bump latency."

Blowfly Adaptation

In blowfly receptors R_1-R_6 three intensity-dependent effects of light adaptation can be distinguished:

1. Very dim background lights (I < 500) quanta absorbed per second) reduce latency, time-to-peak, and response half-width, but do not significantly alter the amplitude of the receptor response elicited by a pulse/probe of fixed intensity (300 quanta absorbed) given 1 s after the cessation of the adapting light. Surprisingly, all responses measured after increasing times of light adaptation (1-120 s) to conditioning-light steps of fixed intensity I cross at an "isosbestic point in time," as shown in Fig. 2.

   These effects can be explained by the following hypothesis. Quantum absorption in a microvillus is assumed to cause local excitation of a membrane area, inducing a relatively slow and large quantum bump. If this same light-adapted membrane area is excited again within a certain period, a faster and smaller bump is produced. Thus after light

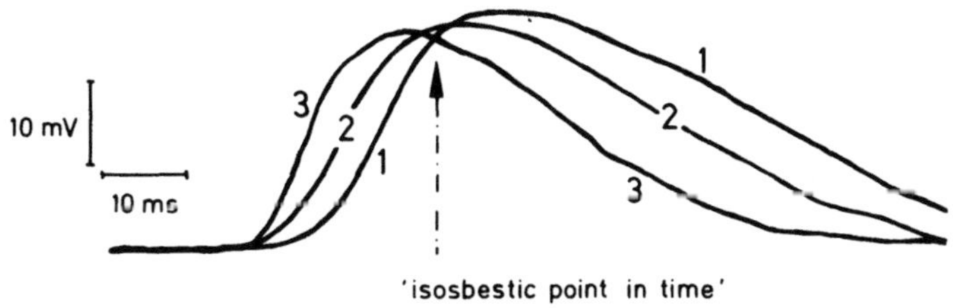

FIG. 2 - Receptor response to a test probe recorded 1 s after the cessation of an adapting light of variable duration. Traces 1-3 correspond to 1, 8, and 32 s adapting time with a steady light which leads to the absorption of 2 x 10^5 light quanta per second.

adaptation two different types of membrane areas exist in the photoreceptor, generating two different types of bumps. Consequently, the time course of probe responses after different times of light adaptation reflects the relative proportion of these types of bumps, and hence the ratio of "light-" and "dark"-adapted membrane areas. In this view, the phenomenon of the "isosbestic point in time" is comparable with the isosbestic point in spectral absorption of a reversible dye system. Because of the better synchronization of the bumps produced by the light-adapted membrane areas, the amplitude of the response to the pulse/probe is almost uninfluenced, although these bumps are smaller in size.

2. After the application of more intense background lights, the amplitude of the response to the pulse probe is reduced. There is evidence that this secondary effect is mainly produced by two mechanisms: a) the "machinery of biochemical transduction" within an individual microvillus becomes exhausted, so that the respective microvillus is refractory for a certain period; and b) the conductivity of the light-regulated ion channels is reduced by calcium.

3. The third phase of adaptation is correlated with "bleaching," i.e., when a significant number of rhodopsin molecules is activated synchronously in the microvillus unit ((19, 20), and see Minke, this volume).

Comment on Formal Models for Limulus Bumps: Their Statistical Analysis

The responses of Limulus ventral photoreceptors to single photons become larger when the cell is very weakly light-adapted and no significant change is found in bump time course under these conditions, when the latter is derived from either a) noise analysis (17) or b) direct measurement (Stieve and Schnakenberg and Keiper, both this volume).

Noise analysis of bumps, however, is limited by some restrictions. Noise analysis of both stationary and nonstationary processes requires a mathematical description of the time course of bumps. It is practical to use a gamma function (characterized by two parameters) with additional distribution parameters for latency and amplitude.

Real bumps do not look like the "average bump," but display a rather large range of shapes with at least four independently fluctuating parameters (24). The mathematical treatment of this problem is tricky. Noise analysis results derived from the simpler gamma curve ought to be considered approximations. Small effects found with such approximate analysis should be viewed with caution. The facilitation effects on bump amplitude at low background intensities (factor of 6-8) found by Grzywacz (17) are of course not in doubt.

At conditioning light intensities I, more than 1 log unit higher than that at which facilitation occurs, bumps become very small and fast. Adaptation hence has at least four different effects: a) on latency, b) on bump slope/amplitude/gain, c) on bump duration, and d) on quantum efficiency.

Kinetic Models of Light Adaptation

Mathematical models of the kinetics of photoreceptor responses can provide mechanistic explanations for observable phenomena and also suggest correspondences between electrophysiological and biochemical mechanisms. Furthermore, <u>quantitative</u> tests of models by experiment force us to confront our ignorance.

Two previous models of visual transduction provide an explanation for light adaptation. These are the FH (Fuortes and Hodgkin (15)) model and the BHL (Baylor, Hodgkin, and Lamb (3)) model. The FH model postulates n stages of amplification and temporal integration. One stage can be diagrammed thusly:

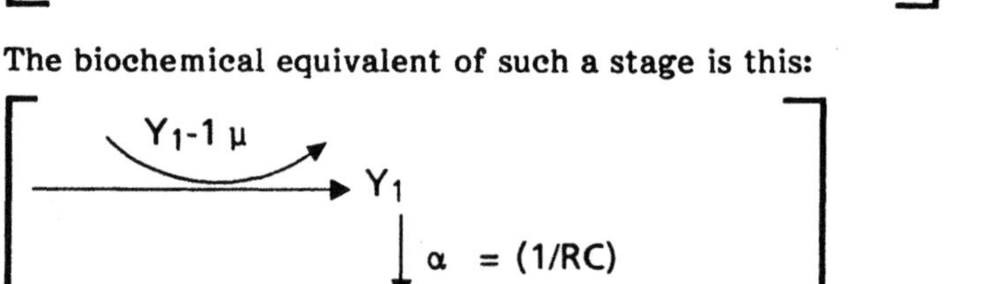

The biochemical equivalent of such a stage is this:

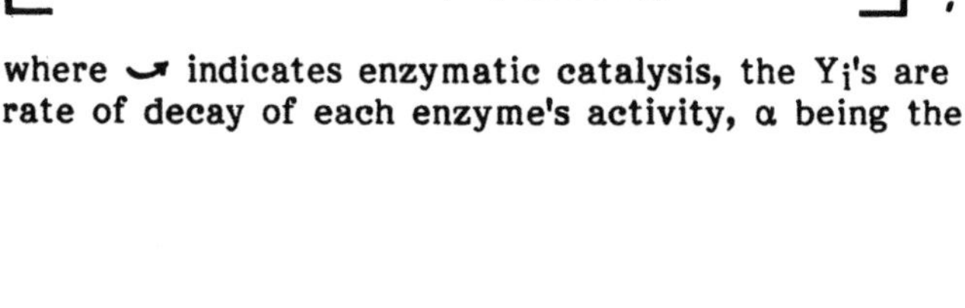

where $\curvearrowright$ indicates enzymatic catalysis, the Y_i's are enzymes, and α is the rate of decay of each enzyme's activity, α being the same for each stage.

The first stage of such a system is special; light input catalyzes the production of Y_1 as follows:

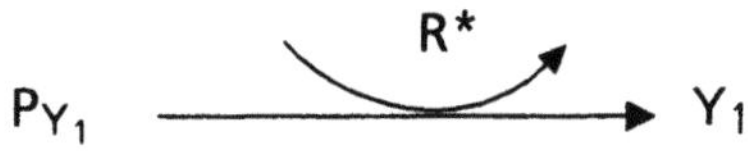

where R_{Y_1} is the precursor of Y_1, and R^* photoactivated pigment. In the FH model, light adaptation is achieved by speeding up α. That is, $\alpha = \alpha_0 f(I)$. If $f(I) = f_0 \exp(I/I_0)$, then the peak of the impulse response obeys Weber's Law (15).

The BHL model is quite different. It can be represented as a linear chain of chemical reactions:

$$R^* \xrightarrow{\alpha} Y_1 \;;\qquad Y_{1}\text{-}1 \xrightarrow{\alpha} Y_1 \;;\qquad \underset{(Z_1)}{Y_n} \xrightarrow{\alpha} Z_2 \,.$$

In this system, increasing α will <u>not</u> affect the peak amplitude of the response to a brief flash (light impulse). Basically, this is because the amount of Y_n (of Z_1) is decreased by just the same factor by which α is increased and the peak amplitude remains constant. Thus, a new mechanism for adaptation to steady light was introduced: autocatalysis by Z_2 of the breakdown of Y_n (or Z_1), the concentration of the transmitter substance that closes the light-sensitive conductance. Thus,

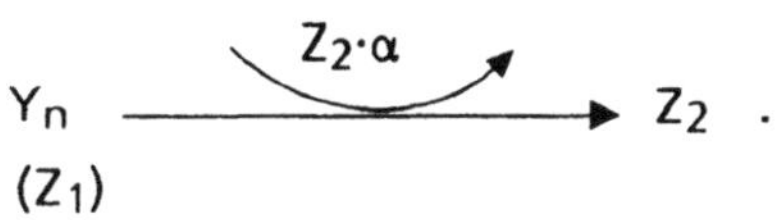

There is reasonable qualitative agreement of this model with quasi-steady-state light adaptation (1 s after step onset) in turtle cone photovoltage recordings (3).

Recent measurements of photovoltages in turtle horizontal and cone receptor cells pertinent to these models have been reported by Tranchina, Gordon, and Shapley (42). They measured the linear temporal transfer function in response to full-field sinusoidal temporal modulation of the illumination around a steady level. Thus, the stimulus was $I_0 + I_1 \sin(2\pi ft)$. The measurements were of the amplitude and phase of the voltage of the cell modulated at each stimulus frequency. The data are <u>not</u> consistent with either the FH or BHL models. As $f \to 0$, the amplitude followed Weber's Law, i.e., $R(f \to 0) = K/I_0$, where K is a constant. As $f \to \infty$ (i.e., $f >$ 10 Hz in cool turtle retina), $R(f \to \infty) = K$, a constant independent of I_0. To state in words, the response amplitude was independent of the mean illumination I_0 at high temporal frequencies. This phenomenon can be "explained" by the following TGS model:

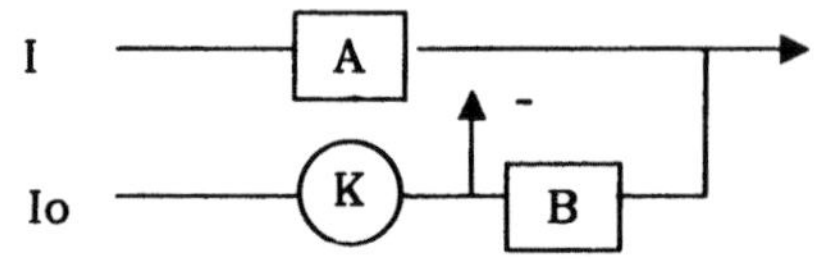

,

which is interpreted to mean that filters A and B (each possibly an FH filter with no built-in adaptation) are arranged in a negative feedback circuit. The strength of the feedback from B onto A is modulated by the mean light level I_0. Tranchina has recently devised a chemical kinetic scheme which represents this model. A and B are to be taken as FH enzyme cascades, as above. The product of the A cascade is enzymatically destroyed by the last stage of the B cascade in a bimolecular reaction involving a cofactor which is a function of the steady level of illumination. Some details need to be worked out, but this model is clearly the one most consistent with turtle cone and horizontal cell kinetics (Shapley, personal communication). Questions for future research concern the nature of the chemical feedback, the existence of the putative cofactor, and its linkage to the steady level of illumination.

BLEACHING ADAPTATION

We next considered a number of issues concerning "bleaching" adaptation, it being kept in mind that most invertebrate pigments do not, strictly speaking, "bleach" but undergo photointerconversion (but see Schwemer, this volume, for discussion of bleaching of some insect visual pigments). An overall scheme comparing phenomena of bleaching adaptation in both vertebrates and invertebrates was proposed by Minke and is included here as Table 1.

This table serves to describe the basic phenomena we subsume under the heading "bleaching adaptation," and to illustrate similarities and differences between vertebrates and invertebrates in adaptation events following exposure to relatively intense lights.

Sensitizing Effect of Exogenous Chromophore in Isolated Retina

The bleaching of rhodopsin in the isolated vertebrate retina ordinarily leads to a sustained elevation of rod threshold. The extent of this "bleaching adaptation" threshold elevation far exceeds that predicted on the basis of reduced quantum catching capability. When externally applied to the retina in this condition of maintained desensitization, 11-cis retinal induces a pronounced decline in threshold - i.e., promotes a resensitization far in excess of that due to restored quantum capturing. The magnitude of this resensitization phenomenon and its dependence on the 11-cis isomeric form of the applied retinal suggest the following: a) the sustained desensitization which occurs after exposure to an intense light arises from some action of the unregenerated pigment molecule itself; and b) other ("downstream") processes in the receptor that mediate the photoresponse remain essentially functional after "bleaching adaptation" in that pigment regeneration per se

rapidly promotes recovery toward the maximally sensitive (fully dark-adapted) state.

TABLE 1- A comparsion of "bleaching adaptation" - the effects of light hitting an appreciable fraction of the visual pigment in vertebrates and invertebrates.

INVERTEBRATES	VERTEBRATES
1 R M induction of the prolonged depolarizing afterpotential (PDA) (see Fig. 5, trace D, Minke, this volume)	R R* induction of the prolonged afterhyperpolarization (or prolonged aftercurrent) (see Fig. 10, trace Bb, Minke, this volume)
2 M R depression of PDA (see Fig. 5, trace B, Minke, this volume)	Regeneration of R by adding 11-<u>cis</u> retinal results in depression of the prolonged aftercurrent (see Fig. 10, trace Bc, Minke, this volume).
3 The PDA is accompanied by a large reduction in sensitivity to light which is much larger than the reduction in the probability to absorb photons. This desensitization is observed also in a Drosophila mutant where R-to-M conversion is not manifested in voltage response. Thus the voltage response and the desensitization can be separated from each other. These effects are reversed by M-to-R conversion.	The prolonged afterhyperpolarization is accompanied by a large reduction in sensitivity to light, a phenomenon which is known as bleaching adaptation. This effect is reversed by regeneration of R.
4 The PDA and the desensitization are not observed when the photopigment concentration is largely reduced by vitamin A deprivation or by mutation. In these preparations sensitivity is depressed only proportional to the loss of pigment.	Bleaching light induces bleaching adaptation in normal Xenopus photoreceptors. However, reduction in pigment concentration by vitamin A deprivation depressed the sensitivity only proportional to the loss of pigment.
5 The PDA is localized to the illuminated area.	The prolonged aftercurrent and the accompanying desensitization are localized to the bleached area.
6 The PDA is accompanied by voltage fluctuations (noise) which are similar to the noise observed during illumination due to superposition of quantum bumps.	The prolonged aftercurrent is accompanied by current fluctuations (noise) which are very similar to the current fluctuations due to absorption of photons.
7 In the fly, when the PDA is maximal the amount of phosphorylated M molecules is minimal. Phosphorylation of M increases with a time course roughly parallel to the decline of the PDA.	Following bleach, phosphorylation level (in the frog) is reduced at the tip of the outer segment relative to the base. The prolonged aftercurrent (toad) was found to be long at the tip relative to the base.

Figure 3 illustrates the basic phenomenon, discussed in several papers by Pepperberg and colleagues (10, 33, 34).

Pepperberg (32) has proposed a model to account for the desensitizations exhibited long after intense light (bleaching adaptation) and during prolonged exposure to weak background light (Weber Law behavior in step/-probe increment threshold experiments). The hypothesis is proposed that, under these conditions, photoactivated and/or bleached forms of the pigment molecule act (in qualitatively similar fashion) to decrease the level of a component of the transduction pathway (i.e., deplete the pool of this substance that is functionally accessible to δR^*). Quantitatively, for steady-state desensitizations of photoreceptors in the isolated retina, the model yields the equation $\sigma = K/E$, where σ is the prevailing value of the semisaturation parameter for the flash response (assuming hyperbolic saturation), K is a constant, and E is the (quasi-) steady-state level of the critical substance that remains available for activation within the lifetime of δR^*.

The model suggests that, both during (steady-state) background adaptation and during the slow (regeneration-dependent) phase of recovery from bleaching light (for example, cf. (13)), desensitizations of proximal visual neurons may depend strongly on the "reduced availability" of a key transduction agent, E, in the photoreceptors.

Equivalent Background (Bleaching-induced Events) in Individual Vertebrate Photoreceptors

Human rod psychophysical dark adaptation has been found consistent with the hypothesis that the "equivalent background" of Stiles and Crawford (41) might arise from events, generated in the dark in a partially bleached photoreceptor, that were identical with light-triggered single-photon responses. Lamb (25) tested the prediction of dark "quantal" events in toad rods. He found qualitative agreement in noise following ca. 1% bleach.

A full quantitative test of equivalence of the bleaching-induced noise events faces various technical difficulties (e.g., recording stability), but at this point we know that a) the power spectrum is correct, b) the event rate declines, and c) the rod itself is desensitized approximately as expected for a background light that would cause the same amount of dark current suppression. Recent work by Baylor, Nunn, and Schnapf (4) measuring the photocurrents of monkey rods has revealed some differences from the results found by Lamb (25) in toad rods. Perhaps the most striking is that during dark adaptation after 3000 isomerizations large, rectangular "discrete" events are observed that last several seconds. These events clearly do not have the form of the normal single-photon response, although about 1/500 flashes given to a totally dark-adapted rod apparently triggers such an event. Thus, although the simplest dark light theory may not obtain for monkey rods, the noise events observed after bleaching may be related to shot events that are occasionally triggered by photons.

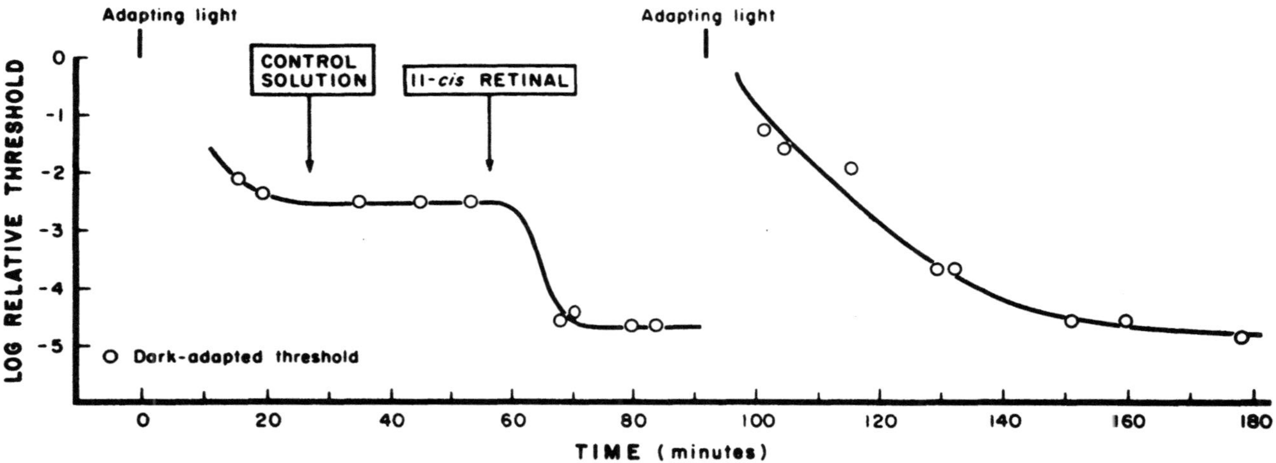

Fig. 3 - Partial bleaching of the isolated skate retina, and sensitization induced by 11-cis retinal. The data show thresholds for the extracellularly recorded flash response of the photoreceptors. The initial adapting irradiation (terminated at time zero) bleached about 42% of the rhodopsin initially present in the receptors. For further details, cf. Fig. 6 and accompanying text of Pepperberg, Brown, Lurie, and Dowling (33). (Reproduced from The Journal of General Physiology (see (33)) by copyright permission of The Rockefeller University Press.)

Summary: Source of Depolarizing Afterpotential (PDA) and Dark Bumps Arising After Intense Conditioning Exposure in Invertebrates Photoreceptors

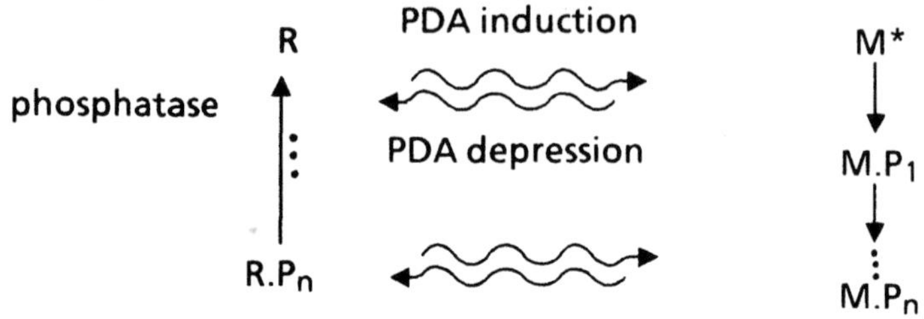

Comments

1. PDA induction is caused by production of a large amount of M*.
2. M* ⇝ R photoreconversion results in depression of the PDA and the accompanying voltage fluctuations (29).
3. According to Hamdorf, Minke, and Paulsen, when a large R-to-M* conversion occurs, the M* molecules remain in an active, non-phosphorylated state (30), giving rise to the PDA and the accompanying voltage fluctuations. However, according to the hypothesis of Lisman (28), all the M* molecules are quickly phosphorylated, and the PDA and accompanying noise are due to dephosphorylation of the various $M.P_i$ states; the decline of the PDA is due to multiple phosphorylations, which reduce the probability of back reactions (i.e., of complete dephosphorylation).
4. RP_n is an inactive state which cannot produce excitation (19). Conversion of MP_n into RP_n initiates a rapid dephosphorylation resulting in the formation of the excitable R state (30, 31).
5. It should be mentioned that recent evidence indicates that invertebrate visual pigments are subjected to turnover. The pigments decay primarily in the state of metarhodopsin (possibly MP_n; single exponential decay of metarhodopsin in the dark with t_{50} of about 2 hours in blowfly and about 10 minutes in butterflies). Biosynthesis of rhodopsin which depends on the availability of the 11-*cis* chromophore replaces the degraded metarhodopsin (reviewed by Schwemer, this volume).

CHEMICAL HYPOTHESES ABOUT LIGHT ADAPTATION

Vertebrates

Work by Bastian and Fain (1), by Hemilä and Reuter (21), and by Lamb, McNaughton, and Yau (26) has established that in amphibian rods which light-adapt, a cytosolic diffusible adaptational transmitter must be produced at relatively low levels of I or Q (a few isomerizations/rod/s). Considerations of volume argue that this transmitter or its disk surface effects must be produced with relatively high numerical gain, the latter being defined as the number of transmitters or effects per isomerization. Having estimated the decay rate of the transmitter, Lamb, McNaughton, and Yau (26) also established an upper limit of about 6 μm to the apparent longitudinal diffusion of the transmitter. It was agreed that the existence, gain, and diffusion of the adaptational transmitter are the most important established constraints on chemical hypotheses about adaptation in vertebrate rods. In vertebrates, calcium has received the most scrutiny as the candidate for adaptational transmitter. The recent work of Bastian and Fain (2) tests this hypothesis in toad rods, as did the work by Bertrand et al. (5) in turtle cones. At present, the weight of evidence seems against the hypothesis.

It must be emphasized, however, that mammalian rods do not light-adapt. Years ago this was shown by Green (16) for the albino rat aspartate-isolateda-wave. More recently, Baylor, Nunn, and Schnapf (4) have demonstrated in their recordings from single monkey rods that these cells do not light-adapt at all. In other words, monkey rod photocurrents have an invariant response characteristic, and once the rate of photon arrival from a steady conditioning light becomes so large that the response is saturated, it remains so indefinitely until the light is turned off. Thus, there is at present no reason to consider a transmitter for light adaptation in mammalian rods.

Invertebrates

A significant body of evidence (reviewed by Brown, this volume) supports the hypothesis that Ca^{2+} is the adaptational transmitter in Limulus. However, its mode of operation remains obscure, and few detailed models exist linking Ca^{2+} to the adaptation properties described by Stieve (this volume).

Stieve reported that whereas the sensitivity of the dark-adapted Limulus photoreceptor did not change when the external calcium was varied between 40 µmol/l and 100 mmol/l, the desensitization caused by identical conditioning illuminations was greater, the higher the external calcium concentration (39, 40).

The amplitude of the light-induced increase in intracellular calcium concentration (monitored by Arsenazo) and even more the rate of its subsequent decline depend on the clamped membrane voltage (23). These findings indicate that the light-induced transient increase in intracellular calcium is to a substantial extent caused by a passive calcium influx from an extracellular compartment and a subsequent electrogenic outward transport (possibly an Na/Ca exchange).

A novel hypothesis of Ca^{2+} desensitization in blowfly has been proposed by Hochstrate and Hamsdorf (22).

Effect of external calcium and magnesium on the photoresponse of the blowfly

Hemisected fly eyes were superfused in the dark with salt solutions which contained various concentrations of calcium, $[Ca^{2+}]_o$, or magnesium, $[Mg^{2+}]_o$. If the eye was superfused with a solution which contained 130 mmol/l NaCl, 0.1 mmol/l Ca^{2+}, and 10 mmol/l Hepes buffer (ph = 7.0), the physiological behavior of the photoreceptors in the cut preparation was in all respects almost the same as in the intact eye. Changing $[Ca^{2+}]_o$ between 10^{-9} and 3×10^{-2} mol/l in the absence of Mg^{2+} did not influence the resting potential or the frequency with which quantum bumps occurred but did lead within 10 s to changes in the latency and amplitude of the response, and most significantly, in the repolarization time, t_r. The plot of t_r vs. $[Ca^{2+}]_o$ revealed that t_r changes significantly in two distinct regions, namely, between 2×10^{-8} and 10^{-7}, and between 10^{-7} and 10^{-4}, respectively. Lowering $[Ca^{2+}]_o$ did not affect the amplitude of the response but did lead to a drastic increase in t_r (by a factor of about 50, independent of light intensity), which was accompanied by an increase in latency and time-to-peak. Raising $[Ca^{2+}]_o$ led to a reduction in the duration and amplitude of the response, which closely corresponds to the effects of light adaptation. The amplitude reduction reveals a loss in sensitivity of the photoreceptor which is dependent on $[Ca^{2+}]_o$. The concentration range in which these effects are observed closely corresponds to the change in the extracellular calcium concentration expected to occur in the intact fly's eye during light adaptation.

It is postulated that two types of calcium-binding sites exist, high-affinity binding sites (HABS) and low-affinity binding sites (LABS), which modulate the functioning of ion channels in the membrane that are activated as a consequence of light absorption. The relationship between the sensitivity of the cell and $[Ca^{2+}]_o$ indicates that the sensitivity is determined by the degree of saturation of the LABS, in that there is a simple titration equilibrium between the LABS and the extracellular calcium, with an equilibrium binding constant of $K = 10^{3.4}$ mol/l.

Similar to the effect of calcium, an increase of $[Mg^{2+}]_o$ also leads to a rapid reduction in sensitivity; but, in contrast to the effect of calcium, magnesium slows down the time course of the response. The relationship between sensitivity and $[Mg^{2+}]_o$ also corresponds to a simple titration equilibrium with $K = 10^3$ mol/l. This and other experiments where mixtures of calcium and magnesium were used to indicate that both ions can compete for the LABS.

These results support the view that the LABS are located on the outside of the membrane. Thus, the effects of light adaptation may be caused mainly by the following mechanism: after light absorption, calcium is released from intracellular stores (see Brown, this volume) and subsequently extruded into the extracellular space, leading to an increase in $[Ca^{2+}]_o$, and in turn to the saturation of the LABS with calcium. When the LABS are saturated the average conductivity of the light-regulated channels is strongly reduced, with the consequence that proportionately more light is needed to evoke a criterion photoresponse - i.e., sensitivity is reduced. In order to also explain the changes in the time course of the receptor response, one may assume that the kinetic behavior of the light-regulated channels is modulated via some calcium-dependent cooperativity between channels. This would explain the fact that at low light levels (i.e., slight degree of LABS saturation) only the time course of the responses is changed, with no significant change in their amplitude (see (22) for details).

Inositol 1,3,4-trisphosphate
Intracellular pressure injection of inositol-trisphosphate (IP_3) has been shown to induce conductance changes of the plasma membrane of Limulus ventral photoreceptors (7). These IP_3-induced conductance changes mimic several aspects of light-induced conductance changes. The injection of IP_3 was found to reduce transiently the sensitivity of the photoreceptor to light; moreover, light was found to reduce transiently the sensitivity to IP_3 injections.

More recently, pressure injection of IP_3 has been found to induce a transient increase in intracellular Ca^{+2} in Limulus ventral photoreceptors, as detected by intracellular aequorin (7, 8, 12). Also, steady luminescence of extracellular aequorin increased after the cell was impaled with a pipette filled with a solution of IP_3 at high concentration (0.5 mmol/l), presumably due to leakage of IP_3 out of the pipette; in this condition, the sensitivity of the cell to light was much reduced. IP_3 may participate in the adaptation process, inasmuch as an increase in intracellular Ca^{2+} has been implicated in light adaptation (cf. Brown, this volume).

QUESTIONS FOR FUTURE RESEARCH
The biochemical insights of cyclic GMP cascade are obviously fundamental and beg to be integrated, at least speculatively, with our discussions on light adaptation. This schematic of the cascade is offered and it incorporates the finding of Fesenko et al. (14) and Yau and Haynes (45) that cGMP appears to be able to open the outer segment light-sensitive conductance ($g_{h\nu}$) by some direct allosteric mechanism.

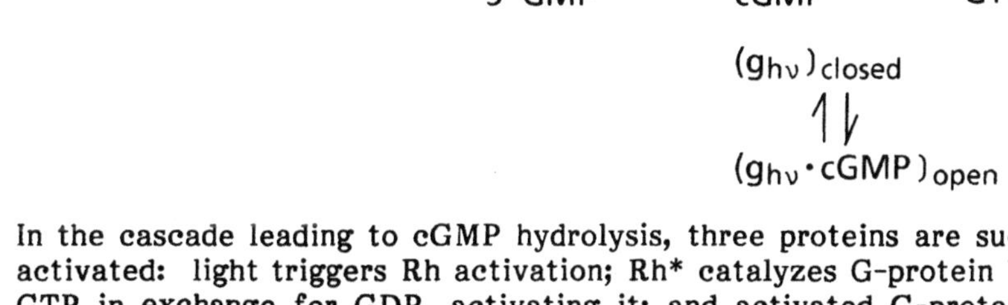

In the cascade leading to cGMP hydrolysis, three proteins are successively activated: light triggers Rh activation; Rh* catalyzes G-protein binding of GTP in exchange for GDP, activating it; and activated G-protein in turn activates PDE. In addition to the three proteins activated by the light-triggered cascade, guanylate cyclase, which restores cGMP levels, must be included in order that all steps leading to regulation of cGMP in the outer segment on the time scale of the light response be represented. Both the activation of G-protein by Rh* and the hydrolysis of cGMP are catalytic steps and involve relatively high gain, estimated at ca. 500/s and 2000/s, respectively, for fully activated Rh* and PDE*. The obvious should perhaps be stated - that if each of these steps is activated in the normal cell in the fashion found in membrane preparations, then there must also exist turnoff mechanisms for each step that render the components inactivate on a time scale briefer than the single-photon response.

Light adaptation has a profound effect on the system gain (photoresponse amplitude/isomerization), and thus it seems that the most likely biochemical stage for adaptation to occur in such a way as to change gain effectively is at the high-gain enzymatic steps of Rh* and PDE*. Reducing the effectiveness of Rh* catalysis of G-protein's activation and reducing the effectiveness of PDE* catalysis of cGMP hydrolysis are thus key points to consider in the theory of adaptation. Among this class of possible mechanisms for desensitization are: phosphorylation of the pigment molecule (27); light-induced increase in the K_m of PDE (35); and light-induced reduction in GBP* formed per isomerization (9). Light adaptation (in nonmammalian vertebrate photoreceptors, see review by Shapley and Enroth-Cugell (36)) also has a profound effect on the response time scale (or bandwidth), and thus it seems that the most potent way to reduce the system gain and decrease the time scale of the response simultaneously would be to cut the lifetime of Rh* and/or GBP*, assuming that the lifetime of PDE* is controlled simply by the lifetime of GBP*. It is interesting to note, however, that whereas changing the lifetime of any of the activated cascade intermediates would immediately produce change in both gain and bandwidth, gain change could also be effected by cutting the effectiveness of the coupling between the steps in the cascade without increase in bandwidth. For example, if the K_m of PDE were regulated by one of the early byproducts of light, profound gain change could be effected without much change in bandwidth.

Questions Whose Answers Might Help Us Understand Light Adaptation

1. Is there any known effect of an early product of the cascade on PDE? Is the PDE inhibitor subunit binding, for example, affected by some byproduct of light?

2. Is the deactivation of GBP* only by GTP-ase catalyzed hydrolysis of the bound GTP, or is there some other much faster and more effective deactivation route? Is this deactivation step (or any other) affected by Ca^{2+}?

3. Yau and Nakatani (46) have evidence that during the light response, intracellular free Ca^{2+} actually goes down. Is the lifetime of Rh* or that of PDE* or GBP* affected by the range of changes in Ca^{2+} expected from Yau and Nakatani's results?

Questions About Dark Adaptation

1. What is the light impulse response during stages of dark adaptation after a bleach? Does it superimpose on the "equivalent background" response?

2. How do the various enzymatic steps above saturate when given an excess of Rh*? For example, does the availability of Rh-kinase, which is present in amounts of only about 1/100 rhodopsin, cause the rate of phosphorylation of rhodopsin in a rod to saturate at bleach levels of ca. 1%?

Acknowledgements. The rapporteur wishes to thank P. Hillman for contributing the introductory section to this report and N.M. Grzywacz, P. Hillman, and B. Knight for the contribution on the dynamic signal noise analysis technique.

REFERENCES

(1) Bastian, B.L., and Fain, G.L. 1979. Light adaptation in toad rods: requirement for an internal messenger which is not calcium. J. Physiol. **297:** 493-520.

(2) Bastian, B.L., and Fain, G.L. 1982. The effects of low calcium and background light on the sensitivity of toad rods. J. Physiol. **330:** 307-329.

(3) Baylor, D.A.; Hodgkin, A.L.; and Lamb, T.D. 1974. The electrical response of turtle cones to flashes and steps of light. J. Physiol. **242:** 729-758.

(4) Baylor, D.A.; Nunn, B.J.; and Schnapf, J. 1984. The photocurrent, noise and spectral sensitivity of rods of the monkey Macaca fascicularis. J. Physiol. **357:** 575-607.

(5) Bertrand, D.; Fuortes, M.G.F.; and Pochobrodsky, J. 1978. Actions of EGTA and high calcium on the cones in the turtle retina. J. Physiol. **275:** 419-437.

(6) Borsellino, A.; Fuortes, M.G.F.; and Smith, T.G. 1965. Visual responses in Limulus. Cold S.H. Symp. Quant. Biol. **30:** 429-443.

(7) Brown, J.E., and Rubin, L.J. 1984. A direct demonstration that inositol trisphosphate induces an increase in intracellular calcium in Limulus photoreceptors. Biochem. Biophys. Res. Commun. **125:** 1137-1142.

(8) Brown, J.E., and Rubin, L.J. 1985. Inositol trisphosphate induces an increase in intracellular ionized calcium in intact and functioning Limulus photoreceptors. Biophys. J. **47:** 38 (Abstract).

(9) Clack, J.W.; Oakley, B., II; and Pepperberg, D.R. 1982. Light-dependent effects of a hydrolysis-resistant analog of GTP on rod photoresponses in the toad retina. Proc. Natl. Acad. Sci. USA **79:** 2690-2694.

(10) Clack, J.W., and Pepperberg, D.R. 1982. Desensitization of skate photoreceptors by bleaching and background light. J. Gen. Physiol. **80:** 863-883.

(11) Claßen-Linke, I., and Stieve, H., 1981. Time course of dark adaptation in the Limuls ventral nerve photoreceptor - measured as constant response amplitude curve - and its dependence upon extracellular calcium. Biophys. Struct. Mech. **7:** 336-337.

(12) Corson, D.W.; Fein, A.; and Payne, R. 1984. Detection of an inositol 1,4,5 trisphosphate-induced rise in intracellular free calcium with aequorin in Limulus ventral photoreceptors. Biol. Bull. **167:** 524 (Abstract).

(13) Dowling, J.E., and Ripps, H. 1970. Visual adaptation in the retina of the skate. J. Gen. Physiol. **56:** 491-520.

(14) Fesenko, E.E.; Kolesnikov, S.S.; and Arkadiy, A.L. 1985. Induction by cyclic GMP of cationic conductance in plasma membrane of retinal rod outer segment. Nature **313:** 310-313.

(15) Fuortes, M.G.F., and Hodgkin, A.L. 1964. Changes in time scale and sensitivity in the ommatidia of Limulus. J. Phyisol. 172: 239-263.

(16) Green, D.G. 1973. Scotopic and photopic components of the rat electroretinogram. J. Physiol. **228:** 781-797.

(17) Grzywacz, N.M. 1985. On individual and interactive properties of the single photon responses in invertebrate photoreceptors. Ph.D. Thesis, Hebrew University, Jerusalem.

(18) Grzywacz, N.M., and Hillman, P, 1985. Statistical test of linearity of photoreceptor transduction process: Limulus passes, others fail. Proc. Natl. Acad. Sci. USA **82:** 232-235.

(19) Hamdorf, K. 1979. The physiology of invertebrate visual pigments. In Handbook of Sensory Physiology. Comparative Physiology and Evolution of Vision in Invertebrates. A. Invertebrate Photoreceptors, ed. H. Autrum, vol. 7, pt. 6A, pp. 145-224. Berlin: Springer-Verlag.

(20) Hamdorf, K., and Razmjoo, S. 1979. Photoconvertible pigment states and excitation in Calliphora: the induction and properties of the prolonged depolarizing afterpotential. Biophys. Struct. Mech. **5:** 137- 161.

(21) Hemilä, S., and Reuter, T. 1981. Longitudinal spread of adaptation in the rods of the frog's retina. J. Physiol. **310:** 501-528.

(22) Hochstrate, P., and Hamdorf, K. 1985. The influence of extracellular calcium on the response of fly photoreceptors. J. Comp. Physiol. A **156:** 53-64.

(23) Ivens, I., and Stieve, H. 1984. Influence of the membrane potential on the intracellular light induced Ca2+ concentration change of the Limulus ventral photoreceptor monitored by Arsenazo III under voltage clamp conditions. Z. Naturforsch. **39c:** 985-992.

(24) Keiper, W.J.M.; Schnakenberg, J.; and Stieve, H. 1984. Statistical analysis of quantum bump parameters in Limulus ventral photoreceptors. Z. Naturforsch. **39c:** 781-790.

(25) Lamb, T.D. 1980. Spontaneous quantal events induced in toad rods by pigment bleaching. Nature **287:** 349-351.

(26) Lamb, T.D.; McNaughton, P.A.; and Yau, K.-W., 1981. Spatial spread of activation and background desensitization in toad rod outer segments. J. Physiol. **319:** 463-496.

(27) Liebman, P.A., and Pugh, E.N., Jr. 1980. ATP mediates rapid reversal of cGMP phosphodiesterase activation in visual receptor membranes. Nature **287:** 734-736.

(28) Lisman, J.E. 1984. Properties of visual pigment off-switch. Inv. Ophthalmol. Vis. Sci. **25:** 157.

(29) Minke, B., and Kirschfeld, K. 1984. Non-local interactions between light-induced processes in Calliphora photoreceptors. J. Comp. Physiol. A **154:** 175-187.

(30) Paulsen, R., and Bentrop, J. 1984. Phosphorylation of opsin induced by irradiation of blowfly retinae. J. Comp. Physiol. A **155:** 39-45.

(31) Paulsen, R.; Bentrop, J.; and Peters, K. 1984. Photochemistry and biochemistry of blowfly photoreceptor membranes. Vision Res. **24:** 1700.

(32) Pepperberg, D.R. 1984. Rhodopsin and visual adaptation: analysis of photoreceptor thresholds in the isolated skate retina. Vision Res. **24:** 357-366.

(33) Pepperberg, D.R.; Brown, P.K.; Lurie, M.; and Dowling, J.E. 1978. Visual pigment and photoreceptor sensitivity in the isolated skate retina. J. Gen. Physiol. **71:** 369-396.

(34) Pepperberg, D.R.; Lurie, M.; Brown, P.K.; and Dowling, J.E. 1976. Visual adaptation: effects of externally applied retinal on the light-adapted, isolated skate retina. Science **191:** 394-396.

(35) Robinson, P.R.; Kawamura, S.; Abramson, B.; and Bownds, M.D. 1980. Control of the cyclic GMP phosphodiesterase of frog photoreceptor membranes. J. Gen. Physiol. **76:** 631-645.

(36) Shapley, R., and Enroth-Cugell, C. 1984. Visual adaptation and retinal gain controls. Prog. Retinal Res. **3:** 263-346.

(37) Stevens, S.S. 1970. Neural events and the psychophysical law. Science **170:** 1043-1050.

(38) Stieve, H., and Bruns, M. 1983. Bump latency distribution and bump adaptation of Limulus ventral nerve photoreceptor in varied extracellular calcium concentration. Biophys. Struct. Mech. **9:** 329-339.

(39) Stieve, H.; Bruns, M.; and Gaube, H. 1984. The sensitivity shift due to light adaptation depending on the extracellular calcium ion concentration in Limulus ventral nerve photoreceptor. Z. Naturforsch. **39c:** 662-679.

(40) Stieve, H., and Klomfaß, J. 1981. Calcium dependence of light evoked membrane current signal and membrane voltage signal and their changes due to light adaptation in Limulus photoreceptor. Biophys. Struct. Mech. **7:** 345.

(41) Stiles, W.S., and Crawford, B.H. 1932. Equivalent adaptation levels in localized retinal areas. **In** Report of a Joint Discussion on Vision, pp. 194-211. Physical Society of London. Cambridge: Cambridge University Press. (Reprinted in Stiles, W.S. 1978. Mechanisms of Colour Vision. London: Academic Press.)

(42) Tranchina, D.; Gordon, J.; and Shapley, R. 1984. Retinal light adaptation - evidence of a feedback mechanism. Nature **310:** 314-316.

(43) Wong, F. 1978. Nature of light-induced conductance changes in ventral photoreceptors of Limulus. Nature **276:** 76-79.

(44) Wong, F.; Knight, B.W.; and Dodge, F.A. 1980. Dispersion of latencies in photoreceptors of Limulus and the adapting bump model. J. Gen. Physiol. **76:** 517-537.

(45) Yau, K.-W., and Haynes, L. 1985. Cyclic GMP-sensitive conductance in outer segment membrane catfish cones. Nature **317:** 62-64.

(46) Yau, K.-W., and Nakatani, K. 1985. Light-induced reduction of cytoplasmic free calcium in retinal rod outer segment. Nature **311:** 661-663.

List of Participants with Fields of Research

APPLEBURY, M.L.
Dept. of Biological Sciences
Purdue University
West Lafayette, IN 47907
USA

Molecular basis of visual transduction in vertebrates

BAER, K.M.
Institut für Neurobiologie der Kernforschungsanlage Jülich GmbH
Postfach 1913
5170 Jülich
F.R. Germany

Protein composition and function of proteins in the photoreceptor of the crayfish, investigation with antibodies

BAYLOR, D.
Dept. of Neurobiology
Stanford Medical School
Stanford, CA 94305
USA

Transduction in vertebrate photoreceptors

BLEST, A.D.
Dept. of Neurobiology
Research School of Biological Sciences, P.O. Box 475
Canberra City, ACT 2601
Australia

Photoreceptor membrane turnover in arthropods, cytoskeletal organization of invertebrate photoreceptors

BOWNDS, M.D.
Laboratory of Molecular Biology
University of Wisconsin
1525 Linden Drive
Madison, WI 53706
USA

Chemistry and physiology of phototransduction

BROWN, J.E.
Dept. of Ophthalmology
Washington University
School of Medicine
660 S. Euclid Avenue
St. Louis, MO 63110
USA

Physiology and biochemistry of transduction in vertebrate and invertebrate photoreceptors

CAVAGGIONI, A.
Istituto di Fisiologia Umana
Universita di Parma
43100 Parma
Italy

Neuropsychology of photoreception

CERVETTO, L.
Istituto di Neurofisiologia, CNR
Via S. Zeno 51
56100 Pisa
Italy

Photoreceptor physiology

CHABRE, M.
Laboratoire de Biologie
Moléculaire et Cellulaire
Département de Recherche
Fondamentale
Grenoble 38041
France

Biophysical studies of the rhodopsin-transducin-phosphodiesterase amplifying cascade in retinal rods

COLES, J.A.
Laboratoire d'Ophtalmologie
Expérimentale
22, rue Alcide-Jentzer
1211 Geneva 4
Switzerland

Ion movements and control of mitochondrial respiration in the drone retina

FAHR, A.
Abt. Biophysik
Fachbereich Physik
Freie Universität Berlin
Arnimallee 33
1000 Berlin (West) 33

Photoelectricity of pigmented membrane proteins

FAIN, G.L.
Dept. of Ophthalmology
Jules Stein Eye Institute
UCLA School of Medicine
Los Angeles, CA 90024
USA

Physiology of photoreceptors and retinal interneurons

FINDLAY, J.B.C.
Dept. of Biochemistry
University of Leeds
Leeds LS2 9JT
England

Studies into the structure and activity of integral membrane proteins

GOLDBERG, N.D.
Dept. of Pharmacology
3-260 Millard Hall
University of Minnesota
Minneapolis, MN 55455
USA

The hydrolysis of cyclic GMP and cyclic AMP as a biochemical event in cellular excitation/response coupling

HAMDORF, K.
Institut für Tierphysiologie
Ruhr-Universität Bochum
Geb. ND 5/33
Postfach 10 21 48
4630 Bochum 1
F.R. Germany

Electrophysiology, microspectrophotometry of photoreceptors

HANKE, W.
Lehrstuhl für Zellphysiologie ND4
Ruhr-Universität Bochum
Postfach 10 21 48
4630 Bochum 1
F.R. Germany

Reconstitution of ionic channels, channels in visual transduction

HILLMAN, P.
Neurobiology Dept.
Institute of Life Sciences
The Hebrew University
Jerusalem 91904
Israel

Measurements and models of single-photon responses and their interactions

HOCHSTRATE, P.
Lehrstuhl für Tierphysiologie
Ruhr-Universität Bochum
Geb. ND5/31
Postfach 10 21 48
4630 Bochum 1
F.R. Germany

Electrophysiology of photoreceptors

KAUPP, U.B.
FB5 Biologie
Abt. Biophysik
Universität Osnabrück
Barbarastraße 11
4500 Osnabrück
F.R. Germany

Biophysics of photoreceptors

KEIPER, W.
Institut für Theoretische Physik
RWTH Aachen
Seffent-Melaten 26C
5100 Aachen
F.R. Germany

Models for the molecular mechanism of an invertebrate photoreceptor (Limulus)

KIRSCHFELD, K.
Max-Planck-Institut für biologische Kybernetik
7400 Tübingen 1
F.R. Germany

Visual-sensitizing-protective pigments in photoreceptors

KORENBROT, J.I.
Dept. of Physiology - 762S
University of California
School of Medicine
San Francisco, CA 94143
USA

Role of Ca ions in retinal rod phototransduction

KÜHN, H.
Institut für Neurobiologie der Kernforschungsanlage Jülich GmbH
Postfach 1913
5170 Jülich
F.R. Germany

Biochemistry of vision

LAMB, T.D.
Physiological Laboratory
University of Cambridge
Downing Street
Cambridge CB2 3EG
England

Electrophysiology of vertebrate photoreceptors

LIEBMAN, P.A.
143 Anatomy-Chemistry Building
University of Pennsylvania
Philadelphia, PA 19104
USA

Binding, kinetics, and mechanism of amplified cyclic nucleotide cascade of vision

LISMAN, J.E.
Dept. of Biology
Brandeis University
Waltham, MA 02254
USA

Electrophysiology of transduction in vertebrate and invertebrate photoreceptors

MacLEISH, P.R.
Laboratory of Neurobiology
The Rockefeller University
1230 York Avenue
Box 138
New York, NY 10021
USA

Phototransduction and retinal synaptic physiology

MILLER, W.H.
Dept. of Ophthalmology and Visual Science
Yale Medical School
New Haven, CT 06510
USA

Molecular mechanisms of phototransduction

MINKE, B.
Dept. of Physiology
The Hebrew University -
Hadassah Medical School
Jerusalem 91010
Israel

Phototransduction in invertebrates

NEUGEBAUER, D.-C.
Lehrstuhl für Neurophysiologie
Zoologisches Institut
Hüfferstraße 1
4400 Münster
F.R. Germany

Morphological and cytochemical characterization of a mechanoreceptor membrane (stereorillus membrane of sensory cells from the inner ear of vertebrates (fish))

OWEN, W.G.
Dept. of Biophysics and
Medical Physics
University of California
Berkeley, CA 94720
USA

Signal transfer between retinal photoreceptors and bipolar cells

PAULSEN, R.
Lehrstuhl für Tierphysiologie
Ruhr-Universität Bochum
Postfach 10 21 48
4630 Bochum 1
F.R. Germany

Biochemistry of phototransduction and adaptation in invertebrate photoreceptors

PEPPERBERG, D.R.
Dept. of Ophthalmology
University of Illinois
College of Medicine
1855 West Taylor Street
Chicago, IL 60680
USA

Visual pigment and photoreceptor adaptation

PRINZ, H.
Max-Planck-Institut für
Ernährungsphysiologie
Rheinlanddamm 201
4600 Dortmund
F.R. Germany

Neuroreceptors

PUGH, E.N., Jr.
Dept. of Psychology
University of Pennsylvania
3815 Walnut Street
Philadelphia, PA 19104
USA

Transduction in vertebrate rods

ROOF, D.J.
Berman-Gund Laboratory
Harvard Medical School
243 Charles Street
Boston, MA 02114
USA

Cytoskeletal proteins related to the vertebrate rod outer segment

SAIBIL, H.R.
Dept. of Zoology
Oxford University
South Parks Road
Oxford OX1 3PS
England

Squid photoreceptor structure and biochemistry

SCHLEICHER, A.
Institut für Biophysik und
Strahlenbiologie der
Albert-Ludwigs-Universität
Albertstraße 23
7800 Freiburg
F.R. Germany

Time-resolved light-scattering on rhodopsin and bacteriorhodopsin

SCHNAKENBERG, J.
Institut für Theoretische Physik
RWTH Aachen
Seffent-Melaten 26C
5100 Aachen
F.R. Germany

Statistical physics applied to biophysical systems, in particular photoreception and nervous excitation

SCHWEMER, J.
Institut für Tierphysiologie
Ruhr-Universität Bochum
Postfach 10 21 48
4630 Bochum 1
F.R. Germany

Molecular mechanism of photoreception in invertebrates

SHAPLEY, R.
Dept. of Biophysics
The Rockefeller University
1230 York Avenue
New York, NY 10021
USA

Neurophysiology of vision, visual perception

STEIN, P.J.
Dept. of Ophthalmology
Yale University
School of Medicine
New Haven, CT 06510
USA

Biochemical mechanisms and phototransduction in photoreceptors

STIEVE, H.
Institut für Biologie II
Verfügungszentrum
RWTH Aachen
Kopernikusstraße 16
5100 Aachen
F.R. Germany

Neurobiology, photoreception

STRYER, L.
Dept. of Structural Biology, D133
Stanford University
Medical School
Stanford, CA 94305
USA

Molecular basis of visual excitation

TORRE, V.
Dipartimento di Fisica
Istituto di Scienze Fisica
Via Dodecaneso 33
16146 Genova
Italy

Biophysics of phototransduction

WONG, F.
Marine Biomedical Institute and
Dept. of Physiology and Biophysics
University of Texas
Medical Branch
200 University Boulevard
Galveston, TX 77550
USA

Molecular mechanisms of visual transduction

YAU, K.-W.
Dept. of Physiology and Biophysics
University of Texas
Medical Branch
Galveston, TX 77550
USA

Mechanism of visual transduction in vertebrate retinal photoreceptors

Subject Index

Author Index

MIX
Papier aus verantwortungsvollen Quellen
Paper from responsible sources
FSC® C105338

If you have any concerns about our products,
you can contact us on
ProductSafety@springernature.com

In case Publisher is established outside the EU,
the EU authorized representative is:
Springer Nature Customer Service Center GmbH
Europaplatz 3, 69115 Heidelberg, Germany

Printed by Libri Plureos GmbH
in Hamburg, Germany